FOOD AND SOCIETY

FOOD AND SOCIETY

Edited by

MARK GIBSON

ELSEVIER

ACADEMIC PRESS

An imprint of Elsevier

Academic Press is an imprint of Elsevier
125 London Wall, London EC2Y 5AS, United Kingdom
525 B Street, Suite 1650, San Diego, CA 92101, United States
50 Hampshire Street, 5th Floor, Cambridge, MA 02139, United States
The Boulevard, Langford Lane, Kidlington, Oxford OX5 1GB, United Kingdom

Library of Congress Cataloging-in-Publication Data
A catalog record for this book is available from the Library of Congress

British Library Cataloguing-in-Publication Data
A catalogue record for this book is available from the British Library

ISBN: 978-0-12-811808-5

For information on all Academic Press publications visit our website at https://www.elsevier.com/books-and-journals

Publisher: Charlotte Cockle
Acquisitions Editor: Patricia Osborn
Editorial Project Manager: Redding Morse
Production Project Manager: Maria Bernard
Cover Designer: Christian J. Bilbow
Illustrated by Pat Newsham

Typeset by TNQ Technologies

Contents

List of figures

List of tables

Table of equations

Contributors

People who have helped shape this book are listed below. Without their valuable contributions, this book would not have been possible. To them, I owe a great deal of gratitude and respect.

Mark Gibson (PhD)
John Paull (PhD)
Mukta Das (PhD)
Loi Kim Ieng (Connie) (PhD)
Lei Weng Si (Clara) (PhD)
Ubaldino Couto (MSc)
David Graham (PhD)
David Wiley (MSc)
Christy Ng Yen Nee (PhD)
Chammy Lai Peng Tai Msc
Ricky On Peng Sou MSc MBA
Cindia Ching Chi Lam (PhD)
Grant O'Bree (MSc)
Miguel Gassmann de Oliveira (MBA)
Marta Hoi Yan Lam (PhD)
Florence Iok San Lam MSc MBA
Alan Wan Lok Cheng Master of Accountancy

Chapters, affiliations, and biographies

Front Matter

*By **Mark Gibson** PhD*
Mark's Bio See Below

Chapter 1: Sociopolitical food and nutrition: pre-20th century

*By **Mark Gibson** PhD*
Mark's Bio See Below

Chapter 2: The 20th century: winds of change (1900-45)

*By **Mark Gibson** PhD*
Mark's Bio See Below

Chapter 3: Food, rights, and politics: the post-war years (1945-60)

*By **Mark Gibson** PhD*
Mark's Bio See Below

Chapter 4: Mixed blessings: a time of hope and crisis (1960-2000)

*By **Mark Gibson** PhD*
Mark's Bio See Below

Chapter 5: The 21st century: ideological convergence

*By **Mark Gibson** PhD*
Mark's Bio See Below

Chapter 6: The beginnings of modern development theory

*By **Mark Gibson** PhD*
Mark's Bio See Below

Chapter 7: Agriculture, forestry, and fisheries

*By **Mark Gibson** PhD*
Mark's Bio See Below

Chapter 8: Organic food and agriculture

*By **John Paull** PhD*
School of Land & Food, University of Tasmania, Australia

Dr. John Paull is an environmental scientist at the University of Tasmania in the College of Science and Engineering. He is the editor of the open-access peer-reviewed Journal of Organics. He has recently been a visiting academic at the University of Oxford. He has degrees in mathematics, psychology, and environmental science. He has presented his research at many international conferences, in Europe, Asia, and Australasia. He has published extensively in international journals with many of his papers available at http://utas.academia.edu/JohnPaull. He is the author of books including *The Value of Eco-labelling (2009)* and book chapters in *Marketing of Organic Products* (2008), *Island Futures* (2011), *Diversifying Food and Diets* (2013), *Organics in the Global Food Chain* (2013), *Good Food for All* (2014), *Sustainable Development of Organic Agriculture* (2017), *Encyclopedia of Food and Agricultural Ethics* (2018), *and Global Leadership Initiatives for Conflict Resolution and Peacebuilding* (2018).

Chapter 9: From agriculture to the global food chain/system

*By **Mark Gibson** PhD*
Mark's Bio See Below

Chapter 10: Global food waste

*By **Mark Gibson** PhD*
Mark's Bio See Below

Chapter 11: The Hunger files: food security

*By **Mark Gibson** PhD*
Mark's Bio See Below

Chapter 12: Food: natural and environmental considerations

By ***Mark Gibson*** *PhD*
Mark's Bio See Below

Chapter 13: Food: population rise and sustainability

By ***Mark Gibson*** *PhD*
Mark's Bio See Below

Chapter 14: Governance and food

By ***Mukta Das*** *PhD and* ***Mark Gibson*** *PhD*

Mukta Das *PhD*
SOAS Food Studies Centre, SOAS University of London

Mukta Das studied Anthropology of Food at SOAS Food Studies Centre, SOAS University of London. Her current research focuses on China's ascendancy as a global power and South Asian migration into the Pearl River Delta, exploring this through ideas of taste and aesthetics, food festivals, and heritage projects. Before this, Mukta worked for several NGOs as well as in the private sector in Asia. She has a master's in the Anthropology of Food and also in History of China and India.

Mark Gibson *PhD*
Mark's Bio See Below

Chapter 15: Food & social trends

By ***Lei Weng Si (Clara)*** *PhD and* ***Loi Kim Ieng (Connie)*** *PhD*
Lei Weng Si (Clara) *PhD*
Institute for Tourism Studies, Macao (IFTM): Colina de Mong-Há Macao, China
Clara is an Assistant Professor at the Institute for Tourism Studies, Macau, China. She received her PhD in International Business from the University of Leeds in the United Kingdom. Her research interests rest on festivals and event management, management education, and cross-cultural food issues and trends. Prior stepping into the academia, Clara worked in the industries for some years and took part mostly in marketing and management. Clara likes traveling and trekking. Her footprint covered six of the seven continents. South America is her favorite continent, where she trekked Torres del Paine, Patagonia, and enjoyed the rich food culture very much.

Loi Kim Ieng (Connie) *PhD*
Institute for Tourism Studies, Macao (IFTM): Colina de Mong-Há Macao, China
A local of Macao and currently the Assistant Professor at Institute for Tourism Studies (IFT) Macao, Connie developed her interest in the tourism and hospitality field at an early age. To pursue this interest, she completed her bachelor degree in Tourism Business Management with a concentration in Hotel Management at IFT. Thereafter, she obtained MSc in Financial Management from the University of London. Her PhD in Tourism at the James

Cook University in Australia strengthens her perspectives on destination marketing, tourism product development, and benchmarking studies. Connie's research interests include entertainment development in tourism destinations; destination positioning and benchmarking; tourist behavior; tourism product diversification; hotel guest satisfaction, service quality, and loyalty; and the like.

Besides academic development, Connie is a big fan of pottery art, which she has been practicing for more than 10 years.

Chapter 16: Trending foods

By ***Lei Weng Si (Clara)*** *PhD*
Clara's Bio – See above

Chapter 17: Food, culture & food tourism

By ***Lei Weng Si (Clara)*** *PhD and* ***Ubaldino Couto*** *MSc*
Clara's Bio – See above

Ubaldino Couto *MSc*
Institute for Tourism Studies, Macao (IFTM): Colina de Mong-Há Macao, China

Ubaldino Couto is a lecturer in event management at the Institute for Tourism Studies in Macao. He is a specialist in festivals, and his research explores the role of festive events in achieving positive social outcomes. Ubaldino is also interested in diaspora communities, particularly the aspects of cultural identity, and how it is manifested to other cultures. He is actively engaged in institute-wide projects and also in the community. He was invited in an expert delegation to Amakusa, Japan, to advise the local government and companies on harnessing the Portuguese-influenced heritage for business and tourism. Ubaldino was also invited to local schools and universities abroad to deliver short courses and seminars on a variety of topics. He is also a mentor and works closely with young aspiring professionals. Ubaldino advocates equality through research, specifically, in race, gender, sexuality, and access issues.

Chapter 18: Feeding the future: challenges and limitations

By ***Mark Gibson*** *PhD*
Mark's Bio See Below

Chapter 19: The sociology of the chef: a new theoretical proposition from the open professional kitchen

By ***David Graham*** *PhD*
Department of Service Sector Management; Sheffield Hallam University

David spent the early years of his career in the restaurant and hotel and licensed retail trade working for large multinational chains, followed by ownership of his own restaurants and coffee shop businesses. On selling his business interests, he moved into further and higher education, teaching and researching at Huddersfield Technical College, Leeds Metropolitan University, and more recently at Sheffield Hallam University.

He is committed Principal Lecturer who has a passion for internationalization in the curriculum and champions the use of technology to facilitate cross-boundary interactions. David has been a leader in the application of video conferencing to enhance the student learning experience.

David holds a teaching qualification, degree, and masters from Huddersfield University and a PhD from Sheffield Hallam University along with other professional qualifications. He has worked on national committees for teaching and learning, validated degree and master's awards for other further and higher education providers, and has externally examined in the United Kingdom and overseas. He has never lost sight of his early career in hospitality and continues to teach and research in this area, presenting and writing articles centered on food and beverage management.

Chapter 20: Food ingredients

*By **David Wiley** MSc and **Christy Ng Yen Nee** PhD*

***David Wiley** MSc*

Macao Institute for Tourism Studies (IFTM), Macao, China

Mr. David C. Wiley, Jr is a lecturer at the Institute for Tourism Studies, Macau. He holds a Master of Science degree in Hospitality Management from Florida International University. Mr. Wiley teaches courses related to Food and Beverage, including Nutrition and Food Safety, and is also a ServSafe-certified instructor and proctor.

***Christy Ng Yen Nee** PhD*

Macao Institute for Tourism Studies (IFTM), Macao, China

Dr. Christy Ng Yen Nee is an Assistant Professor at the Institute for Tourism Studies, Macau. She holds a PhD in Hospitality Administration from the Oklahoma State University, USA. She is a certified proctor and has taught Food Safety (ServSafe) and Nutrition classes for more than 8 years. Dr. Ng has research interests in relationship marketing, customer satisfaction, job satisfaction, and service quality.

Chapter 21: The role of information technology in the food industry

*By **Chammy Lai Peng Tai** Msc, **Ricky On Peng Sou** MSc MBA and **Cindia Ching Chi Lam** PhD*

***Chammy Lai Peng Tai** Msc*

Institute for Tourism Studies, Macao (IFTM): Colina de Mong-Há Macao, China

Chammy Tai received her Master Degree in Public Administration from the Peking University. She has broad experience in the IT industry lasting for more than 20 years and focusing on system development areas and academic projects in tertiary education.

***Ricky On Peng Sou** MSc MBA*

General Manager (IT and Operation) of TEDMEV Macau EDI Van, S.A. Macao, China

Ricky On Peng Sou has over 19 years' experience in the field of information technology, he is experienced in setting up and managing an entire IT system for several enterprises. He obtained his Master Degrees in Computer Science at the Macau University of Science and Technology and University of Macau in Business Administration, respectively. Mr. Sou has comprehensive knowledge in information security area, and he is a Certified Information System Auditor and has good experience in auditing ISO27001 system. Mr. Sou possesses remarkable experience in BPR (Business Process Reengineering) projects, possessing strong knowledge of Electronic Data Interchange system. He is also familiar with Web-based applications and Smart Phone Apps development projects and has provided successful eCustom solutions to various governmental departments.

Cindia Ching Chi Lam *PhD*
Institute for Tourism Studies, Macao (IFTM): Colina de Mong-Há Macao, China

Cindia Ching Chi Lam is the Assistant Professor and Academic Coordinator of the Evening Degree Programmes of Institute for Tourism Studies, Macau, and has been a member of the governing committee, Technical and Scientific Council, for years. Dr. Lam is a local from Macao, graduated from University of Southern Queensland (Australia) with a Bachelor of Science, completed her master degree at Heriot-Watt University (England), and earned her PhD from the Catholic University of Portugal (Portugal). Dr. Lam specializes in teaching service quality and accounting-related courses. Her research focuses on tourism and hospitality areas that spans from the development of tourism products (including medical tourism) and destination positioning to service quality, customer behavior, and corporate social responsibility. Her research works have been published in different books and journals, including Journal of Hospitality and Tourism Management, Journal of China Tourism Research, Tourism Review, etc, and she also serves as the reviewer for numerous international journals. Dr. Lam has worked on numerous government policy and commercial consultancy projects, including "Construction of Macao into a liveable and leisure city" (commissioned by Conselho Executivo of the Macao Special Administrative Region), "A Study on the Perceptions and Cognition of Fair Trade in Macao" (commissioned by Oxfam), "A Study on the Current Situation of Parental Bonding in Macao" (commissioned by Gabinete de Apoio ao Ensino Superior/Kiang Wu Nursing College), etc.

Chapter 22: Food service in hospitality management

By ***Grant O'Bree*** *MSc*
Institute for Tourism Studies Macao (IFTM): Colina de Mong-Há Macao, China

Following a career spanning over 20 years in the tourism and management industries within the Asia-Pacific and Middle Eastern regions, Grant turned his hand to higher education. Since becoming a lecturer, Grant has educated a diverse background of students in Human Resource Management, Digital Marketing, Organisational Behaviour, Hotel Operations, and F&B Cost Control and Lodging Management. Grant has held roles at the University of Queensland teaching in the Master of Hotel Management program and Think education as lecturer. Grant has an extensive range of international experiences within academia and industry, enabling him to deliver courses utilizing a holistic and well-rounded perspective. Grant currently works at the Institute for Tourism Studies, Macao SAR, where he lectures in various hotel and core management subjects.

Chapter 23: The power of service quality: front-of-house service skills

By ***Florence Iok San Lam*** *MSc,* ***Alan Wan Lok Cheng*** *Master of Accountancy and* ***Cindia Ching Chi Lam*** *PhD*
Florence Iok San Lam *MSc*
Institute for Tourism Studies, Macao (IFTM): Colina de Mong-Há Macao, China

Florence Iok San Lam is the adjunct Lecturer at the Institute for Tourism Studies (IFT). She has taught Leadership, Human Resources Management, and Principles of Management for the Institute's Daytime Degree Programmes, as well as Principles of Management and Tourism Product Management for the Evening Degree Programmes. She is a local from Macao, graduated from IFT with a Bachelor of Arts and the Excellence Award, and completed her master degree at the University of Groningen, the Netherlands. She has lived, studied, and worked in Macao, the Netherlands, Scotland, London (UK), Lisbon (Portugal), and Poland, having worked in the oil and gas, entertainment, and arts and cultural sector. Currently, she is the Director of Iao Hin Gallery, Macao, and a contributing writer for the Opinion pages of Macau Closer. As an active member of Macao's society supporting social progress and development, she has been included in Macau News as one of the city's movers and shakers.

Alan Wan Lok Cheng *Master of Accountancy*
Institute for Tourism Studies, Macao (IFTM): Colina de Mong-Há Macao, China
Alan Cheng is an instructor in front office operations at the Institute for Tourism Studies, Macau SAR. He has been in this position teaching front desk theory and practicum since August 2012. Alan commenced his hospitality career in 1999 when he was offered an opportunity to assist with starting a new motel in Sydney, Australia. Then, his position as the general manager at the property gave him the opportunity of providing on-the-job training to the front office staff.

Cindia Ching Chi Lam *PhD*
Institute for Tourism Studies, Macao (IFTM): Colina de Mong-Há Macao, China
Cindia's Bio – See above

Chapter 24: The financial environment: culinary math

By ***Miguel Gassmann de Oliveira*** *MBA*
Institute for Tourism Studies, Macao (IFTM): Colina de Mong-Há Macao, China
Miguel Gassmann de Oliveira has been fortunate to work most of his career in an area he is passionate about; Hotels and Food and Beverage operations. He holds a degree in Hotel management from Escola de Hotelaria e Turismo do Porto, a Bachelor's degree in Tourism Management from IFT, and a Master of Business Administration with a Major in Finance from the Instituto Inter-Universitário de Macau and the Universidade Católica de Lisboa. With over 10 years' experience in the many sectors of the local hotel industry with a focus on Food and Beverage as well as spending the last 11 years teaching Hotel and F&B-related subjects, he has had the opportunity to share this passion with his students. At present, Miguel is lecturing on Theory and Practice of Restaurant and Bar, F&B Cost Control, Hotel Operations, Introduction to Gastronomy, and Food Product Knowledge II at the IFT

Chapter 25: Controlling costs and purchasing food

By ***Miguel Gassmann de Oliveira*** *MBA*
Miguel's Bio – See above

Chapter 26: F&B spatial design and functionality

*By **Marta Hoi Yan Lam** PhD*
Institute for Tourism Studies, Macao (IFTM): Colina de Mong-Há Macao, China

Marta Lam obtained her PhD, titled "Thermal Performance of Deciduous Climbing Plants on Glazed Building Façades' in 2007 at University of Brighton, UK. Her interest on environmentally friendly building designs leads her to the engineering and construction field in one of the fastest growing cities, Macau. Prior Marta become an educator in Institute For Tourism Studies (IFT), she worked in the industry on decoration and project management in various fit-out projects of Venetian Cotai Resorts.

With increasing variety of subsidiary services provided within buildings, the sector of facilities management becoming more and more important in the hospitality field. Marta started teaching facilities management in Institute for Tourism Studies since year 2011. Later, when Culinary Arts program was also introduced in IFT, she also took up the role in teaching kitchen and restaurant design. Marta wishes to help students to understand how services are planned, designed, layout, and maintained within the superstructure.

Summary: What role food?

*By **Mark Gibson** PhD*
Mark's Bio See Below

Appendices

General economic theory

*By **Mark Gibson** PhD*
Mark's Bio See Below

Millennium Development Goals (MDG)

*By **Mark Gibson** PhD*
Mark's Bio See Below

Nutritional guidelines

*By **Mark Gibson** PhD*
Mark's Bio See Below

Glossary

*By **Mark Gibson** PhD*
Mark's Bio See Below

Biography

Mark Gibson

Before he began his academic career, Invited Assistant Professor Dr. Mark Gibson worked in the food and beverage industry for more than 20 years in various roles including chef, consultancy, and front of house manager. After so many years in the industry, Mark had a change of direction and undertook both his Masters and PhD in Food Management, Environmental Sustainability, and Food Security at Manchester Metropolitan University (United Kingdom).

Dr. Gibson has just finished an 8-year mid- to senior-level posting at the Institute for Tourism Studies, Macao. As part of the coordinating team, he helped to develop several new postgraduate Master's degrees: two of which include Food and Beverage Management and in Issues of Gastronomy. Mark also sat on several academic bodies from the CTC through to the pedagogical council as well as the coordinator's panel. After his stint at IFTM, Mark is now taking a break to finish another book after which he will consider a new role as an author, educator, researcher, consultant, environmentalist, food and society practitioner, and gastronomic social scientist among other things.

Some notable publications include several books, papers, and articles about food and its many surrounding issues. In addition to writing books and articles, Dr. Gibson is also an associate editor and reviewer for several food-related journals. He also helped with "Foodsource"—a research and policy unit based in the Social Sciences Division of the University of Oxford's, Environmental Change Institute. Amid all this, Mark continues to write books on the broad subjects of food and food-related material while acting as an ad-hoc consultant for the United Nations.

Foreword

By 2050, the population of the Earth is expected to top out at just over 9 billion. Such numbers will need to be supported in terms of clothing, housing, and being fed all from a single resource: the Earth. From the perspective of food, this challenge is exacerbated especially when considering that presently, and for many years now, there still exists a large cohort of the population that remain persistently undernourished. This figure varies a lot – and ranges from between 8-900 million people. With this in mind, the global community is without doubt in agreement with the need to increase the global food supply. However, any increase in the food supply has to meet persistent present shortfalls as well as future projected needs. Such a rise in production has been estimated to be in the region of around 70%. Of course, this is perhaps an unforgiveable oversimplification, as in truth there currently does exist sufficient production to feed the existing population on a daily basis; the problem is one of food security; in essence there are many issues of access and stability among other things that cloud the food issue. Having said that, despite such issues, there is a need, a very real need to increase food availability by a whopping 70%.

Without getting into too much detail for the moment, the elephant in the room here is the use of the word availability. That is to say, for a while there was much talk by the United Nations and others of this need for a 70% increase in food production—only now are people beginning to accept alternative views that rely less on big production increases but more toward a suite of initiatives that include efficiency savings from technology, etc., as well as measures to decrease a shamefully large component of food wastage.

Of course and once again, this is the simplified version, and one goal of this book is to highlight the many considerations, arguments, and areas of acrimony that besets a holistic view of food and culture as well as many other areas vis-a-vis supply and demand, environmentalism, globalization, population sustainability, and many others. Moreover, food, being the theme of this book, is discussed at every level and from many conceivable angles.

To give some sort of overview, some perspective, or background if you like, the first few chapters delve briefly into the recent history of the growing importance of food from the sole perspective of sustenance through to food and the environment and many other subjects in between. As a result, these few sections evaluate the definitive growth and maturation in the worldwide aspects of food and by extension the multilateral landscape of social accountability. Achievements unfold and are examined and evaluated to understand contemporary influences or forces in action. Starting with communal or social collectivization, a narrative unfolds, whereby agronomy and technological and scientific advances are shown to adapt to the ever-increasing population numbers.

A tall order without a doubt and one in which the book tackles in a very overarching way. Lastly, this book aims to be academic in nature: both rigorous and supported by solid data. Much of the data is empirical in nature and, where supporting data cannot be found, alternative acceptable estimates, guesstimates, and considered opinions will be sought in lieu. In essence, the core of this work looks to bring together or to assimilate if you like as much of the concept of food from many different perspectives as is possible.

Preface: What role food?

Society has always had a unique and special relationship with food—not solely because of the need for sustenance but also in the way people have developed "cultural" values and traditions surrounding the way we cherish and treat food and feeding practices. Such values, whether knowingly or not, affect all manners of race, religions, cultures, and subcultures. Indeed, in this regard, individuals, groups, races, and cultures have formed practices and habits that have dominated the food landscape for millennia. Undeniably, practices that have origins in personal or shared values have, for the most part, transformed food into occasions, into feasts and ceremonies, or more simply into pleasurable past times. Furthermore, the role of food reflects very personal groups' or individuals' values and, as such, one person or culture's food practices might not sit well with another's set of "food values." From this, it can be seen that the role of food in society varies greatly and never more so than in today's globalized world. Nowadays, food choices are increasingly based on considerations of more than just basic survival, instead they go on to include considerations of health, ethics and morality, cultural identity, price, geographic location, and naturally personal preferences. This in turn invokes considerations of politics, economics, technologies, and other variables—all of which affects or is affected by globalization, world trade, and cultural evolution among many other concepts. The following chapters in part-one aim to explore the role of food within some of these contexts; in particular, emphasis is placed on trends that have generally influenced today's food choices as well as the drivers of change within the food we consume. It also serves as a general overview—a barometer of social constructs and paradigms that have evolved over the years—often, but not always, from the perspective of three thematic areas: those of food and politics; food and economics; and food and sociocultural perspectives.

Acknowledgments

Writing this book has been an incredible journey. After having been a chef for over 20 years and after my Masters, PhD, and 10-year worth of studies, I thought I was especially well versed in the subject of food from cooking to culture. The things I never knew and indeed the things I wish I had known could have without doubt made me a more knowledgeable chef. And yet, the more I research, the more I find of what there is to know. Of course, this has had a lot to do with the pure breadth along with the many various disciplines within the Food and Society field. In this regard I have been very lucky in working alongside some professional and enthusiastic individuals who have provided both encouragement and the occasional epiphany. These range from the Institute's President Dr. Fanny Vong through to Connie and Florence.

There are a few other people too, who have helped in this journey and to whom I owe a great deal of gratitude. I would also like to thank Pat Newsham who was particularly generous with his time and who has provided the illustrations for this book. Another individual who I would like to thank is Flora for her patience and understanding; indeed, despite her busy schedule, she always made herself accessible for advice, as a mentor or simply for a chat; in this I am truly grateful. I would lastly like to thank others including my students for their wit and energy, my colleagues for the neverending banter, which is always a welcome aside. Collectively these people have provided me with insight, clarification, and direction as well as the occasional "push."

I would especially like to thank certain close friends, colleagues, and loved ones who have also been enduringly supportive and inspirational—Hugo, Christy, Jerome, Jeffrey, Marieta, Miguel, David G, David Wong, Hans, Helen, Jamie-Leigh, Alan, Leona, Mikey, Dell, Paul Jones, Flora and Awan Wan, Paul and Vee Parsons, Florita Alves, Widy Astutik, Riri and of course the contributors of the book who are introduced later as well as many more other people who have provided constant support and a little nudge now and again.

Abbreviations

APEC	Asia-Pacific Economic Cooperation
ASEAN	Association of Southeast Asian Nations
BMI	Body mass index
BMR	Basal metabolic rate
CDC	Centers for Disease Control and Prevention
CEC	Commission of the European Community
CGIAR	Consultative Group on International Agriculture Research
CH_4	Methane
CIS	Commonwealth of Independent States
CITES	Convention on International Trade in Endangered Species of Wild Fauna and Flora
CO_2	Carbon dioxide
CPI	Consumer Price Index
CSD	Commission on Sustainable Development called for in Agenda 21 and established by ECOSOC
CSSD	Consultative Subcommittee on Surplus Disposal
CST	Committee on Science and Technology as a subsidiary body established under the United Nations Convention to Combat Desertification
DEFRA	UK Department for Environment, Food and Rural Affairs
DES	Dietary energy supply
DESA	Department of Economic and Social Affairs
EBRD	European Bank for Reconstruction and Development
EC	European Community
ECA	Economic Commission for Africa
ECCAS	Economic Community of Central African States
ECE	Economic Commission for Europe
ECOSOC	UN Economic and Social Council
EDF	European Development Fund
EEA	European Environment Agency
EFTA	European Free Trade Association
EIA	Environmental Impact Assessment
ENSO	El Niño–Southern Oscillation
EPA	Environmental Protection Agency
ESA	European Space Agency
EU	European Union
FAC	Food Aid Convention
FAO	UN Food and Agriculture Organization

FAOSTAT	FAO Statistical Database
FBS	Food Balance Sheet
FEWSNET	USAIDs Food Early Warning Systems Network
FIVIMS	Food Insecurity and Vulnerability Information and Mapping Systems
GATT	General Agreement on Tariffs and Trade
GDP	Gross domestic product
GEC	Global Environmental Change
GEO	Global Environment Outlook
GHG	Greenhouse gases
GIEWS	Global Information and Early Warning System
GLP	Global Land Project
GMO	Genetically modified organism
GNI	Gross national income
GRID	Global Resource Information Database
GWP	Global warming potential
HDI	Human Development Index
IBRD	International Bank for Reconstruction and Development aka. World Bank
ICCP	International Climate Change Partnership
IDA	International Development Agency
IFAD	International Fund for Agricultural Development
IFPRI	International Food Policy Research Institute
IGBP	International Geosphere-Biosphere Program
IGC	International Grains Council
IGO	Intergovernmental organization
IMF	International Monetary Fund established in 1945
IMR	Infant mortality rate
IPCC	Intergovernmental Panel on Climate Change
IRC	International Rice Commission
LDCs	Least developed countries
LIFDC	Low-income food-deficit country
MDGs	Millennium Development Goals
MENA	Middle East and North Africa
NAFTA	The North American Free Trade Agreement
NASA	National Aeronautics and Space Administration
NEC	Not elsewhere classified; used in the UN FAO food balance sheet reporting
NGO	Non-governmental organization
OCHA	Office for the Coordination of Humanitarian Affairs
ODA	Official development assistance
OECD	Organization of Economic and Cooperative Development
OHCHR	Office of the United Nations High Commissioner for Human Rights
OPEC	Organization of the Petroleum Exporting Countries
POPs	Persistent organic pollutants
PPP	Purchasing Power Parity
SAC	Scientific Advisory Committee

SAP	Structural Adjustment Program
SCN	United Nations System Standing Committee on Nutrition
SOFI	State of Food Insecurity
UN	United Nations
UNCCD	UN Convention to Combat Desertification
UNCTAD	UN Conference on Trade and Development
UNDG	United Nations Development Group
UNDP	United Nations Development Program
UNEP	United Nations Environment Program
UNESCO	UN Educational Scientific and Cultural Organization
UNFCCC	UN Framework Convention on Climate Change
UNFPA	United Nations Population Fund
UNHCHR	United Nations High Commissioner for Human Rights
UNHCR	United Nations High Commissioner for Refugees
UNICEF	United Nations Children's Fund
UNU	United Nations University
USAID	US Agency for International Development
USDA	United States Department of Agriculture
WFP	World Food Program
WFS	World Food Summit
WHO	World Health Organization; UN specialized agency
WMO	World Meteorological Organization; UN specialized agency
WRI	World Resources Institute
WTO	World Trade Organization

PART ONE

The Historic Political, Economic and Social Constructs of Food

Tis education forms the common mind,
Just as the twig is bent, the tree's inclined
Alexander Pope
Moral Essays

CHAPTER

1

Sociopolitical food and nutrition: pre-20th century

OUTLINE

Populations will only increase if and when they have sufficient basic means to do so. One of these basic needs involves access to a plentiful supply of natural resources such as food and water, not to mention the collective organization and governance of its people. Although the latter is not necessarily quite so obvious, we will look at this aspect in more detail later. In this way, with adequate governance, provisions, time, and a smattering of luck, in due course, social evolution and collective ideology have been shaped and reshaped to define

Food and Society
https://doi.org/10.1016/B978-0-12-811808-5.00001-5

humanity, most often for the better. However, this evolutionary process requires a critical mass of people, the majority of which need to be moving collectively in the same direction for things to get started. Yet, the reality for millennia was that population numbers had remained static at around the 200 million mark. That was, until a certain set of circumstances came together, prevailed and which then collectively encouraged the increase in population figures. Such circumstances included the industrial revolution, the industrialization of agriculture, and an age of enlightenment among others.

Ever since civilizations first began to participate in sedentary cultivation rather than hunting and gathering, they were naturally inclined to share the fortunes and failures of their labors with the rest of the community (next section). As such a hitherto unknown concept of food security emerged, adapted, and evolved. Such is its importance, though that no book about food and society would be complete without something in the way of explanation of the food security concept. Consequently, this and many other food-related topics are included. In this way, the whole picture of food and society comes together from the macro- to the microenvironment of food.

1.1 The growth of agricultural civilizations

Pinning down the birth of modern humanity is fraught with disagreement and contention. The problem is there are numerous theories, scholarly and otherwise, that try to describe the specific origins of the agricultural evolution and the subsequent impetus that gave rise to the complete transition from hunter-gatherer to the very origins of agriculture as we know it today. Many such theories include the demographic, oasis, evolutionary, or socioeconomic hypotheses among others (Childe, 1936; Sauer, 1952; Binford, 1968; Rindos, 1987; Hayden, 1992, 1995; Weisdorf, 2005; Rosen, 2007). However, such theories aside it was widely understood or at least acknowledged that the genuine origins of farming evolved somewhat simultaneously and somewhat independently around 10,000 years ago at several sites around the world. This change (or growth if one prefers) has been said to have been largely fueled as a direct result of the abundance of freshwater—important and undoubtedly of great overriding geologic influence. This was collectively encouraged as previous bands of mobile hunter-gatherers came to understand the numerous benefits of living a static existence nearby to rivers, springs, and lakes (Miller, 1980; Gopher et al., 2001; Weisdorf, 2005; Guisepi, 2009; UOR, 2009). Consequent growth from this mini agrarian revolution, coined in 1935 by philologist turned archaeologist V. Gordon Childe as the "Neolithic Revolution," was understood to be pivotal, or the primary mover in "sedentary cultivation." Directly resulting from such a fundamental change in living styles, Weisdorf (2005) claims that the transition from hunter-gatherer to agriculturalist at this juncture has been universally recognized as one of the most crucial moments in human civilization. Indeed, stemming from this and subsequent population increases resulted in the establishment and growth of many early civilizations (Greene, 1999; Guisepi, 2009). From this point, empires and great civilizations were established, including the Chinese Empire along the Yangtze, Huang, and Yellow Rivers (UOR, 2009); the Mehrgarh and Harappan civilizations along the Indus river; and many Mesopotamian and Near East cultures which eventually settled along the Tigris, Euphrates, and the Nile Delta regions (areas which later collectively became known as the "Fertile Crescent") (Time, 1936). It was also understood that this life-changing transition from hunter-gatherers to farmers was also more than likely a slow one, with early settlers more inclined to still hunt and gather while supplementing their efforts with crop plantations of foods gathered in the wild. All this before the full conversion to pure domestic agriculture-based economies was finally embraced (UOR, 2009). Furthermore, such plant and animal domestication also stood out as a pivotal time for mankind—one in which humans became increasingly dependent on the environment and ultimately (perhaps a little unwittingly) constrained by the environment too (Furon, 1958; Weisdorf, 2005; UOR, 2009).

So, what fueled this new agricultural paradigm, apart from the aforementioned settlements in advantageous geographic locations? Well, technological innovation for one, which although might seem rudimentary by today's standards, in fact, evolved in great leaps in contemporary times. For a start, early farming tools were made of wood and stone. This included the stone "adz" (an axe-like tool); the sickle or reaping knife (which was used to gather grain); the digging stick (later adapted as a spade or hoe); and a rudimentary plow (later adapted to be pulled by oxen). As civilization progressed, new and improved practices and metal tools of bronze and iron greatly improved cultivation with things like the cast-iron moldboard plow. New irrigation systems too, powered by wind and water mills, allowed hitherto unyielding or previous unproductive land to be brought back into cultivation. Productivity was also increased with the introduction of "biowaste" or more commonly animal manure fertilizers. On top of this, the practice of crop rotation and leaving land to fallow was a great leap in understanding of the sustainability and extant land use. Storage methods were also refined; granaries where jars, dry cisterns, silos, and bins all containing stored grains popped up providing an all-round food supply for increasingly needy populations.

1.2 Modern agriculture: on the shoulders of giants we prosper

With the previous model, as mentioned, the population levels still remained fairly static. Clearly there was something missing that allowed for such numbers to increase as they did so around the 19th century. The answer, although evident in the increasing pseudo-societal groups, was the slow evolvement and advantages of collectivization. The next few millennia, for instance, witnessed the sociocultural transformation of small dynamic, mobile egalitarian groups of hunter-gatherers into sedentary agriculturist societies while simultaneously allowing for increased social, political, and technological complexity (Johnson, 1997; Guisepi, 2009). This new complex societal and political paradigm by nature further encouraged the reorganization necessary for social and economic development, as well as the diversification of nonagricultural trades such as craftsmen, politicians, and priests, etc. Moreover, in certain regions, the division of society did not stop at the introduction of a diversity of trades but was also responsible, in certain places, for the introduction of social class systems (Bender, 1975; Price and Gebauer, 1995; Weisdorf, 2005; Agropolis Museum, 2009; Guisepi, 2009). Apart from class-based society, such a marked change in the way society easily adapted to, and adopted a fledgling market-based system of food and nonfood goods, went a long way to accepting a formal basis of modern economic civilization (Cavalli-Sforza et al., 1993; Agropolis Museum, 2009). However, on the flip side, in some quarters, talk and rumor of this new collectivization brought with it fears of insufficient food provision and access for the neediest (more later). Some notable influences over the centuries gave agriculture and by extension society more than a boost.

1.2.1 The eighth-century Muslim Agricultural Revolution

Today the current agricultural economy resembles what it is largely thanks to the collective knowledge of our forefathers. The oft overlooked "Muslim Agricultural Revolution" (MAR) of the eighth century, for example, was privy to many new and "borrowed" techniques and advancements. Yet by the Middle Ages, the MAR were the first to disseminate their collective gained knowledge through detailed written agricultural practices, procedures, and technologies across the Islamic world (Glick, 1977; Al-Hassani et al., 2007). The Romans too built lucrative trade and export businesses based on techniques pioneered by the Sumerians. Jointly, as a consequence of these long-learned lessons and advances, a healthy economy across the Old World was spreading, which over time further facilitated the distribution of replicable farming techniques.

1.2.2 The Columbian Exchange 1492 AD

Frequently cited as the second prodigious food revolution, the Columbian Exchange began with Christopher Columbus and flourished with his successors. Christopher Columbus had an inquisitive mind and in the agricultural field this meant introducing new people to new foods and with it subsequent increased trade in terms of crops, livestock, and, of course, disease between the "Old" and the "New" worlds. Consequently, as a direct result of the Columbian Exchange, new crops such as American corn (maize), the humble potato, and tomato, among others, were introduced into Western Europe, while North America benefited from new livestock including cattle, sheep and pigs. Grains (particularly wheat) were also a point of significance, as it also found its way into the Americas and other places.

Moreover, the Columbian exchange aside, Africa and Asia via Western traders and slave routes became important conduits for labor, rice, onions, olives, grapes, coffee beans, bananas, and sugarcane—all at one point or another put down roots in the New World (Agropolis Museum, 2009). Also, as agricultural trade brought with it increased economic benefits, the rise of "plantation economy" flourished and grew. These large agricultural estates producing bulk demanding crops such as cocoa, cotton, sisal, sugarcane, coffee, banana, citrus trees, palm trees, and indigo were introduced. These measures described above all contributed to the ability of populations to grow through greatly increasing agricultural productivity, albeit on the back of much slave labor (Agropolis Museum, 2009).

1.2.3 Colonial agriculturalism

Around the time of 15th and 16th century Europe, new economic theories of growth emerged based largely on the creation of wealth. This abutted and was perhaps further fueled by this new period of global exploration and colonization. The result was a rich harvest in which exotic or hard to find food and other wealth-creating opportunities for the colonizing home countries were created. International trade was truly born. To this point, the sum collection of agricultural know-how had generally been passed down from generation to generation and from culture to culture—ultimately from civilizations' forefathers. By now, agricultural trade did not only stop at "goods and services" but also through a healthy new trade in knowledge transfer (Danhof, 1949; Johnson, 1997). This substantial flow of knowledge across time did not begin with the growth from the advancement and sophisticated techniques traded during the periods described above. Instead its humble beginnings started before the Neolithic period and subsequently included the MAR and the Columbian exchange (Crosby, 1972; Watson, 1974; Glick, 1977; Salvaggio, 1992; McNeill, 2003; Al-Hassani et al., 2007; Agropolis Museum, 2009). This flow of knowledge cannot be stressed enough. In fact, Justus Von Liebig remarked rather eloquently in his book of the period that

> One of the most remarkable features of modern times is the combination of large numbers of individuals representing the whole intelligence of nations, for the express purpose of advancing science by their united efforts, of learning its progress and communicating new discoveries ... ***(Playfair, 1847, pg 3).***

And so, it was that this gathering of knowledge set the backdrop from which the relatively more modern agricultural and industrial revolutions both grew.

1.3 The agricultural and industrial revolutions

Perhaps the two main revolutions from about 1650 AD onward, in terms of agricultural led population growth, were the agricultural and industrial revolutions, with perhaps a smattering of the soon-to-grow enlightenment revolution.

1.3.1 Agricultural revolution

True agricultural pioneering research is often cited as beginning from around the 16th and 17th century periods (Johnson, 1997). Moreover, it is often upheld that the revolution began first in England before spreading to America and beyond into Europe and elsewhere by the mid-17th century, after which momentum continued with the industrial revolution of the early to mid-18th century (Frey, 1996; UNEP, 1996; Johnson, 2000).

It was boom time; alongside flourishing trade and the ongoing transfer of crops from lands of origin to all corners of the world, Old and New, progress was also being made in areas of selective breeding of both plants and livestock. This was also a period that witnessed the first real attempts at pest control management using poisons and certain biological controls. Further noteworthy production output and rising yields were also buoyed through the heavy use of fertilizers and reclamation of land. And it did not stop there, the knowledge gap was closing; with extensive irrigation projects, proper and adequate drainage, improved and/or increased animal husbandry, and the increasing use of hormones and antibiotics, the foundation of the industrialization of agriculture was established (Johnson 1997, 2000). All in all, such practices led to the specialization and empowerment of the sector. Consequently, it was not long before the great cities of North Africa including the Near East, Europe, and Asia and the Americas were all supporting technically advanced agricultural systems including crop rotation, irrigation, and pest control measures and fertilization use (Cohen, 1995; UOR, 2009). The same period, especially throughout Europe, also saw the widespread practice of feudalism in the agricultural sector where the "... lord of the manor (or liege) presided over his vassals" (Gibson, 2016). Few people at the time understood the ramifications of increasing agricultural efficiency and by extension land productivity.

Increased agricultural output tapped directly into the subsequent increases in the carrying capacity of the earth and the growth in populations too (UOR, 2009). A great deal of anticipation was expected from new and improved agri-science and what effectively started with advances in agricultural practices, experimentation and scientific application continued to encompass all elements of the food chain, especially in the areas of transportation and mechanization (Durand, 1916). This also paved the way for inadvertent consequences in the shape of competition and the displacement of previous long-established trading associates as well as the displacement of suppliers. One more advantage of the global agricultural revolution was the increase in yields per unit area, thus permitting more of the populace to be fed from the same area of land (UOR, 2009). Alas, from such collective advances in agricultural knowledge and practice, this period witnessed the considerable growth in populations and their standards of living especially among the industrialized nations. Notable too were the marked changes for the better in local economies, income growth, and distribution of the labor force, among other things (Watson 1974, 1983; Gardner, 2002).

1.3.2 Industrial revolution

Swiftly on the back of the agricultural revolution came the industrial revolution. This was a period of rapid industrial growth beginning in England toward the second quarter/half of the 18th century (1725—50 AD), which then moved throughout the Europe and the United States. The early part of the revolution observed great leaps of invention, mechanical innovation, and general improvement in labor efficiency, especially on the agricultural playing field among many other sectors of industry. Among many advances in agriculture were the wooden plow, new horse-drawn threshers, grain and grass cutters, cultivators, rakes, and the labor-saving corn shellers and the like. Many in turn were superseded or improved on with arrival of the industrial revolution. Add to this the arrival and application of steam power during this time, which eventually led to the mechanization and industrialization of agriculture. This further led to the commercialization of food and ultimately to the beginning of the food-processing industries (UOR, 2009). Table 1.1 records some of these advances or seminal moments of the agricultural and industrial revolutions. However, during this period, the real coup was undoubtedly the invention of the internal combustion engine in the 1850s. This effectively freed up large agricultural labor forces, allowing millions for the first time to migrate to urban employment (Johnson 1997, 2000).

1.4 The enlightenment: sociocultural movements

As humanity evolves, cultural values exist in a state of flux, constantly adapting out of necessity and circumstance. Yet, far from being fickle, as the description might suggest, cultural values in this context are just the values that constantly evolve. And values do change,

TABLE 1.1 Key innovations of the agricultural and industrial revolutions.

Date	Innovation and invention
1698 The first steam engine	The first practical incarnation of a steam-powered engine was the water pump. However, developed by Thomas Savery, it was not particularly efficient and was readily prone to explosions.
1701 Seed drill	Created by the famous agrarian Jethro Tull, the seed drill allowed for more efficient and workable seeding.
1712 Improved steam engines	Hopping on the Bandwagon, Thomas Newcomen developed a steam engine that was more robust and reliable while operating at atmospheric pressure.
1730 The iron (Rotherham) plow	The first real success in commercial iron plows was the Rotherham plow patented by Joseph Foljambe in 1730.
1732–86 The first threshing machine	Building on unsuccessful threshing machines like Michael Menzies (1732) and Mr. Stirlings machine (1758) (which only threshed wheat), Andrew Meikle in 1786 devised the first successful mechanized threshing machine.
1775 James Watt steam engine	In partnership with Matthew Boulton and James Watt together they improved on previous engines with up to 75% reduction in fuel consumption.
1794-98 Plow	After many improvements on previous plows, innovations by Thomas Jefferson allowed for deeper and more efficient pulling of the plow.
1799 High-pressure steam engine	Around 1800, improvements of atmospheric engines witnessed new engines using high-pressure steam engines first introduced by Richard Trevithick. These were more powerful and smaller in design than those previous.
1800–31 Mechanical reaper	After many unsuccessful attempts between 1800 and 1831, the first useful mechanical reapers were introduced in 1830–34 by McCormack and Hussey.
1804–10 Sealed containers and canning	During this period, advances in technology saw the improvement of hermetically sealed foods for preservation by Francois Appert and canning by Peter Durand in 1810.
1837 Steel plow	Steel plow was invented by John Deere in 1837.
1840s Fertilizer manufacture	During this period saw the introduction of manufactured chemical fertilizers by Baron Justis Von Liebig in the 1840s.
1841 First portable steam threshers	Ransomes first introduced the portable steam threshing machines.
1850s–78 Internal combustion engine	This period witnessed the first successful gas-fired internal combustion engine developed by Etienne Lenoir (1859) and refined by Nikolaus Otto (1878), (Britaninca, 2019).
1871 Pasteurization	Pasteurization is invented by Louis Pasteur.
1890s–1910 Tractors	Engine technology was constantly being pushed to new limits. Benjamin Holts early steam traction engines of the 1900s and the internal combustion engines of the 1850s eventually paved the way for the first internal combustion tractor of 1910.
1888/95 Pneumatic tires	John Dunlop invented the first air-filled pneumatic tires in 1888 for bicycles. However, in 1895, André Michelin was the first to use pneumatic tires on automobiles.

(Continued)

TABLE 1.1 Key innovations of the agricultural and industrial revolutions.—cont'd

Date	Innovation and invention
1895 Refrigeration	While refrigeration had been around by now for 40 years or so, it was Carl Von Linde who developed the first safe domestic refrigerators in 1895.
1899 Artificial insemination (AI)	Pioneering work, built on previous efforts by Spallanzani (1784), Heape (1897), (Francis and Jolly, 1906), and others, led E. I. Ivanow to establish AI as a practical procedure in Russia.

Compiled from Tull 1762; de Graffigny, 1898; Fouts, 1921; Ogburn and Thomas, 1922; Morris, 1933; Kuo-Chün, 1958; Olmstead, 1975; Rasmussen, 1977; Powell, 1988; Hills, 1989; Martin, 1991; Fox, 1993; McMichael, 1995; Brunt, 2003; Heldman, 2003; Kauffman, 2003; Nuvolar, 2004; Elliott, 2008; Britannica, 2009.

whether over time or through circumstance. What is important today might not have been on the agenda yesterday; and by the same token, values which our forefathers fought hard to defend might seem trivial under the microscope of the 21st century. Yet in all of these instances, such values have shaped and forged a sort of collective social glue—one that promotes common ethical and moral frameworks and perceptions of right and wrong. Social cultural values then are a "Force Majeure" and one that is no less evident than in "values" that have in some cases, grown up around the food we eat or which in some cultures, have actually shaped the food we eat.

Food has always been central to human societies; in early cultures, the desire to share food was strong and could be defined as one of the many leading markers of growing humanity. As communities became more complex, so food became entrenched in traditions, habits, and meanings that varied widely over time, geographic location, and cultural practices. From these humble beginnings, it can be seen that no culture is ever static and cultural adaptation in ever-changing environments is a constant feature of world history. This is especially so when it comes to changes in the food supply. For instance, there was no bigger upheaval in this field than in Europe and the Middle East between the 15th and 18th centuries. This 300-year or so period saw sweeping cultural and intellectual developments, most notably the "scientific" and "industrial" revolutions, as well as the "enlightenment," which once it entered the fray spurred on both the cultural and agricultural revolutions, which in turn helped redefine the global food landscape. At the same time, increased movement of labor and goods between large empires (such as the Mughal and Ottoman) encouraged a growing globalized society, carrying with it a wave of changing cultural patterns with economic, demographic, and environmental consequences. Importantly too, globalization, in the form of trade, movement of labor, and/or the changing political environment (especially colonialism), was a two-way affair. That is to say, whether economically, politically, or socially, both sides generally benefitted from this mutual interaction. However, this is not to suggest benefits were equally shared—not by a long shot. However, this mixing of values was especially evident at the cultural level. The sharing of foods helped assimilate global cuisines, as more and more societies came to rely on similar staples such as potatoes, corn, or rice to supplement their diets. This had profound social, cultural, and economic consequences. Sometimes, the introduction and exchange of foods altered existing cultural practices, and in other instances, food was reworked into existing traditions. In fact, in many food cultures of today, the food and cultural histories of many countries and regions cannot be fully understood without reference to the wider world at large. Food then, from an historical and cultural perspective, not only shapes regional identities but also expresses cultural values. Importantly too, food also conveys significant messages about the individual as the 19th-century French philosopher and gourmand Jean Anthelme Brillat-Savarin (1826) noted

> Dis-moi ce que tu manges, je te dirai ce que tu es ***(Brillat-Savarin 1826)***

which translates as "Tell me what you eat, and I will tell you who you are"—put another way "we are what we eat." In this little nugget, Brillat-Savarin clearly believes culture in relation to food and defines the individual. However, from an historical perspective, one could legitimately ask whether food indeed shaped culture or whether culture has in fact defined traditions, foods, and habits. In reality there is probably not one answer; it is more likely a mix of the two, a sort of "back and forth" exchange whereby, depending on time and circumstances, one comes to dominate and define the other and on occasion vice versa. In view of present day situations though, there are several instances where food is certainly and overtly shaping the cultural mix. This is perhaps more evident in three areas than in any others—that of population, globalization, and wealth creation. For background purposes, the first and second items, population and globalization as well as changing patterns of wealth are discussed later in the book.

While much came before, the real emergence of the science, for instance, in terms of tradition and nutrition, leap-frogged on the back of previous knowledge, shaping the beginning of the 20th century. However, even after Thomas Malthus' and others' prophetic warnings, the powers are still managed to relegate hunger (or malnutrition) to a mere technical anomaly of food provisioning (Anon, 1798; Malthus, 1803). The inference being that such conditions could be avoided altogether through sufficient and adequate planning. This collective thinking essentially concreted hunger's social and political trajectories. However, having said that it was the waning enthusiasm of this position (in the early 1910s) that people realized there was more to hunger than a mere "sideshow." Instead and in large part due to the people's collective social awakening, spurred on by a growing body of sensationalist journalism, hunger issues in general were now coming to the forefront of the minds of scientific and lay people alike. At this juncture, things changed and a great deal of this new-found attention helped fully politicize the notion of hunger and starvation; in particular interventionist policies were now being spurned on in many countries at both the government and institutional levels (Aronson, 1982; Vernon, 2007).

Once again, while acknowledging this position, interventionism was not something new, rather, governments, particularly Western governments, had for a prolonged period of time been involved (or actively inserted itself) into the agricultural market in the form of the Corn Laws in the United Kingdom, and elsewhere in the Western world through sometimes controversial use of political and economic strategies in the form of local and international tariffs and protectionism (Barnes, 2006; Vaidya, 2006). However, such measures (the adequate provision of food) were still largely seen as an issue of production and supply. Although directly as a result of the sidelining the notion of adequate food for all, humanitarianism was gaining social and political traction which ultimately squeezed it into the political arena of the time.

Meanwhile, on the subject of health and nutrition, 19th century was a very active time.

1.5 Health and nutrition: an emerging discipline

Through the 19th century, particularly coinciding with better railway links, the newly found mobility of individuals encouraged the spread of communicable diseases across Europe. The problem was recognized at the international level and initial efforts were slow and beset with poor international coordination and agreement. One exception was the International Sanitary Conference held in Paris in 1851 (WHO, 2010). In the initiative, several European countries came together in the effort to tackle the issue of contagion in a way that would not overly hinder trade, etc. Protocols were introduced and quarantine precautions were set in an effort to slow or stop the spread of cholera in this instance.

A few other sanitary conferences were held between 1851 and the close of the century, yet there was a fundamental problem. That is to say that while in principle the sanitary code was agreed upon by its members, it was never actually properly ratified. Although this and a few other coordinated efforts by other international initiatives offered little in the way of successes, collectively, the efforts served to set significant precedents in the introduction of global governance in matters of global welfare. Interestingly despite limited success, the initiative to work together on issues of health was taken up at the fourth International Sanitary Conference (ISC) in Vienna in 1874. The conference discussion ensued about the need for a permanent international agency that would exclusively tackle the question of health. However, in spite of agreements and recommendations, it was only in 1903 that 12 European countries joined together to create the Paris-based Office International d'Hygiène Publique (OIHP) in 1907 that the sanitary conferences' visions were fulfilled. The OIHP continued the work of the international sanitary conferences by adopting their conventions and directing studies on epidemiological diseases and the implementation of quarantine protocols to avoid the spread of the likes of the plague and cholera among others. Thus the OIHP became the first real international health organization (EoN, 2009).

In Britain about this time despite previous limited success regard to health initiatives, the nation's health policy was moved over to the President of the Local Government Board. In operation this ultimately ensured that by the early 20th century, public health amenities (plus sanitation and environmental health) now fell under the auspices of the "Poor Law." It was a relatively hit or miss affair whereby through assistance from voluntary hospitals, workhouses, and the like, only some but not all access to such services, was available. Furthermore, services were patchy in quality and application. Consequently, in 1919 the responsibility of the nation's health fell to the Ministry of health (more later).

1.5.1 The chemical revolution

Further linking health and nutrition in the diet were many scholars around the turn of the 18th century. At this point many had already set their sights on understanding how the body worked physiologically and how food/nutrition was being utilized in the process. However, it was one man—Antoine Lavoisier during what has been described as the "chemical" revolution—who paved the way for further insight (Nichols and Reeds, 1991; Carpenter 1994, 2003; Weaver, 2006).

1.5.1.1 Lavoisier, Antoine

Lavoisier, often referred to as the father of modern chemistry, had a certain interest in metabolism. As an influential man in his field, he had already determined that excreted carbon dioxide from the body was the end product of oxidative activity proposing the connection between oxygen absorbed by the body and the excreted carbon dioxide. He had achieved this by putting guinea pigs into a calorimeter and measuring the amount of melted ice in a process called direct calorimetry. In doing so, Lavoisier working with Pierre-Simon Laplace illustrated that the ice melt was quantitatively related to the amount of carbon dioxide expired by the guinea pig (Passmore, 1982). In other words, Lavoisier established oxygen's role in animal and plant respiration as basically a slow combustion of organic material. In Lavoisier's mind, the conclusion was obvious—animals created energy through a "combustive reaction" involving oxygen. These respiratory experiments were the first studies that would eventually lead to a fuller, more holistic understanding of the metabolic process (Lavoisier, 1780; Lavoisier and Laplace, 1784; Lusk, 1906; Underwood, 1944; Nichols and Reeds, 1991; Bensaude-Vincent, 1996; Gibson, 2016).

1.5.1.2 Boussingault, Jean Baptiste

Having established the respiratory functions of animals, all eyes now turned to energy inputs and outputs. Building on François Magendie's food experiments with dogs, Boussingault had already suggested that foods which *did not* contain any nitrogenous compounds could not help support life. From this it was therefore understood that the nutritional value of a food such as vegetables could be found in the nitrogenous gluten and vegetable (Carpenter, 2003). Buoyed by such progress Boussingault began experimenting in the 1830s. Experimenting first with plants and then animals, he learned two things: firstly, that atmospheric nitrogen could only be synthesized by plants albeit indirectly; and secondly, through the examination of animal excreta, Boussingault determined that the nitrogen content of the animals ingested food was sufficient to meet their needs. In this way, there was no necessity to obtain nitrogen directly from the Earth's atmosphere (Dumas and Boussingault, 1844).

1.5.1.3 Liebig, Justus von

Without doubt, Liebig was the leading German organic chemist of the day. And building on Lavoisier's methodology vis-à-vis organic analysis, Liebig managed to improve the systematic approach to experimentation within the field. A new discipline organic chemistry was born. Liebig did not waste any time either; he questioned Jean Baptiste Dumas's findings about animals, food, and oxidation, suggesting that animals must also be able to convert carbohydrates to fat, requiring reductive changes (later metabolism) rather than solely through oxidation (Carpenter, 2003). Building on this and other similar research in the field, Liebig added and indeed improved upon this work and further hypothesized that energy was actually created through the metabolization of proteid (protein), carbohydrates, and fat (Lusk, 1906). It was also understood that protein contained nitrogen, and Liebig further postulated, based on the conservation of energy, that it might well be conceivable to establish equilibrium between nitrogen intake and excretion from urine and feces (Nichols and Reeds, 1991).

1.5.1.4 Voit, Carl

Such hypotheses and experiments were not without value. They had implications beyond scientific research alone to those of practical real-life nutritional values. In 1857, Carl Voit (a student of Liebig's) determined beyond a shadow of doubt that Liebig's proposed nitrogenous equilibrium could be met. In doing so Voit also recognized that any imbalance in nitrogen intake against that of excreta was reflected in a corresponding loss or gain in body tissue (Lusk, 1906). In 1866, additional experiments, this time by Voit and Pettenkofer, using respiration chambers focused on carbon dioxide and nitrogen waste helped confirm Liebig's theory postulating that not only protein but also carbohydrate and fat were broken down or metabolized and used as energy (Passmore, 1982). We must not forget too that it was Lavoisier who put forward the idea that a person's body heat derives from the oxidation of ingested substances in the body (Ziegler, 1922). Furthermore, while "metabolization" was not fully understood at the time, it was nevertheless proposed that energy released this way produces heat and as such it was offered that

> … heat may become a measure of the total activity of the body. ***(Lusk 1906, pg31).***

1.6 The first dietary studies

By now the sciences and knowledge began to converge and create a new discipline called nutrition. Experiments, hypotheses, and leaps of insight by a plethora of some of the most skilled minds of the era were now coming together, lending scientific credibility to a new and exciting subject—one that was moving stridently in the public and political spheres. By the mid-19th century, many were working on dietary standards such was the leap in knowledge. Out of this, the first formal action by any government, in this case the British Privy Council in 1862/3 introduced the first public policy of dietary recommendations, firmly grounded in scientific principles. These first recommendations arose out of the pioneering efforts of Edward Smith who calculated a minimum basic daily requirement sufficient to avert starvation and disease (Acheson, 1986; Carpenter, 1991; McArdle et al., 1999). From his food intake surveys among Northern England's low-income groups, Smith considered that a minimum basic daily requirement should be equal to about 4300 grains of carbon and 200 grains of nitrogen (equivalent to about 2800–3000 calories) (Tomlinson, 1978; Oddy, 1983; Harper, 1985). He also suggested reduced rations for women by about 10% were required because of their smaller stature and their perceived reduced needs.

1.6.1 Frankland, Edward

Momentum was growing and among the many practitioners, Voit, after visiting Edward Frankland, returned to Munich with a Thompson calorimeter to aid in his studies. That was 1860, and by 1866, Frankland managed to combust organic material through oxidizing it with a mixture of potassium chlorate and potassium nitrate in a calorimeter and then measuring the heat produced. By utilizing this method, the values of 29 foodstuffs were calculated in "heat units" (carrying equal value as calories). These experiments successfully established the first direct measurement of heat from food energy and in the process introduced the quantifiable concept of food energy (Frankland, 1866; McLoed, 1905; Ensminger et al., 1993/4; Russell, 1996; Carpenter, 1998).

1.6.2 Voit, Carl

Not having missed a beat, Voit and Pettenkofer, using the calorimeter, experimented on a fasting man and established how much food was burned in the body and whether or not there was any net gain or loss. From these experiments, in 1866, Voit organized a table in which he calculated the metabolic rate of man to be between 2.25 and 2.4 million small calories—equal to about 2250–2400 Kcal. In achieving this, Carl Voit was the first man to routinely define the energy potential of food in caloric terms (Hargrove, 2006).

1.6.3 Rubner, Max

Additional experiments by Voit began as he worked on the interchangeability in the diet of carbohydrates and fats. The idea was further taken up by a student of Voit's called Max Rubner. Rubner had, by 1884, determined the per gram caloric conversion rates of the macronutrients, establishing energy equivalent factors for fat at 9.3 kcal/g, protein at 4.1 kcal/g, and carbohydrates at 4.1 kcal/g. Such was the importance of further experimentation that it became the foundation of Rubner's 1884 isodynamic law, which specified that foodstuffs replaced each other: "... in accordance with their heat-producing value ..."; in other words the changeability of the main food groups was now established (Rubner, 1885; Lusk, 1906; Mudry, 1974; Nichols, 1992; Ensminger et al., 1993/4; Hwalla and Koleilat, 2004).

1.6.4 Atwater, Wilbur Olin

Atwater, working with Max Rubner in Voit's laboratory, acquired a great deal of knowledge which he then took back to America where he carried on working with calorimetry. From this knowledge, the work of calorimetry grew and being at the pinnacle of an already eminent career built the first calorimeter that could take a human in 1892. Built at Wesleyan University in collaboration with Edward Bennett Rosa, the pair also built a bomb calorimeter for gauging energy values in foods. Originally referred to as the Atwater-Rosa calorimeter, it was later refined by Francis Benedict who also worked with Atwater shortly after this time. Consequently, the final device became known as the Atwater-Rosa-Benedict calorimeter (Chambers, 1952). Shortly after between 1894 and 1900, Atwater also worked on Rubner's earlier energy conversion factors. In doing so Atwater factored in the potential losses due to metabolism and digestion itself and considered conversion factors for protein, fat, and carbohydrate. Working with different animal and plant foods, Atwater based his on calculations as averages equating to 5.65 kcal/g for protein, 9.4 kcal/g for fat, and 4.15 kcal/g for carbohydrates. Of note is the notion that these figures signified actual energy conversion factors of the food rather than what the body absorbed. In fact, later because of the substantial variability within each grouping of macronutrients, Atwater ended up with two main conversion factor tables. By way of example, it was noted that protein energy within potatoes was poor and only equal to 2.78 kcal/g while the higher-quality protein found in eggs contained energy equivalent to 4.36 kcal/g.

Such were the differences according to Atwater, the only reasonable thing to do to compile a table of specific group values which fully replicated the different food conversion factors. The other general table of conversion factors applied to all food groups regardless of a specific food nutrients actual composition. In this way, the previous figures of 5.65, 9.4, and 4.15 kcal/g were recalibrated characterizing actual metabolizable energy, i.e., which was actually absorbed by the body. Such general figures equated to 3.9 kcal/g for both protein and carbohydrates and 8.99 kcal/g for fats (Atwater and Woods, 1896; FAO, 2003). The importance of Atwater's work is still as relevant today as it was toward the end of the 19th century. In fact, many caloric composition tables being used today are still based on the rounded-out Atwater values of 4 kcal/g for both protein and carbohydrates and 9 kcal/g for fats (Nichols, 1992; Ensminger et al., 1993/4; Gibson, 2016).

References

Acheson, E.D., 1986. Tenth Boyd Orr memorial lecture: food policy, nutrition and government. Proc. Nutr. Soc. 45, 131–138.

Agropolis Museum, 2009. History of Food & Agriculture: Beginning and Development of Agriculture. Agropolis Museum.

Al-Hassani, S., et al., 2007. Muslim Heritage in Our World. U.K, Foundation for Science Technology and Civilisation Publishing.

Anon, 1798. Essay on the Principle of Population As It Affects the Future Improvement of Society, with Remarks on the Speculations of Mr. Godwin, M. Condorcet and Other Writers (1 ed.), (London, J. Johnson in St Paul's Church-yard).

Aronson, N., 1982. Nutrition as a social problem: a case study of entrepreneurial strategy in science. Soc. Probl. 29 (5), 474–487.

Atwater, W.O., Woods, C.D., 1896. The Chemical Composition of American Food Materials. US Office of Experiment Stations Bulletin(28).

Barnes, D.G., 2006. A History of English Corn Laws: From 1660–1846. Routledge, London.

Bender, B., 1975. Farming in Prehistory: From Hunter-Gatherer to Food Producer. John Baker Ltd, London.

Bensaude-Vincent, B., 1996. Between history and memory: centennial and bicentennial images of Lavoisier. Isis 87 (3), 481–499.

Binford, L., 1968. Post-Pleistocene Adaptations. New Perspectives in Archaeology. In: Binford, R., Binford, L. (Eds.). Aldine Publishing Company, Chicago.

Brillat-Savarin, A., 1826. Physiologie du goût, ou méditations de gastronomie transcendante. Charpentier, Libraire-Editeur, Paris.

Britannica, 2009. Encyclopædia Britannica Online.

Britannica, Encyclopædia Britannica, 2019. History of technology 2019. Retrieved 2nd April 2019 from. https://www.britannica.com/technology/history-of-technology/Electricity#ref10451.

Brunt, L., 2003. Mechanical innovation in the industrial revolution: the case of plough design. Econ. Hist. Rev. 56 (3), 444–477.

Carpenter, K.J., 1991. Biographical article: Edward Smith (1819–1874). J. Nutr. 121 (3), 1515–1521.

Carpenter, K.J., 1994. Protein and Energy: A Study of Changing Ideas in Nutrition. Cambridge University Press, Cambridge.

Carpenter, K.J., 1998. Early ideas on the nutritional significance of lipids. J. Nutr. 128 (2), 423S–426S.

Carpenter, K.J., March 2003. A short history of nutritional science: Part 1 (1785–1885). J. Nutr. 133, 638–645.

Cavalli-Sforza, L.L., et al., 1993. Demic expansions and human evolution. Science 259 (5095), 639–646.

Chambers, W.H., 1952. Max Rubner. J. Nutr. 48 (Suppl.), 3–12.

Childe, G., 1936. Man Makes Himself. Oxford University Press, UK.

Cohen, J.E., 1995. How Many People Can the Earth Support? W.W. Norton, New York and London.

Crosby, A.W., 1972. The Columbian Exchange: Biological and Cultural Consequences of 1492. Greenwood Publishing Co, Westport, Connecticut.

Danhof, C.H., 1949. American evaluations of European agriculture. J. Econ. Hist. 9 (Suppl.), 61–71. The Tasks of Economic History.

de Graffigny, H., 1898. Gas and Petroleum Engines. Whittaker and Co, London.

Dumas, M.J., Boussingault, M.J.B., 1844. The Chemical and Physiological Balance of Organic Nature. H. Bailliere Publisher, London.

Durand, E.D., 1916. Some problems of population growth. Publ. Am. Stat. Assoc. 15 (114), 129–148.

Elliott, A.G., 2008. Gas and Petroleum Engines. BiblioBazaar, Charleston USA.

Ensminger, A.H., et al., 1993/4. Foods & Nutrition Encyclopedia. CRC Press, London.

EoN, 2009. Economic and Social Development: Regional Commissions the Encyclopedia of the Nations. United Nations, New York.

FAO, 2003. Food Energy – Methods of Analysis and Conversion Factors. FAO FOOD AND NUTRITION PAPER 77. Food and Agriculture Organisation.

Fouts, L.X., 1921. Jefferson the inventor and his relation to the patent system. J. Pat. Trademark Off. Soc. 4 (316).

Fox, P.F., 1993. Cheese: Chemistry, Physics and Microbiology. Chapman and Hall, New York.

Francis, H.A.M., Jolly, W., 1906. Contributions to the Physiology of Mammalian Reproduction, Part I, The Œstrous Cycle in the Dog, Part II, The Ovary as an Organ of Internal Secretion, 1906 in Philosophical Transactions of the Royal Society of London, Series B, Containing Papers of a Biological Character, 198, 99–141. Retrieved from: http://www.jstor.org/stable/91944.

Frankland, E., 1866. On the origin of muscular power. London, Edinburgh and Dublin Philoso. Mag. 4th series (31), 182–199.

Frey, S., 1996. A Glossary of Agriculture, Environment, and Sustainable Development. Retrieved Accessed 3 July 2019, from. https://www.ksre.k-state.edu/historicpublications/pubs/SB661.pdf.

Furon, R., 1958. Manuel de prehistoire general, geologie prehistorique: evolution de Phumanit, areheologie, prehistorique, les metaux et la protohistoire. Payor, Paris.

Gardner, B.L., 2002. American Agriculture in the Twentieth Century: How it Flourished and What it Cost. Harvard University Press, Cambridge, Massachusetts.

Gibson, M., 2016. The Feeding of Nations: Re-Defining Food Security for the 21st Century. CRC Press, Boca Raton, Florida.

Glick, T., 1977. Noria pots in Spain. Technol. Cult. 18 (4), 644–650.

Gopher, A., et al., 2001. The "when", the "where" and the "why" of the Neolithic revolution in the Levant. Doc. Praehistorica 28, 49–61.

Greene, K., 1999. V. Gordon Childe and the vocabulary of revolutionary change. (archaeologist and author). Kevin Greene. Antiquity 73 (279), 97–109.

Guisepi, R.A., 2009. World Civilizations: The Origins of Civilizations: The Agrarian Revolution and the Birth of Civilization. International World History Project. Online. http://history-world.org/neolithic1.htm.

Hargrove, J.L., 2006. History of the calorie in nutrition. J. Nutr. 136, 2957–2961.

Harper, A.E., 1985. Origin of recommended dietary allowances – an historic overview. Am. J. Clin. Nutr. 41, 140–148.

Hayden, B., 1992. In: Gebauer, A., Price Madison, T. (Eds.), Models of Domestication: Transitions to Agriculture. Prehistory. Prehistory Press.

Hayden, B., 1995. In: Gebauer, T., Gebauer, A. (Eds.), A New Overview of Domestication. Last Hunters, First Farmers. School of American Research Press, Santa Fe.

Heldman, D.R., 2003. Encyclopedia of Agricultural, Food, and Biological Engineering. Marcel Dekker, New York.

Hills, R.L., 1989. Power From Steam: A History of the Stationary Steam Engine. Cambridge University Press, Cambridge.

Hwalla, N., Koleilat, M., 2004. Dietetic practice: the past, present and future. Health J. 10 (6), 716–730.

Johnson, D.G., 1997. Agriculture and the wealth of. Nations. Am. Econ. Rev. 87 (2), 1–12.

Johnson, D.G., 2000. Population, food, and knowledge. Am. Econ. Rev. 90 (1), 1–14.

Kauffman, K.D. (Ed.), 2003. Advances in Agricultural Economic History. Elsevier Science Ltd, Oxford.

Kuo-Chün, C., 1958. Organized leadership and agricultural technology in modern China. Agric. Hist. 32 (1), 25–31.

Lavoisier, A., 1780. Expériences sur la respiration des animaux et sur les changements qui arrivent à l'air par leur poumon. Histoire de l'Académie Royale des Sciences, avec les Mémoires de Mathématique et de Physique pour le meme Annee. Acad. R. Sci. 1777, 185–194. Paris, Académie royale des sciences.

Lavoisier, A., Laplace, P.-S., 1784. Mémoire Sur La Chaleur. Histoire de l'Académie Royale des Sciences, avec les Mémoires de Mathématique et de Physique pour le meme Annee. Académie royale des sciences. Académie royale des sciences, Paris, pp. 355–408, 1780.

Lusk, G., 1906. The Elements of the Science of Nutrition. W.B. Saunders, Philadelphia, London.

Malthus, T.R., 1803. An Essay on the Principle of Population; or, a View of its Past and Present Effects on Human Happiness; with an Enquiry into Our Prospects Respecting the Future Removal or Mitigation of the Evils Which it Occasions. (London, Printed for J. Johnson, in St. Paul's Church-Yard).

Malthus, T.R., 1798. An Essay on the Principle of Population, as it Affects the Future Improvement of Society with Remarks on the Speculations of Mr. Godwin, M. Condorcet, and Other Writers (London, Printed for J. Johnson, in St. Paul's Church-Yard).

Martin, J.H., 1991. Iron, Liebigs law, and the greenhouse. Oceanography 4 (2), 52–55.

McArdle, W.D., et al., 1999. Sports and Exercise Nutrition. Williams and Wilkins, Baltimore.

McLoed, H., 1905. Edward Frankland – Born jan 18th 1825, died August 9th 1899. J. Chem. Soc. Trans. 87, 565–618.

McMichael, P., 1995. Food and Agrarian Orders in the World-Economy. Greenwood Publishing Group, Oxford.

McNeill, J.R., 2003. Europe's place in the global history of biological exchange. Landsc. Res. 28 (1), 33–39.

Miller, R., 1980. Water use in Syria and Palestine from the neolithic to the bronze age. World Archaeol. 11 (3), 331–341. Water Management.

Morris, T.N., 1933. Principles of Fruit Preservation Jam Making,canning and Drying. Chapman and Hall, London.

Mudry, J.J., 1974. Measured Meals: Nutrition in America. University of New York, Albany.

Nichols, B.L., 1992. A History of Nutrition Research Reflected in the USDA Tables of Food Composition. 17th National Nutrient Databank Conference. National Nutrient Databank, Baltimore, Maryland.

Nichols, B.L., Reeds, P.J., 1991. History of nutrition: history and current status of research in human energy metabolism. J. Nutr. (121), 1889–1890.

Nuvolar, A., 2004. Collective invention during the British Industrial Revolution: the case of the Cornish pumping engine. Camb. J. Econ. 28 (3), 347–363.

Oddy, D.J., 1983. Urban famine in nineteenth-century Britain: the effect of the Landcashire cotton famine on working-class diet and health. Econ. Hist. Rev. New Series 36 (1), 68–86.

Ogburn, W.F., Thomas, D., 1922. Are inventions inevitable? A note on social evolution. Political Sci. Q. 37 (1), 83–98.

Olmstead, A.L., 1975. The mechanization of reaping and mowing in American agriculture, 1833–1870. J. Econ. Hist. 35 (2), 327–352.

Passmore, R., 1982. Reflexions on energy balance. Proc. Nutr. Soc. 41, 161–165.

Playfair, L. (Ed.), 1847. Chemistry in its Application to Agriculture and Physiology: By Justus Liebig. T.B. Peterson, Philadelphia.

Powell, B., 1988. Scottish Agricultural Implements. C.I. Thomas and Sons, UK.

Price, T., Gebauer, A., 1995. Last hunters-first farmers. In: Gebauer, T., Gebauer, A. (Eds.), New Perspectives on the Prehistoric Transition to Agriculture. New Perspectives on the Transition to Agriculture. School of American Research Press, Santa Fe, pp. 3–20.

Rasmussen, W.D., 1977. Agriculture in the United States: A Documentary History. Greenwood Press, USA.

Rindos, D., 1987. The Origins of Agriculture: An Evolutionary Perspective. Academic Press, New York.

Rosen, A.M., 2007. Civilizing Climate: Social Responses to Climate Change in the Ancient Near East. Altamira Press, Plymouth: USA.

Rubner, M., 1885. Calorimetrische Untersuchungen I and II. Z. Biol. 21, 250–334 and 337–410.

Russell, C.A., 1996. Edward Frankland: Chemistry, Controversy and Conspiracy in Victorian England. Cambridge University Press, Cambridge.

Salvaggio, J.E., 1992. Fauna, flora, fowl, and fruit: effects of the Columbian Exchange on the allergic response of New and Old World inhabitants, 13 (6), 335–344.

Sauer, C., 1952. Agricultural Origins and Dispersals. American Geographical Society, New York.

Time, 1936. After Breasted. Time Magazine Online.

Tomlinson, H., 1978. 'Not an instrument of punishment': prison diet in the mid-nineteenth century. Int. J. Consum. Stud. 2 (1), 15–26.

Tull, J., 1762. Horse-Hoeing Husbandry or, an Essay on the Principles of Vegetation and Tillage London. Printed for A. Millar.

Underwood, E.A., 1944. Lavoisier and the history of respiration. Proc. R. Soc. Med. 37 (6), 247–262.

UNEP, 1996. Glossary of Environmental Terms. Retrieved 10 August 2019, from. https://www.unenvironment.org/explore-topics/resource-efficiency/why-does-resource-efficiency-matter/glossary.

UOR, 2009. University of Reading: Agriculture, Policy and Development, History of Agriculture. Retrieved 4 June 2018, from. http://www.ecifm.rdg.ac.uk/history.htm.

Vaidya, A.K. (Ed.), 2006. Globalization: Encyclopedia of Trade, Labor, and Politics. ABC-CLIO, Santa Barbara.

Vernon, J., 2007. Hunger: A Modern History. Belknap Press, Cambridge, Mass.

Watson, A.M., 1974. The Arab agricultural revolution and its diffusion, 700–1100. J. Econ. Hist. 34 (1), 8–35. The Tasks of Economic History.

Watson, A., 1983. Agricultural Innovation in the Early Islamic World. Cambridge University Press, U.K.

Weaver, L.T., 2006. The emergence of our modern understanding of infant nutrition and feeding 1750–1900. Curr. Paediatr. 16 (5), 342–347.

Weisdorf, J.L., 2005. From foraging to farming: explaining the neolithic revolution. J. Econ. Surv. 19 (4), 561–586.

WHO, 2010. Website of the World Health Organisation. Retrieved 2nd April 2019, from. http://www.who.int/.

Ziegler, M., 1922. The History of the Calorie in Nutrition 1922. The Scientific Monthly 15 (6), 520–526. Retrieved 2nd December 2018, from. http://www.jstor.org/stable/6661.

CHAPTER

2

The 20th century: winds of change (1900–45)

OUTLINE

Food and Society
https://doi.org/10.1016/B978-0-12-811808-5.00002-7

Along with the above, and for the benefit of background, the next few sections focus on certain of these dynamics such as: politics and governance; health and nutrition; science and technology; and a smattering of social awakening to tell the story of food's growing public and political importance. This is important from the stance of food and society, as it tells the story of the beginnings of so many governmental and nongovernmental issues and initiatives, the advancement of society, food science, and several attempts at global governance. The same story also describes the successes and failures while at the same time introducing key prominent people, events, and innovations along the way.

By the 20th century, things were beginning to move quite rapidly. For instance, steam, diesel, gasoline, and electric power were becoming more widely accessible. On top of this, improvements in transportation infrastructures including those of roads, canals, ships, and of course rail improved the lot for agriculturalists, empowering farmers and allowing for greater freedom. Such improvements also meant that supplies existing far away from markets (and those closer of course) could now conceivably be sown, harvested, and brought to market in a timelier fashion. Thus, in essence, the period known as the industrial revolution can be characterized in many ways, in particular three ways might best surmise the period: that of the energy source used (fossil fuels), the greater use of artificial fertilizers and pesticides, and lastly, the improvement and implementation of targeted scientific agricultural research. Coinciding with this was the increased abundance of refrigeration, mechanization leading to increased yields, and labor productivity. This period also saw the introduction of extensive irrigation, significant increased animal husbandry, and the greater application of antibiotics and hormones. Concomitantly, a decline in family-based farming and the simultaneous growth of agribusiness's (large-scale food production and processing industries) helped create greater employment in new start-up industries in new areas of agriculture, such as the agricultural chemical sector and the biotechnology sectors along with a few others, etc. (UOR, 2009). Consequently, such was the influence of the agricultural and the industrial revolutions, which profound changes in the way agri-business was conducted, combined, they allowed for the extensive, long-distance transportation of food and commodities facilitating commodity imports and exports. Thus, efficiently expanding markets and in due course ushered in an era of agricultural globalization. In turn, this new agricultural paradigm was unequivocally instrumental of the foundation of world trade realignment. By extension large production and less lead time from farm to market saw the lowering of the costs all round especially in production: furthermore, commodity prices too saw great benefits in the form of cheaper market prices, more accessible to the laymen.

2.1 The agricultural landscape

From around the late 19th century to the turn of the 20th century, agriculture was facing a mixed bag of challenges (some good, some not so). From stagnating yields and national interventionist policies, which hindered rather than helped, and from agricultural specialization and the industrialization of agriculture combined to alter the landscape of the contemporary agricultural model. All this coincided with an extended period of stagnation in terms of lowered farm salaries particularly within the United States (Gardner, 2002). This also coincided with a spurt of rapid increase in population size and yet in spite of previously mentioned advances in both the agricultural and industrial sectors it appeared as if a commodity yield increases plateaued. Consequently, it came to be seen that technology alone could not arrest the tide of stagnating crop yields. In fact, the situation, not surprisingly, drew great attention to agricultures' potential limitations in providing for the increasing rate of population growth (see Malthus, carrying capacity 13.2 & 13.3).

As an aside, further negativity was also being fostered about international trade too. One of the biggest concerns at the time was the seemingly overreliance on this new international trading platform in general. However, the same phenomenon was also noted in the agricultural and general food chain arena's too (Michie and Smith, 2000). The concern suggested some was that trading food internationally threatened to increase overdependence while simultaneously exposing large net importing/exporting nations to massive "trade" or "economic" shocks. This was because, fundamentally speaking, while keeping up with current economic concepts of the time—net imports appeared to be, more often than not, as a negative "pull" on the national balance sheets, especially among the foreign earnings balance. In doing so, this was seen as some suggested to large "trade gaps" (Michie and Smith, 2000). By contrast, increasing net exports were thought to improve a nations overall wealth (Michie and Smith, 2000) (see also balance of payments in the appendices).

At about this time too much lip service was being paid to increasing national self-sufficiency in agricultural production. The voices were heard and this shepherded a time of much needed improvement. The United States is a good case in point. Up to this point, the United States relied quite heavily on the United Kingdom for the knowledge transfer especially in the areas of technology and agriculture (seed and animal husbandry). As a consequence, society's opinions and expectations began to increase and after a lengthy period of government lobbying. The end result was the creation of dedicated government agencies of agriculture. And it was successful. Once created, with new-found government intervention, US farmers began to invest in the new technologies and innovations in the form of new land and equipment. A similar phenomenon too swept across Britain and some parts of Europe as they too set up specialist dedicated agricultural agencies. Furthermore, accompanying these increasing agencies came greater responsibility and accountability. Legislative instruments of late 19th to early 20th centuries emphasized production and the protection of agricultural goods in the form of mechanisms such as greater tariffs through both export and import taxes. Globally, this opened the door to the creation of the international organization whose sole responsibility was agricultural trade and other issues—International Institute of Agriculture (IIA) founded by a merchant of Polish-American origin in 1905 called David Lubin. The IIA operated more like a global clearing agency among other things. Having said that, the institute's greatest legacy, among other things, was to raise the profile on an international level of the need for increased worldwide awareness of a more formalized agricultural profile from production to legislation (Britannica, 2009).

However, there were many casualties of both agricultural and industrial revolutions and their associated policies as well as the resultant fledgling, fast-growing international trading platform. In particular, the extensive dislocation of time-honored trading partners did not go unnoticed. Such displacement came to be, in part, due to the adoption of more efficient use of existing resources and through specialization. Alas food commodities' increased mobility together with fairer trade barriers observed a sort of geo-displacement among many customary agrarian suppliers (Michie and Smith, 2000). By way of illustration, as grain became less profitable for, say Europe-based agriculturalists, so the North Americans and Australians filled the vacuum by providing alternate cheap and reliable substitutes. Out of this and other opportune circumstances, numerous European farmers responded by specializing too. Many turned their attention to dairying, cheesemaking, and other globally competitive products (Clayton and Black, 1943). Consequently, with increasing global demand, supply (raw materials) and specialization increased or was further developed, thus herein laying the foundation of the worldwide realignment of not only the agrarian sector but also world trade in general.

With this, in terms of policy, in particular agricultural policy, the main aim of agricultural policy-makers in the industrialized world was to expend a good deal of effort, came through in the form of protectionism of sorts whereby legislation provided sufficient agricultural/provisions for not only their own countries but also of their close trading partners as well, and of course for their valuable allies.

2.1.1 Agricultural productivity

By way of supplementary information, before the period mentioned above, improvements in agricultural productivity, more frequently than not, was attained by way of land acquisition and/or reclamation and greater use of fertilization. This helped in the drive for greater cultivatable land and total increased in harvest per acre. Furthermore, before grain yields could markedly increase, one other barrier had to be navigated: that of productivity. Presently, labor-intensive agriculture still relied on heavy manpower per acreage, particularly during harvest time. In fact, agricultural methods had not changed much from about the 14th century to the late 19th and early 20th centuries. Time advances in technology, particularly in America and the United Kingdom, were such that while growth through land expansion or reclamation was slowly decreasing subsequent yields were increasing (Johnson, 2000; Gardner, 2002). Also, as has been touched on, the increased intensification and industrialization of agriculture meant the resultant productivity gains allowed more farm and ancillary workers to pursue nonagricultural employment in both the manufacturing/industry and the service sectors. Not surprisingly too, such migration out of agricultural employment further fueled urbanized living, which, in circular logic, additionally reinforced the needs of the industrialization process and which subsequently also contributed to the increase in prosperity of countries or states (Johnson 1997, 2000; Weisdorf, 2005). No wonder, with such increases in food productivity and better, faster to market commodity, coupled with commodity mobility, that population figures markedly increased (Cohen, 1995; Zhang, 2008). Such increases were also responsible for noticeable changes in the dynamics and makeup in society. That is to say, from a long history of farm-centric work and life, reduced labor, due to mechanization (in both production and harvesting) and the overall industrialization process, societal changes meant there were more people moving to urban areas in the hope of finding work in alternative trades, many of which had never existed before this period. This had a knock-on effect in terms of the creation of national wealth (Watson 1974, 1983).

Concurrently, as the populace amplified due in large part to more than favourable conditions mentioned previously, certain individuals previously concerned about population numbers vis-a-vis the Earth and its potential to supply mankind's needs resurfaced with a vengeance (see Section 13.3).

2.2 Health and diet: the final link

Meanwhile while others were concentrating on runaway population numbers, during the early part of the century, independent caloric studies conducted around the world, as Chittenden writes, many seemed to back up the requirement of 3000 calories per day which was advised by Voit some 20 years or so previously (Chittenden 1906, 1907). That being said, there were still many other studies that showed that no universal consensus leaving Chittenden to lament over what he believed was a general lack of "scientific" evidence in many trials. He also bemoaned certain "apparent" scientific studies where one man's habit or craving vis-à-vis food requirements did not signify sufficient methodical approach to warrant any serious consideration (Chittenden 1904, 1906, 1907; Carpenter et al., 1997). Chittenden further contemplated just how such studies passed muster and truly reflected the reality on the ground (Gibson, 2016). Supporting this argument, Chittenden cites his own predictive studies which revealed a person's physiologic requirements would easily be catered for with fewer calories compared with a growing body of 3000 kcal proffered by many other studies (Chittenden, 1904; Carpenter et al., 1997). Even Lusk, a student of

TABLE 2.1 Early daily nutritional values as offered by Lusk and Chittenden.

Date/Authority	Load (default man)	Protein	Fat	Carbo-hydrate	Calories
1904–07 Russell Chittenden					
Russel Chittenden's research determined that balance for the so-called "brain worker" (slightly more active than the sedentary man) was sustainable on fewer than the often-raised daily figure of 3000 calories. For Chittenden, as he saw himself as fully sedentary, he considered 2000 calories sufficient for his own needs					
		>50 g			2500–2600 calories
1906 Graham Lusk					
					2488–2562 calories

Source: Hutchinson (1916); Chittenden (1904, 1906, 1907); Lusk (1906).

Voit and somebody who did more to raise the profile in the field of nutrition than almost any other during this period, like Chittenden similarly viewed the generally accepted 3000 daily calories as being just a tad excessive. Chittenden and Lusk's recommendations are outlined in Table 2.1.

2.3 Technology and biotechnology close the gap

In the field of technology around the beginning of the 20th century, things were moving very fast. It was not uncommon for what was the previous few months' technical or scientific innovation to become policy in the next few months. Such was the rate of scientific/technical/or industrial development, especially in areas of steam, diesel, and electric power, that there was quick and wide uptake of these new technologies. This facilitated improved transport, infrastructure, and other advances like refrigeration. In fact, improved methods of preservation and procedures effectively permitted prior, far-from-market suppliers, easier access to trade globally.

Biotechnology (although the term had not been coined as of yet) represented the notion that some materials or products could be made with the help of live organisms (Fári and Kralovánszky, 2006). As a result, industries were now using the latest biotechnologies to help in such issues as deficiencies of food and other societal needs. Besides the food component, biotechnologies were also used for manufacturing substances such as acetone, butanol, and glycerol. Indeed, the technology was especially welcomed in Germany where plant produced acetone was employed to produce paint solvents or more dangerously to make bombs. Furthermore, in both Germany and Britain, community wastewater was decontaminated or purification plants, founded on bacteriological yeast action, and were used in overcrowded cities where major outbreaks of disease occurred (Bud, 1993). In addition, groundbreaking scientific investigation by Max Delbrück and team opened up the possibility to use excess brewer's yeast as a food supplement in animal husbandry. Later this became extremely advantageous for Germany, especially during the First World War as they successfully substituted nearly half of their imported animal feed protein requirements with eukaryotic, single-celled microorganism (aka brewer's yeast) (Ugaldea and Castrillob, 2002). There were other major breakthroughs during this early period too (Table 2.2).

TABLE 2.2 Key early dates in biotechnology.

Date	Advancements made in the early 20th century
1900	Mendel's work on genetics had stagnated for a few years until they were rediscovered by Hugo DeVries, Erich Von Tschermak, and Carl Correns reexamined Mendel's work.
1901	In 1901, Beijerinck discovers oxygen-dependent (aerobic) nitrogen fixers.
1902	In this year "immunology" is first used.
1902	Walter Stanborough Sutton suggested that Mendel's so-called "factors" (genes) were transported via chromosomes and tentatively suggested they might also transport heredity from one generation to the next. A few years later, Sutton gave Mendel's "factors" the more familiar name used today "genes."
1902	Archibald Garrod further strengthened the link made between Mendel's genetic heredity and the biochemical pathways of reproduction systems in organisms.
1903	Both Walter Sutton and Theodor Boveri both postulated that every egg or sperm cell comprised one chromosome pairs (one of each). And in doing so joined or enhanced Mendel's gene heredity pathways and with it the fertilization of the egg and by extension heredity.
1905	Two other scientists, Edmund Wilson and Nellie Stevens, postulated that both X and Y chromosomes together determined the gender of an organism.
1906	While used in descriptive above, the actual term "genetics" was first introduced into the scientific world.
1907	In this year, the first in vivo culture of animal cells commenced.
1909	Further work on genes discovered that genes were also associated with hereditary disorders.
1909	Adding to the relatively new field of biotechnologies terminology, Wilhelm Johannsen coined the words "gene," the transporter of heredity, and "genotype" to determine the composition (genetically wise) of flora and fauna, while "phenotype" was given to mean the organism itself.
1912	In the evolution or maturation of X-ray technology, Lawrence Bragg was able to research the structures of modest crystalline materials at the molecular level, thus paving the way to the growth of X-ray crystallography.
1915	Phages also known as "bacterial viruses" were unearthed.
1917	In the field of terminology, "biotechnology" finally came of age to earn its own respective field and title created by Károly (Karl) Ereky, a Hungarian agricultural engineer.
1918	Using acetone produced by plants, the German army used it to make bombs.

Compiled from multiple Sources: Bud (1993); Thackray (1998); Fári et al. (2001); Fári and Kralovánszky (2006); BioTech Institute (2010).

2.3.1 Plant hybridization

Biotechnology can also be applied to the selective breeding of plants and livestock. In fact, for millennia unbeknownst to practitioners aided in the production and betterment of both livestock and crops. Before about 1910, for instance, selective breeding gave agriculturists an ability to select the best examples of the harvest, which was then utilized as seedlings in the round of planting. Having said that, the downside of this particular method (based on the visual aspect alone) had its limitations—this was biotechnology in use. This, however, changed when a trio of geneticists—George Shull, Edward East, and East's student Donald Jones, independently began working on corn genetics. Selective breeding by planting the previous year's best corn, for example, was flawed. Instead, Shull, Edward, East, and East's student not only showed that although hybridization (the cross-breeding of corn offspring) was a more effective guarantee of the next crop quality but also noted that the crop often surpassed the original corn plants. Unfortunately, however, celebrations were short-lived as further experiments exhibited a certain amount of deterioration of superiority in the hybrid's subsequent first-generation examples. Consequently, it was found that in assurance of continuous increasing and better yielding harvests, it seemed one solution was to revert back to the original lineage—the unique parental combination. This notion meant that if farmers were to continue down the hybridization of crops, they would now be tied to seed companies for their higher yielding quality product on an annual basis. However, this new reliance on seed companies was weighed against the potential benefits farmers climbed on board.

2.4 Governance and early food production

Before the "Great" or First World War, Britain's existing food chain (see Food Chain, Chapter 9, Section 9.2) was largely stable. Indeed, it was not uncommon for Britain to routinely be fairly self-sufficient growing about 20% of the required wheat and about 50% of required beef, mutton, and bacon, etc. However, Britain still needed to import foods to the tune of approximately 13 million tons annually. For this bounty, Britain relied heavily on the United States and Canadian markets for what amounted to close to 65%–70% of total nutritional requirements (Pollen, 1917; Pinchot, 1918). Such a situation, however, led Britain (up to half-way through the Great War) to adopt a "let it be" policy. This lackadaisical attitude toward policies concerning nutritional issues was counterproductive at best and foolish beyond that. Granted though, Britain was not suffering to the same extent of food shortages as Europe was at this time—but Europe at this time did not readily benefit a great deal from the same "special" relationship with America and others that Britain shared (Starling, 1919a; Dewey, 1989). Then, around the middle- to late 1916, a general, less than stellar, crop yield in Britain and a progressively successful U-boat campaign by the Germans topped off with reducing allied harvests, so even Britain had to eventually adapt and change. In fact, by this time food shortages had already begun to disrupt Britain's everyday routine, which was further made difficult by profiteers and hoarding food.

Then came the First World War, and British and American governments were, by this time (end of the 19th century), involved in the regulatory policies concerning nutrition and health requirements. However, since the beginning of the century, such policies reflected a light hand, with no real teeth to them. Up to now, it appeared that required food supply was left largely to existing trade routes satisfying supply and demand with limited input from regulatory policies. In fact, things only really changed with the actual onset of the Great War (Beardsworth and Keil, 1997).

Furthermore, up to now, focus was on quantity of food rather than the quality of the diet itself (Beardsworth and Keil, 1997). Moreover, Starling (a well-known physiologist of his time) noted in 1919 that to leave the nations nourishment to market forces (especially during war periods) was simply too much of a gamble (Starling, 1919a). Indeed, it was during this period that governments came to realize the immense complications of adequately feeding nations while ensuring sufficient quality. Starling's rationale was predicated or based on the notion that while many of history's previous skirmishes/wars or the like usually seemed only to affect small regional areas and populations, however, despite the size of the skirmishes, there was always seemed to be continuous lingering issues of hunger and disease. Not surprisingly, Starling also inferred that by ramping up these small wars as was the likes of World War One, when once every suitable man was enlisted so the major proportion of the remaining population were fully immersed in industries supporting the war effort, governments quickly realized the hugely damaging possibilities of doing nothing, especially in the area of food production and supply, etc. (Starling, 1919a).

Consequently, the arrival of agricultural specialization together with the commencement of the First World War provided both problems and opportunities in the areas of food production, supply, security, and incomes. One foreseeable problem though and one which was largely ignored was that as able-bodied farmers were diverted to war duties, farm equipment and soil suffered. That is to say that the remaining agricultural equipment was left improperly or at worst badly maintained while fertilizers became scarce due to lower levels of production. In continental Europe, fertilizer production was especially low (up to between 60% and 70%) reduction in some areas (Pollen, 1917; Pinchot, 1918; Dewey, 1989).

During this period, across the pond in America, however, the American agricultural sector prospered and again found itself in the position of global purveyors of food, especially for the nations at war (Encyclopedia, 2006). Momentum for America's success in this area came from interventionism in the free market; free trade was effectively suspended as the government bought up majority of agricultural grain futures, effectively gaining control over the whole sector. This allowed for a fixed supply of grain at steady prices (Santos, 2006). This was great news for the allied forces.

2.4.1 Interventionism: rationing

Back in Britain in 1914, the Defense of the Realm Act (DORA) passed into law. Originally put in place as a way of reducing the worst excesses logistical security, however, it was an act that saw the government effectively taking control of key industry logistics, particularly in the supply of goods and materials. DORA also came to play a significant role in the area of food supply. This fit snugly with the then Prime Minister Herbert Asquith and the coalition government of 1915, which also favored more proactive interventionist food policies. In the coalitions view, more land was to be reserved for cultivation and putting in price supports as extra incentives. However, it became clear that DORA was failing. However, according to Dewey, these and other measures were not fully implemented until the appointment of the new Prime Minister Lloyd George in 1916. It was then that the full measures of interventionism with a slew of policies to kick-start or stimulate existing production, through mechanisms like guaranteeing prices and, if needed, compulsory powers to enforce cultivation, came into full force (Clynes, 1920; Dewey, 1989). Indeed, in response to the failing DORA, the Government in 1917 formed the Food Ministry (Clynes, 1920) with the laudable goal of not only regulating the production of the local food supply but also consumption. In the United States, similar controls were implemented under the Lever

Food Control Act of 1917, whereby the Department of Agriculture retained control of all on-farm activities while market organization and pricing became the preserve of the Food Administration (Clayton and Black, 1943). Britain and America were not alone in this predicament either, as many other countries in the continents of Europe and Africa were also being hard hit in this new war of attrition (Lysaght et al., 1917; Starling, 1919a; Cary, 1920; Bennett, 1949). Indeed, many other countries, including Germany itself, suffered widespread disease and famine. Whether people died from policy failure, disease, war, or famine itself is difficult to determine as many people died from opportunistic diseases such as tuberculosis, rickets, influenza, dysentery, scurvy, keratomalacia, and hunger. Indeed, there were even reports of exhumed bodies being cannibalized (Lysaght et al., 1917; Starling, 1919a; Cary, 1920; Bennett, 1949). Directly as a result of such experiences, it has been observed that, particularly during wartime, food and agriculture has come to be seen as an industry of strategic economic and social importance (Tanner, 2004).

Even so, knowing this, Britain and other countries were still slow in taking the initiative. Primarily, the UK government organized the Royal Sugar Commission in August 1914, and this was followed in October 1916 by the Royal Commission on Wheat Supplies and in December 1916 by the Ministry of Food (Table 2.3). Nonetheless, it was not until the appointment of Lord Rhondda in 1917 as Food Controller did scientific principles of food and nutrition take their proper place in the determination of policy. Yet, in spite of all this interventionist effort, it was still not enough, and eventually it became necessary, toward the end of the war, to, at first, introduce a voluntary rationing scheme, but which was then later made compulsory in January 1918 (Starling, 1919a,b; Howard, 2002; National Archives, 2011). The United States also could not avoid a similar fate, especially after feeding their servicemen and honoring their export commitment to the allied forces. As a result, even Americans were eventually advised of the timely need of maintaining balance and moderation in their diets (Armour, 1917; Lusk, 1918).

In fact, it can be seen from Table 2.3 this was a busy period full of advancement in many areas especially in the political and scientific arenas.

2.5 Nutrition grows up

Around this time, vitamin and other nutrient research allowed researchers to state that a person's health itself was related to the food one ate. This was supported by the fact that science was tentatively showing that child development was distinctively correlated child's consumption of food. In 1921, building on this, a report originating from the Minister of Health in Britain drew attention to a notion of a balanced diet, i.e., in addition to the main sources of energy, an assortment of vitamins and minerals was also required for the continuing health of the population. It was also noted that fruit, vegetables, and milk were especially rich in nutrients and the recommendation was that they should be made more accessible to the population. It was reports like these in both England and the United States that safeguarded the continuance of good practices such as national milk distribution programs and school feeding and other such programs up to the 1930s. Many such programs' impetus came from active organisations like the Children's Minimum Council

TABLE 2.3 Key dates of the period: early 20th century.

Key dates of the period	
1905 Education Act (The Provision of Meals)	The UK Education Act (otherwise known as the "Provision of Meals act") was the culmination of the work of 365 private, charitable organizations that came together in an effort to improve child nutrition.
1908 International Institute of Agriculture **(IIA)**	David Lubin's (an agricultural reformer) political efforts led to the inception of the "International Institute of Agriculture" (IIA). Its role was to enrich the settings of rural life while also acting as a world clearing-house on information for crops, prices, and cross-border trade.
1909 The Haber–Bosch Process	Developed and first utilized industrially in 1913 the Fritz Haber and Carl Bosch[1] process of making fertilizer took synthesized ammonia, which was directly made from hydrogen gas and atmospheric nitrogen. As said, originally intended to be employed in the manufacture of fertilizer[2]. It was later adapted by the German government into a process of bomb making during WW1.
1914 The UK Defense of the Realm Act (DORA)	The Defense of the Realm Act or DORA as it was known was introduced in 1914. Its mission was to disseminate information on everything people were not allowed to do in times of war—firstly in terms of security and secondly to prevent food shortages. As WW1 progressed, DORA was seen to be failing. With a healthy black market and panic buying together distorted the equitable distribution of food. Consequently, DORA was modified. In a bold move the government took over several million acres of farm land to provide food for the nation. Like others, this initiative also failed and in a desperate measure introduced rationing through the Ministry of Food in 1918 and as a result, no one starved.
1914 Royal Society's War Committee	Formed in 1914 to forge a plan to blockade Germany's food provisions the Royal Society's Physiological War Committee was born. Slightly later, in 1916, a sub-committee, the "Food (War) Committee" was introduced in an effort to advise the government on matters of food and nutrition.
1916 Ministry of Food	Founded by a Royal Charter in 1793 as the "Board or Society for the Encouragement of Agriculture and Internal Improvement" lasted until it was finally dissolved in 1822. Whereupon in 1889, the Board of Agriculture Act led to the creation of the Board of Agriculture (later the Ministry of Agriculture, Fisheries and Food (MAFF)). However, by 1916, this too was further superseded by the Ministry of Food whose main aim was to regulate the consumption and supply of food while also to encourage the overall production of food (Starling, 1919a,b).
1916 US Federal Farm Loan Act	The Federal Farm Loan System was the USA's first massive intervention policy in the United States Agriculture industry.
1917 The Lever Food Act	Founded in 1793 by Royal Charter the "Board or Society for the Encouragement of Agriculture and Internal Improvement" (the Board of Agriculture Act) lasted until it 1822 whereupon it was finally dissolved. By 1889, the same Act led to the creation of the Board of Agriculture, (later the Ministry of Agriculture, Fisheries and Food (MAFF)). However, by 1916, this too was further superseded by the Ministry of Food whose main aim was to regulate the consumption and supply of food while also encouraging the overall production of food.

(*Continued*)

TABLE 2.3 Key dates of the period: early 20th century.—cont'd

Key dates of the period	
1917 Inter-allied Scientific Food Commission	The members of Inter-allied Scientific Food Commission from Great Britain, United States, Italy, and France were given the task to examine the scientific aspect of the inter-allied food problem and to recommend any suitable measures that might improve the dire situation (Starling, 1919a).
1917 Biotechnology	Even though biotechnology was already established as a discipline, the term was finally coined by a Hungarian agricultural engineer named Károly (Karl) Ereky in 1917. The word was found in two publications: "Food production and agriculture" and "large-scale development of pig fattening" (Rao, 2008; Fári and Kralovánszky, 2006).

[1]*Haber–Bosch process was also the work of great insight. Playing with the method of mixing nitrogen from the atmosphere with hydrogen and enormous amounts of heat, the mixture was then burned leaving behind a nitrogen–hydrogen mixture. Furthermore, this was then compressed and heated to between 200 and 700°C in the presence of a catalyst (metal oxide) with the end result being ammonia for further use in making fertilizers.*

[2]*Fertilizers—plants rely on certain nutrients for healthy development; these include but are not limited to nitrogen, phosphates, and other minute amounts of trace elements. Furthermore, plants can only uptake nitrogen in its nitrate form. This means that nitrogen in the atmosphere is not an option for plants. Instead, plants utilize microorganisms which convert atmospheric nitrogen into nitrate (that plants can use) within the soil. Therefore, soil pH (acidity or alkalinity) must be carefully monitored and adjusted to ensure the optimal growing conditions for each particular crop. Fertilizers then are very useful as they contain the right nutrients in the right quantities for the budding plant. This is why one of the major developments of the Green Revolution was the use of chemical fertilizers, not only to adjust soil pH balances but also to achieve the right levels of all the important trace elements and compounds such as nitrates and phosphates.*

Compiled from Baldwin (1907); Kates (1912); Starling (1919a,b); Gunderson (2003); Moore (2008); Smith (2009).

(later the Children's Nutrition Council) and the Committee Against Malnutrition whose collective aims were to improve the nation's overall nutrition, particularly so when it came to children (BMJ, 1938; Lloyd and Shore, 1938; BMJ, 1954; Acheson, 1986).

Further research in this area continued to give scientists more insight into the various nutrients and their respective roles within the diet. As understanding grew, the methodologies of data gathering also improved. For example, evidence for "requirement" needs was modified from predominantly observational research of so-called healthy people to tentative studies whereby nutrient intakes were varied to determine minimum needs (Beaton, 1992). It was this period that the United Kingdom really shone, stood out, and was at the forefront of global nutritional understanding. In fact, Britain, at this time, was definitely at the forefront of a global, ad hoc nutritional movement (Ruxin, 1996).

2.5.1 Metabolic studies

Aligned with nutrition, studies on BMR (basal metabolic rates), i.e., is the rate at which the body absorbs and utilizes the various energy giving nutrients while at rest to keep vital functions working, for instance, breathing and keeping warm. While calorimetry thus far had been the method of choice for researchers to date, two men in particular, Arthur Harris and Francis Benedict together with the Carnegie Institute (CI), were to change all this. Joining forces with the Carnegie Institute of Washington to work together on the analyses of particular biometric measurements to do with BMR, Harris, Benedict, and the CI established a predictive formula for BMR rates without the need for calorimetry. Their subsequent paper published in 1918 still stands up today, after 100 years, and is still employed in some areas around the world even now. Harris and Benedict's extensive statistical analysis of the data for over 100 men and women each, together with 94 newborn babies, was such that they firmly establish a link between the surface area of the whole body, a person's height, and also their weight. The formula, using multiple regression equations, presented BMR values equal to (Equation 2.1):

Equation 2.1 Predictive BMR formula

$$\text{For men, BMR} = 66.4730 + 13.7516\ \text{w} + 5.0033\ \text{s} - 6.7550\ \text{a}$$

$$\text{For women, BMR} = 655.0955 + 9.5634\ \text{w} + 1.8496\ \text{s} - 4.6756\ \text{a}$$

where BMR = total heat production per 24 h, w = weight in kilograms, s = stature in centimeters, and a = age in years (Harris and Benedict, 1918).

Source : Harris, Benedict and the CI.

This research was groundbreaking, and shortly afterward among ongoing concerns over the malnourishment of the people, the Royal Society (of Britain) founded the Physiological War Committee whose main goal was to counsel the then government on questions of nutrition (Table 2.3) (Royal Society, 1917). After much advice, the president of the Board of Trade asked the Royal Society to draw up recommendations for Minimum Dietary Requirements (MDR). As a result, in 1916, a subcommittee of the Royal Society's War Committee—The Food Committee—was inaugurated (Starling, 1919a,b; Beardsworth and Keil, 1997; Moore, 2008). And following advice and research of its own, the committee determined that

> … [We] are convinced that the dietary requirements of a nation engaged on active work cannot 'be satisfactorily met on a less supply … than 100 grams protein, 100 grams fat, 500 grams carbohydrate, equal approximately to 3,400 calories per "man" per day. ***Royal Society (1917), {Table 2.4}.***

However, and once again for an average woman doing an average days' work, less calories were recommended as the Society allotted a total of 2400 calories for women (Starling, 1919a,b).

From this point on, in Britain at least, the impetus for national dietary standards was formed, humanitarianism notwithstanding (Beaton, 1992). Nutritional standards gained further inertia when it was decided in 1917 that all allied nations would pool their nutritional supplies to achieve a comprehensive unity of nutritional supplies (Starling, 1919a,b). Indeed, to this end, an inter-allied conference at Versailles in November 1917 created the Scientific

TABLE 2.4 The Royal Society's daily recommended food provision.

Date/ Authority	Load (default man)	Protein (g)	Fat	Carbohydrate	Calories
1917/18 Royal Society of London					
The Royal Society noted that approximately 3–3400 calories per man based on modified Atwater's figures were sufficient to maintain energy equilibrium on a daily basis					
		100	100	500	3000–3400

Compiled from Royal Society (1917); Starling (1919a,b); DuBois (1940).

Food Commission, in which their task was to study, among other things, the nutritional requirements of allied country's with the objective of planning and controlling food shipments from North America to Europe (Starling, 1919a,b; Beaton, 1992; Iacobbo, 2004; Gibson, 2016). From this point, in Paris 1918, the Inter-Allied Scientific Food Commission gathered to further the abovementioned objectives. In doing so, the committee drew largely on previous work of the Royal Society and implemented many of their food requirement recommendations. However, it was not an easy task at all. The first challenge the committee foresaw was just how to gauge the nutritional needs of apparent heterogeneous populations, especially where energy needs varied greatly inter alia, age, gender, occupation, and size. Aided by the US representatives of the Commission, Chittenden and Lusk (G.R.C. 1944), the answer was to tackle the problem from two perspectives—firstly by employing an average person's BMR as a baseline and then adding further requirements based on the type of work that was to be undertaken. The commission calculated an average person's baseline BMR requirement to be in the region of 1687 calories per day. Then considering an average days' work for the average male and the requirements amounted to just over 3000–3136.8 calories to be precise. However, an important caveat here is that this final calculation did not factor in the variable occupations of people. The following Table 2.5 is from the commission's recommendations.

At about this time, two fledgling movements, the bio-dynamic and organic movements, owed their existence to the popular ideas of Rudolf Steiner.

TABLE 2.5 Daily calorie requirements as proposed by the Inter-Allied Scientific Food Commissions.

Group and work levels	Calories	
	Gross intake (allowing for waste)	**Net**
Male		
- Sedentary	2750	2500
- Heavy	3500–5500	3200–5000
- Soldier		3700–3800
Average male	3300–3400	3000
Female		
- Laundress		3291
- Typewriter	2100	1900
Average female	2650	2400
Children		
0–6	1650	1500
6–10	2100	2310
10–13	2500	2750
14 and above	As average adults	

Source: Adapted from Starling's 'The Feeding of Nations' (London, Longmans, Green & Co., 1919).

2.6 Bio-dynamics and organic farming

(See also Organic Farming, Chapter 8)

As an aside, for as long as early civilized man ceased hunting and gathering and began cultivating his own food, organic agriculture was the norm. This lasted millennia but then was abruptly altered from about 1845 onward. This was because of the agricultural and chemical revolutions, fronted by Liebig and others. The revolutions advocated a whole new industrial approach to the agri-business industry. In fact, a few of the more prevailing approaches advocated the intensification, the industrialization of agricultural output on an unprecedented scale. Not surprisingly, the new approach relied heavily on the use of mechanization of difficult or routine tasks, also with the increased usage of synthetic or chemical pesticides, herbicides, and fertilizers. However, regarding the use of industrial pesticides, fertilizers and the like had its downsides. This led some in the 1920s, particularly the European farmers, to begin to turn away from such products, disillusioned with what they viewed as reduced seed vitality of crops and a general decline in animal health; there was also the issues of pests as well as, to some, a general loss of flavor.

Addressing the problem, a group of land owners sought council from Dr. Rudolf Steiner, an Austrian responsible for the Steiner schools' movement. Steiner immediately embraced the challenge and a short time later, he delivered a series of eight lectures in which the fundamentals of a new agricultural method called *"bio-dynamics"* was put forth as one of the several solutions. Steiner's unique point of view embraced a holistic spiritual, almost ethereal approach to agriculture fully embraced in the natural cycle of life. Thus, in Steiner's view, by treating the Earth as a single organism, the regeneration of life through agriculture was a simple process of aligning the multifarious components of increasing industrialization/chemical usage congruently with the elements of nature. Perhaps a more telling proposition was delivered in his fifth speech in which he suggests

> Spiritual Science always tries to look into the effects of living things on a large scale ... the wide circumference of Nature's workings — that is the talk of Spiritual Science. But we must first know how to penetrate into these wider workings of Nature. ***Steiner (1958).***

Thus, as can be seen, even from the birth of Steiner's biodynamic viewpoint, he, at once, advocated an holistic practice of both agricultural needs and an understanding of the complexities of the interrelated relationships at work in nature (Steiner, 1958; Bradshaw, 2003).

In the field of politics though, especially around the time of the First World War, things were rather bumpy.

2.7 Socio-political tectonics

Following the aftermath of the First World War, socio-political and economic environments had to change and adapt. This resulted in the changing of long-established monarchies, feudalism, and other outdated forms of governance, especially within Europe. This left a vacuum which was quickly replaced with a gamut of emergent communist, fascist as well as democratic, autocratic and other successor states. Political borders were also being negotiated and redrawn resulting in a number of smaller states, especially as a result of the newly dispersed German, Austro-Hungarian, and Ottoman empires. In Russia too, under the pressure of widespread economic disruption throughout the country because of the war effort, it too saw its own revolutions in the 1917 Russian and subsequent Bolshevik uprisings.

An aside to all this, one direct consequence of the cost of war, particularly in Europe (which devastated many country's internal infrastructures), an economic weaker Europe emerged. While in America, economic opportunities to help rebuild and renegotiate the terms of trade with European authorities led the way for the United States to emerge as the economic superpower of the day. Also, in the area of politics and governance of food, the cessation of World War One hostilities proved to be a busy time in the new emerging "food for all" notion. With the advent of WW1, food, unwittingly took on greater importance due to shortages in many areas. In this sense the war almost became a war of attrition. Consequently, in the end, it was quickly noted that post-war reconstruction of local economies and general infrastructure was not the only challenge ahead for many developed countries. Rather, existing and exacerbated economic and social issues (including food) had to be tackled on an unparalleled worldwide platform. It was about this time too that another area—the pressure of population control—also resurfaced and began receiving more and more interest. After fading attention in the optimum carrying capacity of populations possibly due to the preoccupation of war, the end of hostilities together with predictions of ongoing food deficiencies, outside of the United States, awoke public debate regarding carrying capacity of earths inhabitants and the food supply (Ostrolenk, 1930; Mukerjee, 1933; Wolfe, 1934; Gibson, 2016).

Out of the ashes of post–World War 1 it was recognized that there was a great need in the affairs of the world as a whole. This took the form of global governance, and the outcome was the creation of a new, far-reaching organization or institution called the League of Nations. This new body aimed solely at equipping and encouraging international collaboration on the many political and social fronts that WW1 had raised.

2.7.1 The League of Nations

This was not a new concept either. For several years before the end of the war, discussions among the allied parties came to the conclusion that an international organization was needed to help maintain global peace and security (Hitchcock, 1919; League of Nations, 1919; Northedge, 1986). As mentioned, over the previous few years, while an institutional framework had been loosely modeled, it was only properly fleshed out under the "Treaty of Versailles" (the official document marking the end of the "Great" war). The main aim was an institutional body that could help maintain global peace and security. Thus, created during the 1919 Paris Peace Conference the new body was officially named the League of Nations. To date, the League of Nations was the most ambitious attempt at global governance concerning matters of peace and security. However, despite this mini victory of international standing, one major economic and political power, the United States failed to sign the accord. The Americans it seemed, were nervous of what was seen as the creation of a global political power that could effectively restrict America's freedom of action. More importantly, it was felt that the United States would very well be subject to control of what was potentially seen as a super-state or supra-national authority. On top of this, it was also considered that the United States might be forced or coerced into fighting wars not of their choosing and on sides not of their control (White, 1919). However, despite America's lack of involvement, the League of Nations operated surprisingly well. In conjunction with the creation of the "League," several other institutions were also created; these included the International Labor Organization; the Permanent Court of International Justice; and the Health Organization.

Through the mandated Covenant of the League of Nations (the Leagues Operational Charter) two overriding goals were paramount. The first was to encourage international cooperation within economic and social fields. And the second, was the creation of a multilateral instrument whereby world nations could seek refuge from cross-border political disagreements or tensions through discourse without recourse to conflict—effectively, a place to secure or preserve peace through dialogue, arbitration, and conciliation (Hitchcock, 1919; League of Nations, 1919; Kohn, 1924; Pipkin, 1933). It was not just about political issues either. Indeed, while just a fledgling body, the League was the first major international organization whereby issues of international economic and social standing could be addressed; it was in these areas that the League really left its lasting legacy. In fact, one of its founding principles, that of "multilateral dialogue" served as a blueprint for its later, successor organization, the United Nations (discussed later).

In the meantime, the league was busy paving the way for many new initiatives, especially in the area of economic and social issues. One such social milestone was initiated by the renown British social reformer Eglantyne Jebb, who, after the creation of the "Save the Children's fund" in 1919, turned to another matter close to her heart - the notion of children's rights. Indeed, in 1923, in an effort to pursue her goals, Eglantyne headed all the way to Geneva, to the International Union. Equipped with a modest but well-defined ideologue attesting to the inherent rights of children Eglantyne witnessed the fruition of her dream, the "Declaration of the Rights of the Child," or the Declaration of Geneva as it became known. The accord was well received and even taken up a year later by the League of Nations itself under the banner of the "World Child Welfare Charter" (UNICEF, 2009). Groundbreaking by standards of the day, the Charter effectively meant that children were agreed to have rights to such things as food, shelter, and health care, protected from exploitation among others to generally ensure a fully rounded package of welfare.

Meanwhile, Western agricultural policies of initial optimism resulting from the allied victory as well as increasing highly integrated and industrialization of agriculture became somewhat subdued as surpluses and then depression swept in.

2.7.2 Surpluses and depression

As a direct consequence of the First World War effort and concomitant increasing demand for agricultural produce as well as into the post-war reconstruction period, rising demand seemed to buoy the agricultural industry. Indeed, numerous Western government policies from price support, interventionism, and protectionism simultaneously increased output while controlling overall costs for the better. However, so good was the bounty that in America profiteering, despite legislation ensuring middle men sold goods at cost plus, a small token of profit for themselves was insufficient to keep the more unscrupulous of merchants under control. This created universal complaint and much ill feeling from the American consumer (Crutchfield, 1919), in spite of numerous antiregulation sentiments from the masses. However, one commentator of the time, Crutchfield, was in favor of a post-war reconstruction world where the role of American industry, especially agriculture, maintained the duality of regulated pricing and controlled demand. This, Crutchfield cites, would help stabilize prices and supply at a time when the:

> ...the whole world is looking to the American market for an undue share of our supplies... *Crutchfield (1919).*

Nonetheless, whether primarily through the "inelasticity" of the agricultural sector or whether through Western policies, growth in agricultural output at the back end of the war and into the next few years increased markedly (Holmes, 1924; Ostrolenk, 1930; Michie and Smith, 2000). Elsewhere though, especially in Eastern Europe and much of the rest of the developing world, it could not be further polarized—poverty and depression continued as agricultural deficits became commonplace. Yet it was not all doom and gloom as the West still provided for many of the undernourished, equating to millions of tons of food aid being shipped to Eastern Europe and other needy countries on a regular basis (Shaw, 2007). This was during 1919–26 AD. This was generally greeted as mixed news, principally for the American agricultural industries, as the continued gains and resultant yield increases from the industrialization of agriculture meant there was plenty to sell. However, a tipping point was reached, which ultimately led to the "West's" first real glut of agricultural surpluses (Ostrolenk, 1930). At the time this glut was seen both as a hindrance and an opportunity. A hindrance as it meant a reduction of farm incomes and an opportunity for producers with the potential of increasing entries into major new export markets (Mayne, 1947). Inevitably perhaps, with the richer getting richer and the poorer getting poorer, the result was the increasing emergence of economic disparity between developed and the underdeveloped nations while an ever-widening trade gap grew (Durand, 1922; Taylor, 1926).

More seriously, lurking around the corner in Russia was full-scale famine. This arose several years after the uprising of 1917 and the succeeding civil war in which wheat and other food crops were decimated. The situation was exacerbated with the onset of drought, in which wholesale crop failure combined to cause widespread hunger and death. Further adding to the death toll was the fact that Lenin refused to acknowledge the famine and as a result no aid was sent. Resulting from the confluence of events it has been estimated that upward or 4 million people died of famine during this particular period (Chamberlin, 1934; Gantt, 1936; Bennett, 1949). Luckily, it did not last long as a quick recovery took place over the next few years before fully recovering to pre-war levels by about 1927/28. However, it seems that few lessons were learned through all of this as over the next few years famine once again reared its head.

This was also a period of innovation when it came to the governance of food and food-related issues including forward-looking health, political, and scientific issues of the day as can be seen in Table 2.6.

2.7.3 The Great Depression

As mentioned above, harking back to the period shortly after the end of the Great War, the 1920s witnessed a steady global recovery in trade in nearly all but the agricultural sectors of

TABLE 2.6 Key dates of the period: 1918 to 30.

Key dates of the period	
1918/19 Influenza Epidemic	During the period between 1918 and 1919, the influenza pandemic spread like wildfire and killed in the region of 20–40 million people worldwide.
1919 The International Labor Organization **(ILO)**	By 1919, the League of Nations established an agency for matters concerning the reconstruction and the protection of labor unions. The ILO ran for a while; however, after the demise of the League, the ILO became a specialized agency under the umbrella of the UN in 1946 (ILO, 2010).
1919 The International Federation of Red Cross and Red Crescent Societies **(IFRC)**	During this time, the Red Cross widened its international activities to include support to emergencies that fell outside of war (IFRC, 2009).
1919 Save the Children	The Save the Children foundation was created to combat intense civilian suffering caused by the continued a blockade against Germany and Austrian-Hungary (Save the Children, 2009).
1919 League of Nations (**LoN**)	Creation of the first international multilateral organization with the mandate for peace and cooperation (League of Nations, 1919).
1920 Save the Children International Union (**SCIU**) (L'Union Internationale de Secours aux Enfants)	In 1920, Eglantyne Jebb and sister Dorothy Buxton founded the "Save the Children" campaign in the United Kingdom (UNICEF, 2005).
1924 World Child Welfare Charter	In 1924, the League of Nations adopted the "Save the Children" (SCIU's) values of the declaration of the rights of the child (UNICEF, 2005).
1927 The (**LoN**) International Economic Conference (Geneva)	In 1927, the League of Nations Assembly's held its first International Economic Conference. It had two main objectives, which were to strengthen international trade laws (arresting widespread adoption of tariff increases) and in the second, to address certain commercial, industrial, and agricultural challenges (UNOG, 2010).
1929 US Agricultural Marketing Act	In 1929, the US Agricultural Marketing Act created a new body, the Federal Farm Board US, which in an attempt to stabilize the agricultural market was allowed to directly intervene to influence price stabilization, crop surpluses, and export prices (Markham, 2002).
1929 The International Association of Agricultural Economists **(IAAE)**	Established in 1929 by 11 countries, the International Association of Agricultural Economists was created to regularly host conferences to share knowledge and to address challenges throughout the world (IAAE, 2008).
1929 British Colonial Development Act	The British Colonial Development Act aimed to provide agricultural and economic development in their many colonies with a view to stimulate both domestic investments while also advancing Britain's own commerce and trade (Garside, 2002).
1930 Wheat Meetings	Between 1930 and 31, there were 16 international meetings held for international wheat importers and exporters to foster agreement of agricultural products; however, while no agreement was reached, the meetings did eventually lead to the first wheat conference in 1931 (IAAE, 2008; IGC, 2009a).

(Continued)

TABLE 2.6 Key dates of the period: 1918 to 30.—cont'd

Key dates of the period	
1930–34 US Grain Stabilization Corporation (**GSC**)	The United States Federal Farm Board's (FFB) set up the Grain Stabilization Corporation to buy up grain and other commodities to stabilize prices of key crops (Markham, 2002).
1930 US Agriculture Marketing Act established the Foreign Agricultural Service Division (**FASD**)	The US FASD was created to help maintain and expand international export opportunities. However, the FASD was eventually superseded by the Foreign Agricultural Service in 1953 (National Archives 1995; Otto, 1999; Swanson, 2003).

Source: Compiled from multiple sources League of Nations (1919); National Archives (1995); Otto (1999); Garside (2002); Markham (2002); Swanson (2003); UNICEF (2005); IAAE (2008); IFRC (2009); IGC (2009a); ILO (2010); UNOG (2010).

industry (Kindleberger, 1986; Michie and Smith, 2000). However, in economic terms, from about the mid 1920s till the end of the decade (1929) the steady recovery became a boom. However, this was short-lived and, due either in part, to the reparations and war debts (Keynesian), or to the poor adherence to the international Gold Standard (Monetarism) (depending on which one you side with) the world fell into what became known as the "Great Depression" (Kindleberger, 1986). This accentuated the already problematic global economic trade markets and furthermore, the Great Depression also had the effect of further depressing agricultural markets by additionally contributing to widespread unemployment (Ostrolenk, 1930). Not surprising, from increasing global governmental interventionist policies, one had to find a way of tackling an overheated economic market further exacerbated by overproduction in the agricultural sectors. In the United States, one attempt to pass a bill (the McNary-Haugen Bill) which would see excess agricultural products off-loaded in foreign countries, while concurrently raising prices in the domestic market failed. Failing this, in 1929 the US government then introduced the Federal Farm Board (FFB). The FFB while claiming some victory over agricultural excesses, ultimately it too, failed in its remit to address the virtually annual problem of "over productivity." Following on from this, the AAA or Agricultural Adjustment Act was passed in 1933 with the ambitious aim of reducing cultivatable acreage while compensating farmers for nonproduction/trade. Another method also used by the AAA was the environmental-based policy that united reduction in commodities while encouraging environmental protectionism via payments to farmers for shifting soil-depleting crops like wheat and corn etc. to soil-conserving plants such as legumes and grasses (Hardin, 1943).

2.8 Continued hybridization

Following on from previous agricultural attempts at increasing yields and thus raising productivity per hectare so, the big breakthrough came with the first commercially available hybrid corn. Progression in this field can best be shared by two people, George S. Carter (1921) and Henry Wallace (1923). Commonly quoted as the first person attributed with such advances was Henry Wallace who in 1913 became interested in the methods that Shull and East had pioneered. Eventually, through the help from the American government. Wallace established one of the first commercial hybrid corns in 1923 (Berlan and Lewontin, 1986). Carter in 1921 meanwhile, also building on the work of Shull, East and others, created the first double-crossed method of hybridization (Cook, 1937). Consequently, by 1923 this new method of breeding, whether cross-breeding or hybridization became profitable for both farmers and industry too (Morrison, 1947; Sprague, 1962; Smith et al., 2004). Furthermore, by the mid 1920s, simultaneous trials on wheat also permitted commercially available hybrids (APS, 2010). The acceptance of these new hybrids, especially in America by the 1930s, was immediate and production expanded to such a degree that grain yields increased significantly, further adding to problem of overproduction mentioned in previous sections.

Meanwhile, in the arena of health, this period carried mixed blessings.

2.9 The health organization of the League of Nations

Following the near end of the Great War (1918/19), influenza cut a swathe of devastation across the world by killing between an estimated 15 and 20 million people. In addition to this, approximately 2 million cases of typhus placed extraordinary burden on the Office International d'Hygiène Publique (OIHP). Separately, in early 1920, a proposal by the League of Nations (LoN) for a permanent international health organization called the "Health Organization of the League of Nations" (HOLN) was accepted. Not long afterward, another proposal saw the LoN proposing to combine the duties of both the League's Health Organization with those of the OIHP, in doing so creating a single international health organization. However, the reality was problematic in that, while the United States was a member of the OIHP and not the League, the proposal meant that any collaboration or joining of forces required that the United States become a member of the League. Sadly, the United States made it clear that this was counter to US interests and so for this and other reasons talks broke down leaving in its wake two separate international health organizations (WHO, 2010). As it turned out, the OIHP's main sphere of activity continued to be focused on the supervision and development of international quarantine measures, while the HOLN among other things continued to collect, collate, and disseminate information on the status of epidemiological diseases of importance. Initially, the HOLN almost exclusively dealt with health issues in the United States and Europe, its remit eventually expanded to encompass surveys taking on board global nutritional standards. Progress was slow, however, causing one commentator Ruxin (1996) to comment:

> ...prevalence of nutritional disorders in the developing world principally remained a scientific point of ignorance. ***Ruxin (1996)***.

This situation quickly righted itself as the HOLN went on to collaborate with many of the top nutritionists of the time, shaping both understanding and eventually policy. However, before the HOLN moved into full gear, independent studies and events were moving along at a fast pace advancing nutritional understanding.

2.9.1 Dietary energy research

During this period (into the 1920s), the debate as to nutritional requirements continued with Britain and America at the forefront.

2.9.1.1 *Britain*

While many studies showed disagreement and divergence when it came to nutritional consensus, there was growing convergence in the dietary makeup of a balanced diet. It is worth noting too that contemporary studies still relied on calorimetry in many methodologies, yet one noteworthy exception was Miss E. M. Bedale. Importantly, while many clinical trials vis-à-vis basal metabolic rate (BMR) still used calorimetry as mentioned above, the first real study using BMR to evaluate human energy (macronutrient) requirements was Bedale's. Indeed, in her 1923 publication of school-age children's energy needs, Bedale explored various children and their activities and determined energy expenditure in relation to their calorie intakes for each grouping using BMR. However, while these studies were of great pioneering effort, they were largely ignored by the wider nutritional community, owing it seems to her gender and ergo her credibility (Bedale, 1923; McNaughton and Cahn, 1970; Henry, 2005).

Meanwhile too, Chittenden's previous skeptical views regarding the nutritional communities over reliance on general nutritional recommendations was much criticized and was not universally accepted (Carpenter et al., 1997). Yet Chittenden's misgivings continued to echo in the early work of Henry Sherman who in 1920 published a sort of metaanalysis of over 100 (109) previous scientists' experiments and found protein recommendations varying a great deal ranging from 21 to 65 g for an average 70 kg man. This equated to an average requirement of 0.635 g of protein per kilogram of body weight equal to 44.4 g per day for an average 70 kg per man appearing to echo Chittenden's views. As such Sherman suggested that:

> Many of the published experiments which were designed to test the amounts of protein required in normal nutrition are now seen to have given misleadingly high results... ***Sherman (1920, p. 98).***

2.9.1.2 *America*

During the same period in the United States, the United States Department of Agriculture's (USDA) nutritionist Hazel Stiebling promoted nutritional research. Indeed, regarding her work at the Sherman laboratory, Stiebling proposed the first comprehensive dietary standards taking into account vitamin and mineral needs too (Harper, 1985). Table 2.7 highlights the original values even though she later partnered with Esther Phipard, further increasing recommendations to include both thiamine and riboflavin.

Importantly, this signified an ideological break from previous attempts to promote values that aimed to prevent starvation to one of inclusivity, taking into account overall optimum requirements for health, nutrition, and well-being (see also appendix Table 29.1 & 29.2 Macro- and Micro-Nutrient Guidelines of the EU, USA and WHO for Adult Men and Women - Parts One and Two) (Harper, 1985).

In the field of population pressures, amid global economic depression and agricultural uncertainty, the perennial argument of population growth and even overcrowding as well as the world's natural resources was brought once again brought to the fore (Holmes, 1924; Merritt-Hawkes, 1928).

TABLE 2.7 USDA recommended daily requirements by Hazel Stiebling (1933).

USDA Recommended daily requirements by Hazel Stiebling 1933	Energy	Protein	Calcium g	Phosphorous g	Iron g	Vitamin A units	Vitamin C units
Child under 4	1200	45	1	1	0.006 −0.009	3000	75
Boy 4–6; girl 4–7	1500	55	1	1	0.008 −0.011	3000	80
Boy 7–8; girl 8–10	2100	65	1	1	0.011 −0.015	3500	85
Boy 9–10; girl 11–13	2400	75	1	1.2	0.012 −0.015	3500	90
Moderately active woman or boy over 11–12, girl over 13	2500	75	0.88	1.2	0.013 −0.015	4000	95
Very active woman or boy 13–15	3000	75	0.88	1.32	0.015	4000	100
Active boy 15+	3–4000	75		1.32	0.015	4000	100
Moderately active man	3000	67	0.68	1.32	0.015	4000	100
Very active man	4500	67	0.68	1.32	0.015	4000	100

Source: Recreated from Stiebling's Food Budgets for Nutrition and Production Programs. (US Department of Agriculture, USDA, 1933).

2.10 Population pressure resurfaces: the First World Population Conference

Under the auspices of the League of nations, a population conference was held in Geneva in 1927. Importantly, this was the first real multilateral attempt to collate information and viewpoints with regard to the population problem (Carr-Saunders, 1927). The World Population Conference was also perhaps the first real international forum to consider the question of *optimum* population capacity of food and other resources, among other things (De Gans, 2002). Professor Raymond Pearl from Johns Hopkins University, one of the key scientists at the conference, utilized previous empirical studies of the US population, together with his work with rats, fruit flies, and yeast cells, extrapolated his findings of a potential burgeoning population to the present-day familiar growth curve that resembles a flattened "S" or "logistic growth" curve as it has become known (Pearl, 1927; De Gans, 2002).

Pearl further hypothesized that this growth was primarily due to fertility and mortality and only secondarily from economic and social factors (Pearl, 1927; De Gans, 2002). Pearl's work also signaled a departure from numerous previous trends predicated on historic growth rates. Instead he asserted the notion of future alternative conditional predictions based on combinations of single or multiple alignments of inter alia socioeconomic variables (Pearl, 1927).

Yet despite such innovation, Pearl's work was received as being divisive and polarizing, although many individuals, including the Secretary General of the League of Nations Sir Eric Drummond, considered it a great success (Carr-Saunders, 1927; Merritt-Hawkes, 1928; Connelly, 2006). Indeed, it was also, according to birth control pioneer and sponsor of the conference Margaret Sanger (1927), the first time that sociologists and biologists collaborated on a project with the sole purpose of using scientific analysis of social problems in a search for solutions to economic dilemmas (Sanger, 1927). Yet, despite this renewed drive in population matters, there was still little broad-ranging consensus over limiting factors that might determine population ceilings.

Shortly after in 1930, another high profiler of population pressures Hiller explored several previous studies that collectively cited limited factors of population growth ranging from temperature, light, air, and nutrition to humidity among others (Hiller, 1930). However, Hiller swiftly dismissed such studies as

> ...invalidated by methods used in their construction ... [that contain] ... logical inadequacies. ***Hiller (1930, p. 524).***

Furthermore, he also fully dismissed the notion of population pressures or predictions as generally untenable suggesting instead that previously used variables such as technological, social and cultural influences that had been used as determinants of population ceilings less than useful as, in his words:

> ...any attempt to predict long time population cycles will probably go as far astray as did the forecasts of Malthus. ***Hiller (1930, p. 550).***

Such concerns caused ripples in the field and were outed by many including Speier and Kähler who countered in their book "The War of Our Time" (1939) when it was noted that:

> ...interdependently social and economic conditions influence population development. ... A study of the world-wide implications of population trends gives us only further proof that it is necessary to build up an international system of social and economic co-operation if we are to prevent the starvation of millions of human beings... ***Staudinger (1939).***

2.11 The gold standard

On the economic front, the pre-1930s global trading platform was based on "relative" free trade backed by a hitherto strong gold standard. However, with the advent of competitive currency devaluations by certain nations topped with strategic manipulation of trade policy, the standard finally collapsed. There were some who came to consider this failing of classical economic paradigms, as paving the road for the new "Keynesian" revolution, which focused more on national or domestic fiscal and monetary policies (Goodrich, 1947). Not long after the collapse of the standard, the spectra of surpluses once again raised its head. Attempting to head off the worst of the economic fallout, the United Kingdom alongside the major wheat exporting countries of the time combined forces in an attempt to address the need to control and dispose of the unbelievably voluminous wheat surpluses (double the wheat exports of all countries in 1938–39). The answer came in the realization of the need to reduce production-altering trade barriers and other production, stock, or quota controls to stabilize supplies and by extension prices (Thompson, 1943).

2.12 Biotechnological advances

The late 20s and early 1930s saw great developmental leaps in the fields of microbiology and fermentation technologies (Zymotechnology). And, while a fuller, more complete understanding of chemical functions of post-war molecular biology was still to come, the early 1930s was nevertheless an exciting time. One such development and perhaps the most important discovery around this time was penicillin, discovered by Alexander Fleming in 1928 (derived from the Penicillium mold); this was undoubtedly a seminal moment in the field of bioscience. Another breakthrough saw the manipulation of certain microorganisms in which they could now be mutated through physical and chemical manipulation to become faster growing and more tolerant of less oxygen. It was this new hybridization that further promoted the development of new technologies, especially in food fermentation—zymotechnology (Shurtleff and Aoyagi, 2007; Gibson, 2016). Such developments led to the subsequent expansion of large-scale fermentation production, principally in the pharmaceutical industry. This propagated a rapid upsurge in interest in life sciences' research, which in turn also saw the eventual generation and the creation of new antibiotics in addition to increasing the range of enzymes and vitamins available (AU/ICE, 1997). Other notable moments in the biotechnology sciences are delineated in the following Table 2.8.

2.13 Green shoots of agricultural evolution

In the field of agriculture, little else has achieved more in the way of increasing crop production than the humble fertilizer. Fertilizers were not new at this time, rather it was the economical industrial synthesis of ammonia (Smil, 2001) that made all the difference.

TABLE 2.8 Key dates: Biotechnology 1919 to 38.

Date	Notable dates in the biotechnology sciences
1921	The first marketable hybrid corn is developed.
1925	Consuming more than 10% of the USDA total budget in 1921, Congress had to cut the decades-old free Seed Distribution Program.
1926	Hermann Muller revealed X-rays encouraged genetic mutations providing researchers a way to induce transformations. Henry Agard Wallace founded a hybrid-corn seed company called "Hi-Bred" known today as Pioneer Hi-Bred International, Inc.
1928	Alexander Fleming discovers Penicillin.
1934	Desmond Bernal illustrated that molecules of protein could be studied using X-ray crystallography. Martin Schlesinger purified bacteriophage and in doing so found equal amounts of DNA and protein.
1938	The term "molecular biology" was first coined.

Source: Compiled from: NHM (1989/90); BioTech Institute (2010).

However, up until about the turn of the century, obtaining commercial fertilizer for agricultural use was not very easy. It was about the same time when global crop yields started to wane and stagnate. Indeed in the United States, crop production, particularly wheat grain, was achieving only about 1.7% increase in 1909 than practically the whole of the previous decade (Webb et al., 2008). Consequently, it was arguable that actual, natural crops physiological limits were being reached.

Accordingly, it was felt, more than understood, that any substantive increase in yields would have to be driven by agricultural technologies—technologies such as fertilizers and the like. Unfortunately, what was holding progress back was "on-the-ground farmers" shortage of nitrogen for fertilizers[1] Indeed, in the first two decades of the 1900s, it was common for farmers in the United States and Europe to lack sufficient credit and purchasing power for the acquisition of nitrates to make these fertilizers. This occurred despite nitrates being one of the top three commodities (by volume), globally shipped. As far as synthesizing artificial nitrogen compounds in the laboratory went, the goal was to make it plentiful and cheaper for the end user. And indeed, a process had previously been developed by two Germans Fritz Haber and Carl Bosch[2] in 1909 that could directly synthesize ammonia from atmospheric nitrogen and hydrogen gas, which could in turn be used in the production of fertilizer. The process was then first used industrially by BASF Limburgerhof agricultural research center in 1913 and then by the government of Germany. Yet during the War, the German government shifted focus and instead of using ammonia for peaceful agricultural purposes they channeled the conversion of ammonia away from fertilizer to the production of high explosives instead. Incidentally, a side note saw both chemists later winning the Nobel Prize for their work in chemistry—Fritz Haber in 1918 and Carl Bosch in 1931. This fixing of nitrogen became known as the Haber–Bosch process and was at first a closely guarded German secret. That was until after the war when a clause of the treaty of Versailles ensured that BASF was obliged to share the knowledge, albeit under license, to a new, purpose-built unit in France. Industrial plants then began to pop up everywhere across the globe taking advantage of this "marvelous" new industrialized technology, which, in turn, opened up plentiful possibilities in the global agricultural sector (Smil, 1999).

Thus, the preindustrialization of fertilizers and their constituent components, which were globally traded among growing financial speculation, crop failures, and trade constraints,

[1]Fertilizers—plants rely on certain nutrients for healthy development; these include but are not limited to nitrogen, phosphates, and other minute amounts of trace elements. Furthermore, plants can only uptake nitrogen in its nitrate form. This means that nitrogen in the atmosphere is not an option for plants. Instead, plants utilize microorganisms which convert atmospheric nitrogen into nitrate (that plants can use) within the soil. Therefore, soil pH (acidity or alkalinity) must be carefully monitored and adjusted to ensure the optimal growing conditions for each particular crop. Fertilizers then are very useful as they contain the right nutrients in the right quantities for the budding plant. This is why one of the major developments of the Green Revolution was the use of chemical fertilizers, not only to adjust soil pH balances but also to achieve the right levels of all the important trace elements and compounds such as nitrates and phosphates.

[2]Haber–Bosch process was also the work of great insight. Playing with the method of mixing nitrogen from the atmosphere with hydrogen and enormous amounts of heat, the mixture was then burned leaving behind a nitrogen–hydrogen mixture. Furthermore, this was then compressed and heated to between 200 and 700°C in the presence of a catalyst (metal oxide) with the end result being ammonia for further use in making fertilizers.

were readily translated into higher food costs and by extension food prices. All this changed with the advent of industrialized fertilizers.

2.14 Health and nutrition studies foster acrimony and collaboration

Back in the field of health and institutional collaboration, it was a mixed bag of pluses and minuses.

2.14.1 Cross-border nutritional collaboration

On a wide-reaching mission, the League of Nations (LoN) and others looked closely at the health and well-being of children first before concentrating on the wider populace. Indeed, in the writings of Périssé (1981), the international community shocked by the overwhelming impact the Great Depression had on people's overall health and living conditions, people and institutions began to take serious interest in the situation and started to make sweeping decisive moves to arrest and prevent further undernourishment (Périssé, 1981). In this endeavor, collective multinational institutional collaboration first began in 1925 with the Health Organization of the LoN, the International Institute of Agriculture (IIA), and the International Labor Organization (ILO). Together they researched nutrition in, vis-à-vis agriculture, social, and economic problems (League of Nations, 1936a,b; Mayne, 1947). This paid dividends and by the 1930s many foreign governments began refining children's nutrition with the establishment of free or subsidized milk as well as school meals. From these small beginnings, one could feel a palpable momentum growing. However, it was not until the consolidation of so many strands of study by the League and others that the LoN finally made real international progress (Carpenter, 2007). It began with the recognition that what was needed was a set of common nutritional requirement standards that could be used as guidance and comparison across regions and nations.

At the same time by 1931, based in Switzerland, the LoN's Health Secretariat's Nutrition Division recruited Wallace Aykroyd as what was perhaps the first ever nutritionist explicitly tasked with international nutritional guidance and responsibility (Carpenter, 2007). Then in 1932, the League organized two conferences of expert: one in Rome and the other in Berlin. Their goal was to assess the fundamentals of acceptable diets while assessing the role of nutrition inter alia the economic crisis (League of Nations, 1936a; Périssé, 1981; Carpenter et al., 1997). This and others acted as forerunners to the League's 1935 Expert Commission on Nutrition Conference, London (see Chapter 2.14.4) (Medical Science, 1936).

2.14.2 The Burnet and Aykroyd report

Meanwhile, on the Yugoslav representative of the HOLN suggestion, information was collected on the food situation in a sample of representative countries. Overseen by the Director, Dr. Frank G. Boudreau, the result was the Burnet–Aykroyd report: "Nutrition and Public Health." The report, drawing on the works of John Boyd Orr (UK) and Hazel K. Stiebling (US), showed that despite plentiful or excessive food production, the poorest of people were still suffering from hunger and deprivation. Furthermore, perhaps of overriding inference, was the fact that the report formally acknowledged that at the center of the issue was the underconsumption of food as a direct result of low purchasing power. This brought poverty into the mix (more below) (Burnet and Aykroyd, 1935; League of Nations, 1936a; Harper, 1985). The report focused on three main points: firstly, nutritional needs; secondly, the resources required to fulfill such needs; and thirdly just how best to employ these resources for the betterment of the general population (Burnet and Aykroyd, 1935; League of Nations, 1936a,b). The report was submitted at the 16th full assembly of the League of Nations in 1935. It was a key piece of work that not only acknowledged the previous ideas of John Boyd Orr (later the first Director-General of the FAO) but was also timely. Timely, in that it brought another level of understanding to an international forum amid a period when other important discoveries were being made in the field of diet and nutrition. In fact, the report was very significant in that it recognized that health had indeed become a problem of social and economic importance and by way of addressing the situation proposed a marriage of health and agriculture (Gibson, 2016).

2.14.3 Joint memorandum of the BMA and MoH

Domestically, in the Britain at the time, regarding dietary recommendations disharmony came about with the publication of the British Medical Association's 1933 report stating that for the average person, a daily health and maintenance regime should comprise 3400 calories. While, previous work by the Ministry of Health (MoH) advocated only 3000 calories (BMJ, 1933; BMJ, 1934), the tabloids picked up on this indiscretion in a heartbeat and implied that the cost of the BMA's recommendations for sufficient basic food for good health could not even be covered with limited income from unemployment pay. Not surprisingly a political furor erupted. Equally not surprising was that the MoH was enraged and the report led to an acrimonious confrontation among the two bodies. Eventually, good sense prevailed and all agreed that the margin had to somehow be decreased. To this end, a conference between the Nutrition committees of both the MoH was held in 1934 to address the bodies various areas of contention. While tense, the meeting nevertheless reached an eventual agreement in the form of a joint memorandum. This memorandum proposed a set of increasing caloric requirements (Table 2.9) based on age sex and activity level. For the purpose of general aggregation, an average calorie consumption of 3000 calories per person was also agreed (BMJ, 1934; Smith, 2003).

Incidentally, Beardsworth and Keil (1997) writes that despite reservations by both parties and although malnutrition was still widespread throughout the United Kingdom, the interwar years actually saw Britain's general dietary health improve quite markedly (Beardsworth and Keil, 1997). Meanwhile in a more global arena, the League of Nations and others were also active in the role of nutrition for the betterment of children and the masses. Further key dates of this period can be found in the following Table 2.10.

Back to the Burnet–Aykroyd report and its many forward-thinking objectives brought renewed enthusiasm to the table.

2.14.4 The Mixed Commission of the LoN

The Burnet–Aykroyd report was of such consequence that it could not be ignored and great discourse among the various parties ensued. Deliberated in detail at the LoN's general

TABLE 2.9 Agreed daily requirements by the BMA and MoH.

Group	Calories
Male	
- Light work	2600–3000
- Medium work	3000–3400
- Heavy work	3400–4000
Female	
- Active	2800–3000
- Housewife	2600–2800
Children	
Boy 14–18	3000–3400
Girl 14–18	2800–3000
Children	
12–14	2800–3000
10–12	2300–2800
8–10	2000–2300
6–8	1700–2000
3–6	1400–1700
2–3	1100–1400
1–2	900–1200
Adult average	3000

Source: Recreated from The Nutrition Question. British Medical Journal 1(3828): 900–901, 1934.

assembly, it stirred up many emotive topics. One of the people enthused by the Burnet–Aykroyd report was the former Australian Prime Minister Stanley Bruce. Bruce, stimulated by the report, addressed the LoN and firmly backed up the notion of a merging agriculture and health responsibilities (FAO, 1970). Consequently, in direct response to the meeting, several initiatives were put in place. Firstly, a technical committee was selected to review the report's physiological foundations of nutrition, while secondly another committee, the Mixed Committee, was inaugurated and tasked with looking into the report vis-à-vis the full wide-ranging issues of nutrition in regard to health and economics (Campbell, 1938). The new Mixed Committee was to be chaired by Lord Astor who was to oversee the nutrition work of the HOLN, the International Institute of Agriculture, and the ILO (Royal Statistical Society, 1936).

The mixed committee produced two reports: an interim and final reports. The interim report titled "Problem of Nutrition" comprised four volumes: An Interim Report of the Mixed

TABLE 2.10 Key dates of the period: 1931 to 34.

Key dates of the period	
1931 International Wheat Conferences	The first few International Wheat Conferences between exporters and importers (Rome, London) failed to reach an agreement; it did, however, eventually lead to the first successful agreement among its cohorts (IGC, 2009a).
1931 The International Relief Association **(IRA)**	Created to aid the adversaries of Hitler and the Nazis movement, the International Relief Association was suspended in 1933 (Montgomery, 2008).
1933 International Rescue Committee **(IRC)**	Despite the cessation of the International relief Association, an American branch of the IRA was actually created the behest of Albert Einstein to continue the previously good work of the IRA (IRC, 2009).
1933 First International Wheat Agreement—Wheat Advisory Committee	After extensive negotiations, the third International Wheat Conference was a good chance for both the importing and exporting countries to sign the first International Wheat Agreement. On top of this, the Wheat Advisory Committee (later to become the IWC in 1942) was born. However, sadly the first International Wheat Agreement failed in the first year (League of Nations, 1933; Meerhaeghe, 1998; IGC, 2009a,b).
1933 US Agricultural Adjustment Act (**AAA**)	Under the auspices of the US Agricultural Adjustment Act (**AAA**) (PL. 73 −10), the United States intervened in the market mechanism to restrict US agricultural production and by extension surpluses. They did this by paying farmers to reduce the cropping area they cultivated and thus reduce and stock, thus raising the value of the actual crops that were grown (Fite, 1962; NAL, 2010).
1933 Kwashiorkor	Inadequate protein consumption in infants usually induced by poor breastfeeding (the sickness of weaning) was given the name "kwashiorkor" in 1933 by a British nurse Cecily Williams working in the Gold Coast. Originally, the term was taken from the GA language of coastal Ghana.
1933 US Commodity Credit Corporation (**CCC**)	The United States Commodity Credit Corporation (**CCC**) became the preferred body for managing food stocks and surpluses while protecting farm income and market prices. The corporation formed the foundation of the first food aid and became part of USDA in 1939 (FSA, 2008).
1934 US Export & Import Bank	US Export & Import Bank was created to promote US exports through the use of concessional loans.

Source: Compiled from multiple reports and committees League of Nations (1933); Fite (1962); Meerhaeghe (1998); FSA (2008); Montgomery (2008); IGC (2009a,b); IRC (2009); NAL (2010).

Committee on the Problem of Nutrition; the report on the Physiological Basis of Nutrition; Nutrition in various countries; and lastly, Statistics of Food Production, Consumption and Price (League of Nations, 1936a,b; Royal Statistical Society, 1936). Collectively, recommendations were made encouraging to both governments in general and the LoN to, inter alia, setup of national nutritional committees to fully promote up to date nutritional policies (Mcdougall, 1940). The report was further revised which eventually led to the Final Report of the Mixed Commission of the League of Nations on the Relation of Nutrition to Health, Agriculture and

Economic Policy (League of Nations, 1937). The report expressly noted the failure of mankind to satisfy normal nutritional requirements (League of Nations, 1937). The report further elaborated on whether it was not the sole duty of public authorities to assume the responsibilities inherent in a nutrition policy, or whether wider stakeholders, including the public, should take up some of the responsibility (League of Nations, 1937).

The Final Report of the Mixed Committee was a remarkable document, which contemporarily was the most detailed and authoritative analysis of nutrition and the broader notions of social economic and agricultural issues ever undertaken to date. The report even became a best seller voted by the New York Times as the most important and influential book of the year (Mcdougall, 1940). While the report and its supplementary technical papers set forth the LoN recommended nutritional requirements albeit without supporting scientific justification, nevertheless, it did propose a blueprint calling for dietary standards: the dietary recommendations for special groups; nutritional surveillance; and global nutritional education (Périssé, 1981; Harper, 1985; Carpenter et al., 1997). Caloric recommendations for the general population were given for adults with various workloads, infants, and children (Table 2.11). Indeed, the guidelines further promoted the notion of optimal nutritional needs rather than minimum nutrition which had been practice for so long (League of Nations, 1936a; League of Nations, 1937; Périssé, 1981).

Moreover, while the report was more focused on industrialized nations, it did, however, look at the developing world too—although in some eyes, not enough. Among the many conclusions and recommendations, some of which are described previously, the report went on to acknowledge the fact that in many countries, grave malnutrition was rife; furthermore, it was also understood that the means and ware withal to address such issues already existed. The report was further decisive on the notion that while ignorance was a big problem in the exacerbation of malnutrition, poverty was undeniably an important central component that needed to be tackled. This further reinforced the idea that optimum health and well-being was not solely the provision of energy through food intake but rather seeing diet as a holistic notion that involved numerous other factors, not least of which was poverty (League of Nations, 1937; Hanekamp and Bast, 2007).

The period between the two wars was very productive in its nutrition and health research and policies. Great momentum was also garnered in terms of social responsibility and scientific understanding. While, in the early 1920s, protein standards appeared to be more predicated on opinion than fact, by the end of the 1930s, experimental nutritional research laid more solid foundations for energy requirements and better health (Hwalla and Koleilat, 2004). Furthermore, despite an ongoing lack of general universal consensus, continued nutritional research nevertheless flourished and increasing convergence began to develop (Table 2.12). It was also understood at about this time that provision of food in adequate quantities and quality would profoundly affect a nation's health more than any other single reform (McDougall, 1943).

In the world of governance during the 30 s, a growing awareness of food and nutrition as well as the consequences of an inadequate diet were beginning to loom large in many countries policy priorities.

TABLE 2.11 The League of Nations nutritional recommendations.

Age	Calories	Protein gm/per kg	Calcium gm	Iron mg
Infants				
0–6 Months	100 (per kg of body weight)			
6–12 Months	90 (per kg of body weight)			
Children				
1–2	840	3.5	1.0	
2–3	1000	3.5	1.0	
3–5	1200	3.0	1.0	
5–7	1440	2.5	1.0	
7–9	1680	2.5	1.0	
9–11	1920	2.5	1.0	
11–12	2160	2.5	1.0	
12–15	2400	2.5	1.0	
15–17	2400	2.0	1.0	
17–21	2400	1.5	1.0	
Men	2400	1.0	0.75	10
Women	2400	1.0	0.75	10+
Pregnant (1–3 months)	2400	1.0	1.5	10+
Pregnant (4–9 months)	2400	1.5	1.5	10+
Nursing	3000	2.0	1.5	10+
	Extra calories to be added per hour of work			
Light work	Up to 75			
Moderate work	75–150			
Hard work	150–300			
Very hard work	300 calories and upwards			

Source: Compiled from mixed reports League of Nations (1936a,b); League of Nations (1936a,b); League of Nations (1937).

2.15 Growing multilateralism

During this period, regardless of Ruxin's (1996) suggestions to the contrary, many national and international research studies and articles surrounding the issues of food and the developing world were undertaken (Firth, 1934; Gilks, 1935; Worthington, 1936; Watson, 1937; Auchter, 1939). Having said that though, in Ruxin's defense, the LoN's Nutrition report of 1935 did in fact highlight a shortage of research in certain developing countries, particularly

TABLE 2.12 Nutritional recommended requirements for adults 1919 to 39.

Date/ Authority	Load (default man)	Protein	Fat	Carbohydrate	Calories
1932 Ministry of Health's Advisory Committee on Nutrition (ACN)					
MoH suggested a standard 3000 calories as being adequate for daily requirements for the average person (Ministry of Health, 1932; BMJ, 1934)					
					3000
		67–75			
1933 British Medical Association (BMA)					
The BMA's 1933 report caused a social outcry by citing 3400 calories as their recommendations for health and maintenance (BMJ, 1933; BMJ, 1934)					
					3400
1933 USDA					
Under the auspices of Hazel Stiebling (an American Nutritionist), the first recommendations were made containing not only energy requirements but also proposed vitamin and mineral allowances for calcium, phosphorous, iron, and vitamins A and C (Stiebling, 1933)					
					3–4500
1934 Joint Ministry of Health and British Medical Association Memorandum					
The disagreement and political consequences caused as a result of the MoH and the BMA's disagreements resulted in collaboration out of which an agreed sliding scale of values was recommended (BMJ, 1934; Smith, 2003)					
	Man				2600–4000
	Woman				2600–3000
1932 League of Nations Conference of Experts					
The LoN met to consolidate several different dietary values so that nutritional status could be compared across and within countries. (League of Nations, 1936a,b; League of Nations, 1937; Hwalla and Koleilat, 2004)					
		Not less than 1 g per KG			2400

Source: Compiled from Multiple Data sets

in Asia. In due course though and building on 10 years of research, the 1935 meeting of the League of nations described previously was in fact the first of its kind. Indeed, it was the first major multilateral forum whereby ideas inter alia food, nutrition, and public health came together synthesizing into one lofty, multilateral global objective to expand principles of good nutrition throughout the world (League of Nations, 1937). It was also noted that food and nutrition vis-à-vis good health and well-being was a multidimensional beast that had to be tackled on many fronts if progress were to be made (Campbell, 1938; McDougall, 1943; FAO, 1970; Shaw, 2007). Furthermore, to tackle such a wide-reaching goal of sufficient food for everyone, it was recognized that there was a need for a multilateral world food security organization that could combine both food and agriculture under one sphere of responsibility. And the LoN was not that body (Gibson, 2016).

2.15.1 Poverty and supply

To date, the problem of sufficient quantity and quality of food had only been considered in terms of production and supply. Yet the Burnet–Aykroyd report opened people's eyes to the idea that amid talks of surpluses and excessive food production, the main hurdle in fact was poverty and low purchasing power that ultimately led to much hunger and malnourishment around the globe (Burnet and Aykroyd, 1935). Much later, the idea of inadequate or low purchasing power was central to Amartya Sen's 1981 essay on deprivation and entitlement and one that incidentally contributed to his 1998 Nobel Prize on Welfare Economics. Yet it, even before poverty, as central to the notion of inadequate food supplies, had its genesis long before. Saying that, however, the idea was perhaps first properly elucidated in the mixed committee's report—the first internationally focused scientific publication that unconditionally placed poverty at the center of the problem:

> Poverty and ignorance remain formidable obstacles to progress in the disparity between food prices and incomes and increases the difficulty experienced by the poorer sections of the community in obtaining an adequate supply of the proper foods ... That this situation can exist in a world in which agricultural resources are so abundant ... remains an outstanding challenge to constructive statesmanship and international cooperation. ***League of Nations (1937).***

2.15.2 The League of Nations: for Whom the Bell Tolls

During its early years, as an institutional body, the League of Nations flourished. Yet despite great leaps in international economic and social cooperation over the years, particularly in the fields of health and nutrition; the inability of the institution to foresee or prevent hostilities of the next World War sounded the death knell for the first truly international organization. Indeed, the league suffered from many shortcomings, the biggest of which was its inability to offer little more than sound bites and moral outrage when hitherto friendly states (in this case, Japan, Italy, and Germany) became aggressors. The stage was set for an almighty confrontation when in 1931 Japan, in defiance of both the Assembly and the Council, waged war against China. Furthermore, in 1935, Italy invaded Ethiopia and the resultant League's sanctions were limited and only half-heartedly supported. Finally, in 1936, the League's impotency became clear as the illegal German reoccupation of the Rhineland left the League looking powerless, unwilling or unable to do anything about it. Indeed, even in 1939, while Hitler's army marched into Poland, not one member of the League raised his/her voice or called for a meeting of the Council or Assembly (Encyclopedia, 2002). In this one move, it had become clear for all to see that the LoN had ultimately failed in its main duty to prevent what was looking to become the outbreak of World War Two. Effectively, the LoN was undermined and was seen as powerless and incapable (Shaw, 2007).

By contrast, up to 1938, the HOLN had, even now, been busy in the spreading of its message far and wide with resultant nutrition committees springing up everywhere (in 21 countries, to be precise). However, despite such progress, there still remained a lack of material or evidence on the gravity or extent of poverty and undernutrition throughout the world. In response, in 1939, the HOLN called for more, in-depth extensive nutritional surveys. Unfortunately, with the outbreak of the Second World War, such ideas had to be put on hold (Ruxin, 1996).

As the debate over good, bad, and adequate nutrition continued, Rudolf Steiner was at the forefront of a new, more gentle method of organic farming.

2.16 Organic farming and environmentalism

(See also Section 2.6 Bio-dynamics and Organic Farming Chapter 7)

As mentioned previously, Rudolf Steiner had by 1924 already acquainted those that were interested with the notion of bio-dynamics and reenforcing the benefits of organic farming. In this, Steiner's central belief in "natural agricultural" bio-dynamic farming became the forerunner of the modern organic movement. One keen practitioner was the British agriculturist come writer Lord Northbourne, born Walter Ernest Christopher James. Northbourne tended his estate—Northbourne Court as well as his Home Farm in accordance with the principles of biodynamic as advocated by Steiner. In doing so, Lord Northbourne avoided the use of all chemical fertilizers, herbicides, and pesticides. Also, in tune with the movement, all organic waste was composted and returned to the soil. In this one can easily argue that Lord Northbourne was indeed one of the first organic farmers in Britain. In fact, such was his importance within the organic movement as in 1938/9—the first known conference on organic farming took place at Northbourne Court. On top of this, in 1940, Northbourne authored an influential book entitled "Look to the Land." In the book he foresaw the many fears that were to preoccupy people both inside and beyond the organic movement. This was also undoubtedly one of the key inspirational works that influenced the then-to-be-famous environmentalist of the 1960s Rachel Carson. In a moment of prophetic clarity, Northbourne wrote

> In the long run, the results of attempting to substitute chemical farming for organic farming will very probably prove far more deleterious than has yet become clear. ***Northbourne (1940, p. 61).***

Additionally, Northbourne has since been credited as first person to use the phrase "organic farming" (James and Fitzgerald, 2008).

Shortly before in 1939, inspired by the work of Sir Albert Howard and Sir Robert McCarrison, Lady Eve Balfour started the famous Haughley Experiment. It began in 1939 on her own farm in Suffolk whereby she intended to investigate the beneficial claims of organic husbandry. She began by managing three side-by-side tracts of land, allowing Balfour to study the full cycle of food chains over successive generations of plants and animals. By 1945, such was the scale and involvement of the experiment that Lady Balfour helped to manage the experiment and correlate the information. This led to the creation of a body called the Founders Committee, which later, in 1946, became the Soil Association (SA). A year or so later under the threat of financial strain, the experiment was fully taken over and administered by the SA. Incidentally, the longitudinal experiment lasted for over 40 years by which time enough information was collated and analyzed to make some firm conclusions. The experiment's results exhibited moderately higher moisture content, as well as higher organic C, and other mineral components N, P, K, and S content within the organic soil (Balfour, 1975; Blakemore, 2000) (Gibson, 2016).

From such modest beginnings, so the world's first organic association - the Australian Organic Farming and Gardening Society (AOFGS) came into being in Sydney in 1944 (Paull, 2008).

During the same period, the advent of the beginning of the Second World War brought with it the specter of more food worries with low provisions, surpluses and rationing.

2.17 Second World War: provisioning and rationing

Just as before, once hostilities of the Second World War broke out, the threat of food shortages once again raised its head. Previous surpluses of previous years had dwindled and it became one of the priorities for the allied powers to scale-up production and provision of food on a massive scale. As one commentator acknowledged

> ...the food and nutrition problems of a nation at war are tremendously complex and involve a wide network of interrelated activities... ***Clayton and Black (1943, p. 105).***

2.17.1 Production of food

So, throughout Europe, not surprisingly, those nations heavily reliant on imported foodstuffs once again had to adopt widespread rationing and other food control measures (Bacon, 1943; Volin, 1943; Whipple, 1943; Lloyd, 1943; Richter, 1943). As needs must, so those more reliant, and economically capable, paid more attention to the long-term vision. Thus, for some, reduced dependence on imports while increased local agricultural acreage helped secure some measure of success. However, not all nations in need faired the same, due to many factors. To offset some of these food shortages and general aid in the War effort, the United States created the Lend Lease Agreement. Introduced in 1941 the lend lease arrangement was a simple measure that allowed countries to purchase goods and services from the United States on a loan basis, but under favorable terms (Clayton and Black, 1943). Procurement was then expedited through the Office of the Lend Lease Administration before being transferred to the respective countries. During this time countries and region's fared quite differently due to adaptation and policy shifts.

2.17.1.1 Britain

As the Second World War broke out, Britain was still in a period of food surpluses. However, as needs grew so surpluses dwindled, or were being managed, once again the UK Ministry of Food introduced rationing in 1940. This move was also the result of genuine worries of a successful bombing and submarine campaign by the German forces (Black, 1943; Lloyd, 1943). This meant that food imports were drastically reduced and increased acreage under production grew by as much as 50%. It also meant that national production was jointly micromanaged by both the Ministry of Food and the Agricultural Department. The Ministry of Food also aimed to keep staples like bread and meat affordable by becoming the nation's sole purchaser of food imports. This also meant employing interventionist policies to aid in the goal of feeding the nation (Clayton and Black, 1943). Thanks to the US lend lease program and Britain's wartime food planning policies atop equitable distribution through rationing meant that Britain's general nutritional status was, quite possibly, better than that accomplished in the United States at that time (Cassels and Hall, 1943; Lloyd, 1943).

2.17.1.2 *Europe*

Generally, before the war broke European countries, particularly Austria, Czechoslovakia, Germany, and Sweden were mostly self-sufficient in the range of 80% caloric production to need (Richter, 1943). However, hostilities took its toll and by about, midway through the war, Belgium and Norway at least had become reliant on imports of grain and other feedstuffs. Imports were meagre and both countries ended up suffering badly from suboptimum diets. This was further compounded by the German acquisition of fish from Norway and meat from Belgium. Alternatively, both Sweden and Switzerland, with continued access to imports, fared better. In Denmark though, they still managed to produce much of its needed national requirement of food, even after exports to Germany. So much, in fact, was produced that

> ...enough was left for the Danes to maintain the highest food allowances in all of rationed Europe. ***Richter (1943).***

In France, Spain, and Italy though, rationing was introduced while a thriving black market and peasant revolts guaranteed it was the urbanized population that mostly suffered from shortages (Bacon, 1943). Contrastingly, with an agricultural workforce largely comprising women and children, Greece was less affected than other highly developed regions (Whipple, 1943).

2.17.1.3 *Africa and Asia*

It was realized that in Central Africa to achieve the goal of sufficient food in areas deficient of sufficient caloric intake, extensive international action would be required. And in some cases, this would have to include considerable financial aid. Overridingly, in Central Africa, the problem of supply was thought to be mainly one of education that would bring together farmers and producers to the table with the express goal of innovating and exploring ways to increase production (McDougall, 1943). In Asia, however, (Japan, China, Manchuria Philippines, Netherland Indies, French Indo-China, British Malaya, Thailand, Burma, and India), the food situation was a little different. The self-sufficiency of rice and other basic staples was not adversely affected by the war and the respective countries nutritional status quo was largely maintained (Ladejinsky, 1943).

2.17.1.4 *The Americas*

During the pre-war abundance of food, Americans were used to thinking in terms of agricultural surpluses rather than shortfalls (Cassels and Hall, 1943). However, that all changed as military needs, the allies lend lease requirements, on top of labor shortages, and challenged the continuation of the status quo (Clayton and Black, 1943). At this juncture, previous agricultural abundance gave way to a system of rationing based more on need rather than a person's ability to pay. To say that food shortages came as a shock to the Americans might be a bit of an understatement but the Americans nonetheless rallied to the challenge by increasing production that helped supply the war effort. Also aiding in this collective war effort was the political inclusion of the Secretary of Agriculture into the Defense Council in May of 1940. This essentially placed agriculture on a war footing. The aims of the Advisory Commission to the Defense Council was in part the maintenance of pricing parity between both industrial and farm prices and the scaling-up of production where possible (Clayton and Black, 1943). However, in spite of such ambitious goals, it quickly became clear that such measures were not working as was hoped. Then, shortly after, in 1941, President Roosevelt inaugurated the Office of Production Management solely accountable for production and purchasing.

In areas of Central and South American countries, despite certain regional variations, particularly in Central America, the region was basically self-sufficient in production, although in some instances, especially in the temperate zones, some countries were producing large surpluses (Almonacid, 1943).

2.17.1.5 Russia

Midway through hostilities Russia collectivized national agriculture and while every effort to increase acreage under cultivation, as well as large-scale heavy mechanization, for one reason or another, policy after policy failed. With such significant failure, it was noted by the allies that Russians were increasingly under grave threat of mass hunger and starvation—so much so that substantial allied assistance was seriously anticipated (Volin, 1943). One major downside with Russia's particular system was seen after the collectivization of agricultural land. However, things did not pan out as some had wanted and so perishable food stocks fluctuated somewhat. Not surprisingly, between 1932 and 1933, this led to either feast of famine on localized and regional scales. For much of this, the blame was laid squarely at the government's feet. Indeed, failures of Soviet central planning were bad in themselves but was further compounded by Stalin's decision to deny or withhold food from the masses, reserving food instead for the Red Army. This, together with the tendency of funding the war machinery through sales of much-needed food (through exportation) of food supplies, led to huge loss of life (Gantt, 1936). Such was the estimated volume of numbers who died through starvation and other opportunistic diseases at the time ranged between 2 and 15 million (Gantt, 1936); however, these figures seem somewhat vague and too broadly ranged, that later estimates asserted the figures were more likely closer at between 3 and 7 million (Dyson and Ó Gráda, 2002).

Thus, rationing and other measures while having a drastic effect on the region or nations population did highlight and focus attention on a person or nations nutritional need as well as directed like-minded people of the need to increase food production on a massive scale (WHO, 1958).

2.17.2 Food and health on a war-footing

In the field of health, with the outbreak of the Second World War nutritional health strategies, especially in the United Kingdom, the United States and a few other countries were becoming increasingly bound up with national health directives (Beardsworth and Keil, 1997). One major consequence of improved scientific knowledge within issues of health was that it allowed rations to be more precisely determined at the national level. But that's not all, interest was also shifting in favor of understanding nutritional needs of disaggregated populations. This effect trickled down the social demographic ladder—firstly directed in favor of population subgroups, then onto household, and then to the individual (Beaton, 1992). However, one major drawback regarding the work of respective National Nutrition Committees was the fact that once the Second World War raised its arms in aggression; it also effectively shut down many domestic nutritional programs throughout several countries; although some slipped the net and were able to carry on their research as well as on the ground policies (McDougall, 1943). In this respect and at this time (1940s), Britain was very much at the forefront of studies such as health and nutrition. At the heart of such policies was the almost zealous belief that nutrition was a wholehearted remedy for many social ills of the day (Ruxin, 1996). So, armed with a renewed zeal, up to date knowledge and a general acceptance of a necessary rationing culture the UK government embarked on several nutritional education programs.

About this time, even before the United States had entered the Second World War, a report by the revered and influential National Research Council (NRC) was passed along, assuring the US government that the American people were in no immediate threat of a shortfall in protein supply. That said, the NRC did acknowledge that, were the country once again called upon to export their high-protein foods to allied nations, etc., then a protein shortage would probably indeed affect the American population. At about this time too, both the US and British governments concerned for their respective troops attempted to establish an ideal nutritional requirement for soldiers in the field. However, as an aside, the NRC and numerous other similar agencies and bodies had, for a few years now, tended to be preoccupied with proteins in the diet. Although, as mentioned, the NRC was in the frontline in this respect, others seemed to follow blindly. The obsession with protein at this time not only fueled an ongoing obsession but also went on to dominate much scientific nutritional literature in the years that followed.

Science and technology in the war years seemed to advance in leaps and bounds gathering momentum as one invention or discovery after another was quickly applied to real-life situations. The timing could not be better either, as much weight was given to the application of existing technologies for the benefit of the war effort.

2.18 Science and technology: increased momentum

This was a time during which the automation and mass production of food really took off. Whether frozen, dehydration, and concentration of foods, the application of science and technology was readily adopted. Perhaps more so during this period as the War encouraged innovation and application. Whether concentrated juice, for example, was perfected and was readily available to the troops. As was flour that had been fortified with iron and vitamins. Furthermore, technology brought advances like aseptic processing and packaging of foods which prolonged shelf life through the treatment of high-temperature, short-time sterilization techniques. Such advances further improved the quality and safety of food through the retention of much needed nutrients (IFT, 1999; Gardner, 2002).

Other events around this time witnessed the large-scale manufacturing of penicillin; at the same time the introduction of electron microscopy was moving quite rapidly, which incidentally was effectively used in the identification of viruses that infect bacteria (bacteriophage).

This was 1942 and within a year separate studies by Salvador Luria and Max Delbruck who independently studies mutation in bacteria. This, among other previous developments in the field, observed the growth of a new field of scientific discipline, bacterial genetics. In the following year, further research led to the uncovering of the fact that intricate protein was shown not to be the pathway of genetic heredity bat rather DNA was the material substance of the gene and responsible for generational transformations (BioTech Institute, 2010).

2.18.1 International standards of dietary nutrition

As increasing involvement in all aspects of food's production and supply by the British government of the day, it was inevitable that a new philosophy evolved that looked to enhance the quality of the diet and in turn the nation's nutritional health. This was partly due to the Boer war of 1899–1902 in which more than one-third of British applicants were rejected on the grounds of poor health. And things did not seem to get much better either. In 1917/18, conscription to fight in the First World War left a further 2 1/2 million Britons graded "C3" meaning unfit for military service. Yet again the government was at a loss and was shocked to find the problems were that of nutrition. This further politicized the public health issues of the day and additionally increased interventionist policies regulating and monitoring suitable and adequate standards of dietary intake (Starling, 1918; Smith, 2009).

This was about the same period when other countries began to recognize the need for long-term monitoring of food issues and to finally take responsibility for the nation's health and well-being. This was especially so with children (Acheson, 1986).

As such, this was a seminal time when many governments from around the world began to adopt nutritional guidelines based on scientific guidelines. As luck would have it, this was also about the same time as that new technology and further understanding was readily forthcoming especially in the field of food microbiology, health, and nutrition particularly when it came to vitamins and minerals (Beardsworth and Keil, 1997). Fortuitously such understanding came at a time when nutritionists were beginning to understand the growing body of work of chemical and physiological understanding of microorganisms (DuBois, 1940; Passmore, 1982; Carpenter, 2004). Concurrent research into diseases of deficiency including beriberi pellagra and scurvy to name but a few led to the realization in 1901 that

> ...there occur[s] in various natural foods, substances which cannot be absent without serious injury to the peripheral nervous system ... *Grijns (1936, p. 38).*

This moment of epiphany led to further pioneering work by leading notable scientists of the day including Eijkman, Grijns, Funk, Hopkins, Osborne, McCollum, Mendel, and others led to the detection of nutrients such as vitamines (as they were then called) and essential minerals and trace elements. Therefore, one can see that clearly such advances began to pull all the strings together upon which foundation of nutrition, health, and by extension food could only grow (Carpenter, 2003). Alas over the next few decades, many more common vitamins were discovered along with the realization of the importance of such nutrients to the subsequent vitality and general health of humans (Table 2.13) (Carpenter, 2003). This era of discovery also became known as "the golden age of nutrition" in which anything seemed possible (Carpenter, 2003; Ordovas and Mooser, 2004; Go et al., 2005).

TABLE 2.13 Timeline of the discovery of particular vitamins.

Vitamin	Year of discovery or proposed existence
Thiamine	1901
Vitamin C	1907
Vitamin A	1915
Vitamin D	1919
Vitamin E	1922
Niacin	1926
Biotin	1926
Vitamin K	1929
Pantothenic acid	1931
Folate	1931
Riboflavin	1933
Vitamin B_6	1934

Source: Based on Carpenter's 'A Short History of Nutritional Science: Part 3' (1912–1944). Journal of Nutrition 133 (Oct.): 3023–3032.

2.19 Economic front

While the previous interwar years were busy in terms of nutritional advancement, governance and multilateral cooperation were coming together on an as-yet unprecedented scale. However, a cautionary note on free trade and the idea of comparative advantage (a staple of international economics at the time) came from Karl Brandt when he noted in "War in our Time" (1939) that in general, nations' seemingly growing overreliance on foreign resources was the result, more through political expediency rather than necessity (Brandt, 1939). Furthermore, in arguing (in the economic sense) against the oft parried phrase "the have's and have-not's," Brandt went on to argue that it was in fact this overdependence and neglect of a country's own resource base that was more likely at the heart of widening economic gaps between inter- and intraregional areas.

In his rationale, Brandt suggested that nations were very adaptable (historically, at least) when talking of food and raw materials in the past. Following on from this Brandt also offered that, rather than pursuing an economic course, heavily dependent on imports in general, a rethink of the whole economic chain was perhaps warranted. That is not to say that Brandt was against free trade either. Instead, he proposed that in times of "supposed" free trade, when protectionism, quotas, and other barriers rendered the international economic playing field uneven, then a fundamental change of focus in favor of domestic reliance might be one answer to the economic trade gaps. Brandt was also insightful when looking at localized resource allocation touching on many of the natural resource issues that became predominant in the 1990s to the present day. Although his main focus had been economics, trade, and foodstuffs, Brandt also advocated thrift in sustainability and consumption. In such issues, he was ahead of his time in the sympathetic use of local natural resources.

Furthermore, it can be found that in his works (again ahead of his time) that more than a hint of the modern-day food sovereignty movement is laid out in principle, especially in his advocacy of self-determination and food self-reliance (Brandt, 1939).

2.20 More surpluses

Going back to interwar food surpluses, on the whole the situation became bad, as not surprisingly, these same surpluses ended up overheating an industry at a time when real prices were falling. On top of this, a once healthy, but now fading, demand from a near bankrupt Europe coincided with increased competition and the reentry of military personnel back into the US civilian labor market. All these factors came at a time of drastic changes in fiscal and monetary policies, principally in America which ultimately brought about prodigious unemployment and further depressed purchasing power—not just in America, but gradually throughout Europe and eventually leading to worldwide economic depression. It goes without saying that the resultant depression was hard for everyone. By now, agriculture, with its increasing mechanization of and growing industrial sector, all led to free-falling real farm incomes (Black, 1942). In an attempt to rebalance the floundering sector in the United States, polices of agricultural parity were introduced. This was an important concept and a sharp tool in the interventionist's arsenal. Of particular note was the Agricultural Adjustment Act (AAA) of 1938, which ensured that congress

> Establish and maintain such balance between the production and consumption of agricultural commodities … as will reestablish prices to farmers at a level that will give agricultural commodities a purchasing power with respect to articles that farmers buy … *USDA (1985).*

Such policies basically aimed to connect farmers' incomes with costs of labor and materials in an effort to register them on par other industries' incomes (Tenny, 1938). This was important from many angles because this particular intervention policy, it was expected, would encourage and maintain real farm incomes at a time when farms and farm laborers were vulnerable. Just as importantly too, such measures helped the industry reflect the sectors growing importance within the overall national economy (Black, 1942; Varnee, 1955). Parity (described in detail later) meant the preservation of farm grown product prices through federal support to make sure that farming remained an essential part of the US economy. Consequently, parity had to reflect a real starting point for the sake of a true income-based approach. As such the golden era of farming (considered to have occurred in the 1910–14 period) was used in the calculations of parity pricing.

In the midst of all of this, there was still many that remained in a situation want. This dichotomy could not be allowed to continue and several initiatives, not least of which was the food stamp program, aimed to address this very situation.

2.21 UK food stamp program

By 1938/39 the UK food stamp program was introduced—this allowed people on relief to buy orange stamps equal to the value of their normal reasonable food expenditure. In turn, they would also receive 50% of the value of their orange stamps in blue stamps. These could then be redeemed for food that has been considered by the US department of agriculture to be surplus (Waugh, 1943). Other social welfare initiatives too, like the school lunch program, also profited from this new government viewpoint. Indeed, such distribution programs prospered during a period when mountains of surpluses were just hanging over the market while many on welfare were living on the bread line. This led Waugh to comment:

> We were confronted with the striking dilemma of want and hunger in the midst of plenty. ***Waugh (1943).***

Luckily, there was a mounting social awareness to the plight of the hungry and needy. This was particularly noticeable during the last half of World War One. It was also recognized that what was needed was collaborative action on the part of the governments and institutions—a necessity for governments and institutions including humanitarian and other aid agencies to work together as a single collective entity with an international remit. And as history has shown (part one), what came after was a new era of cooperation with goals and objectives firmly sighted on the international platform (Belshaw, 1947; Encyclopedia, 2006).

In the social arena around this time, things were changing too. Spurred on by a new emerging socio-cultural ideology that embraced artistic, literary, philosophy, musical, and cultural movements, people were taking more notice of what was happening around them.

Surpluses aside, the fight for hunger, famine, and food for all was advanced in leaps and bounds with the creation of the United Nations.

2.22 The inception of the United Nations

With the collapse of the League of Nations in 1946, it became clear that what was needed was an international body with stronger executive powers (UNOG, 2009). In fact, the League's failure to prevent cross-border hostilities in 1939 did not put paid to the overall belief in the concept of a multinational organization with the remit of international peace. On the contrary, the new body resolved to learn and adapt from previous mistakes and to shape a new multinational body, more able and more willing to maintain international peace envisioned. Indeed, so strong was this belief that, even before the League of Nations itself came to an end, a new multilateral body had been fully conceived. Setting the foundation for the new body to be called the "United Nations" was the inter-allied assertion "the London Declaration" of June 1941. This declaration came about from several meetings, although the most important was the meeting at St. James's Palace in London where representatives of major powers including Australia, Canada, New Zealand, the United Kingdom, and the Union of South Africa, as well as governments in exile like Belgium, Czechoslovakia, France, Greece, Luxembourg, the Netherlands, Norway, Poland, and Yugoslavia each pledging to not sign a separate or independent peace document with Germany and collectively declared

> The only true basis of enduring peace is the willing cooperation of free peoples in a world in which, relieved of the menace of aggression, all may enjoy economic and social security.... ***UN (2006).***

A few months after the above meeting, aboard the warship USS Augusta off the coast of Newfoundland, a declaration outlining the key articles of agreement between Prime Minister Winston Churchill and President Roosevelt was issued. Although not a treaty per se, the document, the "Atlantic Charter" of August 1941, as it became known was further strengthened in 1941 by Roosevelt's congressional State of the Union speech in which the Four Freedoms concept was made known and elucidated upon. The Four Freedoms were (1) the freedom of speech, (2) the freedom of worship, (3) the freedom from want, and ultimately (4) the freedom from fear. Such freedoms, Roosevelt insisted, were central to the growth of humankind throughout the world (Roosevelt, 1941; UN, 1945). Collectively, these four notions established a broad and succinct permanent ideology of all-round social security.

After further talks, this led to the signing of the United Nations Declaration by 27 nations on January 1 and 2, 1942. By signing the declaration, governments were effectively bound to any future war effort and the inability to unilaterally sign any separate peace deal with aggressor powers (UN, 2006). However, while declarations of interest had been signed and sealed by 1943, the foundation for the anticipated world organization was still not in place. Instead, and once again, after great discussion, the cause was further strengthened by the Foreign Ministers of the United States, Great Britain, and the Soviet Union along with a Chinese ambassador who together drew up the Four-Power Moscow Declaration of October 1943, which, among other things, recognized the need to for a global organization responsible for the preservation of international peace and security. So, while the initial ideologies were now created, it was during the Dumbarton Oaks (a private mansion in Washington, D.C.) talks that the ideologies were forged into workable goals. Thus, by October 1944, the structure of the proposed United Nations four bodies was hammered out. These included a General Assembly, comprising provision for an Economic and Social Council; a Security Council composed of 11 members; an International Court of Justice; and of course, the *Secretariat*. However, in 1945, before fleshing out the final text, President Roosevelt suddenly passed away, yet despite this, the scheduled United Nations Conference on International Organization still convened in San Francisco. At the conference 50/51 (depends on who you read) representatives from various countries met to finally hammer out the final text that inevitably became the Charter of the United Nations. In 1945, the United Nations was finally born.

When it came to food and the United Nations (UN), one astute observer, Frank McDougall, a member of the Economic Committee of the League of Nations said of the UN's responsibilities at the time, said:

> If the United Nations decide to make freedom from want of food ... will require national action in every country and international action to assist countries lacking technical knowledge and financial means to secure improvements in food production ... [therefore] the more economically advanced United Nations should pledge themselves, first to institute policies designed to ensure that the foods required for diets adequate for health are within the purchasing power of their own citizens, and second to provide technical and financial aid to enable the less advanced nations to progress towards the accomplishment of this objective. ***McDougall (1943).***

While this was what the UN had planned for itself, how it fared in this endeavor is unraveled in the course of the book.

2.22.1 The Food and Agriculture Organization is born

With all that had happened vis-à-vis food and the two World Wars; and predating the San Francisco Conference that helped shape the yet to be created UN by 2 years, President Roosevelt convened the first UN conference on Food and Agriculture at Hot Springs, Virginia, in May and June of 1943 (UN, 1943). This came to be, in no small part, due to the efforts of Frank McDougall who wrote a report in 1942 predating the above quote, titled "Draft memorandum on a United Nations Program for Freedom from Want of Food." The report gained the attention of President Roosevelt, the two men finally met. In this and subsequent discussions, McDougall counseled the president on the need for the new, as yet unborn UN to tackle the problem of food as their first priority. Out of these discussions, Roosevelt, in the following year, called for a United Nations Conference on Food and Agriculture (FAO, 1995). The conference was a success and recommendations were made re-iterating the 1935 League of Nations commission for universal, multilateral action to reduce instances of hunger and malnutrition. This would be (as both bodies had envisaged) through the creation of a permanent body responsible solely for the domains of food and agriculture (Parran, 1943; Thompson, 1943; Evang and McDougall, 1944). Essentially, what was being proposed was an organization able to act as a central information clearing house, while promoting stability through international commodity instruments; the maintenance of sufficient reserves; and when necessary the disposal of any surpluses (Shaw, 2007). This pre-UN Conference on Food and Agriculture was nevertheless held under the banner of the UN, which further led to the formation of a Provisional Commission on Food and Agriculture. The Commission even went as far as to draft a Constitution for a new permanent body to be known as the Food and Agriculture Organization (UN, 1943; Phillips, 1981).

2.23 Freedom from want of food

The 1943 conference (UN) was also a turning point in the history of food security (a concept dealt with in 11). Furthermore, using Roosevelt's notion of "freedom from want," the conference defined—as one of its main objective—the "freedom from want of food suitable and adequate for the health and strength of all peoples" (Thompson, 1943). In addition, the conference also acknowledged that

> The first cause of hunger and malnutrition is poverty. It is useless to produce more food unless men and nations provide the markets to absorb it. There must be an expansion of the whole world economy to provide the purchasing power sufficient to maintain an adequate diet for all. ***The freedom from want of food UN (1943).***

This was groundbreaking as the conference covered three main arenas of operation. These included the improvement of the national nutritional health and diets of the populace, increased food production by means of global agricultural expansion, and lastly, better support in terms of resources for the accomplishment of such aims. The talks also noted several important matters regarding the accomplishment of sufficient food for all:

- The first freedom, the "Freedom from Want," among other things, would ensure a secure, suitable, and nutritious supply of food for every man, woman, and child on the earth;
- The conference also aimed to create an enabling environment for the growth of national nutritional organizations and to collect, collate, and disseminate information and experience;
- The conference also advocated essential ideals about sustainability in its aim to provide sufficient nutritious food while conserving lands and water resources;
- It was also understood the interdependence of both producers and consumers and that coordinated agricultural policies must be considered with this in mind;
- It was also encouraged to create cooperative movements to create of preserve land tenure, education, and research;
- The conference, having brought together so much material research, concluded certain connections were evidently clear of many dietary deficiencies and diseases;
- The recognition of special or vulnerable groups such as the young, pregnant women, or the marginalized must be given equal if not greater protection by placing these groups at the top of food security.

The conference stimulated much chatter among individuals and institutions and the resounding implicit feeling was of the acceptance of an overarching multilateral body that could bring the full might of its power in the search for solutions (UN, 1943; Carlson, 1944; Evang and McDougall, 1944; Schultz, 1945; Phillips, 1981). Indeed, in this endeavor, the conference could be succinctly described by one of its own resolutions which stipulated that

> There has never been enough food for the health of all people. This is justified neither by ignorance nor by the harshness of nature. Production of food must be greatly expanded; ... [and] requires imagination and firm will on the part of each government ... to make use of that knowledge *UN (1943)*.

In accomplishing this, much work, understanding, collaboration, and the creation of new bodies was required—which should include the cooperation of existing, on the ground agencies; the introduction of nationwide nutritional organizations; and more agricultural investment opportunities like credit agencies; increasing cooperative movements; international commodity agreements; more equitable distribution measures. Even at the macroscale, financial and commercial instruments were needed to secure expanding world economic growth, which would naturally filter down to the agricultural sector. And this was just the beginning: improved land tenure was needed for resident security; humanitarian aid policies were required in times of emergencies and for fallback safety nets. But above all, what was needed most was improved education and research, particularly in the areas of sustainability and areas of conservation (McDougall, 1943; Thompson, 1943; UN, 1943; Phillips, 1981). The urgency of these needs was not lost on McDougall, as poor social and nutritional conditions across much of the developing world meant that as much as 60% of the global working population was employed in agriculture:

> ...on the basis of numbers, the social condition of the peasants, sharecroppers, and other farm workers is the [single, most] outstanding social problem of the world ***McDougall (1943, p. 126).***

The flip side of the above, as outlined in the conference, was the very real issue of farmers' incomes, sustainability, and consumers spending power. In real terms at the time, it was shown that farmers could not rely on sufficient returns to lead a normal healthy nutritional life, or would consumers be in a position where they can purchase sufficient food for themselves and their families. The answer, as suggested, was that real progress was needed and not at the national/regional level; but rather, change was needed at the international level, and the best way to achieve this was through increased employment across all sectors in all countries, i.e., the general worldwide expansion of economic activity (UN, 1943). Such a huge project would be achieved through increased regulation policies of production and other instruments of development (Thompson, 1943). Such forward thinking placed food and agriculture at the heart of the first United Nations conference as said by the then Surgeon General of the United States Thomas Parran (Parran, 1943) to have been a rare moment of epiphany by Roosevelt. Consequently, in 1945 and not surprisingly, the preofficial UN conference (1943) became the forerunner of two new bodies: the new bodies of the United Nations and the Food and Agriculture Organization. The first meeting was held in Quebec that year and was headed by British nutritionist Sir John Boyd Orr.

2.23.1 Bretton Woods Agreement

Another of the fledgling United Nations founding conferences was the UN Monetary and Financial Conference, which quickly came to be known as the Bretton Woods Conference of 1944. The need for certain financial agreements was evident in the previous years and, in part, in response to practices that heralded the 1930s era of depression. Such practices saw certain nation's attempts to gain a competitive edge (usually economic) by increasing obstacles to "fair" foreign trade. Having said that, certain practices, such as the devaluation of currencies to better compete in export markets, ironically became counterproductive. In the end, a rapid slowdown in world trade affected much of the world's living standards, which dropped significantly. One aim of the Bretton Woods agreements saw new financial instruments designed ostensibly to regulate international monetary and financial directives as well as aiding in the post-war reconstruction of many countries (Bordo and Eichengreen, 1993; IMF, 2010b). Some success out of Bretton Woods agreement came into being when in 1945 the International Monetary Fund (IMF) formally came into existence beginning operations in 1947. Its remit was simple—to oversee the international monetary system—a system of exchange rates and international capital and current payments with the role to facilitate the free flow of goods and services between countries. The IMF ensured cross-border trade through stability of the exchange rate and aimed to achieve this using the par value system. On top of this, the IMF acted as the international lender of last resort.

Next from the Bretton woods agreement came the International Bank for Reconstruction and Development (IBRD), which was originally created to finance international post-war reconstruction. Although in recent times its goals have been extended to help with the extension of long-term investments for the benefit of development, especially in structural improvements in underdeveloped nations, as well as to fight poverty, today, collectively the IMF and the IBRD are known as the World Bank.

2.23.2 UN Relief and Rehabilitation Administration

Meanwhile as the war came to an end, there became greater need for international action and coordination. This need was put before the fledgling United Nations and resulted in the creation of the *United Nations Relief and Rehabilitation Administration* (UNRRA) in 1943 (Foreign Office, 1943; House of Commons, 1944; League of Nations, 1946). This new body incorporated another newly formed office of the US State Department (1942) *Office of Foreign Relief and Rehabilitation* (OFRRO) (Williams, 2005). Ultimately the UNRRA's remit was both broad as it was sweeping (League of Nations, 1946). That is to say, in the UNRRA's brief 3-year term, it aimed to manage

> ... relief of [the] victims of war in any area under the control of any of the United Nations through the provision of food, fuel, clothing, shelter and other basic necessities, medical and other essential services ***League of Nations (1946, p. 92)***.

2.23.3 Aid, welfare, and surpluses in the United States

Originally, in the beginning (1930s), the United States department of Agriculture (USDA) began buying up surplus commodity stocks that appeared on the market almost on a regular basis and which were then passed along to state welfare agencies. Initially, this altruistic measure was a convenient way of controlling farm prices. And while a good idea in principle, the distributed food relied heavily on the surpluses available rather than any attempt to provide a variety or satisfactory nutritional diet to the needy. That was the early 30s. Later (the mid-to late 1930s), the US Department of Agriculture (USDA) was becoming busy once again with fluctuating surpluses. Its main aim this time around was twofold—first, to adjust pricing policies, effectively aiming to depress surpluses, and second, where surpluses existed, the USDA utilized these surpluses in a nationwide effort for the benefit of school children and low-income families (Waugh, 1943). Ultimately though, this had to change as it was recognized that the two problems of farm incomes and food prices were intricately related. This meant that with restrictions, poverty, and other social barriers, millions of people were still unable to afford to buy farmers' products. Consequently, farmers' incomes suffered while people went hungry. As a result, it became clear that larger nationwide, welfare feeding programs were required. But this time the focus was more holistic with more consideration of the diet and nutritional requirements of the nation.

2.23.4 Inter-allied committee on post-war food and relief

In 1940 in a House of Commons speech by the then Prime Minister Churchill, it was agreed that at the end of hostilities, the peoples of Europe would receive food and relief

(House of Commons, 1944). Not long after in 1941 at St. James's Palace, the Allied powers met to consider relief and rehabilitation passing a resolution stating:

> … the only true basis of enduring peace is the willing co-operation of free peoples in a world, in which, relieved of the menace of aggression, all may enjoy economic and social security; and that it is their intention to work together, and with other free peoples, both in war and peace to this end. ***House of Commons (1944).***

This meeting, before the signing of the Atlantic Charter, reaffirmed the mutual cooperation of the Allied forces. In another meeting in September, a resolution was passed securing supplies of food and raw materials for post-war liberated countries. It was through these committees that the Inter-Allied Committee (IAC) was set up to gather relevant information in which to formulate policy to this end (Foreign Office, 1943; House of Commons, 1944; League of Nations, 1946). Furthermore, it was because of these efforts that the committee advanced the use of food balance sheets (FBS), which remained in common use long after the war had ended (Foreign Office, 1943; FAO, 2001). Eventually, the same committee, because of the diligent work and ideology, was later transformed to become the European advisory Board (EAC) (Williams, 2005).

Meanwhile back in the new scientific discipline of nutritional research the first ever recommended dietary allowances (RDA) were made.

2.23.5 Recommended dietary allowances

Thanks to the work of Mitchell Stiebling and Roberts, who worked for what began in 1940 as the Committee on Food and Nutrition, under the auspices of the National Research Council of the US Federal Government, which was later to become the Food and Nutrition Board under the US research Council of the National Academy of Sciences produced, the first widely accepted RDA in 1941. This was a milestone in nutritional science and was built on 50 years of collaborative research and analyses of countless scientists and individuals.

Indeed, despite the dillydallying, the Board of Nutrition finally presented their recommended allowances at the Spring Meeting of the American Institute of Nutrition (1941) (Editorial, 1943; Harper, 1985; Rosenberg, 1994; Aggett et al., 1997; McArdle et al., 1999; Harper, 2003), and once again despite further procrastination, these recommended allowances were not published by the National Academy until 1943. Nonetheless, the table did appear in the Journal of the American Dietetic Association in 1941 forming the basic needs of the diverse population (Table 2.14) (ADA, 1941; NRC, 1943; Harper, 1985; Mayer, 1986; Harper, 2003). Furthermore, since then, this groundbreaking work was built on and

TABLE 2.14 The Food and Nutrition Board's 1941 recommended dietary allowances.

Energy kcal	Protein g	Calcium g	Phosphorus g	Iron mg	Vitamin A IU	Vitamin B1 IU2	Vitamin C mg	Riboflavin mg	Nicotinic acid mg	Vitamin D IU
2775	66	0.91	—	12	4696	516	71	2.3	15.5	210

Source: Recommended Dietary Allowances: Report of the Food and Nutrition Board, Reprint and Circular Series No. 115. (Washington, DC, National Research Council. 6, 1943).

expanded with many countries adopting, adapting, and "borrowing" entire recommendations (see also Appendix Table 30.1).

Meanwhile, a few key dates (Table 2.15) help place much of the previous information in context.

TABLE 2.15 Key dates of the period: 1939 to 45.

Key dates of the period	
1939 US Food Stamp Program	Around this time many food distribution programs were developed by the United States. Coinciding with massive unemployment and surpluses the first few programs later became known as the Supplemental Nutrition in Assistance (SNAP) (Waugh, 1943; Dimitri et al., 2005; Landers, 2007; National Archives, 2009).
1941 The term "Genetic Engineering" is first used	Danish microbiologist A. Jost coined the term "genetic engineering" in a lecture on sexual reproduction in yeast.
1941 Recommended Dietary Allowances **(RDA)**	Building on previous nutritional research work by the League of Nations as well as Burnet and Ackroyd among others, the US Research Council of the National Academy of Sciences set the Committee on Food and Nutrition. Consequently, it was not long before the US research council provided the first Recommended Dietary Allowances in 1941, which was subsequently published in 1943. The committee was renamed the Food and Nutrition Board in 1941 (Harper 1985, 2003; Mayer, 1986; Rosenberg, 1994).
1941 The Atlantic Charter	Premiers Churchill and President Roosevelt met aboard the USS Atlanta and signed a declaration indicating to many that the two powers would work toward the creation of a new global organization as soon as peace once again prevailed (UN, 1943; UNOG, 2009).
1941 US Lend Lease Act	The US Lend Lease Act was responsible for essential food supplies and other vital war materials to the allied nations (Ficker, 1967; Hoffmann and Maier, 1984).
1942 International Wheat Council **(IWC)**	The International Wheat Council (IWC) superseded the Wheat Advisory Committee of 1933. The wheat advisory committee, headquartered in Washington, finally established an international seat of IWC in London in 1949 (League of Nations, 1933; IGC, 2009a).
1942 OXFAM	OXFAM began in Oxford by an enthusiastic group dedicated to famine relief. The acronym translates as Oxford Committee for Famine Relief (OXFAM, 2007).
1943 United Nations Relief and Rehabilitation Administration **(UNRRA)**	The UN Relief and Rehabilitation Administration began work in 1943 providing health care and relief to regions liberated from the Axis powers after the cessation of World War II. The UNRRA further provided billions of dollars in aid packages, before ceasing operations in 1947 (Europe) and in Asia in 1949. UNRRA was ultimately disbanded in 1946 and its functional duties transferred to the Interim Commission of the World Health Organization (WHO) (WHO, 2009).
1944 Norman Borlaug and the Rockefeller Foundation	Norman Borlaug is often referred to as the father of the green revolution because of his work with the Rockefeller Foundation as well as his other studies in the field of crop hybridization during the 50s and 60s led him to the Nobel Peace Prize in 1970 (Herdt, 1998; USAID, 2009).

TABLE 2.15 Key dates of the period: 1939 to 45.—cont'd

Key dates of the period	
1944 UN Monetary and Financial Conference—The Bretton Woods Agreement	The monetary and financial conference of the United Nations (more commonly referred to as the Bretton Woods conference) signatories signed an agreement regulating international monetary and financial order after the end of WW2. Agreements also established the International Monetary Fund (IMF) and the International Bank for Reconstruction and Development (IBRD) (IMF, 2010a,b).
1943 Plant Breeding as Foreign Aid	In the first of its kind, the collaboration with the Rockefeller Foundation and the Mexican government on the Mexican Agricultural Program became the first incarnation of plant breeding as foreign aid.
1945 CARE	CARE, the Cooperative for American Remittances to Europe later renamed Cooperative for Assistance and Relief Everywhere, was inaugurated (CARE, 2009).
1945 United Nations **(UN)**	The incorporation of the United Nations in 1945 was a milestone in international cooperation with the remit covering economic, political, judicial/security, and social governance (UN, 1945; UN, 2000).
1945 Food and Agriculture Organization **(FAO)**	The FAO, a branch of the United Nations, was responsible for the overall economic and social development of food and agriculture. It dealt with all the major issues of agriculture, forestry, and fisheries via both its field programs and its regular programs (FAO, 1946; Phillips, 1981; FAO, 2010). (Phillips, 1981).
1945 World Bank Group **(WB)** International Bank for Reconstruction and Development **(IBRD)**—International Monetary Fund **(IMF)**	The International Bank for Reconstruction and Development (IBRD) was created to finance post-war reconstruction. Nowadays, however, its remit has been changed somewhat to deal with poverty and development issues. While the financial aspect, the orderly currency practices in international trade has been handed over to the IMF (IMF, 2010a,b).
1945 UN Educational, Scientific and Cultural Organization **(UNESCO)**	UN Educational, Scientific and Cultural Organization (UNESCO) was formed in 1945 as a specialized United Nations agency. The organization serves as a clearinghouse or repository of ideas while importantly, ensuring the sharing and dissemination of information and knowledge (UNESCO, 2009).

Source: Derived from multiple sources League of Nations (1933); UN (1943); Waugh (1943); UN (1945); FAO (1946); Ficker (1967); Phillips (1981); Hoffmann and Maier (1984); Harper (1985); Mayer (1986); Rosenberg (1994); Herdt (1998); UN (2000); Harper (2003); Dimitri et al. (2005); Landers (2007); OXFAM (2007); CARE (2009); IGC (2009a,b); National Archives (2009); UNESCO (2009); UNOG (2009); USAID (2009); WHO (2009); FAO (2010); IMF (2010a,b).

References

Acheson, E.D., 1986. Tenth Boyd Orr memorial lecture: food policy, nutrition and government. Proc. Nutr. Soc. 45, 131–138.

ADA, 1941. Recommended allowances for the various dietary essentials. J. Am. Diet. Assoc. 17, 565–567.

Aggett, P.J., et al., 1997. Recommended dietary allowances (RDAs), recommended dietary intakes (RDIs), recommended nutrient intakes (RNIs), and population reference intakes (PRIs) are not "recommended intakes". J. Pediatr. Gastroenterol. Nutr. 25 (2), 236–224.

Almonacid, P.N., 1943. The food situation in a world at war: the other Americas. Ann. Am. Acad. Pol. Soc. Sci. 225, 93–95.

APS, 2010. Population Genetics of Plant Pathogens: Interactions Among Evolutionary Forces and the Genetic Structure of Pathogen Populations. Retrieved from: https://www.apsnet.org/edcenter/disimpactmngmnt/topc/PopGenetics/Pages/InteractionsGeneticStructure.aspx.

Armour, J.O., 1917. Food shortage: an appeal to physicians. J. Am. Assoc. LXVIII (18), 1339–1340.

AU/ICE, 1997. ICE Cases Studies – Intellectual Property Rights & Biotechnology: Indian Seed Conflicts. Inventory of Conflict and Environment (ICE).

Auchter, E.C., 1939. The interrelation of soils and plant animals. Science 89 (2315), 421–426.

Bacon, L., 1943. The food situation in a world at war: southwestern Europe. Ann. Am. Acad. Pol. Soc. Sci. 225, 87–88.

Baldwin, S.E., 1907. The international congresses and conferences of the last century as forces working toward the solidarity of the world. Am. J. Int. Law 1 (3), 808–829.

Balfour, L.E.B., 1975. The Living Soil and the Haughley Experiment. Faber and Faber, London.

Beardsworth, A., Keil, T., 1997. Sociology on the Menu: An Invitation to the Study of Food and Society. Routledge, London.

Beaton, G.H., 1992. Human nutrient requirement estimates: derivation, interpretation and application in evolutionary perspective. Food Nutr. Agric. (2/3).

Bedale, E.M., 1923. Energy expenditure and food requirements of children at school. Proc. R. Soc. Lond. – Ser. B Contain. Pap. a Biol. Char. 94 (662), 368–404.

Belshaw, H., 1947. The food and agriculture organization of the united nations. Int. Organ. 1 (2), 291–306.

Bennett, M.K., 1949. Food and agriculture in the Soviet union, 1917-48. J. Political Econ. 57 (3), 185–198.

Berlan, J.-P., Lewontin, R.C., July–August 1986. The political economy of hybrid corn. Mon. Rev. 38, 35–47.

BioTech Institute, 2010. Timeline of Biotechnology 20th Century. Retrieved from: http://www.biotechinstitute.org/go.cfm?do=Page.View&pid=22.

Black, J.D., 1942. Parity, Parity, Parity. The Harvard Committee On Research In The Social Sciences, Cambridge, MA.

Black, J.D., 1943. Food: war and postwar. Ann. Am. Acad. Pol. Soc. Sci. 225, 1–5.

Blakemore, R.J., 2000. Ecology of earthworms under the 'Haughley experiment' of organic and conventional management regimes. Biolo. Agric. Hortic. 18, 141–159.

BMJ, 1933. British medical association. Report of committee on nutrition. Br. Med. J (Suppl), 1–16.

BMJ, 1934. The nutrition question. Br. Med. J. 1 (3828), 900–901.

BMJ, 1938. Malnutrition among school children. Br. Med. J. 2 (4053), 585–586.

BMJ, 1954. Correspondence: women and child nutrition. Br. Med. J. 1 (4874), 1322.

Bordo, M.D., Eichengreen, B., 1993. The Bretton Woods International Monetary System: An Historical Overview. A Retrospective on the Bretton Woods System: Lessons for International Monetary Reform. Chicago University of Chicago Press.

Bradshaw, J., 2003. A brief history of bio-dynamics – an Australian perspective. Biodynamic Grow. 1 (4–6).

Brandt, K., 1939. Foodstuffs and Raw Materials War in Our Time. H. Speier and A. Kahler. WW Norton, New York.

Britannica, 2009. Encyclopædia Britannica Online.

Bud, R., 1993. The Uses of Life: A Hstory of Biotechnology. Cambridge University Press, Cambridge.

Burnet, E., Aykroyd, W.R., 1935. Nutrition and public health. Q. Bull. Health Organ. IV (2), 152.

Campbell, J.M., 1938. The nutrition report. Int. Aff. 17 (2), 251–253 (Royal Institute of International Affairs 1931-1939).

CARE, 2009. History of CARE International. Retrieved from: https://www.careinternational.org.uk/who-we-are/about-care.

Carlson, A.J., 1944. Symposium on civilian wartime problems in nutrition: from the standpoint of the physician: the importance of food in wartime. Calif. West. Med. 61 (6), 281–285.

Carpenter, K.J., 2003. A short history of nutritional science: part 3 (1912–1944). J. Nutr. 133 (Oct.), 3023–3032.

Carpenter, K.J., 2004. The Nobel Prize and the Discovery of Vitamins. Nobel Prize.

Carpenter, K.J., 2007. Biographical article: the work of Wallace Aykroyd: international nutritionist and author. J. Nutr. 137, 873–878.

Carpenter, K.J., et al., 1997. Experiments that changed nutritional thinking. J. Nutr. 127 (5), 1017S–1053S.

Carr-Saunders, A.M., 1927. The population conference at Geneva. Econ. J. 37 (148), 670–672.

Cary, R.L., 1920. Social and industrial conditions in the Germany of today. Ann. Am. Acad. Pol. Soc. Sci. 92, 157–162.

Cassels, J.M., Hall, F.L., 1943. Food supplies for our civilian population. Ann. Am. Acad. Pol. Soc. Sci. 225, 106–115.

Chamberlin, W.H., 1934. The ordeal of the Russian peasantry. Foreign Aff. 12 (3), 495–507.

Chittenden, R.H., 1904. Physiological Economy in Nutrition: With Special Reference to the Minimal Proteid Requirement of the Healthy Man. An Experimental Study, New York, NY (Frederick A. Stokes Co).

Chittenden, R.H., 1906. A discussion on over-nutrition and under-nutrition. Br. Med. J. 2 (2391), 1100–1103.

Chittenden, R.H., 1907. Physiological Economy in Nutrition, with Special Reference to the Minimal Proteid Requirement of the Healthy Man. F. A. Stokes Co, New York.

Clayton, C.F., Black, J.D., 1943. The food situation in a world at war: Wartime Food Administration-U. S. A. Ann. Am. Acad. Pol. Soc. Sci. 225, 96–105.

Clynes, J.R., 1920. Food control in war and peace. Econ. J. 30 (118), 147–155.

Cohen, J.E., 1995. How Many People Can the Earth Support? W.W. Norton, New York and London.

Connelly, M., 2006. Seeing beyond the state: the population control movement and the problem of sovereignty. Past Present 193, 197–233.

Cook, R., 1937. Yearbook of Agriculture: A Chronology of Genetics. . Department of Agriculture, Washington, U. S.

Crutchfield, J.S., 1919. Food in the reconstruction period. Ann. Am. Acad. Pol. Soc. Sci. 82, 7–10.

De Gans, H.A., 2002. Law or Speculation?. A debate on the method of forecasting population size in the 1920s. Population 57 (1), 83–108.

Dewey, P.E., 1989. British Agriculture in the First World War. Routledge, London.

Dimitri, C., et al., 2005. Economic Information Bulletin: Number 3: The 20th CenturyTransformation of U.S. Agriculture and Farm Policy. United States Department of Agriculture, USDA.

DuBois, E.F., 1940. Biographical memoir of Graham Lusk, 1866-1932. In: National Academy of Sciences Biographical Memoirs. 3d Memoir, Biographical Memoirs (National Academy of Sciences (U.S.)), vol. xxi. National academy of sciences, Washington, D.C, pp. 95–142.

Durand, E.D., 1922. Agriculture in eastern Europe. Q. J. Econ. 36 (2), 169–196.

Dyson, T., Ó Gráda, C. (Eds.), 2002. Famine Demography: Perspectives from the Past and Present UK. Oxford University Press.

Editorial, 1943. The marriage of public health and agriculture (United Nations' conference on food and agriculture). Am. J. Public Health 33 (7), 847–848.

Encyclopedia, 2002. Science of Everyday Things. Encyclopedia.com. Retrieved from: http://www.encyclopedia.com.

Encyclopedia, 2006. Agriculture. Funk & Wagnalls New Encyclopedia.

Evang, K., McDougall, F.L., 1944. The hot springs conference. Proc. Nutr. Soc. 2 (3–4), 163–176.

FAO, 1946. Report of the First Session of the Conference Held at the City of Quebec, Canada, 16 October to 1 November, 1945, p. 89. Washington, FAO.

FAO, 1970. Report of the Conference of FAO: Annex D - Commemorative Address by Professor M. Cépède, Independent Chairman of the FAO Council. Food and Agriculture Organisation, Rome.

FAO, 1995. Dimensions of Need: An Atlas of Food and Agriculture. Food and Agriculture organisation, Rome.

FAO, 2001. Food Balance Sheets: A Handbook. Food and Agriculture Organisation, Rome.

FAO, 2010. The Food and Agriculture Organization of the United Nations. Retrieved from: http://www.fao.org/about/en/.

Fári, M.G., et al., 2001. History of the Term Biotechnology: K. Ereky & his Contribution: Presentation at the Fourth Congress of Redbio Encuentro Latinoamericano de Biotecnologia Vegetal (Goiânia, Brazil).

Fári, M.G., Kralovánszky, U.P., 2006. The founding father of biotechnology: Károly (Karl) Ereky. Int. J. Hortic. Sci. 12 (1), 9–12.

Ficker, H., 1967. Fifty Years of Foreign Loans and Foreign Aid by the United States, 1917–1967. Library of Congress, Legislative Reference Service, Wasington, DC, p. 40, 1967.

Firth, R., 1934. The sociological study of native diet. Afr. J. Int. Afr. Inst. 7 (4), 401–414.

Fite, G.C., 1962. Farmer Opinion and the Agricultural Adjustment Act, 1933. Miss. Val. Hist. Rev. 48 (4), 656–673.

Foreign Office, 1943. Report to Allied Governments by the Inter-allied Committee on Post-War Requirements. UK Foreign Office, HMSO.

FSA, 2008. About FSA: About the Commodity Credit Corporation. Retrieved from: http://www.fsa.usda.gov/FSA/webapp?area=about&subject=landing&topic=sao-cc.

G.R.C., 1944. Russel Henry Chittenden: February 18 1856- December 26, 1943: an appreciation. J. Nutr. 28 (1), 2–6.

Gantt, W.H., 1936. A medical review of Soviet Russia: results of the First Five Year Plan. Br. Med. J. 2 (3939), 19–22.

Gardner, B.L., 2002. American Agriculture in the Twentieth Century: How it Flourished and what it Cost. Harvard University Press, Cambridge, Massachusetts.

Garside, W.R., 2002. British Unemployment 1919-1939: A Study in Public Policy. Cambridge University, Cambridge.
Gibson, M., 2016. The Feeding of Nations: Re-defining Food Security for the 21st Century. CRC Press, Boca Raton, Florida.
Gilks, J.L., 1935. The relation of economic development to public health in rural Africa. J. R. Afr. Soc. 34 (134), 31–40.
Go, V.L.W., et al., 2005. Nutrient-gene interaction: metabolic genotype-phenotype relationship. J. Nutr. 135, 3016S–3020S. Supplement: International Conference on Diet, Nutrition, and Cancer.
Goodrich, L.M., 1947. From League of Nations to United Nations. Int. Organ. 1 (1), 3–21.
Grijns, G., 1936. Prof. Dr. G. Grijns' researches on Vitamins, 1900–1911, and his thesis on the physiology of the N. opticus, translated and reedited by a Committee of Honour on occasion of his 70th birthday. J. Am. Med. Assoc. 107 (6).
Gunderson, G.W., 2003. The National School Lunch Program: Background and Development. ova Science Publishers, New York.
Hanekamp, J.c., Bast, A., 2007. New recommended daily allowances: benchmarking healthy European micronutrient regulations: let governments take care of safety. J. Environ. Liabil. 4, 155–162.
Hardin, C.M., 1943. The food production programs of the United States Department of Agriculture. Ann. Am. Acad. Pol. Soc. Sci. 225, 191–200.
Harper, A.E., 1985. Origin of recommended dietary allowances – an historic overview. Am. J.Clin. Nutr. 41, 140–148.
Harper, A.E., 2003. Symposium: historically important contributions of women in the nutrition society: contributions of women scientists in the U.S. to the development of recommended dietary allowances. Am. Soc. Nut. Sci. 133, 3698–3702.
Harris, J.A., Benedict, F.G., 1918. A biometric study of human basal metabolism. Proc. Natl. Acad. Sci. U.S.A 4 (12), 370–373.
Henry, C., 2005. Basal metabolic rate studies in humans: measurement and development of new equations. Public Health Nutr. 8 (7A), 1133–1152.
Herdt, R.W., 1998. The Rockefeller Foundation: The Life and Work of Norman Borlaug, Nobel Laureate. The Rockefeller Foundation, New York.
Hiller, E.T., 1930. A culture theory of population trends. J. Political Econ. 38 (5), 523–550.
Hitchcock, G.M., 1919. In Defense of the League of Nations. Ann. Am. Acad. Pol. Soc. Sci. 84, 201–207.
Hoffmann, S., Maier, C. (Eds.), 1984. The Marshall Plan: A Retrospective. Westview Press, Boulder.
Holmes, C.L., 1924. The economic future of our agriculture. J. Political Econ. 32 (5), 505–525.
House of Commons, 1944. House of Commons Debates, 25 January 1944: United Nations (Relief Administration), vol. 396. House of Commons, pp. cc567–632.
Howard, M.E., 2002. The First World War: A Very Short Introduction. Oxford University Press, New York.
Hutchinson, R., 1916. Food and the Principles of Diatetics. W. Wood and Co, New York.
Hwalla, N., Koleilat, M., 2004. Dietetic practice: the past, present and future. Health J. 10 (6), 716–730.
IAAE, 2008. About IAAE: History. Retrieved from: http://www.iaae-agecon.org/about-iaae/history.
Iacobbo, M., 2004. Vegetarian America: A History. Praeger Publishers, Westport Connecticut.
IFRC, 2009. Red Cross and Red Crescent Movement: History. Retrieved from: https://media.ifrc.org/ifrc/who-we-are/history/.
IFT, 1999. 20th Century Marks Achievements in Food Science from. https://www.ift.org/about-ift.
IGC, 2009a. 60 Years of Successive Agreements: Before 1949: The Early Years. International Grains Council, London, p. 4.
IGC, 2009b. Grains Trade and Food Security Cooperation: The International Grains Agreement. Retrieved from: http://www.igc.org.uk/en/aboutus/default.aspx#igc.
ILO, 2010. ILO: Origins and History. Retrieved from: http://www.ilo.org/global/About_the_ILO/Origins_and_history/lang–en/index.htm.
IMF, 2010a. Cooperation and Reconstruction (1944–71). Retrieved from: https://www.imf.org/en/About.
IMF, 2010b. International Montetary Fund Website. Retrieved from: https://www.igc.int/en/default.aspx.
IRC, 2009. Who We Are. Retrieved from: https://www.rescue.org/who-we-are.
James, C., Fitzgerald, J.A. (Eds.), 2008. Of the Land and the Spirit: The Essential Lord Northbourne on Ecology & Religion. World Wisdom Inc, Bloomington, Indiana.
Johnson, D.G., 1997. Agriculture and the Wealth of Nations. Am. Econ. Rev. 87 (2), 1–12.
Johnson, D.G., 2000. Population, food, and knowledge. Am. Econ. Rev. 90 (1), 1–14.

Kates, C.S., 1912. Origin and growth of rural conferences. Ann. Am. Soc. Sci. Acad. Pol. Sci. 40, 110–116.
Kindleberger, C.P., 1986. The World in Depression, 1929–1939. University of California Press, Berkely and Los Angeles, California.
Kohn, G.F., 1924. The organization and the work of the League of Nations. Ann. Am. Acad. Pol. Soc. Sci. 114, 5–77.
Ladejinsky, W.I., 1943. The food situation in a world at war: the food situation in Asia. Ann. Am. Acad. Pol. Soc. Sci. 225, 91–93.
Landers, P.S., 2007. The food stamp program: history, nutrition education, and impact. J. Am. Diet. Assoc. 107 (11).
League of Nations, 1919. Covenant of the League of Nations, 28 April 1919. League of Nations.
League of Nations, 1933. Final Act of the Conference of Wheat Exporting and Importing Countries, held in London from August 21 to 25, 1933, with Appendices and Minutes of Final Meeting. Signed at London, August 25, 1933. League of Nations, League of Nations, 3262, pp. 73–88.
League of Nations, 1936a. The Problem of Nutrition: Volume I Interim Report of the Mixed Committee on the Problem of Nutrition. League of Nations, Geneva.
League of Nations, 1936b. The Problem of Nutrition: Volume Il Report on the Physiological Bases of Nutrition. Technical Commission of the Health Committee of the League of Nations, Geneva.
League of Nations, 1937. Nutrition: Final Report of the Mixed Commision of the League of Nations on the Relation of Nutrition to Health, Agriculture and Economic Policy (Geneva, League of Nations).
League of Nations, 1946. Transit Department: Food, Famine and Relief 1940-1946 (Geneva, League of Nations).
Lloyd, D.J., Shore, A., 1938. Chemistry of the Proteins. J. and A. Churchill, London.
Lloyd, E.M.H., 1943. The food situation in a world at war: in the United Kingdom. Ann. Am. Acad. Pol. Soc. Sci. 225, 83–85.
Lusk, G., 1906. The Elements of the Science of Nutrition. W.B. Saunders, Philadelphia, London.
Lusk, G., 1918. Food in War Time. W.B. Saunders, Philadelphia and London.
Lysaght, E.E., et al., 1917. The farmers and the food problem. Ir. Q. Rev. 6 (21), 21–34.
Markham, J.W., 2002. A Financial History of the United States: From JP Morgan to the Institutional Investor (1900-1970. M.E. Sharpe, Armonk, NY, and London.
Mayer, J., 1986. Social responsibilities of nutritionists. J. Nutr. 116, 714–717.
Mayne, J.B., 1947. FAO the history. Rev. Mark Agric. Econ. 15 (11).
McArdle, W.D., et al., 1999. Sports and Exercise Nutrition. Williams and Wilkins, Baltimore.
Mcdougall, F.L., September 1940. Food for a Hungry World. Th Rotarian. US., pp. 11–14
McDougall, F.L., 1943. The food situation in a world at war: international aspects of postwar food and agriculture. Ann. Am. Acad. Pol. Soc. Sci. 225, 122–127.
McNaughton, J.W., Cahn, A.J., 1970. A study of the food intake and activity of a group of urban adolescents. Br. J. Nutr. 24, 331–344.
Medical Science, 1936. League of Nations and nutrition. Ir. J. Med. Sci. 11 (1), 1.
Meerhaeghe, M.A. G.v., 1998. International Economic Institutions. Kluwer Academic Publishers, Massachusetts.
Merritt-Hawkes, O.A., 1928. The population conference at Geneva. J. Hered. 19, 313–315.
Michie, J., Smith, J.G. (Eds.), 2000. Managing the Global Economy. Oxford University Press, New York.
Ministry of Health, 1932. Advisory Committee on Nutrition Report to the Minister of Health on the Criticism and Improvement of Diets. Ministry of Health, London. HMSO, 1932 London.
Montgomery, B., 2008. $E=mc^2$, the Theory of Relativity, and IRC: Einstein's Great Ideas. Retrieved from: http://www.einstein150.com/timeline/#/.
Moore, K., 2008. The Royal Society's War Committee on Engineering 1914–19. Notes Rec. R. Soc. Lond. 62 (3), 315–319.
Morrison, G., 1947. Hybrid corn: science in practice. Econ. Bot. 1 (2), 5–19.
Mukerjee, R., 1933. The criterion of optimum population. Am. J. Sociol. 38 (5), 688–698.
NAL, 2010. Agricultural Adjustment Act of 1933: Definition. N. A. Library, USDA The National Agricultural Library. Agriculture Fact Book, USDA.
National Archives, 1995. Federal Records: Records of the Foreign Agricultural Service. National Archives of the United States, US National Archives.
National Archives, 2009. Federal Register: Food and Nutrition Service, vol. 74. National Archives of the United States, US National Archives.

National Archives, 2011. Britain 1906-1918: Defence of the Realm Act (DORA). Retrieved from: http://www.nationalarchives.gov.uk/education/britain1906to1918/g5/background.htm.
NHM, 1989/1990. Where Did Biotechnology Begin? Access Excellence.
Northbourne, L., 1940. Look to the Land. J. M. Dent & Sons, London.
Northedge, F.S., 1986. The League of Nations: Its Life and Times. University Press, Leicester, Leicester, pp. 1920–1946.
NRC, 1943. Recommended Dietary Allowances: Report of the Food and Nutrition Board, Reprint and Circular Series No. 115. National Research Council, Washington, DC, p. 6.
Ordovas, J.M., Mooser, V., 2004. Current opinion in lipidology. Genet. Mol. Biol. 15 (2), 101–108.
Ostrolenk, B., 1930. The surplus farm lands. Ann. Am. Acad. Pol. Soc. Sci. 148, 207–211.
Otto, J.S., 1999. The Final Frontiers, 1880-1930: Settling the Southern Bottomlands. Greenwood Press, Westport USA.
OXFAM, 2007. How Did Oxfam Start? Retrieved from: https://www.oxfam.org/en/our-history.
Parran, T., 1943. A blueprint for the conquest of hunger. Public Health Rep. 58 (24), 893–899.
Passmore, R., 1982. Reflexions on energy balance. Proc. Nutr. Soc. 41, 161–165.
Paull, J., 2008. The lost history of organic farming in Australia. J. Org. Syst. 3 (2), 2–17.
Pearl, R., 1927. The biology of population growth. In: Sanger, M. (Ed.), Proceedings of The World Population Conference Held at the Salle Centrale, Geneva, August 29th to September 3rd, 1927. Edward Arnold, London.
Périssé, J., 1981. Joint FAO/WHO/UNU Expert Consultation on Energy and Protein Requirements: Past Work and Future Prospects at the International Level. Food and Agriculture Organisation, Rome.
Phillips, R.W., 1981. FAO: Its Origins, Formation and Evolution 1945–1981. Food and Agriculture Organisation, Rome.
Pinchot, G., 1918. Essentials to a food program for next year. Ann. Am. Acad. Pol. Soc. Sci. 78, 156–163.
Pipkin, C.W., 1933. Relations with the League of Nations. Ann. Am. Acad. Pol. Soc. Sci. 166, 124–134.
Pollen, A., 1917. The food problem of Great Britain; the shipping problem of the world. Ann. Am. Acad. Pol. Soc. Sci. 74, 91–94.
Rao, C.K., 2008. Who coined the terms 'biotechnology' and 'genetic engineering', and when? Curr. Sci. 95 (11).
Richter, J.H., 1943. The food situation in a world at war: central and northern Europe. Ann. Am. Acad. Pol. Soc. Sci. 225, 85–87.
Roosevelt, F.D., 1941. XXII Annual Message to the Congress, 1941. Department of State, United States Government Printing Office.
Rosenberg, I.H., 1994. Nutrient requirements for optimal health: what does that mean? J. Nutr. 124 (9 Suppl. l), 1777S–1779S.
Royal Society, 1917. The Food Supply of the United Kingdom: A Report Drawn up by a Committee of the Royal Society. Royal Society, London.
Royal Statistical Society, 1936. Current notes. J. R. Stat. Soc. 99 (4), 830–835.
Ruxin, J.N., 1996. Hunger, Science, and Politics: FAO, WHO, and Unicef Nutrition Policies, 1945 -1978, Chapter II the Backdrop of UN Nutrition Agencies, by Joshua Nalibow Ruxin. University College London. **PhD**, London.
Sanger, M., 1927. Proceedings of the World Population Conference, Geneva, August 29th-September 3, 1927. Edward Arnold, London.
Santos, J., 2006. Political Economy of Enterprise: Going against the Grain: Why Did Wheat Marketing in the United States and Canada Evolve So Differently? Business and Economic History Annual Conference. Business History Conference, Toronto.
Save the Children, 2009. History: Creating the Foundation. Retrieved from: http://www.savethechildren.org/about/mission/our-history/.
Schultz, T.W. (Ed.), 1945. Food for the World. University of Chicago Press, Chicago.
Shaw, J., 2007. World Food Security: A History since 1945. Palgrave Macmillan, UK.
Sherman, H.C., 1920. Protein requirement of maintenance in man and the nutritive efficiency of bread protein. J. Biol. Chem. 41, 97–109.
Shurtleff, W., Aoyagi, A., 2007. History of Soybeans and Soyfoods: 1100 B.C. To the 1980s. Lafayette, California. Soyinfo Center.
Smil, V., 1999. Detonator of the population explosion. Nature 400 (6743), 415–416.
Smil, V., 2001. Enriching the Earth: Fritz Haber, Carl Bosch, and the Transformation of the World Food Production. Massachusetts Institute od Technology, Massachusetts.

Smith, C.W., et al., 2004. Corn: Origin, History, Technology, and Production. John Wiley and Son, New Jersey.

Smith, D.F., 2003. Commentary: the context and outcome of nutrition campaigning in 1934. Int. J. Epidemiol. 32, 500–502.

Smith, R., 2009. The emergence of vitamins as bio-political objects during world war I. Stud. Hist. Philos. Sci. C Stud. Hist. Philos. Biol. Biomed. Sci. 40 (3), 179–189.

Sprague, G.F., 1962. Hybrid Corn. After a Hundred Years – Yearbook of Agriculture, 1962. Department of Agriculture, Washington, U.S., pp. 106–107

Starling, E.H., 1918. The significance of fats in the diet. Br. Med. J. 2 (3005), 105–107.

Starling, E.H., 1919a. The Feeding of Nations. Green & Co, London, Longmans.

Starling, E.H., 1919b. The Oliver-Sharpey Lectures on the Feeding of Nations: A Study in Applied Physiology. Longmans Green and Co, London.

Staudinger, H., 1939. Problems of Population. War in Our Time. H. Speier and A. Kähler. W.W. Norton & Co, New York.

Steiner, R., 1958. The Agriculture Course. Bio-Dynamic Agricultural Association, London.

Stiebling, H.K., 1933. Food Budgets for Nutrition and Production Programs. Miscellaneous Publication No.183 US Department of Agriculture, USDA.

Swanson, R., 2003. Foreign Agricultural Service Act of 1930. U. Congress. US, USDA Foreign Agricultural Service, p. 8.

Tanner, J., 2004. Incorporated Knowledge and the Making of the Consumer: Nutritional Science and Food Habits in the USA, Germany and Switzerland (1930s to 50s). Knowing Consumers: Actors, Images, Identities in Modern History. ZIF (Zentrum für Interdisziplinäre Forschung/Centre for Interdisciplinary Research), Bielefeld, Germany.

Taylor, A.E., 1926. World food resources. J. Foreign Aff. 5 (1), 18–32.

Tenny, L.S., 1938. The Agricultural Adjustment Act of 1938: a symposium. J. Land Public Util. Econ. 14 (2), 162–166.

Thackray, A., 1998. Private Science: Biotechnology and the Rise of the Molecular Sciences. University of Pensilvania Press, Pensilvania.

Thompson, R.J., 1943. The United Nations conference on food and agriculture. J. R. Stat. Soc. 106 (3), 273–276.

Ugaldea, U.O., Castrillob, J.I., 2002. Single cell proteins from fungi and yeasts. Appl. Mycol. Biotechnol. 2, 123–149.

UN, 1943. United Nations Conference on Food and Agriculture, May 18-June 3, 1943 : Final Act and Section Reports/ Department of State, United States of America. Washington, D.C. - U.S. G.P.O. United Nations Conference on Food and Agriculture. United Nations Government Printing Office, Washington D.C.

UN, 1945. Charter of the United Nations, 24 October 1945, 1 UNTS XVI. U.N., United Nations.

UN, 2000. U.N. History. Retrieved from: https://www.un.org/en/about-un/.

UN, 2006. The Declaration of St. James's Palace. Retrieved from: https://www.un.org/en/charter-united-nations/index.html.

UNESCO, 2009. About UNESCO: What is it? What does it do? Retrieved from: https://en.unesco.org/.

UNICEF, 2005. United Nations Childrens Fund. Retrieved from: http://www.unicef.org/about/who/index_history.html.

UNICEF, 2009. The Convention on the Rights of the Child. Retrieved from: http://www.unicef.org/rightsite/sowc/.

UNOG, 2009. The End of the League of Nations. Retrieved from: https://www.unog.ch/80256EE60057F2B7/(httpHomepages)/5562355D4417A43F80256F04007174DB?OpenDocument.

UNOG, 2010. UNOG Registry, Records and Archives Unit, 1870- (Archive). Retrieved from: http://biblio-archive.unog.ch/detail.aspx?ID=404.

UOR, 2009. University of Reading: Agriculture, Policy and Development, History of Agriculture. Retrieved from: http://www.ecifm.rdg.ac.uk/history.htm.

USAID, 2009. FrontLines: Borlaug, Father of Green Revolution, Dies. Retrieved from: https://www.usaid.gov/sites/default/files/documents/1867/USAID-Legacy-in-Agricultural-Development.PDF.

USDA, 1985. Possible economic consequences of reverting to permanent legislation or eliminating price and income supports. Agric. Econ. Rep. Natl. Agric. Libr. 526, 99.

Varnee, D.B., 1955. A new concept of parity. J. Dairy Sci. 38 (8), 935–939.

Volin, L., 1943. The food situation in a world at war: the Russian food situation. Ann. Am. Acad. Pol. Soc. Sci. 225, 89–91.

Watson, A., 1983. Agricultural Innovation in the Early Islamic World. Cambridge University Press, U.K.

Watson, A.M., 1974. The Arab agricultural revolution and its diffusion, 700-1100. J. Econ. Hist. 34 (1), 8–35. The Tasks of Economic History.
Watson, M., 1937. Malaria and nutrition in Africa. J. R. Afr. Soc. XXXVI (CXLV), 405–420.
Waugh, F.V., 1943. The food distribution programs of the Agricultural Marketing Administration. Ann. Am. Acad. Pol. Soc. Sci. 225, 169–176.
Webb, P., et al., 2008. More Food, but Not yet Enough: 20th Century Successes in Agriculture Growth and 21st Century Challenges. Food Policy and Applied Nutrition Program. School of Nutrition Science and Policy, Tufts University, Boston, Massachusetts.
Weisdorf, J.L., 2005. From foraging to farming: explaining the neolithic revolution. J. Econ. Surv. 19 (4), 561–586.
Whipple, C.E., 1943. The food situation in a world at war: southeastern Europe. Ann. Am. Acad. Pol. Soc. Sci. 225, 88–89.
White, T.R., 1919. The amended covenant of the League of Nations. Ann. Am. Acad. Pol. Soc. Sci. 84, 177–193.
WHO, 1958. The First Ten Years of the World Health Organization: Ch 22 – Nutrition. WHO, Geneva.
WHO, 2009. Archives of the United Nations Relief and Rehabilitation Administration (UNRRA). Retrieved from: http://www.who.int/archives/fonds_collections/bytitle/fonds_2/en/index.html.
WHO, 2010. Website of the World Health Organisation. Retrieved from: http://www.who.int/.
Williams, A.J., 2005. Reconstruction' before the Marshall plan. Rev. Int. Stud. 31 (3), 541–558.
Wolfe, A.B., 1934. On the criterion of optimum population. Am. J. Sociol. 39 (5), 585–599.
Worthington, E.B., 1936. On the food and nutrition of African natives. Afr. J. Int. Afr. Inst. 9 (2), 150–165. Problems of African Native Diet.
Zhang, W., 2008. A forecast analysis on world population and urbanization process. Environ. Dev. Sustain. 10 (6), 717–730.

CHAPTER

3

Food, rights, and politics: the post-war years (1945–60)

OUTLINE

Food and Society

https://doi.org/10.1016/B978-0-12-811808-5.00003-9

3.1 A new gold standard

Ravaged by the Second World War, the world's financial markets were in chaos. Consequently, regulation and stability had to be brought back; as a result the job was given over to the United Nations and the as-yet untested World Bank Group (the IMF and the IBRD). Luckily, one of the outcomes of the Bretton Woods agreement was to establish an alternative to the previous difficulties of maintaining a stable gold standard. The difficulty with the original gold standard system of 1875–1914 was that countries had to maintain voluminous gold reserves in case of common currency volatility. This was found to be very fragile and restrictive at best and it was not long before countries were dropping out of the system. A new system had to be found, and harking back to the Bretton Woods talks was the agreement of a fixed monetary exchange system. Similar to the previous mentioned gold standard whereby a country's monetary system was fixed at a specific rate to that of gold, the new system would fix or peg other countries' currencies to to that of the US dollar. This new par value 1 system, as it was called, then pegged the US dollar to gold at a rate of $35 per ounce, the inference now being that US dollar would become the new gold standard, and large hordes of gold were no longer needed at the country level. Initial response to this system was positive and helped nervous global economy to settle down and stabilize. This brought back some measure of stability to the exchange rate system and facilitated the return of stable international trade without the fear of wild swings in the terms of trade.

Separately, on the "right to food" front, the fledgling United Nations set up the Human Rights Commission ITO (1948) chaired by Roosevelt's widow Eleanor. This was a pivotal moment in the history of food as a right.

3.2 The United Nations human rights commission of 1948

Considering the disparate cultural values, ideologies, and political systems, as well as the religious beliefs of the then 58 member states, the fact that any cross-cultural Declaration was agreed upon was a minor miracle in itself. Representing a remarkable achievement of cooperation spanning 18 months, the resultant document during a period where war weary nations sought direction and focus on both physical and social matters. Through the incorporation of a common set of values as described above, the Universal Declaration of Human Rights (UDHR) represented a unified vision of an equitable world sharing more than just a simple set of common ideologies; it was a forward-looking philosophical instrument beyond its years.

The declaration was finally drafted on December 10 of that year (UN, 1948), comprising 30 articles spanning the right to freedom (Article 1) to the right to education (Article 26). The declaration can be simply divided into two main parts whereby articles 3 to 21 dealt almost exclusively with civil and political rights from torture, arbitrary arrest together with slavery and the freedom of movement. The second part, articles 22 to 27, dealt once again almost exclusively with the social, economic, and cultural rights viz-a-viz work community and leisure and education. However, in this mix was also an important milestone in what was called the food security; in article 25 specifically parts one and two, every man woman and child's right to food was recognized. Once again this was a remarkable achievement considering the makeup both politically and philosophically of the member group (Gibson, 2016). Article 25 went on to describe

(1) That everyone had a right to a standard of living that would maintain the health and well-being of oneself and their families, comprising housing, medical care, social services including unemployment, sickness, old age, and of course the right to sufficient and adequate food.
(2) Moreover, it was determined both mothers and children were entitled to extra special attention, especially if the child was born out of wedlock, in which case both should enjoy the same governing principles, in terms of social and welfare protection (UN, 1948).

Building on experience of the First World War, food was once again a hot topic.

3.3 The Marshall plan

Indeed, at the outset of the Second World War, Allied governments colluded and introduced preferential bilateral and multilateral trade agreements promoting domestic production over the importation foodstuffs. This made perfect sense because of the experience of the First World War whereby the German U-boat campaign was relatively successful in blockading some of the Allied countries, thus making it difficult to receive food imports. Yet, whichever way you looked at it, this did not sit well with certain nations who despite the war relied on international trade to support economic growth. Compounding the problem with the allied forces continuing price support policies which not surprisingly ended up with large food surpluses. In fact, it was foreseen that the surpluses would continue after the war with one contemporary observer noting:

> the [FAO] … will soon be faced with the problem of disposing of supplies which cannot be marketed at profitable prices … [if] the goal of freedom from want is to be brought within even approximate reach, the problem of expanding world purchasing power must be solved ….A progressive expansion of world purchasing power, associated with national policies which facilitate rather than impede necessary transfers of resources, provides the setting in which surpluses will be least depressing on price and most easy to handle … ***Belshaw (1947), (insertions by Gibson (2016)).***

This was not so for everybody as some countries were still food impoverished finding it difficult to import necessary supplies in the face of insufficient foreign reserves.

However, not everything fell on the wrong side of the balance sheet. In line with the avowed global cooperation philosophy and aid, the 1947–1948 US Foreign Assistance Act permitted post-war European economic recovery to begin. In turn, this act opened the doors for Truman's Secretary of State George C. Marshall's European Recovery Program (ERP) or more commonly the "Marshall Plan" (Borchard, 1947). The stated goal was not unlike the previous United Nations Relief and Rehabilitation Administration (*UNRRA*) goals of earlier times in which international relief was administered to those that required it. While the previous efforts of the UNRRA acted as a blueprint for the Marshall plan, the Marshall plan was not without political agenda (Williams, 2005).

So, what was the Marshall plan? Overall, it involved multibillion-dollar program of financial, technical, and food aid over a 4-year period to facilitate the redevelopment of Europe. Yet despite its flaws it came to an end in 1951 and was hailed as one of the most remarkable achievements of humanitarian aid ever to have taken place. It was an eye-opener too; in that it was the first full realization of the social and economic interdependence of nations. Globalization had truly begun (Borchard, 1947; Crawford, 1949; Scott, 1951; Weir, 1952; Hoffmann and Maier, 1984; FAS, 2018; USAID, 2009b; USAID, 2009c).

Meanwhile, at around this time, the idea of more equitable terms of trade for everybody stepped out of this theoretical phase and entered the spotlight.

3.4 General agreement on tariffs and trade

As a result of the large number of international barriers that flew in the face of the underlying previously agreed principles of global free trade and equitable cooperation, something had to be done. Amid loud calls for change the United Nations Economic and Social Council (ECOSOC) tasked The New International Trade Organization (ITO) with the coordination of more equitable multilateral trade policies. That was in 1946, and by 1947 an international treaty called the General Agreement on Tariffs and Trade (GATT) was initiated and ready to operate under the ITO. However, the ITO never got off the ground, yet the GATT continued and ending up being an important instrument that would prove its worth many times over the ensuing years (Gibson, 2016).

Post-war reconstruction was not just about infrastructure or economic development, it was also about food and nutritional construction.

3.5 Post-war nutritional reconstruction

As a direct result of the outbreak of World War II, great deal of work by both the League of Nations' (HOLN) health units and the Office International d'Hygiène Publique (OIHP) was put on hold. Yet very real fears of the potential for widespread epidemic rallied the UN's UNRRA placing health among its primary responsibilities.

Immediately preceding the war, the task development and reconstruction provided the first real challenge of international cooperation. Leading the multilateral effort was the United States Marshall plan and the newly created United Nations. Initially, the British dominated

Allied Committee on Post-War Reconstruction Requirements (the Leith Ross Committee) began planning the relief requirements required of each nation; however, it was a slow process and lacked direction. As a result, it was superseded by the Office of Foreign Relief and Rehabilitation Operations (OFRRO) set up by Roosevelt in 1942 and headed by Herbert Lehman was a little more proactive in the mission. Together with the UN Relief and Rehabilitation Administration (UNRRA) (also headed by Lehman) worked tirelessly providing much needed relief supplies. And recognizing the role of mothers and children in the rehabilitation process milk and other relief supplies became the preferred choice in existing infant school feeding programs. However, when it came to the refugees, who also fell under the umbrella of the UNRRA, it was just one job too many and was eventually delegated to the International Refugee Organization (IRO) (House of Commons, 1946; USHMM, 2010) (Broughton, 1997).

About this time too, there was much debate over the future of certain international health policies among loosely related bodies, in particular the FAO; WHO; and UNICEF. Increasing infighting took place among the rival organizations; the first on the scene was the FAO.

3.5.1 Food and agriculture organization

Although the League of Nations held its final meeting in April 1946, the organization was all but essentially laid to rest with the United Nations more or less taking over where the league left off (Sweetser, 1946; Goodrich, 1947). Proposed in 1943 and officially created in 1945, one of the first bodies of the new United Nations was the Food and Agricultural Organization (FAO). Its new Director General was none other than the nutritional scientist Sir John Boyd Orr [2], whose first task was to appoint Dr. W. R. Aykroyd as Director of the FAO's Nutrition Division. Early on while its remit was specifically to improve the lot of global agriculture, forestry, and fisheries, Boyd Orr, Aykroyd, and others saw their role as taking in a more broader aspect to include general welfare: "... to end hunger and raise general standards of nutrition." (Gillespie, 2003). The next on the scene was the WHO.

3.5.2 The world health organization

In the first meeting of the ECOSOC Economic and Social Council (1947), the creation of a permanent specialized health organization was considered, after which much wrangling went on to establish the World Health Organization. Its mission was to carry on the previous functions of the League of Nations as well as the OIHP (EoN, 2009). In the meantime, the Interim Commission of the proposed WHO continued the work of health functions of the UNRRA's until the WHO's constitution was finally ratified ITO, 1948. This was a departure from what was expected of the WHO as in its initial inception there was little room for the WHO to be involved in such cases or programs such as child and maternal health care. This was counter to what some people expected, or as seen by some an effort to expand the remit of the WHO beyond a mere institution concerned with hygiene and little else. However, fierce opposition to a dumbed-down body prevailed and the final draft of the WHO ensured it became a much-needed universal public health organization (WHO, 1946).

Then came UNICEF.

3.5.3 The UN international children's emergency fund

Before the final meeting of UNRRA in 1946, the United States Army's film makers made a short 19-minute documentary called the "Seeds of Destiny." Shockingly the film was full of harrowing images of children in hospitals and orphanages and others foraging and begging for food. By highlighting the plight of postwar children, the short film was aired at the last meeting of UNRRA's governing council. However, the stark reality of the production was both moving and disturbing. Consequently, the Council proposed to the United Nations that its considerable surplus funds be used to continue and enhance the relief effort for needy children (UNICEF, 2005). As a direct result of the meeting, a new child-welfare organization—the UN International Children's Emergency Fund (UNICEF) —was established to provide much-needed relief assistance to children of post-war Europe. Maurice Pate, the enthusiastic and pragmatic head of the new body, only accepted the post on provision that UNICEF would mandatorily include all needy children, no matter creed or color and, more importantly, the political persuasion of the children's parent government. It was a controversial decision that sought to place children outside of institutional politics, especially among a growing number of committees and reports, etc. (UNICEF, 2005)

It was also eloquently pointed out in a lecture given in honor of UNICEF receiving the 1965 Nobel Peace Prize by the then chairman Zena Harman who suggested

> Child suffering could not be distinguished by virtue of its cause or origin. Children in desperate need anywhere and everywhere required help and attention. (Nobel Prize, 2010).

At the outset, UNICEF was only supposed to be a temporary affair with an original mandate that recognized children's health to be of primary importance. The basic needs of such children were considered to include adequate nutritional food; protection against disease; shelter and clothing; clean and safe water; and a suitable environment in which children can grow both emotionally and socially. It was already understood that schools were an exceptional means by which distribution and monitoring programs aimed at health care and nutritionally deficient children could fairly easily be achieved. As a result, by the end of WW2, UNICEF was primarily dedicated to supplementary feeding programs of school-age children.

3.5.4 FAO first world food survey (1946)

The first world food survey of the FAO in 1946 involved 70 countries which included many nutritionally savvy governments and people's from across the globe. During the conference, there was widespread agreement regarding the existence of the problem of global hunger and malnutrition, yet what was lacking was the fact there was very little evidence to back this up. This was summarized by the FAO at the time when commenting on the issue:

> ... it is well known that there is much starvation and malnutrition in the world [yet] vague knowledge that this situation exists is not enough; facts and figures are needed if the nations are to attempt to do away with famine and malnutrition (FAO, 1995).

However, due in large part to the cessation or the breakdown of statistical data gathering during the war, final estimates relied heavily on estimations and imprecise data. Nevertheless, by providing a pre-war baseline of total and per-capita calorie availability, this marked the beginning of ongoing longitudinal research for the FAO. In this sense, the report provided solid semiscientific evidence indorsing long-held beliefs that global hunger and malnourishment did indeed exist, and at staggering levels too. The report indicated that more than half of the global population survives on less than 2250 calories per day (FAO, 1995).

Having established, not only the existence of hunger and malnourishment but also the report identified the extent to which it prevailed; there were now no excuses for doing nothing. The next step then was to establish long-range nutritional targets to 1960 based on apparent needs while also taking into account a population increase in the range of 25% (FAO, 1946; McMmillan, 1946; Belshaw, 1947).

By 1948, the FAO had fully recognized the importance of food, energy, and nutrition as central to its remit within the United Nations (Weisell, 1995).

3.5.5 International nutritional requirements (1949)

In order that nutritional requirements are met, it is incumbent on the provider to fully understand the needs and requirements of humanity (FAO, 2002). Thus, continuing where the League of Nations left off, the UN's Food and Agricultural Organization's Food and Nutrition Division convened an expert committee (Committee on Calorie Requirements) took up the challenge in 1949; the remap was to undertake the first of several nutritional-related human energy and nutrient requirement reviews (Aykroyd, 1956; Périssé, 1981; FAO, 2003). The outcome was based on the 1947 report of the FAO's Food Composition Tables, in particular, the Calorie Conversion Factors in which the Committee on Calorie Requirements authored the first real international food composition tables (Chatfield, 1949; FAO, 1950; Aykroyd, 1956). These tables and the following 1949 revisions (published in 1950) were designed using the early pioneering work of Atwater's food energy conversion factors around the turn of the century (Wu Leung, Busson et al., 1968). The first incarnation of the tables expressed food values in calorie, protein, fat, and carbohydrate content (Chatfield, 1949); however, the tables were subsequently revised and extended to incorporate vitamins and minerals, firstly in 1954 and then in 1955 by Merrill and Watt (Chatfield, 1954; WHO, 1955; Aykroyd, 1956; Wu Leung, Busson et al., 1968; Weisell, 1995). By this time, a country's food balance sheets noticeably took on more importance. Their ability in analyzing a country's food situation was indeed well understood, so much so in fact that the FAO ITO, 1948 recommended that nations develop their own food balance sheets have published regularly. The FAO's "Handbook for the Preparation of Food Balance Sheets" first published in 1949 is still published periodically to this day (FAO, 2001).

3.5.6 Kwashiorkor and protein energy malnutrition

During the 1950s, children's nutritional welfare was still very much high on the agenda (WHO, 1950; WHO, 1951; WHO, 1953; WHO, 1955). The First Joint FAO/WHO Expert Committee on Nutrition observed that high mortality rates of infants and children, particularly in the tropical and subtropical regions, were blamed on widespread disorders, the most prevalent of which was kwashiorkor (a condition that was thought resulted from a lack of protein intake) (WHO, 1950; WHO, 1953). It was also noted that once the infant came off weaning and went on to solids comprising cereals and other foods containing very little milk, ergo more protein (PEM) was seen to take place (WHO, 1950, pg 16). Furthermore, the FAO/WHO's report was further enhanced by Brock and Autret' s report further detailed the extent of kwashiorkor in Africa at the time (Brock and Autret, 1952). As we know now, the observations were slightly off key. However, contemporary 1950s collective conscience was still under the assumption that PEM was a major contributor to malnutrition and many new policies and humanitarian aid was based on this assumption. Back to the 1950s and in answer to the perceived ongoing problem, the joint Committee advocated more education for professionals and lay people alike. During the same time, Joint FAO/WHO Expert Committee on Nutrition launched a worldwide investigation to research and assess the extent of the malady (WHO, 1951; WHO, 1958; Weisell, 1995). Complementing these proactive measures, increased research into nutritional requirements was undertaken in support of the FAOs 1949 energy and calorie requirement tables (WHO, 1951). This indirectly led to the second world food survey (see below).

In terms of terminology, although "kwashiorkor" was first coined by Cecilly Williams during the early 1930s, it had not by the late 1950s been adopted as a catchall for protein energy malnutrition. However, in 1952, the expert committee met in Gambia and used the term kwashiorkor to better distinguish this type of protein energy malnutrition to other types of protein malnutrition—from this point forward, kwashiorkor became synonymous with protein malnutrition.

3.5.7 Protein energy requirements

While a considerable body of work was collected, collated, and analyzed, it was the research in respect of specific energy producing nutrients for the individual that took on greater importance (FAO, 2002). Consequently, during the early 1950s, scientists were becoming increasingly aware that aiming for optimal energy giving nutrients (in particular an obsession with protein) alone was done so at the expense of other nutrient and trace element research. In other words, sufficient energy alone did not necessarily provide adequate nutrition (Allen, 2003; Gibson, 2016). This obsession with protein as the main nutrient was further studied. This included Nevin Scrimshaw and others in their research on infants and breastmilk among other things, which in the end revealed that protein was not the single or sole source of food. It was first realized through the work of Nevin Scrimshaw and others that protein deficiency in infants first developed when breast milk was no longer the single or sole source of food. Thus, cheap alternatives to milk protein were sought in an effort to slow down and prevent kwashiorkor as it was known in weaning infants.

3.6 FAO second world food survey 1952

The second World Food Survey revealed global estimates of calorie and protein supply and concluded that the situation was even worse than the first survey (1946). The survey averaged per capita calorie availability (after adjusting for losses, seed, and feed), for total world output grounded on a daily supply of 2600 calories/capita. This stark reality serves to express the fact that with these figures, over two-thirds of the global population was actually undernourished.

Such results led to the credence that kwashiorkor was indeed one of the most serious and nutritional disorder in the world at the time (Goldsmith, 1955; Aykroyd, 1956).

Buoyed by the report, the FAO and WHO alongside the Josiah Macy Jr. Foundation held a conference in June 1955 to discuss such matters. Following this the WHO established the Protein Advisory Group (PAG), drawn from some of the most revered experts of the time (WHO, 1958). The PAG's official role was adviser to the FAO, the WHO, and UNICEF. The group ultimately recommended that protein be the foundation of most nutritional programs from that point on. Supported by top nutritionists from developing countries the PAG became very influential, although the PAG was eventually rolled into a tripartite program in 1961 (FAO/UNICEF/WHO) (Aykroyd, 1956; PAG, 1967; WHO, 1972; Carpenter, 1994; Weisell, 1995).

Around the middle of the 1950s, further bodies and committees came together to recommend a suite of protein requirements for the token adult or child (FAO, 1955). This is despite the fact that such research had already been undertaken several times before by others including the League of Nations. For adults, the FAO's protein findings were calculated to be 0.35 g/kg of body weight, which is a great difference compared to values given by the League of Nations at 1 g/kg in its 1936 report. While for children and adolescents, despite limited data meant that results were largely estimated at 0.7–0.8 g/kg. Although tentative, the report did encourage further research, which was taken up by the 1963 FAO/WHO Expert Group (FAO/WHO, 1973; WHO, 1973; Allen, 2000).

3.7 Gender and work rates

By 1956, the second FAO Calorie Committee reconfirmed the practicality of the notion of a certain number of calories per day for a "reference" male and female. Although to this there was some minor revisions made to the overall daily calorie allowances:

> ... a range of energy expenditure between 2400 and 4000 kcal/day for men and between 1600 and 3000 for women would appear to include most human beings ***FAO (1957), pg 16.***

The second Committee also reacknowledged the difficulty in allowances due to the difference in physical activity levels by which the reference male or female person is modified. For such purposes, the committee referred to the Christensen classification of work (Table 3.1). That said, Passmore (1964) in his review was troubled by the fact that there were a multitude of alternative classifications in use at the time and strongly suggested that it would be more scientifically prudent if a single agreed scale were used for the various work rates (Passmore, 1964).

There was also one more important difference in FAO's calculations of energy requirements. This was contrary to the work of Voit, Atwater, Lusk, NRC, and others who traditionally based advisory energy requirements on food intake alone. By contrast, the new FAO 1957

TABLE 3.1 Christensen's energy expenditure rates at differing work levels.

Work rate	kcal/min
Very light	Less than 2.5
Light	2.5–4.9
Moderate	5.0–7.4
Heavy	7.5–9.9
Very heavy	More than 10.0

Source: Christensen's Physiological Valuation of Work in the Nykroppa Iron Works. Ergonomics Society Symposium on Fatigue. Ed's W. F. Floyd and A. T. Welford (London, Lewis. 93–108, 1953).

report encouraged the use of energy expenditure to calculate energy requirements (FAO, 1957; Henry, 2005).

3.8 The link between malnutrition and infection

People had long since postured a connection between health and infection. By the mid- to late 1950s, two successive Joint FAO/WHO Expert Committee's on Nutrition showed concern for the lack of knowledge regarding any interaction between nutrition and infection. It was finally through the work of Scrimshaw, Taylor, and Gordon that light shed on this potential link. However, it was not until 1959, in a paper written for the American Journal of the Medical Sciences, that Scrimshaw and colleagues noted a scientific correlation between both malnutrition and diarrhea. Through extensive research, the authors firmly suggested that susceptibility to infection was one outcome of poor malnutrition. Furthermore, it was seen that any infection contributed to a synergistic relationship, which in turn caused further deterioration of the person's nutritional status. Such interactions too it seemed were further exacerbated when malnutrition and infection were present simultaneously. Combined this seemed to amplify the individual effects of either one alone. As a result, the study's message seemed to suggest that to break the vicious cycle, it was not merely enough to intervene to break the cycle, but rather it was preferable to tackle both nutrition and infection at the same time if an optimal outcome was to be achieved (Scrimshaw et al., 1959; Keusch, 2003).

3.9 Genetic engineering further milestones

This was also a period in which genetic engineering really stood out. By 1946, in the discipline of genetics, both Edward Tatum and Joshua Lederberg observed that bacteria could occasionally exchange genetic material. This was quite a leap, yet in the same year, independently, Max Delbruck and Alfred Day Hershey revealed that during the process of genetic recombination, genetic material from different viruses could indeed be united forming a new type of virus. Adjacently, progress in the field of biology witnessed the isolation and growth of animal cells in vitro. 1950 was also the time that the first instance of artificial insemination within livestock using frozen semen was used; this was tremendous news especially for the farmers. Furthermore, among scientists, the boundaries of electron microscopy were being pushed further and further and by 1952, microscopes were now able to see the tiny well-formed anatomical structures or organelles that existed inside eukaryotic cells.

A little time earlier the transfer of genetic material from parent to child was still only a matter of guesswork. It took Ostwald T Avery's work with bacteria in which he postulated that it was the nucleic acid, the DNA of the cells and not proteins (as was often suggested) that actually contained the genetic material of life. This theory paved the way for Watson and Cricks later discovery. In fact, on the subject of Watson and Crick, without doubt, in terms of biological advancement, the discovery in 1953, of the molecular structure of DNA by James Watson and Francis (Watson and Crick, 1953). In their observation's DNA molecules comprised a double-stranded, helical model. To simply state that this was of great importance is an understatement in itself; however, such a major breakthrough in genetics was remarkable as, to this point the method of transfer of genetic from parents to offspring till this point had only been conjecture. What Francis and Crick showed, was a biological vehicle that allowed for the transfer of genetic material from adult to offspring. This effectively opens the door to the potential of DNA, which proved to be extremely complex, and indeed it was able to carry the complete coded recipe for heredity. This further strengthened the belief in the "DNA-gene" transfer theory and ultimately the principles of Mendelian inheritance.

Such a discovery opened up untold possibilities of introducing into the gene pool cures for human diseases and other significant uses. What perhaps Watson and Crick did not foresee was the ramifications such a breakthrough was to have on the way we looked the food we consume. In one respect it opens up a whole: "Aladdin's cave of possibilities whilst on the other it threatens a Pandora's box of ills" (Gibson, 2016). Key moments of this era can be seen in the following Table 3.2:

3.10 Surpluses and the agricultural trade development and assistance act

In the United States, as the Marshall Plan slowed in 1949/50, stocks of grain once again began to rise. This led to ongoing storage problems including costs and also in some instances to depressed producer prices. Overproduction of food in the absence of previously strong European aid exports was again turning into a serious problem for the United States. Domestically, the United States increased government interventionist policies in an attempt to slow down the overheating of the agricultural market. By this time too, Eisenhower managed to push through the Agricultural Trade Development and Assistance Act in 1954. The act also called the Public Law 480 (PL 480) and contained four titles (Table 3.3), which paved the way for direct donation of US agricultural commodities to be used in emergency and nonemergency programs alike. However, it was met with mixed blessings; on the one hand it was seen as an altruistic gesture on the part of the Americans to be used as aid in developing countries as part of their developmental agenda (Sullivan, 1970). On the other hand, others saw this altruism as politically charged and has often been cited agricultural surpluses were being transformed into a US foreign policy program (Toma, 1967; Sullivan, 1970; USAID, 2009a).

TABLE 3.2 Key dates of the period: 1946 to 49.

Date	Key dates of the period
1946 Administrative Committee on Coordination (**ACC**)	The Administrative Committee on Coordination (ACC) was established in 1946. Its main functions were to help achieve the coordination of UN programs and to aid in the implementation of agreements between the United Nations and the specialized agencies. A number of subcommittees also evolved with a similar focus on coordination and cooperation of work programs throughout the UN system. In the end, the ACC became the CEB in 2001 (Mezzalama et al. 1999; CEB 2010; IANWGE 2010).
1946 Nutrition Division of FAO	Predicated predominantly on increasing food supplies, the FAO was launched in 1946. Its other remit was to raise levels of nutrition throughout the world. The FAO was aided by the Standing Advisory Committee on Nutrition, which met three times: 1946, 1947, and 1948, which was itself replaced by the Joint FAO/WHO Expert Committee on Nutrition who continued to advise the FAO until the joint committees' end in 1954 (WHO 1950; Aykroyd 1956).
1946 International Trade Organization **(ITO)**	The UN Economic and Social Committee (1946) began talks on a conference to draft a charter for an International Trade Organization (ITO), which would become a specialized agency of the UN. Completed in 1948, the Charter set out the fundamental rules for international trade. The ITO Charter, however, was not approved; instead, GATT took its place (ITO 1948; WTO 2010).
1946 International Emergency Food Council **(IEFC)**	After WW2, the International Emergency Food Council (IEFC) was organized by FAO to mitigate problems regarding the ensuing world food crisis. It ran until 1948 at which point the FAO absorbed and renamed it the International Emergency Food Committee (IEFC) (FAO 1947; NAL 2010).
1946 The UN International Children's Emergency Fund **(UNICEF)**	UNICEF provided emergency relief assistance to the children of post-war Europe and China but then changed focus to concentrate on long-term benefits to children everywhere, especially those in developing countries. Consequently, the name was changed to the UN Children's Fund, but the acronym "UNICEF" was kept (UNICEF 2005).
1946 The Division for the Advancement of Women **(DAW)**	The Division for the Advancement of Women was established as a Section on the Status of Women in the Human Rights Division of the Department of Social Affairs (DAW 2009).
1947 UK's Agriculture Act	The AA's support for using mostly deficiency payments guaranteeing a certain market price for food became challenging as consumers continued to pay market prices, while farmers were protected irrespective of market forces. This was supposed to be a policy predicated on stability and efficiency (Bikistow 1998; OPSI 2007).
1947 General Agreement on Tariffs and Trade **(GATT)**	The GATT was supposed to have been a provisional agreement intended as a stop gap while the ITO was being drafted. Yet, with the collapse of the ITO, provisional agreements for GATT to operate in its place lasted from 1947–94 after which the WTO took up the mantel. The GATTs role was to substantially reduce tariffs between member nations and thus facilitate better world trade. There were eight rounds of trade talks spanning more than 30 years (WTO 2010).
1947 The Economic Commission for Europe **(ECE)**	One of the five regional commissions of ECE headquartered in Geneva was to establish help and mobilize action for the economic reconstruction of post-war Europe

TABLE 3.2 Key dates of the period: 1946 to 49.—cont'd

Date	Key dates of the period
	and to increase European economic activity both within and outside the region (UNECE 2010).
1947 Economic and Social Commission for Asia and the Pacific **(ESCAP)**	Established in 1947 UN ECOSOC, headquartered in Bangkok, with a Pacific Operations Centre based in Port Vila, Vanuatu. ECOSOC essentially operated in regions that contained more than half the world's population (ESCAP 2010).
1946 UN Population Commission	The Population Commission was established by ECOSOC as a subsidiary body to research and advise the General Council on the size, demographics, and other changes in the world population (POPIN 1994).
1946 Meals for Millions	One man Clifford E. Clinton had the vision to end world hunger; he formed Meals for Millions in 1946, which joined forces in 1960 with Kennedy's Freedom from Hunger Foundation (Freedom from Hunger 2010).
1947 Interim Coordinating Committee for International Commodity Arrangements **(ICCICA)**	ECOSOC recommended the establishment of Interim Coordinating Committee for International Commodity Arrangements (ICCICA) to help with trade and tariff reductions (GATT 1952; Khan 1982).
1947/48 US Foreign Assistance Act aka Marshall Plan	The Marshall's European Recovery Program (ERR) involved a 4-year, multi-billion-dollar program of financial, technical, and food aid for the redevelopment of Europe.
1947 GATT 1st Round Geneva	In 1948, the agreement signed by 23 countries allowed the GATT and members to promote over 45,000 tariff reductions affecting over $10 billion worth of trade, amounting to 20% of the total global market at the time (Morrison 1986; WTO 2010).
1948 FAO Standing Advisory Committee	The Standing Advisory Committee advised that "...the problem of assessing the calorie and nutrient requirements of human beings, with the greatest possible degree of accuracy, is of basic importance to FAO" (Weisell 2002).
1948 The World Health Organization **(WHO)**	During this period having taken over from the LoN's Health Organization, The WHO, a specialized agency of the UN, was tasked with coordinating and disseminating info on international public health issues (WHO 1998; WHO 2010).
1948 Universal Declaration of Human Rights	The Declaration, adopted by the UN General Assembly, arose out of the ashes of the experiences of WW2 representing the first global appearance of a 'rights' based approach (UN 1948).
1949 Committee on Commodity Problems (CCP)	The FAO set up the Committee on Commodity Problems (CCP) as a committee of the FAO council its responsibilities were to continually review any international commodity problems that might affect production, trade, distribution, consumption and related economic matters and relate this info back to the FAO (CCP 2005a,b).
1949 International Wheat Council (IWC)	The 1949 International seat of the Wheat Council (IWC) was finally established in London (IGC 2009).
1949 GATT 2nd Round Annecy	In the second GATT rounds were around 5000 tariff reductions (Morrison 1986; WTO 2010).

Compiled from multiple sources: FAO (1947); ITO (1948); WHO (1950); Aykroyd (1956); Mezzalama et al. (1999); UNICEF (2005); CEB (2010); ESCAP (2010); IANWGE (2010); NAL (2010); WTO (2010); ITO (1948); UN (1948); WHO (1950); GATT (1952); Aykroyd (1956); Khan (1982); Morrison (1986); POPIN (1994); Bikistow (1998); Weisell (2002); OPSI (2007); DAW (2009); IGC (2009); ESCAP (2010); Freedom from Hunger (2010); UNECE (2010); WTO (2010).

TABLE 3.3 Agricultural trade development assistance Act: PL 480.

Title	Title objectives	Administered by:
Title 1	Trade and Development Assistance: Providing for government-to-government sales of US agricultural commodities specially to developing countries on favorable credit or grant terms	USDA
Title 2	Emergency and Development Assistance Program	USAID
Title 3	Food for Development	USAID
Title 5	Farmer to Farmer	USAID

Source: Compiled from multiple sources Toma (1967); Eulau (1968); Sullivan 1970; USDA (2009).

While the PL480 was underway, there were parallel debates over government interventionism in general and the regulation of prices, productivity, and fair trade as a matter of principle (Dimitri et al., 2005). Moreover, the United Nations Committee on Commodity Problems (CCP) failed in its 1946/7 efforts to achieve an International Commodity Clearing House (ICCH) in an attempt to address a proposed World Food Reserve. This was because a report concluded that technical and financial difficulties, rather than physical world food shortages, were larger impediments in the relief of emergency famine or other humanitarian aids. However, while some argue this is true, others see the failure of the clearing house as being attributed to the reluctance of United states and United Kingdom to devolve such power to a multilateral organization over which both the United States and United Kingdom had no control over (UN, 1945; FAO, 1957; FAO, 1958; GATT, 1960).

That said, the passing of the Agricultural Act in 1956 (also known as the soil bank program) allowed the Sec of Agriculture to offer surplus farm products to humanitarian/voluntary agencies that were registered with the Department of State's Advisory Committee on Voluntary Foreign Aid. On the face of it, this was a positive move in the right direction as surplus US grain stocks were distributed around the world. However, while this was in line with America's foreign policy aim, others argued that it better served America's domestic policy by opening up new global markets. In reality, it was probably a little of both especially at a time when choice was a scarce commodity (Sullivan, 1970; Webb et al., 2008). Yet, as with many good things, this was set to change, with the expansion of humanitarian aid from not only governments but also through NGOs like CARE international and Catholic Relief Services to name a few, so people began to ask questions. The lines were being blurred as some argued that domestic agricultural policies were being favored over foreign policy objectives. In fact, that aid, in the form of grain and other nonnative foods, was taking priority over recipients traditional and cultural food ways (Sullivan, 1970). And so, as the debate rumbled between political and humanitarian goals, one commentator Eulau (1968) took a more philosophical view. In his opinion, the dual often ambiguous quandary appeared to benefit both domestic and foreign policy agendas of the donating country irrespective of motive, as he succinctly summarized:

> If congressmen are willing to support a program such as Food for Peace, their private motivations or constituency interests, however selfish, do not defile, tarnish or befoul its humanitarian consequences. ***Eulau (1968).***

However, having said all, there was some good news. It seemed that powers that he did in fact understand the reality of releasing all these surplus on the international markets unimpeded by government interventionism. As such, it felt that the threat was so great that the CCP established a Consultative Subcommittee on Surplus Disposal (CSD) to closely watch the surpluses and consider their potential effects on the global market whether for good or bad (ActionAid, 2003; CCP, 2005a,b).

Elsewhere in the world, the late 1950s witnessed the birth of the European Community and what was quite possibly the worst ever famine in human history—The Chinese Great Leap Forward Famine.

3.11 The European economic community

The establishment of the European Economic Community was established in 1957. It was a remarkable moment as what it showed was the willingness of Member States to relinquish part of their sovereignty in favor of a supranational Europe. This was encouraging and negotiations the Messina Conference in 1955 was followed by a draft report on the creation of a European common market (1956). The creation of both the EEC and the Common Market was predicated on two main goals. The first was to reduce trade barriers within the community allowing the free movement of goods and services across borders. The second goal was simply looking to construct a politically unified Europe. These goals or objectives were further fleshed out over the years through a set of common policies. In the economic world, the treaty (the Treaty of Rome) aimed to promote the smooth development of European economic activity founded on free competition. This was predicated on the EEC's four freedoms including fifth—the free movement of people, services, goods, and capita (Europa 2007, 2010).

In the arena of food for all the Chinese fourth failed in a big way. During the period 1958–61, a famine ensued, and it lasted 3 years and has been quoted as being the biggest single event of its kind ever recorded in history.

3.12 The great leap forward famine (1958–61)

Billed as the largest famine of the 20th century and potentially of history the Chinese Great Leap Forward Famine lasted from 1958 to 61. It certainly was not a precedent either, as up to this point during the 20th century, China had already suffered several large famines over which it lost millions of its people. However, rather than being prepared or asking for external assistance, the political arena with its lack of transparency suppressed the full extent of the unfolding horror until it was just too late. Some say, at the heart of the problem was Chairman Mao Zedong's attempt to transform China's economic model from that of a traditional agricultural nation to one of a more industrial footing. To this end, peasants were instructed to leave their smallholdings and go to collectively work in large production farms. Furthermore, surplus workers were also instructed to produce steel and other necessities in small foundries. Indeed, such collectivization witnessed huge social upheaval and coupled with unfavorable weather conditions, as well as the overconsumption of foods in some quarters and the underinvestment in agriculture in others—the scene was set for widespread famine on an unprecedented level. Fueling this tinderbox was the communist party's natural suppression of information. Couple this with the tendency to report only good news, the scale of the disaster ultimately kept the rest of the world in the dark. Eventually, when the leadership learned of the full scale of the problem it did little in mitigating events instead opting to further clamp down on the information flow. The tight control of the news flow was so effective and almost absolute that it took a further 20 years for the full extent of the atrocity to be told. The total death toll is difficult to estimate but some have placed estimates of at between 20 and 40 million lives. This results in the single greatest peacetime disaster of the 20th century and perhaps of history (Watson, 1999; Devereux, 2000; Ó Gráda, 2009).

Fully 35 years after the League of Nations incorporated the Declaration of the Rights of the Child, the United Nations finally revisited the idea.

3.13 Declaration of the rights of the child (1959)

After a decade or so of increasing activity regarding child welfare as well as the full realization of children's nutritional well-being (especially of the under-fives), the UN General Assembly adopted the Declaration of the Rights of the Child in 1959. This was the name given to a series of related proclamations or declarations over the years that recognized and accepted children's rights. It began in 1923 with Eglantyne Jebb's (Save the Children International Union (SCIU)) declaration of children's rights, which was ultimately taken by the League of Nations in 1924. Then in 1946, the SCIU merged with the International Union of Child Welfare. After that, in 1946, UNICEF replaced the United Nations Relief and Rehabilitation Administration (UNRRA) and he then continued to work for children's rights. And lastly in 1948, the special rights of children were once again articulated in Article 25(2) of the Universal Declaration of Human Rights. Thus, in 1959, the United Nations decided to formalize all these instruments into a special, nonbinding declaration, pronouncing that every child, no matter their family race, sex, religion, social origin, or political views, should be entitled to protection of rights vis-à-vis health, education, food, freedom, dignity, and discrimination among others (UN, 1959; UNICEF, 2009).

3.14 The freedom from hunger campaign

The creation of FAO's then Director-General B. R. Sen, the Freedom from Hunger Campaign (FFHC) was rolled out on a 5-year provisional mandate aimed to draw attention to the suffering of hunger and malnutrition in developing countries. It began in 1960 at a time when progress had been slow. The FFHC was designed to be a focal point for cooperation for international organizations, governments, and various NGOs alike. In this, it was successful and in March of 1963, the campaign concluded in the "World Freedom from Hunger Week" to attract even greater worldwide attention of the injustice of hunger and to the repeated concerns of population growth. A year later, at the request of President Kennedy, the American Freedom from Hunger Foundation was created to help mobilize support of the FAO's FFHC. The Foundation again by joining forces with Clifford E. Clinton's Meals for Millions organization and were both eventually combined (1979) to form the group Freedom from Hunger (Freedom from Hunger, 2010) (UN, 1963; SOFA, 2000) (Table 3.4).

TABLE 3.4 Key dates of the period: 1950 to 59.

Date	Key dates of the period
1950 The UN High Commissioner for Refugees (UNHCR)	The UN High Commissioner responsible for refugees elicited the knowledge and expertise of other UN organizations such as food production (FAO), education (UNESCO), health measures (WHO), and child welfare (UNICEF). It also cooperated closely with the World Food Program in providing for refugees (UNHCR, 2009).
1950 World Vision	World Vision founded by the Reverend Bob Pierce initially focused on orphans and other children in need South Korea. It then expanded throughout Asia then into more than 90 countries by the 2000s (World Vision, 2009).
1951 GATT 3rd round Torquay	The third GATT round in the UK (Torquay) tackled about 8700 tariff concessions (WTO, 2010).
1953 Foreign Agricultural Service (FAS) of the US Department of Agriculture (USDA)	Partnered with USAID, the FAS predominantly concentrated on the agribusiness aspect of humanitarian aid (National Archives, 1995; Otto, 1999; FAS, 2018).
1954 US Agricultural Trade Development and Assistance Act (Food for Peace)	US Federal Law established Food for Peace, which became the primary US overseas food assistance program (FAS, 2018).
1954 US Agency for International Development (USAID)	USAID is the US governments primary agency for humanitarian aid purposes and is the top provider of food aid in the world (UOVS, 2008; USAID, 2009a; USAID, 2010).
1954 The FAO Consultative Sub-Committee on Surplus Disposal (CSSD)	Established in 1954 as a subcommittee of the CCP, the CSSD's role was designated as that of oversight to monitor the disposal of agricultural surpluses used as food aid (ActionAid, 2003). Its remit also include disposing of surpluses in a way that avoided unnecessary displacement of commercial imports while simultaneously safeguarding export interests (ActionAid, 2003; CCP, 2005a,b).
1955–56 GATT 4th Round Geneva	In the 4th round of the GATT negations, they managed over $2.5 billion in eliminated or reduced tariffs (WTO, 2010).
1957 The European Economic Community (EEC)	The European Economic Community (EEC) renamed the European Community EC in 1993 was established by the Treaty of Rome in 1957 (Europa 2007, 2010).
1957 The Joint Research Centre (JRC)	In 1957, while the Treaty of Rome established both the European Economic Community (EEC) and the European Atomic Energy Community (Euratom), it also created the Joint Research Centre, which was founded under the Euratom treaty. The JRC, however, expanded to include other fields it felt important to policy making. These included sciences, energy, security, and consumer protection offering research-based policy support, etc. (JRC, 2009).
1957 UK's Agriculture Act	As with the 1947 Act, the UK's 1957 Agriculture Act set out long-term pledges of price stability (UOR, 2009).
1958 Economic Commission for Africa (ECA)	The ECE was the first intergovernmental agency in Africa. It covered countries whose economic and social environment differed widely from each other and where many countries and dependent territories were among the poorest in the world (UNECE, 2000; EoN, 2009).

(Continued)

TABLE 3.4 Key dates of the period: 1950 to 59.—cont'd

Date	Key dates of the period
1959 UN Declaration of the Rights of the Child	In 1959, the UN General Assembly accepted the Declaration of the Rights of the Child, in turn offering special protection, facilities, and opportunities to develop mentally, physically, spiritually, morally, and in conditions of freedom and dignity (UN, 1959).

Source: Compiled from multiple sources: UN (1959); National Archives (1995); Otto (1999); UNECE (2000); Action Aid (2003); FAS (2018); CCP (2005a,b); UOVS (2008); EoN (2009); JRC (2009); UNHCR (2009); UOR (2009); USAID (2009a,b,c); USAID (2010); WTO (2010).

Meanwhile in 1960, the FAO had commissioned a study looking at utilizing global food surpluses as humanitarian aid. Meeting in Rome the following year there was much discussion on this and other similar issues. In the end, the director of the United States Food for Peace Program, George McGovern, suggested to trial the multilateral food aid program over a 3-year period. The result was the World Food Program (WFP) (ODI, 2000; CCP, 2005a,b; Mousseau, 2005; Rucker, 2007; Walker, 2008).

3.15 World food program

Sponsored jointly by both the FAO and the United Nations, the World Food Program immediately took responsibility for humanitarian aid in the context of world hunger, thus becoming the frontline arm of the US international food aid program. In fact, more than this, the creation of a new body marked the start of a global multilateral food organization whose multiple aims looked to combat hunger, develop economic and social well-being, and provide humanitarian assistance in times of emergencies (WFP, 2008). So important was this work that in 1965 via parallel resolutions of both the FAO and the UN General Assembly established the World Food Program on an ongoing basis.

Previous to this, the WFP subsumed UNICEF's milk supplement programs among a few others. As a result, UNICEF adapted taking on relatively new nutrition programs such as the Applied Nutrition Projects (ANP). However, progress was slow and their effectiveness was duly questioned. The effectiveness of the individual projects approach was placed in doubt and as such program changed tack and attempted to apply a country-specific approach. However, once again it was found that too many similar projects lacked country-specific insight needed to implement programs more effectively (Ruxin, 1996).

3.15.1 The returning burden of surplus

During the early part of the 1960s, the United States Department of Agriculture (USDA) once again had to face the specter of agricultural overproduction. Continuing the practice which began in the late 1940s and which continued into the 50s, it was clear that the American government was facing a losing battle. The soil bank program of the previous decade seemed to do little more than take 28 million acres of farmland out of production while simultaneously paying farmers subsidies in lieu of grain crops on the land. However, the farmers reinvested these subsidies into mechanization which resulted in the end in yet more surpluses. Consequently, under both administrations of Presidents Kennedy and Johnson, overproduction once again became the primary objective of US farm policy. Policies were enacted to achieve multiple goals with the end result of finally dealing with the agricultural surpluses. In a two-pronged attack, more land was taken out of production, while measures were put in place in attempt to expand export markets for American agricultural products. It was also time that saw farmer's income grow from 50% to 75% compared with non-farmer's income (Reinhardt and Ganzel, 2003).

3.16 Hope holds out

The 1960s was considered a time of hope especially in the fight against hunger and malnutrition. Even agrarian reform was initially thought to be going well too. Yet there was a nagging feeling by certain people that the issues of malnutrition as well as economic and social development of the poorer nations were not going so well (SOFA, 2000).

In other news, the United Nations organization, in particular the WHO, FAO, and UNICEF, was complemented by new international players, initiatives, and policies that would unite all the players into one common cause. Indeed, the decade continued with two campaigns: firstly, the Freedom from Hunger Campaign and the World Food Program. Hot on the heels of these programs was the European Common Agricultural Policy and the United Nations World Food Conference.

3.17 The green revolution

The green revolution of the 1960s onward (discussed later) was in fact extension of the 19th century's chemical and industrial revolutions (Swaminathan, 1990; Borlaug 2000; Webb et al., 2008). What progressed to make the 1960s focal point the green revolution were the numerous advances in technology, mechanization, and genetic engineering. It was also the government's conducive policies of the time that made available inorganic fertilizers and advances in irrigation facilities. Not only were crop yields increasing quite considerably but also better soil management through new soil nutrient technologies aided by successful plant hybridization (Khush, 1999). Dwarf grain and rice varieties were grown to increase their yield compared with their biomass; this led to an increase in the harvest index, i.e., grain to straw ratio which saw prerevolution indices of 0.3 (30% grain and 70% straw) increase to green revolution indices of 0.5.

Meanwhile science saw new disciplines emerge, mature, and converge in the ever-growing field of biotechnology.

3.18 Single-cell proteins

The search for cheap high-protein foods was a challenge the United Nations and aid agencies hoped would come to fruition in the war against world hunger. Although the agencies did not find an immediate solution, it did, however, expedite an increasing enthusiasm for alternative, somewhat unconventional sources of protein. Since ancient times, people had harvested Spirulina, the microbial filamentous blue green alga from lakes. It was then dried in the sun and used it for food. The focus in the 1950–60s, however, centered on other microbial sources of protein-rich food including yeast, algae, fungi, and bacteria. Indeed, yeasts importance as an alternative animal feed supplement was already commonly known. And indeed, during the First World War, the German government had already successfully used imported yeast to replace about half of the much-needed animal protein. Also during the Second World War, Germany had already successfully integrated yeast-based foods (yeast contains between 45% and 55% protein) such as sausages, soups, and the like into the nation's diet (Najafpour, 2007).

The idea was simple and elegant, if a way could be found to produce say yeast, for example, on a cheap, commercially practical scale and then the potential for yeast-based foods was limitless. However, yeast can feed on a number of foods (substrates) and the challenge was to find low cost substrates. Efforts focused on by-products such as molasses, cheese whey, starch, ethanol, and even hydrocarbon-based substrates including oil and natural gas (Ugaldea and Castrillob, 2002). This is not as far-fetched as it sounds. Indeed, researchers at British Petroleum (BP) led by Alfred Champagnat at BP's Lavera refinery in France (1962) developed a way that allowed yeast to feed on waxy n-paraffins (linear saturated hydrocarbons). Toprina, as the yeast protein was called, was cleared after nutritional and toxicological tests and by 1967 was being produced on an industrial scale supplying the agricultural industry with alternative poultry and cattle feed (Najafpour, 2007). In separate but related experiments, BP used methanol made from methane conversion (natural gas) and marketed it as "Pruteen," which in turn was used as a "milk" alternative in calf feeding. By now, the idea caught on around the world and several refineries or plants around the world had taken up the challenge and were between them producing around a quarter of a million tons of food yeast (Ward, 1977).

After much wrangling over the years regarding protein research, it was time to question the respective roles of kwashiorkor and marasmus.

3.19 Protein and the growing kwashiorkor–marasmus debate

By early 1960s, the Protein Advisory Group (PAG), already a somewhat influential group, always insisted that their interest lies in both protein and calorie intake. The reality, however, was slightly different in that the PAG tended to put protein research way above that of calorie and much to the chagrin of many nutritionists (Mclaren, 1966). Cracks too were beginning to show up in the debate that protein was the primary protagonist in the problems of nutritional kwashiorkor. This was further evidenced in the Sixth Joint FAO/WHO Expert Committee on Nutrition in 1962 when its concerns were expressed that perhaps too much focus on protein was becoming a problem. The committee suggested that not only kwashiorkor but all aspects of the problem of protein calorie deficiency diseases would in fact benefit from a wider research focus (FAO, 1962). In strengthening this viewpoint, the Expert Committee talked of protein calorie malnutrition and also designated the term "protein calorie deficiency diseases" to include kwashiorkor and marasmus and all its sequelae. Further debunking kwashiorkor's role of primacy in all things protein related was a study in Southern India supported by the WHO which found that marasmus essentially was twice as prevalent as kwashiorkor. Even Scrimshaw, a long-time proponent of proteins role in kwashiorkor and who had previously spent a great deal of effort and time in researching the problem of kwashiorkor, shifted focus toward marasmus and marasmic kwashiorkor (Scrimshaw, 2010).

However, despite these words of concern and shift in focus, protein research is still going strong. Building on this and the PAG's momentum, the FAO Expert Group met again in 1963 to once again reexamine protein requirements. Yet despite a plethora of studies, intake levels for children still remained divisive. Ultimately, not fully satisfied the committee finally adopted the factorial approach to calculating protein requirements. This involved the summation of base BMR values and then allowance for daily individual workload variability. Although it was later concluded by the 1971 expert group, armed with new information that the 1963 protein requirements had perhaps been incorrectly assessed due to the overestimation of nitrogen losses (Passmore, 1964; FAO/WHO, 1965; FAO/WHO, 1973; WHO, 1973).

At about this time, the third World Food Survey was completed and once again highlighted what appeared to be political malaise.

3.20 The third world food survey

After examining the previous World Food Survey 1952, there was general consensus by many FAO statisticians that the notion that nearly two-thirds of the population were malnourished had in fact vastly exaggerated the problem. Yet despite this exaggeration, the third World Food Survey (1963) still suggested that the global nutritional status was in fact still a major problem, with as much as between 10/15% and 50% considered to be suffering from hunger and malnutrition (FAO, 1963). This was still unacceptably high.

References

ActionAid, 2003. Food Aid: An ActionAid Briefing Paper (ActionAid).

Allen, L.H., 2000. Ending hidden hunger: the history of micronutrient deficiency control. In: Background Analysis for the World Bank-UNICEF Nutrition Assessment Project. World Bank, Washington, DC, pp. 111–130.

Allen, L.H., 2003. Interventions for micronutrient deficiency control in developing countries: past, present and future. The Jornal of Nutrition 133, 3875S–3878S. Supplement: Animal Source Foods to Improve Micronutrient Nutrition in Developing Countries.

National Archives, 1995. Federal Records: Records of the Foreign Agricultural Service. National Archives of the United States, US National Archives.

Aykroyd, W.R., 1956. FAO and nutrition. Proc. Nutr. Soc. 15 (01), 4–13.

Belshaw, H., 1947. The food and agriculture organization of the united nations. Int. Organ. 1 (2), 291–306.

Bikistow, G., 1998. Measuring regional variation in farm support: Wales and the UK, 1947–72. Agric. Hist. Rev. 46 (1), 81–98.

Borchard, E., 1947. Intervention–The truman doctrine and the Marshall plan. Am. J. Int. Law 41 (4), 885–888.

Borlaug, N.E., 2000. The Green Revolution Revisited and the Road Ahead. 30th Anniversary. Lecture. The Norwegian Nobel Institute, Oslo.

Brock, J.F., Autret, M., 1952. Kwashiorkor in Africa. Bull. World Health Organ. 5 (1), 1–71.

Broughton, L., 1997. Lester B. Pearson's Role in the Formation of the Food and Agriculture Organization (FAO) and in Other United Nations Activities. United Nations Agency in Canada, Canada.

Carpenter, K.J., 1994. Protein and Energy: A Study of Changing Ideas in Nutrition. Cambridge University Press, Cambridge.
CCP, 2005a. Consultative SubCommittee on surplus disposal: fortieth report to the CCP. In: Committee on Commodity Problems: Sixty-Fifth Session Rome, Italy, 11–13 April 2005. Food and Agriculture Organisation, Rome.
CCP, 2005b. A historical background on food aid and key milestones. In: Committee on Commodity Problems: Sixty-Fifth Session. Food and Agriculte Organisation, Rome.
CEB, 2010. The Chief Executives Board (CEB): Overview. Retrieved 23rd March 2019, from: http://www.unsceb.org/.
Chatfield, C., 1949. Food Composition Tables for International Use: FAO Nutritional Studies No. 3. Nutrition Division, FAO UN, Washington, DC, p. 56.
Chatfield, C., 1954. Food Composition Tables Minerals and Vitamins for International Use: FAO Nutritional Studies No. 11. Nutrition Division, FAO UN, Washington, DC, p. 56.
Crawford, D.M., 1949. United States foreign assistance legislation, 1947-1948. Yale Law J. 58 (6), 871–922.
DAW, 2009. Division for the Advancement of Women. Retrieved 23rd Feb 2019, from: https://www.un.org/development/desa/publications/2009-world-survey-on-the-role-of-women-in-development.html.
Devereux, S., 2000. Famine in the Twentieth Century. Brighton, Istitute for Development Studies.
Dimitri, C., et al., 2005. The 20th Century Transformation of U.S. Agriculture and Farm Policy. Electronic Information Bulletin Number 3.
ECOSOC Economic and Social Council, 1947. 1946–1947 First to Fifth Sessions 1947. Retrieved 1st June 2019, from: https://library.un.org/sites/library.un.org/files/itp/1430-201404101302204694093_0.pdf.
EoN, 2009. Economic and Social Development: Regional Commissions the Encyclopedia of the Nations. United Nations, New York.
ESCAP, 2010. United Nations Economic and Social Commission for Asia and the Pacific: General Description. Retrieved 28th March 2019, from: https://www.unescap.org/about.
Eulau, H., 1968. Book reviews: the politics of food for Peace. By peter A. Toma. Political Res. Q. 21, 530–532.
Europa, 2007. Treaty Establishing the European Economic Community, EEC Treaty. EU PO, European Union Publications Office.
Europa, 2010. Website of the European Union. The EU at a Glance. Retrieved 21st June 2019, from: http://europa.eu/index_en.htm.
FAO, 1946. World Food Survey. Food and Agriculture Organisation, Washington, D.C.
FAO, 1947. Report of the First Session of Council of FAO 4-11 November 1947. Food and Agriculture Organisation, Washington D.C, p. 35.
FAO, 1950. Calorie Requirements. Report of the Committee on Calorie Requirements 1949: FAO Nutritional Studies No. 5. Food and Agriculture Organisation, Washington, DC.
FAO, 1955. Protein Requirements. Report of the F.A.O. Committee. FAO. Nutritional. Studies No. 16. Food and Agriculture Organisation, Rome.
FAO, 1957a. Calorie Requirements. Report of the Second Committee on Calorie Requirements: FAO Nutritional Studies No. 15. Food and Agriculture Organisation, Rome.
FAO, 1957b. Report of the Conference of FAO. Conference Reports. Food and Agriculture Organisation, Rome.
FAO, 1958. National Food Reserve Policies in Underdeveloped Countries. Food and Agriculture Organisation, Rome.
FAO, 1962. Joint FAO/WHO Expert Committee on Nutrition. Sixth Report. World Health Organization Technical Report Series No. 245. Food and Agriculture Organisation/World Health Organization, Rome, Geneva, p. 68.
FAO, 1963. Third World Food Survey, FFHC Basic Study No. 11. Food and Agriculture Organisation, Rome, p. 102.
FAO, 1995. Dimensions of Need: An Atlas of Food and Agriculture. Food and Agriculture organisation, Rome.
FAO, 2001. Food Balance Sheets: A Handbook. Food and Agriculture Organisation, Rome.
FAO, 2002. Food, Nutrition and Agriculture. FAO Food and Nutrition Division, Rome.
FAO, 2003. Food Energy – Methods of Analysis and Conversion Factors. Fao Food and Nutrition Paper 77. Food and Agriculture Organisation.
FAO/WHO, 1965. Joint FAO/WHO Expert Committee on Protein Requirements - Nutrition Studies No. 37. FAO/WHO, Rome.
FAO/WHO, 1973. FAO/WHO AD HOC Committee of Experts on Energy and Protein: Requirements and Recommended Intakes. Food and Agriculture Organisation, Rome.
FAS, 2018. Timeline of U.S. Agricultural Trade and Development. Retrieved 12 Jan 2019, from: https://www.fas.usda.gov/about-fas.

Freedom from Hunger, 2010. Website of the Freedom from Hunger. Retrieved 15th June 2019, 15th June 2010, from: http://www.freedomfromhunger.org/.
GATT, 1952. The History Structure and Functions of the ICCICA. World Trade Organisation, Washington DC.
GATT, 1960. General Agreement on Tariffs and Trade: Seventeenth Session; Disposal of Commodity Surpluses. UN General Agreement on Tarrifs and Trade, New York.
Gibson, M., 2016. The Feeding of Nations: Re-defining Food Security for the 21st Century. CRC Press, Boca Raton, Florida.
Gillespie, J.A., 2003. International organizations and the problem of child health, 1945-1960. Dynamis 23, 115–142.
Goldsmith, G.A., 1955. The kwashiorkor syndrome. Am. J. Clin. Nutr. 3 (4), 337–338.
Goodrich, L.M., 1947. From league of nations to united nations. Int. Organ. 1 (1), 3–21.
Henry, C., 2005. Basal metabolic rate studies in humans: measurement and development of new equations. Public Health Nutr. 8 (7A), 1133–1152.
Hoffmann, S., Maier, C. (Eds.), 1984. The Marshall Plan: A Retrospective. Westview Press, Boulder.
House of Commons, 1946. House of Commons Debates, 25 January 1944: UNRRA Winding up. House of Commons, 430: cc1979-2018.
IANWGE, 2010. UN Interagency Collaboration: History of the Collaboration on Women and Gender Equality from: https://womenwatch.unwomen.org/.
IGC, 2009. Grains Trade and Food Security Cooperation: The International Grains Agreement. Retrieved 21st Feb 2019, from: https://www.igc.int/en/.
ITO, 1948. ITO press release: speech to be delivered to the chief delegate of China at the closing session of the havana conference. In: UN Conference on Trade and the Employment. International Trade Organisation, Havana.
JRC, 2009. European Commission Joint Research Centre: History. Retrieved 22nd June, 2019.
Keusch, G.T., 2003. The history of nutrition: malnutrition, infection and immunity. J. Nutr. 133, 336S–340S.
Khan, K., 1982. The Law and Organisation of International Commodity Agreements. Martinus Nijhoff, Hague.
Khush, G.S., 1999. Green revolution: preparing for the 21st century. Genome 42 (4), 646–655.
Mclaren, D.S., 1966. A fresh look at protein-calorie malnutrition. The Lancet 288 (7461), 485–488.
McMmillan, R.B., 1946. World food survey" - a report from F.A.O. Rev. Mark Agric. Econ. 14 (10), 372–378.
Mezzalama, F., et al., 1999. Review of the Administrative Committee on Coordination and its Machinery. UN Joint Inspection Unit, Geneva, p. 31.
Morrison, A.V., 1986. GATT's Seven Rounds of Trade Talks Span More than Thirty Years - General Agreement on Tariffs and Trade. Business America, Swanton USA, pp. 8–10.
Mousseau, F., 2005. Food Aid or Food Sovereignty? Ending World Hunger in Our Time. Oakland Institute, USA.
Najafpour, G.D., 2007. Biochemical Engineering and Biotechnology. Elevier, Netherlands/Oxford.
NAL, 2010. Consumers Guide. N. A. Library, USDA the National Agricultural Library. Agriculture Fact Book, USDA.
ODI, 2000. Briefing paper. In: Reforming Food Aid: Time to Grasp the Nettle. Overseas Development Institute, London.
Ó Gráda, C., 2009. Famine: A Short History. University Presses of California, Columbia and Princeton. Princeton.
OPSI, 2007. Agriculture Act 1947. UK Parliament. HMSO: Office of Public Sector Information.
Otto, J.S., 1999. The Final Frontiers, 1880-1930: Settling the Southern Bottomlands. Greenwood Press, Westport USA.
PAG, 1967. PAG Bulletin Number 7: II. Articles - Development of Supplementary Food Mixture (CSM) for Children. World Health Organisation, Geneva, p. 24. Protein Advisory Group News Bulletin.
Passmore, R., 1964. An Assessment of the Report of the Second Committee on Calorie Requirements (FAO, 1957). Food and Agricultire Organisation, Rome.
Périssé, J., 1981. Joint FAO/WHO/UNU Expert Consultation on Energy and Protein Requirements: Past Work and Future Prospects at the International Level. Food and Agriculture Organisation, Rome.
POPIN, 1994. Background Document on the Population Programme of the UN. United Nations Population Fund, Geneva.
Nobel Prize, 2010. UNICEF: Achievement and Challenge. Retrieved 2019, from: http://nobelprize.org/nobel_prizes/peace/laureates/1965/unicef-lecture.html.
Reinhardt, C., Ganzel, B., 2003. A History of Farming. Living History Farm.
Rucker, A., 2007. Key players in food aid. eJournal USA.

Ruxin, J.N., 1996. Hunger, Science, and Politics: FAO, WHO, and Unicef Nutrition Policies, 1945 -1978, Chapter II the Backdrop of UN Nutrition Agencies, by Joshua Nalibow Ruxin. University College London. PhD, London.

Scott, S.L., 1951. The military aid program. Ann. Am. Acad. Pol. Soc. Sci. 278 (47).

Scrimshaw, N.S., 2010. History and early development of INCAP. J. Nutr. 140 (2), 394–396.

Scrimshaw, N.S., et al., 1959. Interactions of nutrition and infection. Am. J. Med. Sci. 237 (3), 367–403.

SOFA, 2000. The State of Food and Agriculture 2000. Food and Agriculture Organisation, Rome.

Sullivan, R.R., 1970. Politics of altruism - introduction to food-for-peace partnership between united-states-government and voluntary relief agencies. West. Political Q. 23 (4), 762–768.

Swaminathan, M.S., 1990. Sir John Crawford Memorial Lecture: Changing Nature of the Food Security Challenge: Implications for Agricultural Research and Policy. Consultative Group on International Agricultural Research, Washington.

Sweetser, A., 1946. From the league to the united nations. Ann. Am. Acad. Pol. Soc. Sci. 246, 1–8.

Toma, P.A., 1967. The Politics of Food for Peace. University of Arizona Press, Tucson.

Ugaldea, U.O., Castrillob, J.I., 2002. Single cell proteins from fungi and yeasts. Applied Mycology and Biotechnology 2, 123–149.

UN, 1945. Charter of the United Nations, 24 October 1945, 1 UNTS XVI. U.N., United Nations.

UN, 1948. Part A of general assembly resolution 217 (III). In: International Bill of Human Rights: Universal Declaration of Human Rights UN. General Assembly (3rd Sess 1948-1949: Paris). United Nations. Resolution 217.

UN, 1959. Declaration of the Rights of the Child. U.N., United Nations. A/RES/1386 (XIV).

UN, 1963. Operational Files of the Secretary-General: U Thant: Speeches, Messages, Statements, and Addresses - Not Issued as Press Releases. United Nations, New York. U. Secretariat.

UNECE (2000). "Overview of the Economic Commission for Africa." Retrieved 23rd June 2019, from: http://www.uneca.org/.

UNECE, 2010. Economic Commission for Europe: Inception. Economic Commission for Europe. Retrieved 21st March 2019, from: http://www.unece.org/info/ece-homepage.html.

UNHCR, 2009. About Us: Basic Facts. Retrieved 13th April 2019, from: http://www.unhcr.org.au/basicfacts.shtml.

UNICEF, 2005. United Nations Childrens Fund. Retrieved 30 Jan 2019, from: http://www.unicef.org/about/who/index_history.html.

UNICEF, 2009. The Convention on the Rights of the Child. Retrieved 30 Jan 2019, from: https://www.unicef.org/about-us/70-years-for-every-child.

UOR, 2009. University of Reading: Agriculture, Policy and Development, History of Agriculture. Retrieved 4 June 2019, from: http://www.ecifm.rdg.ac.uk/history.htm.

UOVS, 2008. Demand and Supply Theory.

USAID, 2009a. Food for Peace. Retrieved 4th Jan 2019, from: https://www.usaid.gov/who-we-are/organization/bureaus/bureau-democracy-conflict-and-humanitarian-assistance/office-food.

USAID, 2009b. The Marshall Plan. Retrieved 12th Jan 2019, from: https://www.usaid.gov/who-we-are/usaid-history.

USAID, 2009c. United States Agency for International Development: History. Retrieved 12th May 2019, from: https://www.usaid.gov/who-we-are/usaid-history. http://www.usaid.gov/about_usaid/usaidhist.html.

USAID, 2010. Our Work: A Better Future for All. US Agency for International Development. Retrieved 12th Jan 2019, from: https://www.usaid.gov/what-we-do.

USDA, 2009. History of the. U.S. Department of Agriculture USDA. Retrieved 12th Feb 2019, from: https://www.nal.usda.gov/topics/usda-history.

USHMM, 2010. US Holocaust Memorial Museum: Politics - UNRRA. Retrieved 21st Feb 2019, from: https://www.ushmm.org/learn.

Walker, P., 2008. The Origins, Development and Future of the International Humanitarian System: Containment, Compassion and Crusades. ISA's 49th Annual Convention, Bridging Multiple Divides. International Studies Association (ISA), San Fransisco, CA.

Ward, W.A., 1977. Calculating Import and Export Parity Prices. Training material of the Economic Development Institute, Washington D.C, World Bank. CN-3.

Watson, F., 1999. One hundred years of famine - a pause for reflection. Field Exchange (8), 20–23.

Watson, J.D., Crick, F.H., 1953. Genetic implications of the structure of deoxyribonucleic acid. Nature 171, 964–967.

Webb, P., et al., 2008. More food, but not yet enough: 20th century successes in agriculture growth and 21st century challenges. In: Food Policy and Applied Nutrition Program. School of Nutrition Science and Policy, Tufts University, Boston, Massachusetts.
Weir, C., 1952. Dollar exports in the Marshall plan period. J. Int. Aff. 28 (1), 9–14.
Weisell, R.C., 1995. In: Lupien, J.R., Richmond, K., Randell, A., et al (Eds.), Food, Nutrition and Agriculture: FAO Celebrates 50 Years - Expert Advice on Energy and Nutrient Requirements - an FAO Tradition. Food and Agriculture Organisation, Rome.
Weisell, R., 2002. Measurement and assessment of food deprivation and undernutrition: Part IV - summary of the draft findings of the joint FAO/WHO/UNU expert consultation on human energy requirements. In: International Scientific Symposium. FAO, Rome.
WFP, 2008. The World Food Programme: The First 45 Years. World Food Programme. Retrieved 25 April 2019, from: https://www.wfp.org/history. https://www.wfp.org/overview.
WHO, 1946. Official Records of the World Health Organisation No1 Minutes of the Technical Prepatory Committee for the International Health Conference. World Health Organisation, Paris, p. 80.
WHO, 1950. JOINT FAO/WHO expert committee on nutrition: report on the first session 1949 - technical report series No. 16. In: World Health Organization Technical Report Series No. 16. World Health organisation, Geneva, p. 24.
WHO, 1951. JOINT FAO/WHO expert committee on nutrition: report on the second session 1951 - technical report series No. 44. In: World Health Organization Technical Report Series No. 44. World Health organisation, Geneva, p. 24.
WHO, 1953. JOINT FAO/WHO expert committee on nutrition: third report 1953 - technical report series No. 72. In: World Health Organization Technical Report Series No. 72. World Health organisation, Geneva, p. 24.
WHO, 1955. JOINT FAO/WHO expert committee on nutrition: fourth report 1953 - technical report series No. 97. In: World Health Organization Technical Report Series No. 97. World Health organisation, Geneva, p. 24.
WHO, 1958. The First Ten Years of the World Health Organization. WHO, Geneva. Ch 22 - Nutrition.
WHO, 1972. Nutrition: a review of the WHO programme. II. WHO (World Health Organ.) Chron. 26 (5), 195–206.
WHO, 1973. Energy and protein requirments: report of a joint FAO/WHO ad hoc expert committee - technical report series No. 522. In: WHO Technical Report Series No. 522 FAO Nutrition Meetings Reports Series No. 52. World Health organisation, Geneva, p. 24.
WHO, 1998. Fifty Years of the World Health Organization in the Western Pacific Region, 1948–1998:Report of the Regional Director to the Regional Committee for the Western Pacific. Forty-Ninth Session. World Health Organization - Regional Office for the Western Pacific, Manila, Philippines.
WHO, 2010. Website of the World Health Organisation. Retrieved 2nd April 2019, from: http://www.who.int/.
Williams, A.J., 2005. Reconstruction' before the Marshall plan. Rev. Int. Stud. 31 (3), 541–558.
World Vision, 2009. Website of World Vision. Retrieved 23rd May 2019, from: http://www.worldvision.org.uk/.
WTO, 2010. Understanding the WTO: Basics - the GATT Years: From Havana to Marrakesh. Retrieved 30 March 2019, from: http://www.wto.org/english/thewto_e/whatis_e/tif_e/fact4_e.htm.
Wu Leung, W.-T., et al., 1968. Food Composition Table for Use in Africa. U.S. Department of Health/FAO, Maryland, Rome.

CHAPTER

4

Mixed blessings: a time of hope and crisis (1960–2000)

OUTLINE

Food and Society
https://doi.org/10.1016/B978-0-12-811808-5.00004-0

Shortly after the war, much was achieved and greater awareness of many of the issues surrounding food and society was becoming more obvious as reflected in the number of active bodies, NGOs, civil societies and a growing number of charities. In the mind of many, perhaps one of the growing issues was that of the environment, as well as issues concerning both social and economic conditions—all this in the few decades leading up to the millennium.

4.1 Rachel Carson and the environment

The 1960s saw the birth and growth of modern environmentalism as we know it today and which continued with Rachel Carson's famous attack on what was seen at the time as questionable agricultural practices. Building on the slow burning yet increasing environmental movement which sprung up around the 1930s, Carson passionately shared opposition to the industrialization of the agricultural sector. Building on this and of previous notions of organic farming by Lord Northbourne, the environmental movement gained incalculable impetus from the tireless work of Rachel Carson.

4.1.1 Environmentalism

It can be argued that the modern environmental movement reached critical mass about this time. Indeed, with the serialization in the New Yorker of an as-yet unpublished work by Rachel Carson more people were becoming aware of the environmental issues of the day. By the time the book of the serialization "Silent Spring" came out in 1962, Rachel Carson, now a well-known writer had also become an equally outspoken social critic. The book Silent Spring, using case studies, discusses the deleterious effects that chemicals and synthetic pesticides were having on the environment. In her writings too, she publicly challenged the many ecological destructive practices. In the eyes of the public, awareness was raised and generated a lot of concern and controversy not least because as Carson so eloquently pointed out, that many of these practices would have untold effect not only on the environment potentially on humans too. Such was the controversy that Carson generated, it could not be avoided either socially or politically. In political terms it led to the creation of environmental legislation as well as the creation of agencies to regulate the use of such chemicals et cetera. Of course, one person up against the entire agricultural industry was a tough challenge and attempts to discredit Carson proved only to strengthen their position (Gibson, 2016).

One commentator, the US Supreme Court Justice William O. Douglas, was of the opinion that serialization was "the most important chronicle of this century for the human race," while another, Loren Eisely, suggested that it was a "devastatingly, heavily documented, relentless attack upon human carelessness, greed and irresponsibility." The gravity Carson's accusations were so worrisome that even President Kennedy became involved when he asked his Science Advisory Committee to investigate. The subsequent report, which came out in 1963, vindicated Carson completely. In fact, in an article written by Bruce Frisch, he commented on the fact that the President's Committee also proclaimed that there were unknown long-range risks to the environment, humans, and animals alike. Consequently, true global environmentalism can be said to have been borne from one woman's cautionary tale, in which enough momentum was built that environmentalism now had a seat at the big table contributing to policies at all levels of government (Carson, 1962; Frisch, 1964; Baker, 2003).

Meanwhile on a related note, even the United States Science Advisory Committee (SAC) to the White House became entangled in the debate on nutrition when, at the bidding of President Johnson, the Science Advisory Committee published a three-volume tome called the "The World Food Problem" in 1967.

4.2 To the White House

Setting the tone and gravity of the problem, the SAC in the preamble of the report declared that

> ... the problem of hunger has lingered on and the shadow of starvation and impending famine has grown ever darker. Hunger's unceasing anguish drains hope, crushes aspirations, and obstructs the generation of programs of self-help. The threat of starvation sets man against man and citizen against government, leading to civil strife and political unrest. ***SAC (1967, p. IV).***

The White House report showed remarkable insight and perception as it elucidated many of the difficulties and challenges faced by the issues of hunger and malnutrition. Furthermore, it openly wondered whether the notion had been oversimplified and whether or not compassion fatigue and improper understanding contributed to what appeared to be a prevailing lack of progress. The report also went on to acknowledge that even though we might proclaim to have some of the answers, this did not always translate directly into successful or even appropriate policies and programs in the developing world. In sum, among the reports many and varied recommendations it advocated the need for more adaptive research and called for increased governmental aid on every level (Tables 4.1–4.4) (SAC, 1967; Gibson, 2016).

Ultimately, the close of the 1960s the development decade as it has been dubbed culminated on a high note for the United Nations with the adoption/creation of the International Bill of Rights to serve and protect against oppression or discrimination.

4.3 International Bill of Human Rights

The International Bill of Human Rights is not a single formal document or legislation, rather it is an informal name given to a collection of three instruments: the first is a United Nations General Assembly Resolution and the remaining two are international treaties:

- Universal Declaration of Human Rights (UDHR) 1948
- International Covenant on Economic, Social and Cultural Rights (ICESCR) 1966
- International Covenant on Civil and Political Rights (ICCPR) 1966

This turned out to be a marathon process spanning nearly 2 decades. On accepting the Declaration of Human Rights in 1948 the Commission on Human Rights then set about

TABLE 4.1 Key dates of the period: 1960 to 65.

Date	Key dates of the period
1960 International Rice Research Institute (IRRI)	The Philippines along with the Ford and Rockefeller Foundations created the IRRI. IRRI was an autonomous, nonprofit rice research organization (IRRI, 2019).
1960 Overseas Development Institute (ODI)	The ODI, was a leading independent think tank that worked with issues of international development and humanitarianism (ODI, 2019).
1960 International Freedom from Hunger Campaign (FFHC)	In 1960, the FAO launched a 5 year campaign of International Freedom from Hunger to mobilize and cash in on non-governmental support (Osmanczyk, 2002; FAO, 2010).
1960 American Freedom from Hunger Foundation (FFHF)	Founded to marshal Americans to get involved in the FAO's 5-year global Freedom from Hunger Campaign. Of note too is the joining of forces in 1979 of the FFHF and the Meal for Millions thus creating the Freedom from Hunger organization (Freedom from Hunger, 2010).
1960–62 GATT "Dillon" 5th Round Geneva	The 5th GATT rounds of trade reduction reduced over $4.9 billion in trade tariffs (WTO, 2019).
1961 Organization for Economic Development (OECD)	OECD worked with democratic governments and the global market economy to, among other things ensure the sustainability for economic growth (OECD, 2019).
1962/3 World Food Program (WFP)	The WFP was proposed in 1961 by George McGovern, then Director of the US Food for Peace Program. The WFP's governing body, the Intergovernmental Committee (IGC) was responsible for supervision over the WFP The IGC was eventually rolled over into the Committee on Food Aid Policies and Programs (CFA).
1962 The European Common Agricultural Policy (CAP)	The EEC, initially set out at the Stresa Conference (1958), where the Common Agricultural Policy (CAP) was created (1962). The main decision-making body of the CAP was the Agricultural Council. The CAP was mainly geared toward increasing agricultural productivity and food self-sufficiency. This they achieved through a combination of price supports and food import protectionism (Europa, 2010a,b).
1962 FAO/WHO Codex Alimentarius Commission	The FAO/WHO Codex Alimentarius Commission was first set up to develop international food standards and guidelines (WHO/FAO, 2006; FAO, 2010).
1963 The United Nations Research Institute for Social Development (UNRISD)	An autonomous UN agency conducting research on contemporary problems of development the UNRISD promoted discussion and contribution to key social development issues (UNRISD, 2010).
1964–67 GATT "Kennedy" 6th Round	The Kennedy Round of the GATT negotiations ushered in the new Anti-Dumping Agreement vis-à-vis the problem of Developing Countries. Furthermore, GATT rules were also more clearly defined. The Kennedy Round was also the first time that negotiations looked beyond Tariff cuts to include non-tariff barriers to trade as well (WTO, 2019).
1964 UN Conference on Trade and Development (UNCTAD)	The UN Conference on Trade and Development met in Geneva To discuss the recommendation of the establishment of a permanent UN body to deal with trade in relation to development. Consequently, The General Assembly established UNCTAD as one of its permanent organs in 1964 (UNCTAD, 2002).

TABLE 4.1 Key dates of the period: 1960 to 65.—cont'd

Date	Key dates of the period
1965 The United Nations Development Fund (UNDP)	With the merging of the UN Expanded Program of Technical Assistance (1949) and together with the UN Special Fund (1958) the UNDP was established in 1965, the UNDP worked on several development issues while promoting to developing countries on how best to attract and use aid effectively (Winderl, 2006; UNDP, 2010).

Source: Compiled from many sources including: Osmanczyk (2002); UNCTAD (2002); WHO/FAO (2006); Winderl (2006); IRRI (2019); Europa (2010a,b); FAO (2010); Freedom from Hunger (2010); ODI (2019); OECD (2019); UNDP (2010); UNRISD (2010); WTO (2019); Gibson (2016).

TABLE 4.2 Key dates of the period: 1966 to 69.

Date	Key dates of the period
1966 The Institute of Development Studies (IDS)	The IDS, a leading organization of international development aimed to understand the world and to apply academic skills to real world challenges (IDS, 2019).
1966 International Maize and Wheat Improvement Center (Centro Internacional de Mejoramiento de Maíz y Trigo, (CIMMYT))	Out of the Rockefeller, Ford and Mexican collaboration research led to the founding of the Office of Special Studies to ensure food security in Mexico and beyond. An agreement, the CIMMYT was signed with Mexico's agricultural minister in 1963 which further developed into a collaboration of international researchers (Hesser, 2006; Byerlee and Dubin, 2008; Ortiz et al., 2008).
1966 UN Covenant on Economic, Social and Cultural Rights and: The International Covenant on Civil and Political Rights	A rights-based approach to identity and culturism was adopted by the UN General Assembly in 1966. I was established that human rights were inherent in the dignity of human beings. Indeed, in Article 1 of each Covenant affirmed that all peoples were free to determine their own political status and have the right to pursue their own economic, social and cultural development (UN, 1966a,b).
1967 The Food Aid Convention (FAC)	The Food Aid Convention (FAC) was agreed as part of the International Grains Agreement. Run by the London-based International Grains Council the FAC was intended to act as a safety net, its primary objective to ensure a minimum availability of food aid to meet emergency requirements in developing countries (IGC, 2009a,b).
1968 International Grains Arrangement (IGA)	The International Grains Arrangement (IGA) comprised two separate instruments: The first was the Wheat Trade Convention, 1967 (WTC), while the 2nd introduced the Food Aid Convention, 1967 (FAC). While the WTC consults on prices, rights and obligations so the FAC provided food aid to developing Countries (Meerhaeghe, 1998; IGC, 2009a,b).
1969 Indicative world Plan for Agriculture (IWP)	The IDC introduced the first detailed report on the present and future problems of agriculture (Fasbender, 1970; Goldsmith et al., 1972).
1968 The First International Conference on Human Rights	The first International Conference on Human Rights was held in Teheran, Iran.
1969 United Nations Population Fund (UNFPA)	The Population Commission recommended to ECOSOC an fuller, more comprehensive and expanded population program fund which was subsequently agreed and placed under the administration of UNDP (POPIN, 1994).

Source: Compiled from many sources including: UN (1966a,b); Fasbender (1970); Goldsmith et al. (1972); POPIN (1994); Meerhaeghe (1998); Hesser (2006); Byerlee and Dubin (2008); Ortiz et al. (2008); IGC (2009a,b); IDS (2019); Gibson (2016).

TABLE 4.3 Key dates of the period: 1970 to 73.

Date	Key dates of the period
1970 Committee on Natural Resources (CNS)	The importance of ownership of natural resources when it comes to a nations sovereignty was emphasized in 1970 when the Economic and Social Council established the Committee on Natural Resources (Osmanczyk, 2003).
1971 Office of the United Nations Disaster Relief Coordinator (UNDRO)	As many nations became independent in the 1960s the UN established the Office of the UN Disaster Relief Coordinator (UNDRO). Headquartered in Geneva, its main role was that of coordinator of donors of aid and services (Osmanczyk, 2003).
1971 Doctors Without Borders/Médecins Sans Frontières (MSF)	An international humanitarian organization was created in 1971 by doctors and journalists to relieve suffering around the world (MSF, 2010).
1971 Earthwatch Institute (EI)	The Earthwatch Institute, an international environmental charity was formed in 1971 because of the need for new scientific research as well as promoting the public's understanding of science of environmentalism (Earthwatch, 2009).
1971 The Consultative Group on International Agricultural Research (CGIAR)	Pioneering work by the Ford-Rockefeller Foundation initially led to four research centers: CIMMYT (Mexico, 1966); IITA (Nigeria, 1967); the CIAT (Colombia, 1967); and IRRI (Philippines, 1960). However, by 1970, the FAO, UNDP, and the World Bank proposed a worldwide network of agricultural research centers under a permanent secretariat. Thus, in 1971, the Consultative Group on International Agricultural Research was formed and immediately subsumed the above 4 groups. By 2006 there was a coalition of 15 International Research Centers around the globe (CGIAR, 2008).
1972 The 1st UN Conference on the Human Environment	Held in Stockholm 1972, the UN's first major conference on international environmental issues was later dubbed the First "Earth Summit." It focused on the enhancement and preservation of the environment. And indeed, on recommendations of the conference the UN created the Environment Program (UNEP) to monitor changes and to encourage and coordinate sound environmental practices (Jackson, 2007).
1972 Bread for the World	A group of Catholics and Protestants came together in 1972 to address the causes of hunger. Initially a small affair the group has grown exponentially together with a global remit (BFW, 2009).
1972–81 The end of the Bretton Woods System	By early 1960s, the US dollar was seen as being overvalued. Under the Bretton Woods agreement however the dollar was pegged against gold. In order to slow down the wild swings in exchange rates President Nixon temporarily suspended the dollar-gold convertibility in 1971. An attempt was made to re-introduce the fixed exchange rate system however global currencies were beginning to free "float" against each other marking the beginning of a floating exchange rate system (IMF, 2010).
1972 ActionAid	ActionAid was founded in 1972 to fight worldwide poverty (Actionaid, 2010).
1972 World Resources Institute (WRI)	Recognizing a need for a credible research organization on the environmental, resource, population and development front on a global scale, so the World Resources Institute came into play (WRI), (WRI, 2010).
1972 The UN Environment Program (UNEP)	Born out of the first so-called Earth Summit (1st UN Conference on the Human Environment) of 1972, UNEP advocated policies of education and promotion for a fair, sustainable global development usage for the environment (UNEP, 2010a,b).

TABLE 4.3 Key dates of the period: 1970 to 73.—cont'd

Date	Key dates of the period
1973 Economic and Social Commission for Western Asia (ESCWA)	In 1973 ECOSOC further developed a regional economic commission for the Middle East. Although first proposed in 1947–48 it did not come to fruition. Instead, the idea of a regional commission was revived in 1973, instead ESCWA was established to supersede the UN Economic and Social Office in Beirut (UNESOB) (EoN, 2009a; UN ESCWA, 2009; UN, 2010).
1973–79 GATT "Tokyo Round" 7th Round	The Tokyo GATT Round saw a sweeping attempt to extend and improve the system of tariffs reductions and the result was the antidumping and subsidies agreement further reducing barriers to trade (Morrison, 1986; WTO, 2019).
1973 Center for Research on the Epidemiology of Disasters (CRED)	CRED was created as a non-profit institution founded by epidemiologist Professor Michel F. Lechat. The focus was on training and research, particularly in the areas of relief, rehabilitation and development. In 1980 the Center was recognized as a World Health Organization (WHO) Collaborating Center (CRED, 2009).
1973 University of United Nations (UNU)	UNU is the educational facility for the United Nations to promote knowledge and disseminating that information through the strengthening of individual and institutional capacities (UNU, 2009).

Source: Compiled from many sources including: Morrison (1986); Osmanczyk (2003); Jackson (2007); CGIAR (2008); BFW (2009); CRED (2009); Earthwatch (2009); EoN (2009a); UN ESCWA (2009); UNU (2009); Actionaid (2010); IMF (2010); MSF (2010); UN (2010); UNEP (2010a,b); WRI (2010); WTO (2019); Gibson (2016).

TABLE 4.4 Key dates of the period: 1974 to 79.

Date	Key dates of the period
1974 World Food Conference (WFC)	In response to the worsening global food situation, together with a general lack of progress in the fight against hunger, the first World Food Conference was held in Rome in 1974. The conference promoted the Universal Declaration for the eradication of hunger and malnutrition within a decade. A tall-order indeed and to help achieve this, there were calls for the creation of a ministerial-level World Food Council (WFC)* as well as recommending the reconstitution of the WFP's governing body as well as the incorporation of the International Fund for Agricultural Development (IFAD) as well as the FAO Committee on World Food Security (CFS) (UN, 1975).
1974 The Worldwatch Institute (TWI)	The Worldwatch Institute was designed as an independent research organization focusing on climate and energy, food and agriculture, and the green economy (Worldwatch, 2018).
1974 World Food Council (WFC)	The World Food Conference called for, and indeed established, the World Food Council to review problems affecting the world's food situation and to use its influence on governments and UN bodies and agencies alike. Eventually, the WFC recommended the creation of an inter-secretariat consultative body promoting cooperation between the four Rome-based food organizations, i.e., the FAO, IFAD, WFC, and WFP. However, in 1996, identifying and rationalizing duplicative efforts, the WFC was absorbed by the FAO and World Food Program (WFP) (CCP, 2005; EoN, 2009b).

(Continued)

TABLE 4.4 Key dates of the period: 1974 to 79.—cont'd

Date	Key dates of the period
1975 Committee on Food Aid Policies and Programs (CFA)	The World Food Conference of 1974 was a major milestone for food and indeed focused much attention on the issues of food and food aid. Consequently, the World Food Council's governing body; the Intergovernmental Committee's (IGC) remit was broadened to include the more general problems of food aid and related policies. The new governing body was named the Committee on Food Aid Policies and Programs (CFA) (Phillips, 1981).
1974 AGRIS	AGRIS is the agricultural sciences information system created by the FAO to promote information exchange (AGRIS, 2010).
1974 The International Fund for Agricultural Development (IFAD)	A specialized agency of the UN, IFAD finances agricultural development projects in the developing countries through low-interest loans and grants (Rucker, 2007; IFAD, 2019).
1974 The 1st World Population Conference (UNFPA)	The first World Population Conference (held in Bucharest, 1974) adopted a World Population Plan of Action (WPPA), thus stressing the relationship between population and economic and social development (UNFPA, 2004).
1974 Committee on World Food Security (CFS)	The World Food Council established the FAO Committee on World Food Security, as a forum for review and follow-up of global policies of world food security. It also promoted the use of food aid to support economic development and food security in vulnerable countries (CFS, 2008a,b; IAAH, 2008; CFS, 2011).
1975 The International Food Policy Research Institute (IFPRI)	IFPRI research's sustainable solutions toward ending hunger and poverty. It is one of the 15 research centers under the umbrella of the Consultative Group on International Agricultural Research (CGIAR) (IFPRI, 2010).
1975 Institute for Food and Development Policy (IFDP) AKA Food First	The IFDP analyses look for solutions to the root causes of hunger, poverty, and environmental degradation (FoodFirst, 2009).
1975 FAO Global information and Early Warning System (GIEWS)	FAO's Global Information and Early Warning System (GIEWS) keeps country-specific assessments of food problems under continuous review and issues reports on the global food situation (ODI, 1997).
1976 FAO's Technical Cooperation Program (TCO)	Supports the FAO by allowing greater flexibility in responding swiftly to urgent situations (TCP, 2009).
1976 International Emergency Food Reserve (IEFR)	In 1974, the WFC led to the establishment of the International Emergency Food Reserve (IEFR) requiring member countries to pledge food target donations, which would be used in quick response to emergencies (ODI, 1997; EoN, 2009b; TCP, 2009).
1977 The International Fund for Agricultural Development (IFAD)	Established out of the 1974 WFC, IFAD is a financial institution and specialized agency of the UN concerned primarily with financing agricultural development projects for food production in the developing countries (IFAD, 2019).
1977 The Standing Committee on Nutrition (SCN)	The Protein Advisory Group was incorporated into the UN's Administrative Coordination Committee's Subcommittee on Nutrition (ACC/SCN), (Allen, 2000). The Standing Sub-Committee on Nutrition (SCN) is a forum for UN agencies, NGOs/CSOs, and partners to coordinate and cooperate on nutrition-related issues (SCN, 2009).

TABLE 4.4 Key dates of the period: 1974 to 79.—cont'd

Date	Key dates of the period
1979 Action Against Hunger (ACF)	An agency born out of the need for a different kind of philanthropist, one who was; just as comfortable fighting the front line as well as in tackling the problem from the other side. It espoused a new brand of humanitarian politics that tries to influence political actors and institutions and not just mitigate the worst emergencies on the ground (ACF, 2010). Proactive, rather than reactive.
1979 International Year of the Child	UNICEF creates the International Year of the Child (UNICEF, 2005).

Source: Compiled from many sources including: UN (1975); Phillips (1981); ODI (1997); CCP (2005); UNICEF (2005); Rucker (2007); CFS (2008a,b); IAAH (2008); Worldwatch (2018); EoN (2009b); FoodFirst (2009); IFAD (2019); SCN (2009); TCP (2009); ACF (2010); IFPRI (2010); CFS (2011); Gibson (2016).

the task of translating the principles into international treaties. However, owing to the sheer scale of the task so the UN decided upon drafting two international agreements representing political, civil, economic social and cultural rights. Then for the proceeding near 2 decades, member states discussed, argued and debated the instruments until in 1966 consensus was finally reached. Yet this was not the end of the road as it took another 10 years to ratify the international covenants in 1976. The Bill effectively set in concrete proposals of the Universal Declaration (1948) and ultimately ushered in a new period of "rights" for all individuals (UN, 1948; UN, 1966b; Perry, 1998). Despite the length of time it took to ratify the instruments human rights were firmly on the right footing giving all self-determination in all things political and religious and is non-discriminatory in its application. This was now binding to those that ratified the treaties and the emphasis was now on the signatories to make sure that the: "… equal right of men and women to the enjoyment of all human rights." were upheld (UN, 2019).

In practicality, the two international treaties firmed up ideals and ideologies espoused in the 1948 UDHR. On the one hand the ICESCR acknowledged the right of self-determination and to:

> … social security (Article 9); their right to family life and the protection of children (Article 10); rights to an adequate standard of living, including food, clothing and housing (Article 11); good health (Article 12); a good standard of Education (Articles 13 and 14); and their full and free participation in cultural life (Article 15) *(UN, 1966a,b)*. On the other hand the ICCPR secured individuals rights to personal security and judicial fairness (article 9); a person's political rights (article 1); and the right of non-discrimination (article 4) ***UN (1966a,b); Gibson (2016).***

Meanwhile further developments through the course of the decade was closing policy gaps and promoting multilateral corporation on many fronts.

4.4 Economic uncertainty

Up to the early part of the 1970s, the Bretton Woods agreement did a good job of stabilizing international currency exchange rates. However, due to US wage rises and large capital exports, the US dollar was heating up and eventually became this overvalued and unmanageable. In response, the United States decided to de-couple from the gold standard ushering in an era of floating exchange rates. In turn nations could now renegotiate their own terms of trade allowing for more competitive global trade and the advent of true economic globalisation (McMichael, 1994). McMichael (1994) also postulated that as a result of the de-coupling, new markets and new trading partners could be established with the relatively new countries and sovereign states that emerged after the disbanding of the Cold War Bloc. This in turn further destabilised the global economy and by extension international trade (McMichael, 1994). Table 4.5 details some of the key events of the period.

Still in the early 1970s momentum for both environmental and economic sustainability had permeated into the collective public and political conscience.

TABLE 4.5 Genetic engineering milestones of the 1970s.

	Biotechnological advances
1972	The first cloning research was undertaken using a restrictive enzyme and ligase to form the first recombinant DNA molecule was undertaken by Paul Berg himself.
1973	Scientists Stanley Cohen, Annie Chang and Herbert Boyer successfully transferred DNA from one organism to another thus creating the very first recombinant DNA organism.
1974	In a letter to the top Journal, the Science Magazine, Paul Berg and others proposed to the National Institutes of Health (NIH) to allow a moratorium on certain DNA research (specifically recombinant DNA) until concerns of safety could be discussed and addressed.
1975	The requested moratorium on recombinant DNA research was agreed at an international meeting at Asilomar, California.
1976	Together Boyer and Swanson co-founded Genentech Inc. to research, develop as well as market products founded on recombinant DNA technology.
1976	The National Institutes of Health (NIH) released the first operational guidelines for recombinant DNA research and experimentation.
1977	Although legislative Bills to regulate recombinant DNA research was proposed in the US, none of the bills actually passed into legislation. Walter Gilbert and Allan Maxam at Harvard University formulated a new approach to the sequencing of DNA using chemicals rather than enzymes. Genentech Inc., developed the first protein – somatostatin – a growth hormone, also found in humans.
1978	Genentech Inc. were also fruitful in the production of human insulin and playing rDNA technology.
1977	Genetically engineered bacteria were used to successfully synthesize human growth proteins.
1978	Scientists Hutchinson and Edgell demonstrated that it was indeed possible to introduce specific mutations at specific sites in a DNA molecule.

Source: Compiled from Multiple Sources Berg et al. (1974); Bud (1993); Peters (1993); Dobson (1995); Najafpour (2007); Nature (2007); Styhre (2009); BioTech Institute (2010).

4.5 Sustainable development

In the sprint for economic development, if left unabated, the fear was that developing countries were in danger of achieving their goals at the expense of the environment. Foreseeing this dichotomy, the first United Nations Conference on the Human Environment (UNCHE) was held in Stockholm in 1972. While it was acknowledged that the wide gap between the developed and developing countries had to be bridged, so the conference urged everybody to do so in a non-detrimental manner to the environment (Earth Summit, 2002).

Moreover, it was also understood that the issues of the environment were of global concern and as such, solutions would require international cooperation. Indeed, so important was the task that one of the outcomes of the conference was the formation of the United Nations Environmental Program (UNEP). Its remit was simple, it was to engage international networking while also encouraging the many UN agencies and NGOs to implement the UN strategies into their respective programs.

About the same time the world food crisis was looming.

4.6 World food crisis

The world food crisis was a complicated affair and precipitating events was a number of factors concerning the political environment, the economic background, and agricultural production issues. It began with the relatively high stocks of grain of the grain exporting countries in the late 1960s. While governments and other organizations tried to rapidly reduce such levels to more realistic ones while also trying to reduce production. While successful in part, the international grain markets were a little bit jittery. Alongside this, in 1972 for approximately 2 years, global crude oil barrel prices increased tenfold, with a knock-on effect on crude oil prices which in turn saw fertilizers increased fivefold which also impacted on grain stocks which, by this time had stabled and reduced slightly. Add to this the fact that the wealthier countries were demanding more, coupled with the exchange rate volatility due to the recent decoupling of the gold standard meant that competition in the markets was fierce. This resulted in increased prices and low production levels which effectively doubled over this period. This was a major blow for the notion of food security for all especially for net importing countries. Over the ensuing 2–3 years the situation did not get any better as global food supplies remained at uncomfortably low levels; stocks were depleting, production was slowing, and an imbalance in the terms of trade of importing countries especially all colluded to make markets, as mentioned, jittery and reactionary (Table 4.1) (Maxwell and Frankenberger, 1992; Mitchell et al., 1997; Gerster-Bentaya and Maunder, 2008; UOR, 2009).

Unfortunately, against this backdrop newly released colonial outposts were now in competition with the developing nations all eager for economic development, food and independence.

4.7 The developing world's debt

At the frontier of the resultant development packages the World Bank institutions began lending money and other resources to those nations affected. Such loans favored infrastructure, industry and agriculture (Carrasco and Berg, 1999a). Yet, growth and investment were uneven at best pushing more people out of agriculture and into the industrialized big cities looking for work and sustenance. Not surprisingly, rural agriculture was further impoverished to the extent that a new approach was needed by the World Bank (Carrasco and Berg, 1999a,b). In the interim period, as was already mentioned, unfair trading practices on the international market only served to create a wider gap in economic development between the developed and developing countries (CFS, 2008b). As a result self-sufficiency in developing countries was reduced drastically and dependence and reliance on the developed countries and foreign debt assistance group only increased (FAO, 1996).

4.7.1 Debt and economic reforms

Bearing this in mind, the World Bank maintained that a country's development, in particular its economic development need not to be traded at the expense of quality of life. As a result, the World Bank and others while still engaged in project lending financially also came up with a loan based on growth and equity. Its aims were laudable, and in line with previous institutional initiatives over the years, the World Bank group and others aim to facilitate social welfare programs with a view to decreasing absolute poverty. These welfare programs included many employment sectors including health, food and agriculture, family planning, and education. Whether intentional or not such programs integrated social welfare with food and development programs which were then intertwined with economic growth (Munasinghe, 1998; Carrasco and Berg, 1999a,b; Slusser, 2006). However, the burden of debt grew as wealth from the Oil Producing Exporting Countries (OPEC), which had seen huge increases in the early 1970s.

In a roundabout sort of way, which is beyond the remit of this book, proceeds resulting from increased oil prices filtered through the system and ended up being used to help promote growth in developing (formerly Third World) countries (Eichengreen and Lindert, 1989; Carrasco and Berg, 1999a,b). While it did help, the debt burden of the developing countries was increasing to the point where servicing the debt alone was not sustainable (Lipson, 1981; Munasinghe, 1998; Slusser, 2006).

As the decade matured, it came as no surprise that international intervention was again needed across the board. This paved the way for the 1974 World Food Conference (FAO, 2003).

4.8 World Food Conference (WFC) 1974

The World Food Conference took place in Rome under the purview of the United Nations FAO. Once again, much lip service was paid to the problems of food production and consumption and yet another Declaration was added to the numerous other proceeding instruments (Maxwell and Frankenberger, 1992; Argeñal, 2007). On this score, the Universal Declaration of the Eradication of Hunger and Malnutrition, openly stated that:

> Time is short. Urgent and sustained action is vital. The Conference, therefore, calls upon all peoples expressing their will as individuals, and through their Governments and non-governmental organizations, to work together to bring about the end of the age-old scourge of hunger. *UN (1974).*

One good thing arising out of the conference was the understanding that the cause of food insecurity and famine were not always the result of failures of production but rather structural issues to do with social deprivation, poverty and equality. This led to the proposition echoed in so many other instruments of policy, declarations and agreements that:

> Every man, woman and child has the inalienable right to be free from hunger and malnutrition in order to develop fully and maintain their physical and mental faculties. *UN (1974).*

During the conference it was also requested that the many disparate bodies of the United Nations such as the FAO, WHO, and UNICEF should be united in their nutritional research efforts.

Other bodies that came out of the conference came with specific and important roles in the modern development of hunger and the security of food. One body in particular, the International Fund for Agricultural Development's (IFAD) remit was the financing of self-sufficient agricultural development projects (IFAD, 2019). Another, the Committee on World Food Security (CFS) was tasked with the responsibility of oversight review in production and people's access to food (CFS, 2009). Others, included the World Food Council (WFC) created to review any problems or policy issues as well as reviewing all United Nations agencies policies relating to world food security in general. The WFC was also expected to co-ordinate all UN agency policies related to food trade, nutrition, food aid as well as other related matters (UN, 1975; FAO, 2003).

All the while, great advances were being seen in the new discipline of genetics, spurred on in no small part to the work of Watson and Crick and their discovery of the structure of DNA.

4.9 Recombinant DNA: the arrival of genetic engineering

Leading the charge in the new pioneering field of research was Cohen and Boyer's discovery in 1973 of recombinant DNA (rDNA), more commonly known as gene splicing. In gene splicing one takes gene segments from one bacterium and transferring it into another thus replicating the jeans traits. The practice of transferring spliced genes into another organism was coined transgenic or in lay terms genetic engineering or genetic modification (GMO). This was real progress. However, irrespective of one's viewpoint genetic engineering would have enormous consequences in food technologies et cetera (Styhre, 2009). Yet, as can be seen not everybody was happy to take this on board. Indeed, many eminent biologists from the

National Research Council led by Paul Berg wrote in the "Science" magazine in 1974 of their concerns in which the group suggested:

> There is serious concern that some of these artificial recombinant DNA molecules could prove biologically hazardous ... recombinant DNA molecules might be more easily disseminated to bacterial populations in humans and other species, and thus possibly increase the incidence of cancer or other diseases. ***Berg et al. (1974).***

Such was the impact of this open letter that it led to the voluntarily moratorium on the work on recombinant DNA. At least that was until such time as a conference could be convened to discuss the consequences or potential ramifications of this line of enquiry. A year later in 1975 a conference was indeed held in California at the Asilomar Conference Center. However, not everybody was happy as the outcome allowed the continuation of this line of research albeit under the strictest of guidelines (Nobel Prize, 2010b). Following the moratorium, it was not long before research came up trumps using rDNA in the production of insulin by Genotech, the newly incorporated biotechnology company belonging to Boyers (Styhre, 2009). Other milestones (Table 4.5) of this time shows the extent and breadth of achievements that was made using this relatively new technology.

4.10 A decade of mixed blessings

The debt crisis of the 70s continued into the 1980s, this gave pause for thought as the borrower's domestic economies were clearly not being improved as a result of the original loans. The answer it seemed came in the form of structural adjustment policies whereby the lenders would give more money on condition that structural changes in domestic fiscal and monetary policies were made by the borrowers. For the right reasons, this neo-liberalist ideology was given much credence in the hope of freeing up the developing world and encouraging inertia, development and growth. A few key moments from this decade can be seen in Tables 4.3 and 4.8.

The 1980s also saw the growth of huge, cavernous food mountains which were being stored at considerable expense. However, of prominence during the early period of the decade seemed to be humanitarian aid.

4.11 Maturing humanitarianism

First generation humanitarian agencies and NGOs of the 40 and 50s tended to operate on the front-line mitigating emergencies as they happened. Second-generation agencies of the 1960s and 70s, whether new or evolved from the previous decades, sought instead engaged in "humanitarian" politics in the hope of influencing policies and political agendas more directly (Nash and Humphrey, 1987). This fundamental organizational shift in the way humanitarian agencies operated revolutionized how they interacted with governments and other stakeholders alike (ACF, 2010). About this time several UN agencies also began to foster new relationships with an ever-increasing number of aid agencies and NGOs. Even His Royal

Highness Prince El Hassan bin Talal who, up to this point, had already been fully engaged in humanitarian work. Being very active and with a long history in the field Prince El Hassan bin Talal was acutely aware of the changing landscape of humanitarianism and the stakeholders involved. To this end at the United Nations General Assembly of 1981 His Royal Highness proposed a new international humanitarian order. The motion was adopted and by the following year the GA asked the former United Nations High Commissioner for Refugees, Saddrudin Aga Khan to look into the subject among other things. Consequently, the UN proposed an independent Commission on International Humanitarian Issues. The commission would operate outside of the remit of the United Nations framework and would be composed of leading personalities in the field (IBHI, 2019). Shortly after in 1983 the Independent Commission on International Humanitarian Issues (ICIHI) and was co-chaired by HRH Prince El Hassan bin Talal and HH Prince Sadruddin Aga Khan. During this 4-year long mandate the work of the commission covered a breadth of issues from children, famine, deforestation, desertification, indigenous peoples, conflict and refugees to name but a few (IBHI, 2019). Consequently, a number of reports were produced touching on a range of humanitarian issues that were disseminated amongst the interested parties. The commission itself was succeeded in 1988 by the Independent Bureau for Humanitarian Issues (IBHI).

The following table (Table 4.6) details other events occurring around this period.

TABLE 4.6 Key dates of the period: 1980 to 88.

Key dates of the period	
1980 Southern Africa Development Community (SADC)	The SADC began as a loose alliance of nine States in Southern Africa known as the Southern African Development Coordination Conference (SADCC). Its main aim was to lessen economic dependence on South Africa. Its role now was to achieve economic development and growth and alleviate poverty (SADC, 2018).
1981 EuronAid	EuronAid was a European network of non-governmental organizations (NGOs). Originally founded in 1981 as a non-profit association using large food surpluses as an instrument of development policy rather than dumping on the Third World. EuronAid ceased operations in 2007 (EuronAid, 2008).
1981 Amartya Sen – "Poverty and Famines"	Economist AmartyaSen's 1981 book "Poverty and Famines" set the stage turning away from contemporary thinking turning away from the one-dimensional paradigm of production and supply to one encompassing "entitlement" theory. He won the Nobel prize in 1998 for his for his contributions to welfare economics (Nobel Prize, 2010a).
1981 World food day	World Food Day (WFD) was established by the FAO at the Organization's Conference in November 1979 to coincide with the date of the FAO's anniversary; 16 October (Argeñal, 2007).
1982 Mercy Corps	The Save the Refugees Fund founded by Dan O'Neill in 1979 was a task force for Cambodian refugees. And teaming up with Ellsworth ("Ells") Culver in 1982 the two founded Mercy Corps with a change in remit to from solely relief assistance to on long-term solutions to poverty and hunger (MercyCorps, 2019).

(Continued)

TABLE 4.6 Key dates of the period: 1980 to 88.—cont'd

Key dates of the period	
1984 European Lomé III Convention	Starting in 1975 the Lomé Convention set out the principles and objectives in respect of aid, trade and politics of cooperation between the European Union and the ACP countries; African, Caribbean and Pacific Group. In 1984 the Lomé III Convention shifted its attention from simple industrial development to development including self-reliance (EC, 2009).
1985 USAID Famine Early Warning System Network (FEWS NET)	USAID-funded FEWS Network (FEWS NET) is collaboration of stakeholders providing appropriate vulnerability and early warning information in emerging food security issues (USAID, 2010).
1985 Global Resource Information Database (GRID)	GRID, a Division of the Early Warning and Assessment, Global Resource Information Database (DEWA/GRID) forms UNEPs global network of environmental information centers which provide access to environmental information and data useful for stakeholders or decision-makers (UN ISDR, 2006; UNEP GRID, 2008).
1986 AGROSTAT	In 1986 AGROSTAT, FAO's comprehensive statistical database covering world agricultural information becomes operational changing its name in the mid 1990s to FAOSTAT (Wu, 2001; FAO, 2010).
1986–93 GATT Uruguay Round 8th Round	During previous GATT rounds, people considered the agricultural exemption left out of the previous 7 rounds to be an obstacle. Therefore the inclusion of agriculture in international trade negotiations of the 8th Uruguay Round was the most substantial trade liberalization agreement in agricultural products in the trade rounds' history (WTO, 2010).
1988 The Monitoring Agricultural Resources Unit Mission (MARS)	Part of the Joint Research Center (JRC) applied space technology in the acquisition of opportune information on crop areas and yields (EC JRC, 2010).
1988 Intergovernmental Panel on Climate Change (IPCC)	The IPCC established by the UNEP and WMO in 1988 was to assess existing knowledge of climate change science. Looking particularly at potential impacts and possible responses. The preparation of the Assessment Reports on Climate Change is the main output of the IPCC (Agrawala, 2004; UNEP, 2010a).

Source: Derived from Multiple Origins: Wu (2001); Agrawala (2004); UN ISDR (2006); Argeñal (2007); EuronAid (2008); UNEP GRID (2008); EC (2009); MercyCorps (2019); EC JRC (2010); FAO (2010); Nobel Prize (2010a); SADC (2018); UNEP (2010a); USAID (2010); WTO (2019).

4.12 The 1984–85 famines in Africa

The 1984/1985 famine of Ethiopia (present day Ethiopia and Eritrea) was brought on by drought and further exacerbated by civil war. The famine started in the North of the country and rapidly spread to the south. However humanitarian efforts were hindered by problems of security and insurgency. The Ethiopian government was not helping itself either, by holding back shipments of international relief from the rebel areas served only to ensure greater death toll. Together with the resultant economic collapse, the country lost over a million lives to famine and further millions more were *made* homeless and destitute. This situation served to illustrate even with sufficient food and motivated international community problems of conflict and security of aid had to be addressed (Watts, 1991; Maxwell and Frankenberger, 1992; Devereux, 2000; Ó Gráda, 2009).

It was early 1980s and once again food surpluses were back only this time the problem seems to be that much worse further highlighting a widening gap between the developed and developing countries.

4.13 Food mountains: the scourge of Europe

The low stocks that had previously exacerbated the global food crisis of the 1970s had finally recovered. Simultaneously, global recession reduced demand for many grains (Mitchell et al., 1997). As far as Europe was concerned this was in part driven by the EU's Common Agricultural Policy (CAP) of increased productivity and price support. While such policies were in place, they were intended to ensure availability of supply at reasonable prices as well as to boost farm incomes (UOR, 2009). However, because of the recession as well as reduced demand so stocks piled up. This was especially prevalent within the EU countries itself. They quickly followed general dissatisfaction and indignation of the millions of tons of food that were being stored and, in some cases, going rotten while there was widespread hunger and starvation across the world. This also gave EU sceptics cause for complaint and further, loud calls for CAP reform (Howarth, 2000; Dobbs and Pretty, 2001; Wakeman, 2003). This was not limited to Europe either, with many other countries suffering a similar fate including India (FAO, 1982; Chapman and Baker, 2001; Ray, 2004). In America too, although not to the same extent as the US agricultural policies were slightly different came to subsidies and farm incomes et cetera (Subramanian, 2003). Other key dates of the period are shown in Table 4.7.

TABLE 4.7 Key dates of the period: 1990 to 95.

Key dates of the period	
1990 UN World Summit for Children	To this point in history the UN Summit for Children claimed the largest gathering of world leaders. The World Summit adopted a Declaration on the Survival, Protection and Development of Children as well as a Plan of Action for implementing the Declaration in the 1990s.
1990 Human Development Report **(HDR)**	Till now, the development of a countries progress was measured largely in monetary Terms i.e., GDP, GNP and a few other metrics. The Human Development Report (HDR) first launched in 1990 was given the remit of placing people back at the heart of the development process. Such metrics were adjusted to include, not only metrics of monetary development but also on a suit of indicators looking at social development goals pertaining to the quality of life.
1992 Department of Humanitarian Affairs **(DHA)**	In 1991, the UN General Assembly wanted to improve the overall effectiveness of their humanitarian operations and, as such, created the DHA in 1992. Also created was the Emergency Relief Coordinator (ERC) which meant coordinating with the UN Disaster Relief Coordinator, UNDRO. The same resolution also created the Inter-Agency, the Standing Committee (IASC), the Consolidated Appeals Process (CAP) (not to be confused with the Common Agricultural Policy (CAP) and the Central

(*Continued*)

TABLE 4.7 Key dates of the period: 1990 to 95.—cont'd

Key dates of the period	
	Emergency Revolving Fund (CERF) as coordinating bodies and tools of the ERC. In 1998, the DHA was incorporated into the OCHA (OCHA, 2011).
1992 United Nations Conference on Environment and Development (**UNCED**) - The Earth Summit (**ES**)	The UN Conference (UNCED) known as the Earth Summit 1992 was a historic meeting. Taking place, 2 decades after the first global environment conference, the Earth Summit sought opinions and ways to halt the degradation of natural resources while reversing the pollution of the planet. From this meeting an 800-page document called Agenda 21, a blueprint for action in achieving sustainable development was published (UN, 1997b).
1992 Commission on Sustainable Development **(CSD)**	As a result of the UN Conference on Environment and Development (UNCED) of 1992, the UN council established the CSD to review progress (UN, 1993).
1992 The Center for Global Food Issues	The center for global food issues conducts research into agriculture and environmental concerns surrounding food production. It is also the Center assessing policies, and in turn heightens awareness of the environmental impacts of various food and farming systems and policies (CGFI, 2019).
1992 European Commission Humanitarian Aid **(ECHO)**	The EC Humanitarian Aid department (ECHO) was established in 1992 to help provide emergency assistance to victims of natural or man-made disasters outside the EU (EC, 1996).
1993 Second World conference on Human Rights	The World Conference on Human Rights was held by the UN. The rights conference since the collapse of the Cold War. The main outcome of the conference was the Vienna Declaration and Program of Action (Lawson and Bertucci, 1996).
1993 Global Policy Forum **(GPF)**	Founded by an international group of citizens, through networks and coalitions the GPF aimed to create a more equitable and sustainable global society (GPF, 2009).
1994 FAO's The Special Program for Food Security **(SPFS)**	The new Program was launched to boost food production through National and Regional Programs (FAO SPFS, 2009).
1994 Food Security Analysis Unit **(FSAU)**	The FSAU sought to provide nutrition and livelihood security, initially for the Somali people. It began in 1994 funded by USAID/OFDA and was managed by WFP (FSAU, 2010).
1995 International Grains Council **(IGC)***	The International Grains Agreement of 1995 linked two Conventions the Grains Trade Convention (GTC) and the Food Aid Convention (FAC). However, by 1995 GTC, the International Wheat Council (IWC) became the International Grains Council (IGC), in full recognition of all coarse grains rather than just wheat (IGC, 2009a,b)
1995 Oxfam International	Founded in 1995, Oxfam International is a collection of approximately13 organizations supporting humanitarian programs throughout the world (Oxfam, 2018).
1995 The World Trade Organization **(WTO)**	In 1993 GATT was updated. The 75 existing members as well as the EC became the founding members of the WTO which took over from GATT. Its new 150 + members were to supervise and further liberalize international trade. Its 150 members account for over 97% of world trade (Wahlberg, 2008; WTO, 2019).

Sources: Compiled from Multiple Origins: UN (1993); EC (1996); Lawson and Bertucci (1996); UN (1997a,b); Oxfam (2018); Wahlberg (2008); FAO SPFS (2009); GPF (2009); IGC (2009a,b); CGFI (2019); FSAU (2010); WTO (2019); OCHA (2011).

4.14 A changing concept of food security: entitlement theory

With the arrival of Nobel prize-winning welfare economist Amartya Sen whose book, "Poverty and Famines" (Sen, 1981a) challenged the status quo when it came to food security, people started to listen. Indeed, in his book he questioned many of the prevailing assumptions and argued that even if the country produces enough food for everybody it is still possible that many thousands or even more will go hungry as a result of poverty and/or with little or no access to food (Sen, 1981b). The notion that poverty acted as a cause of hunger was not a new idea in fact the UN had several anti – poverty programs aiming to combat this very same problem. However, what was new was that Sen posited his own Entitlement Theory rather than the seemingly singular preoccupation by the UN and some aid agencies on education and supply-side policies. Sen's entitlement theory places more emphasis on socio-behavioral coping and response strategies that are sometimes more firmly based in cultural social systems than in the availability of food (Maxwell and Frankenberger, 1992; CFS, 2007). In his theory, Sen further elaborates on the entitlement theory suggesting that access to food can be established in one of several ways. Firstly, is Directly through either growing the food for themselves through farm, marginal lands or local gardens. Secondly, it is through Exchange, this refers to the idea of bartering or trading and even buying food with sufficient funds. Lastly, is Transfer, which results in the receipt of food aid, or gifted food by friends or extended families or through inheritance whether food or finance. Thus, in Sen's view, hunger and famine are not solely about the aggregate supply of available food or even poverty alone, instead his notion suggests it's a lack of Entitlement that is at the center of food insecurity (FAO, 1970; Maxwell and Frankenberger, 1992; Sen, 1997; FAO, 2003; CFS, 2007). This notion was to entirely change the way that food security was viewed from then on, so much so that while the 1974 World Food Conference's original fears rested on volume and stability of food supplies so with Sen's insight the FAO expanded their notion of food for all to incorporate the concept of access into their definition of food security:

> ... ensuring that all people at all times have both physical and economic access to the basic food that they need *FAO (1983)*.

The World Bank also incorporated the same concept into their definition suggesting that people should have:

> ... access of all people at all times to enough food for an active, healthy life ... ***World Bank (1986)***.

Consequently, access now became a central tenet in the fight against hunger and famine. Put another way Sen's entitlement theory was a way of establishing people's capability of converting their various entitlements; financial, land, social etc. into food.

4.15 Hidden hunger

Hidden or Silent Hunger as it is sometimes called are terms that are both used to define circumstances in which the intake of food is sufficient enough so as not to allow for the detection of clinical symptoms. But is still insufficient for a good and healthy life style, with it the lack of energy intake or a lack of other nutritional deficiencies of micronutrients et cetera, silent hunger can go undetected and go on to cause major problems in later life (Kruif, 1926, Breckenridge, 1942; Macy and Williams, 1945; McGovern, 1969; Allen, 2000; Scrimshaw, 2003). As has been discussed, during the 1970s it had already been established that iron, vitamin A and iodine were all more prevalent than protein energy malnutrition however the problem did not receive enough attention and subsequently remained on the policy fringes.

By now, the term hidden or silent hunger was firmly established as a disorder of insufficient micronutrients and was a term also used to describe the millions of silent sufferers around the world. By the mid-1980s it was time however to redress the balance and place micronutrients firmly in the mix of international nutrition programs. Although iodized salt had been in play for a long time especially in the efforts to combat cretinism and goiter, it seemed that initial enthusiasm had waned somewhat and efforts to keep Up with the iodized salt program had to be renewed in order to keep iodide and deficiency disorders (IDD) at bay. Vitamin A was another one that was on the radar of aid agencies, especially so after a massive study of 25,000 children by Sommer and colleagues in Aceh, Indonesia concluded that a huge 34% reduction in child mortality rates was attributed to vitamin A supplements alone (Sommer et al., 1986; UNICEF, 2004).

4.16 Single cell protein's partial success

In previous decades, the promise of single cell proteins (SCP) and the fact they might be able to become a mainstream source of protein became a moot point in the 1980s. This was because marked improvements in crop production and plant breeding witnessed increasing agricultural output. Couple this with the lowering of the trade barriers, especially after the Uruguay round of the General Agreement on Tariffs and Trade (GATT), the first time in the many decades of GATT talks that agriculture was in fact included. As barriers to international trade were lowered so new trading partners helped reshape the global food supply chain. Back to SCP's, and the coupling of both the increased agricultural output, crossbreeding, and globalized trading meant that the market price for plant proteins had dropped to the point where the idea of mainstream single cell proteins became virtually obsolete. However, there was one success that made it all the way to the market and that was Quorn. During the 1960s and 70s Rank Hovis McDougall's research team found they had a lot of waste starch. This proved to be very useful as a substrate to grow single cell proteins and after much research they came across the fungus mold, Fusarium venenatum. After much rigorous testing Quorn was given a safe bill of health in 1985 and by 1994 it was launched onto the market as a meat free alternative mycoprotein (Ugaldea and Castrillob, 2002).

While this was a small success for single cell protein alternatives to meat so the growing biotech industry witnessed the first genetically modified organisms (GMO's) to help improve certain traits in plants.

4.17 Genetic engineering gains momentum

The decade of the 1980s was a good time for genetic engineering. What really helped was a new method of genetic amplification called the Polymerase Chain Reaction (PCR). It was a great breakthrough as it expedited the process of modern genetic engineering. Traditionally bacteria were used to amplify certain DNA sequences creating millions of copies, a necessary process in the work of genetic engineering. However, in 1983 Kary Mullis developed the Polymerase Chain Reaction which achieved the same results in a few short hours thus negating the need for bacterial alternatives.

This method of DNA cloning was built on decades of research and helped expedite rDNA technologies where the insertion of DNA material from one organism into another was simplified. In turn, the private sector was not stupid either and it didn't take them long before they could see the full potential of genetic engineering variously described as transgenetic, transgenic, genetic modified organism GMO's etc.

The first GMO plants to benefit from this new technology were the tobacco plants. It was discovered in 1983 that by transgenically splicing Agrobacterium tumefaciens vectors into the tobacco plants, more robust cell growth was witnessed (An, 1985; Vlasák and Ondřej, 1992). On the strength of the research and successes of the previous few years so private companies were now working at tremendous pace donating lots of resources into researching the commercial potential benefits of the sector. This is borne out of the fact that in America alone between 1987 and 90 the USDA issued close to 100 permits for the field testing of genetically modified crops. Indeed, under testing at that time were new or improved varieties of alfalfa, cantaloupe melons, corn, cotton, cucumbers, potatoes, soybeans, rice, squash, tomatoes, tobacco, and walnuts (Table 4.8) (Swaminathan, 1990).

4.18 Agriculture and the GATT rounds

Promoting international trade through the reduction of tariffs and quotas and other barriers to trade the general Agreement on Tariffs and Trade (GATT) came together periodically since 1949, as shown below:

- Annecy Round: 1949
- Torquay Round: 1951
- Geneva Round: 1955–56
- Dillon Round: 1960–62
- Kennedy Round: 1962–67
- Tokyo Round: 1973–79
- Uruguay Round: 1986–94

The meetings were held periodically in an effort to roll back oppressive instruments of international fair trade in an effort to even out the playing field – especially for the developing countries. The Uruguay round of negotiations finally came to an end in 1994 after 8 years of negotiations with the result that saw for the first time the full integration of agricultural trade issues. This effectively opened the door to a whole new batch of global opportunities. At the

TABLE 4.8 Genetic engineering milestones of the 1980s.

Date	Advances in genetics
1980	The US Supreme Court decided that genetically altered life forms (organisms) can in fact be patented.
1981	Scientists at Ohio University fashioned the first transgenic animals by splicing genes from animals into mice. Simultaneously the first gene-synthesizing machines were now being developed.
1982	Humulin, the name given to Genentech's insulin drug for humans was produced. It was also a milestone in the fact that it was the first biotech drug to be approved by the Food and Drug Administration for human use.
1983	The Polymerase chain reaction (PCR) was a process invented by Kary Mullis and others at Cetus Corporation in Berkeley, California. Rather than using traditional bacteria to cultivate multiple copies of DNA genes Mullis and colleagues pioneered the use of heat and enzymes to produce unlimited copies of genes and gene fragments – in the process becoming an important tool in biotech research and development worldwide.
1984	The DNA fingerprinting technique was created by Alec Jeffrey's to identify individuals. By now too the first genetically engineered vaccine was also developed.
1986	The first field trials of GMO tobacco plants (tobacco) were conducted also the first biotech-derived interferon drugs for the treatment of cancer, are produced.
1987	Advanced Genetic Sciences' Frostban, was the first authorized use of genetically altered bacterium to inhibit the formation of frost on plants like strawberry's and potatoes were field tested. In other news, Calgene, Inc. received a patent for the DNA sequence that is employed to extend the shelf-life of fruit and vegetables.
1988	Geneticists Philip Leder and Timothy Stewart applied and were given the first ever patent for a genetically altered animal, (mouse).
1989	The Human Genome Project started. It was an enormous effort to map out the sequence of the human genetic code. Also, the mapping of other animals' genomes was organized by two bodies - the U.S. Department of Energy and the National Institutes of Health. Scientists also created a recombinant vaccine to guard against the deadly rinderpest virus, which to date was responsible for the death of millions of cattle in developing countries.

Source: Compile from multiple sources Bud (1993); Peters (1993); Dobson (1995); Najafpour (2007); Nature (2007); Styhre (2009); BioTech Institute (2010).

same time, GATT itself was being updated or phased out being replaced, rolled into a new body – the World Trade Organization (WTO) in a concerted effort to further liberalize international trade (Warnock, 1997).

The timing couldn't be better in light of some ongoing undesirable practices such as continuing to support price support and protectionist policies. The common agricultural policy (CAP), as previously discussed in relation to Food Mountains, cost the CAP in the region of 20% of the entire CAP's annual budget of around 30 billion Euros at the time. And that's not all a further 28% was earmarked for export subsidies.

However, amid loud and public criticisms of the CAP's use of resources, so the EU CAP proposed certain reforms. Beginning in 1992 and continuing into the decade to 1997 the European Commission eventually offered a roadmap outlining the future direction of policies pertaining to agriculture, including the CAP among other reforms.

Indeed, this was in line with the growing liberalization of world trade as well as the many obstacles or challenges of the European Union's (EU) eastward expansion. The roadmap was packaged as an action plan—Agenda 2000 to provide a new financial framework in order to strengthen policies and to continue with agricultural reform which had incidentally begun in 1988 and 1992 (EU, 2010).

Meanwhile, the continuing inconsiderate use of the earth's resources in regards to global climate change, deforestation and an overall general disregard for the sustainability of the natural resource base brought the notion of environmentalism, once again to the fore.

4.19 Earth Summit: environmentalism

People were so concerned in fact that it led to calls, at the highest levels, to take action. The result was the UN Conference on Environment and Development (UNCED) in 1992, also known as the Rio Earth Summit. As it happened, the Summit turned out to be the first in a series of global conferences concentrating on economic and social development through sustainable solutions (Earth Summit, 2002). Solutions that would halt, slow-down or even stop and reverse the deterioration of the Earth's natural resources. Lessening the impact of humanity's toll on the environment while maintaining development goals, the summit promoted a sustainable development program - Agenda 21, to be overseen by the Commission on Sustainable Development (CSD) (UN, 1993; UN, 1997b; Earth Summit, 2002; Jackson, 2007). Agenda 21 was an idea that espoused cooperative worldwide responsibility for the environment irrespective of blame in which, acting locally, on both large and small projects, would collectively introduce worldwide improvements. In the field of food, concern was given to the potential of further degeneration of land, marine waters, biota and biodiversity. The conference was an epic effort of determination to address major concerns of a growing environmentalist movement and with its Agenda 21 document, it essentially bound food in particular that food security to untold future policies of sustainability and environmentalism (UN, 1992). The major outcomes of Agenda 21 touched on many socio-economic aspects of life in general. These included the cooperation between developing and developed countries in an effort to reduce poverty and to protect and promote health and vitality among peoples. It also played a role in conservation of natural resources by vowing to protect Earth's many environmental ecosystems through an integrated approach to better land resource management, more sustainable agriculture, the protection of biological diversity, to combat deforestation as well as protecting and managing fresh water and marine resources. With such laudable aims it was understood that between the United Nations and the various humanitarian groups and other NGOs that they had to engage in the frontline of things which invariably involved working closely with indigenous populations, inclusive of women and children (UN, 1992). Another underreported outcome of the Rio Earth Summit were the NGOs and a noticeable growth in their powerbase. Historically, NGO's came in all flavors, shapes and sizes. Their budgets, mandates and accountability were all varied, yet as a collective they roused sufficient support between themselves and the influence policy will help ensure governments adopted agreements et cetera.

4.20 Genetic engineering comes of age

With the many advances made in genetic engineering in the previous decade the 1990s finally bore fruit for the fledgling commercial technology. In 1994, for instance the first commercial transgenic product was Monsanto's rDNA version of the natural cow growth hormone Bovine Somatotropin (BST). The product was an immediate success.

rBST, also known as bovine growth hormone (BGH) helped boost milk production by an amazing 25%. Another product, this time by the US company Calgene (later acquired by Monsanto), introduced a genetically modified tomato which had a longer shelf life by effectively slowing down the ripening process. This new tomato called the Flavr Savr became the first commercially grown product to be granted a license for human consumption (Hoban, 1995; Schneider and Schneider, 2010). Two years later, Monsanto transgenically introduced the bacterium bacillus thuringiensis (Bt) into corn and by doing so boosted corn's own internal pest deterrent. In this way, Monsanto introduced the first Bt corn requiring little or no pesticide during growth. However, perhaps Monsanto's most famous genetically engineered product of the time, were the corn and soya plants whereby Monsanto had altered to be resistant to glyphosate. Roundup, or Roundup Ready, was Monsanto's trade name for Glyphosate. Glyphosate is an herbicide that is usually sprayed around plants and was particularly good at killing a broad spectrum of weeds. Thus, by introducing the genes into soya and corn, farmers could now effectively save time and money by spraying the whole field indiscriminately (Roberts, 2008).

However, all was not well, as enthusiasm, from the general public, was hampered by the growing fear that came with every new patent.

Consequently, people began to protest, and it was not a local affair, all around the world people were raising public awareness and fear of such foods hampering their general acceptance. Worse to come, that is and articles began to appear regularly in newspapers et cetera worldwide (O'Neil, 1992). As the backlash became stronger, so one adventurous group of innovative German protesters led a walking protest with a banner containing a very interesting menu comprising genetically altered foods:

Starters
Smoked trout fillets with the gene for human growth hormone
Tomato salad with flounder-fish gene
Main course
Grilled chicken with bovine growth hormone gene
and baked potato with scorpion gene
Dessert
Melon with virus gene

This menu was not a figment of their overactive imaginations either. As it turned out, all these foods had had in fact been developed by genetic engineers (Blythman, 1993). Indeed, such was the feeling among the general public that it led one observer to suggest:

> If they want to sell us Frankenfood, perhaps it's time to gather the villagers, light some torches and head to the castle. *Lewin (1992)*.

TABLE 4.9 Genetically modified organisms, come of age.

1990	Calgene Inc. saw the first field trials of genetically engineered cotton plants to endure use of the herbicide Bromoxynil GenPharm international, Inc. created the first transgenic dairy cow able to produce human milk proteins for infant formula
1993	The Food and Drug Administration (FDA) of the United States declared that genetically engineered foods were not inherently dangerous and subsequently did not require special regulation. Meanwhile George at Washington University researchers successfully Kloehn human embryos provoking protests from politicians, ethicists and critics around the world
1994	The first to market genetically engineered food product, Calgene's Flavr Savr tomato was approved by the US FDA
1995	A European research team identified a human genetic defect which appeared to underlie the most common cause of deafness
1996	UK government announces that 10 people may have become infected with the BSE agent through exposure to beef
1997	Scottish scientists clone Dolly the sheep using DNA from adult sheep cells
1998	Scientists at Japan's Kinki University clone eight identical calves using cells taken from a single adult cow

Source: Compiled from multiple sources Nature (2007); Huckett (2009); BioTech Institute (2010); Dennis et al. (2010).

Furthermore, while Paul Lewin was the first person to label such foods as "Frankenfoods"; nevertheless, it was a turn of phrase that was repeatedly seen in the tabloids time and time again. By 1998 this backlash had reached the European Union and in response to fears and public demand several EU countries held an unofficial moratorium on the importation and approval of new varieties of GM crops (Table 4.9) (EC, 2000).

In the meantime, food-related issues were gathering increasing momentum and in 1996 the FAO held the World Food Summit to discuss global hunger.

4.21 World Food Summit (WFS)

At the time of the 1996 World Food Summit (WFS), the global food situation, in terms of hunger and malnutrition was considered tight, that is to say, supply and demand were almost in equilibrium. Couple that with reduced stock holdings, higher prices and a subsequent decline in food aid and it comes as no surprise that hunger and malnutrition persisted (FAO, 1996). It was also thought that once again, by bringing together heads of State further suffering could be alleviated. Yet, as with many such conferences the WFS aimed to renew their aims and objectives with commitment at the highest political level. Furthermore, it also intended to raise more awareness within the media and the public once again the effort to highlight the sheer scale of the global problem. The Summit was not a wasted affair however as it drew on a great deal of empirical evidence that had been accumulated over the previous decades. Supplemented with research from Sen and others and finally, in the minds of many, established the full complexity and multi-dimensionality of the food situation as it is understood today. The notion that hunger and malnutrition could be dealt with in isolation of wider national or global issues had long since been debunked yet for some it was only now that the full concept was only just being fully realized. In short measure the Summit essentially established the fundamental fact that above all else, poverty eradication was incumbent on improving the food situation. The summit also skillfully underscored the notion that a rights-based approach to food, one that engaged not only in economic or issues of resource resources, was the only way forward (Food Summit, 1996; UN, 1997a; Argeñal, 2007).

Consequently, in 1996 World Food Summit adopted both the Rome Declaration as well as the Plan of Action. Lastly, one of the explicit aims of the summit was to halve the total number of undernourished people by 2015 – and considering the estimated number of undernourished people during the specified timeframe topped 800 million people, it was a promise that never materialized (Table 4.7) (FAO, 2003).

4.22 Footprints and ecological accounting

As with promises of progress in any other field so there has to be some kind of benchmarking in the environmental sense too. Indeed, the idea of environmental accounting was becoming mainstream during this period as more and more people, governments, and NGOs needed to benchmark their promises against actual progress. It was intuitive too, taking stock either quantitatively or qualitatively of natural resources, as well as any social, physical or economic impacts is inherent in any business model. And looking after the environment was now big business. Beginning in the 1970s amidst concerns over the environment:

> was the simple epiphany that all renewable resources came from the Earth and that as yet there was no tally or accountability as to what was being taken, replenished or destroyed. *Gibson (2016).*

Consequently, books, articles and academic papers on proposals and methods of ecological accounting began to appear with more and more frequency (Krebs, 1972; Hannon 1973, 1985; Stearns and Montag, 1974). From here the notion of ecological accounting took several parts. However, of the more familiar methods was the "foot printing" or more precisely, ecological foot-printing method popularized by Mathis Wackernagel and William Rees in 1990 and further established in their 1996 book on the subject (Rees, 1992; Wackernagel and Rees, 1996). In the foot-printing method of accounting, calculations can be made at the individual, country or globally and can be defined as:

> ... how much biologically productive land and water an individual, population or activity requires to produce all the resources it consumes and to absorb the waste it generates using prevailing technology and resource management practices ... *GFN (2010).*

Using this method, the ecological footprint is measured in units of global hectares on a per person per year basis. To arrive at this figure the demands placed on the Earth's resources are divided by the capacity of the Earth to support such demands - on a per hectare basis. From here the popularity of this simple, but definable notion helped found the Global Footprint Network in 2003. Indeed, in the words of Gibson (2016):

> "...similar measures spring boarding from such concepts were also the water and carbon footprints which collectively combine to form a powerful arsenal in the field of ecological accounting. ***Gibson (2016).***

On political level too, there were increasing calls to incorporate such accounts into a country's macroeconomic System of National Accounts (SNA). This was not a new call-to-arms either as back as 1976 ideas for a system of national environmental accounting had already been proposed (Peskin, 1976). In other areas, Table 4.10 highlights other things going on in the period surrounding the mid- to late 1990s.

TABLE 4.10 Key dates of the period: mid-late 1990s.

Key dates of the period	
1996 World Food Summit (**WFS**)	It was generally understood for the 1974 World Food Conference goals that they would need to be accelerated. Once again as in so many conferences the 1996 World Food Summit renewed its promises for global commitment to eliminate hunger and malnutrition. It would achieve this by raising awareness among decision-makers such as, the media and with the public too. The Summit culminated in the Rome Declaration on World Food Security and the World Food Summit Plan of Action. It also pledged to cut the number of hungry to about 400 million by 2015 (UN, 1997a,b).
1996 ReliefWeb	Spurred by the idea that timely and reliable information during any particular crises was critical to improving and maximizing responses. ReliefWeb came onto the scene. The group also operated on the premise that transparency and accountability were paramount in building trust (which some agencies seem to have forgotten) among the humanitarian relief agencies, NGOs and the (ReliefWeb, 2019).
1997 The UN Development Group **(UNDG)**	In 1997 the UN Development Group (UNDG) took on the role of coordinating and harmonizing the 32 UN funds, programs, agencies, departments, as well as offices that independently or co-dependently worked for the social and economic development community (UNDG, 2009).
1998 The Office for the Coordination of Humanitarian Affairs **(OCHA)**	In 1998 the DHA was reorganized becoming the OCHA. Its remit was expanded to coordinate policy development and humanitarian responses of the UN agencies, funds and NGOs. It primarily works via the Inter-Agency Standing Committee (IASC) chaired by the ERC (OCHA, 2011).
1998–2008 Food and Nutrition Technical Assistance **(FANTA)**	The Food and Nutrition Technical Assistance (FANTA) project was a 10-year project to improve knowledge, program development, policy, strategy, implementation and monitoring within the field of food and nutrition. The Project was funded by USAID and was renewed in 2008 for a further 5 years (Copeland et al., 2002; POPLINE, 2010).
1999–2004 Food Insecurity and Vulnerability Information and Mapping **(FIVIMS)**	FIVIMS helps countries comprehend, through sectoral analyses as well as evidence-based analyses, the underlying causes of food insecurity. This is further aided by the KIDS Database. The database is a set of interlinked international databases which is used for information exchange networking called - The Key Indicators Database System (KIDS). The KIDS database effectively allowed the dissemination and analyses of FIVIMS among numerous other datasets (CFS, 1998; FAO/GIEWS, 2006; FIVMS, 2008).

Sources: Compiled from Multiple Origins: UN (1997a,b); CFS (1998); Copeland et al. (2002); FAO/GIEWS (2006); FIVMS (2008); ReliefWeb (2019); POPLINE (2010); OCHA (2011).

References

ACF, 2010. Action Against Hunger International: Who We Are. Retrieved from: http://www.actionagainsthunger.org/.

Actionaid, 2010. Actionaid. Retrieved from: https://www.actionaid.org.uk/about-us.

Agrawala, S., 2004. Context and early origins of the intergovernmental panel on climate change. J. Clim. Change 39 (4), 1998.

AGRIS, 2010. About FAO's AGRIS. Retrieved from: http://agris.fao.org/agris-search/info.action.

Allen, L.H., 2000. Ending hidden hunger: the history of micronutrient deficiency control. In: Background Analysis for the World Bank-UNICEF Nutrition Assessment Project. World Bank, Washington, DC, pp. 111–130.

An, G., 1985. High efficiency transformation of cultured tobacco cells. J. Plant Physiol. 79 (2), 568–570.

Argeñal, 2007. History of Food Security in International Development: Background Paper Developed for Christian Children's Fund Honduras. Christian Children's Fund, Honduras.

Baker, R., 2003. Rachel Carson's Silent Spring and the Beginning of the Environmental Movement in the United States. Environmental Policy Analysis. Indiana University, Indiana.

Berg, P., et al., 1974. Potential biohazards of recombinant DNA molecules. Sci. New Ser. 185 (4148), 303.

BFW, 2009. Bread for the World: Our History. Retrieved from: https://www.bread.org/about.

BioTech Institute, 2010. Timeline of Biotechnology 20th Century. Retrieved from: http://www.biotechinstitute.org/go.cfm?do=Page.View&pid=22.

Blythman, J., 1993. Food & drink: do we have to eat our genes?: we will soon be offered genetically engineered food. In: Joanna Blythman Does Not Think We Should Accept it the Independent. London.

Breckenridge, M., 1942. The preschool child's nutrition. Am. J. Nurs. 42 (5), 533–537.

Bud, R., 1993. The Uses of Life: A History of Biotechnology. Cambridge University Press, Cambridge.

Byerlee, D., Dubin, H.J., 2008. Crop improvement in the CGIAR as a global success story of open access and international collaboration. In: Microbial Commons International Conference. Microbial Commons, Ghent, Belgium.

Carrasco, E.R., Berg, K.J., 1999a. The E-Book on International Finance and Development: Part One – V. The 1980's: The Debt Crisis and the Lost Decade. The University of Iowa Center for International Finance and Development (UICIFD).

Carrasco, E.R., Berg, K.J., 1999b. The E-Book on International Finance and Development: Part One IV - the 1960s and 1970s: The World Bank Attacks Poverty. Developing Countries Attack the IMF The University of Iowa Center for International Finance and Development (UICIFD).

Carson, R., 1962. Silent Spring. Houghton Mifflin, new york.

CCP, 2005. A historical background on food aid and key milestones. In: Committee on Commodity Problems: Sixty-Fifth Session. Food and Agriculture Organisation, Rome.

CFS, 1998. Guidelines for national food insecurity and vulnerability information and mapping systems (fivims): background and principles. In: Committee on World Food Security: Twenty-Fourth Session. Committee on World Food Security, Rome.

CFS, 2007. Assessment of the world food security situation. In: Committee on World Food Security: Thirty-Third Session. Food and Agriculture Organisation, Rome.

CFS, 2008a. Agenda item II: assessment of the world food security and nutrition situation. In: Committee on World Food Security: Thirty-Fourth Session. Food and Agriculture Organisation, Rome.

CFS, 2008b. Report of the CFS: Statement by the Deputy Director-General. Committee on World Food Security: Thirty-Fourth Session. Food and Agriculture Organisation, Rome.

CFS, 2009. Report of the Thirty-Fifth Session of the Committee on World Food Security (CFS). Committee on World Food Security, Rome.

CFS, 2011. FAO Committee on World Food Security (CFS). Retrieved from: http://www.fao.org/economic/cfs09/cfs-home/en/.

CGFI, 2019. About the Center for Global Food Issues. Retrieved from: http://www.cgfi.org/.

CGIAR, 2008. Consultative Group on International Agricultural Research: Who We Are: History of the CGIAR. Retrieved from:

Chapman, G., Baker, K.M., 2001. The Changing Geography of Asia. Routledge, London.

Copeland, R., et al., 2002. Food and Nutrition Technical Assistance Project Assessment. The United States Agency for International Development, Washington, DC.

CRED, 2009. The Centre for Research on the Epidemiology of Disasters: History. Retrieved from: https://www.cred.be/about.

Dennis, R.H., et al. (Eds.), 2010. Encyclopedia of Biotechnology in Agriculture and Food. CRC Press, Taylor & Francis Group, Boca Raton.

Devereux, S., 2000. Famine in the Twentieth Century. Institute for Development Studies, Brighton.

Dobbs, T., Pretty, J., 2001. Economics Research Report: Future Directions for Joint Agricultural-Environmental Policies: Implications of the United Kingdom Experience for Europe and the United States. South Dakota State University, Dakota.

Dobson, A., 1995. Biocentrism and genetic engineering. Environ. Values 4 (3), 227–239.

Earth Summit, 2002. Decision Making: Briefing Sheet. Retrieved from: http://www.earthsummit2002.org/es/life/Decision-making.pdf.

Earthwatch, 2009. Earthwatch Our History. Retrieved from: http://www.earthwatch.org/europe/aboutus/history/.

EC, 1996. Council Regulation (EC) No 1257/96 of 20 June 1996 Concerning Humanitarian Aid. European Commission. EU. 1257/96.

EC, 2000. Commission Takes Initiative to Restore Confidence in GMO Approval Process. Press release. European Commission, Brussels.

EC, 2009. The Lomé Convention. Retrieved from: https://ec.europa.eu/europeaid/search/site/The%20Lom%C3%A9%20Convention_en.

EC JRC, 2010. European Commission Joint Research Centre - MARS: About Us. Retrieved from: http://mars.jrc.it/mars/About-us.

Eichengreen, B., Lindert, P.H., 1989. The International Debt Crisis in Historical Perspective. MIT Press, Cambridge, Mass.

EoN, 2009a. Economic and Social Development: Regional Commissions the Encyclopedia of the Nations. United Nations, New York,.

EoN, 2009b. Economic and Social Development: World Food Council. The Encyclopedia of the Nations. United Nations, New York.

EU, 2010. CAP Reform a Policy for the Future. Retrieved from: https://ec.europa.eu/info/food-farming-fisheries/key-policies/common-agricultural-policy/cap-glance_en.

EuronAid, 2008. EuronAid: Code of Conduct. Retrieved from: https://odihpn.org/magazine/euronaid-general-assembly-adopts-code-of-conduct-on-food-aid-and-food-security/.

Europa, 2010a. Agriculture: Meeting the Needs of Farmers and Consumers " the EU at a Glance. Retrieved from: http://europa.eu/pol/agr/index_en.htm.

Europa, 2010b. European Union History: 1960 – 1969 the 'Swinging Sixties' – a Period of Economic Growth. The EU at a Glance. Retrieved from: http://europa.eu/abc/history/1960-1969/index_en.htm.

FAO, 1970. Report of the Conference of FAO: Annex D - Commemorative Address by Professor M. Cépède, Independent Chairman of the FAO Council. Food and Agriculture Organisation, Rome.

FAO, 1982. Report of the Council of FAO: Eighty-Second Session. Food and Agriculture Organisation, Rome.

FAO, 1983. World Food Security: A Reappraisal of the Concepts and Approaches. . Director General's Report. FAO, Rome.

FAO, 1996. World Food Summit – Technical Background Documents 1-5: Volume 1: 3. Socio-Political and Economic Environment for Food Security. World Food Summit. Food and Agriculture Organisation, Rome.

FAO, 2003. Trade Reforms and Food Security: Conceptualizing the Linkages. Food and Agriculture of the United Nations, Rome.

FAO, 2010. The Food and Agriculture Organization of the United Nations. Retrieved from: http://www.fao.org/about/en/.

FAO SPFS, 2009. Special Programme for Food Security. Retrieved from: http://www.fao.org/3/w9990e/w9990e09.htm.

FAO/GIEWS, 2006. The Use of Geospatial Data for Food Insecurity and Agricultural Drought Monitoring and Asessment by the FAO GIEWS and Asia FIVIMS. Seventeenth United Nations Regional Cartographic Conference for Asia and the Pacific, Bangkok, United Nations.

Fasbender, K., 1970. The FAO indicative world plan. J. Intereconomics 5 (6), 190–192.

FIVMS, 2008. The FIVIMS Initiative: Food Insecurity and Vulnerability Information and Mapping Systems. Retrieved from: https://www.fao.org/FOCUS/E/SOFI/fiv-e.htm http://www.fao.org/3/W8500e/W8500e.htm.

Food Summit, 1996. Rome Declaration on World Food Security World Food Summit. Food and Agriculture Organisation, Rome.
FoodFirst, 2009. About the Institute for Food and Development Policy. Retrieved from: http://www.foodfirst.org/.
Freedom from Hunger, 2010. Website of the Freedom from Hunger. Retrieved from: http://www.freedomfromhunger.org/.
Frisch, B.H., 1964. Was Rachel Carson right? Sci. Dig. 39–45.
FSAU, 2010. Food Security Analysis Unit for Somalia: FAQ. Retrieved from: http://www.fao.org/somalia/programmes-and-projects/fsnau/en/.
Gerster-Bentaya, M., Maunder, N., 2008. Food Security Concepts and Frameworks: What Is Food Security? EC-FAO.
GFN, 2010. Global Footprint Network. Retrieved from: http://www.footprintnetwork.org/en/index.php/GFN/.
Gibson, M., 2016. The Feeding of Nations: Re-defining Food Security for the 21st Century. CRC Press, Boca Raton, Florida.
Goldsmith, E., et al., 1972. Introduction: the need for change. Ecologist 2 (1).
GPF, 2009. Global Policy Forum: About Us. Retrieved from: http://www.globalpolicy.org/about-gpf-mm/introduction.html.
Hannon, B., 1973. The structure of ecosystems. J. Theor. Biol. 41 (3).
Hannon, B., 1985. Linear dynamic ecosystems. J. Theor. Biol. 116 (1).
Hesser, L.F., 2006. The Man Who Fed the World: Nobel Peace Prize Laureate Norman Borlaug and His Battle to End World Hunger. Durban House Publishing, Dallas, Texas.
Hoban, T.J., 1995. The Construction of Food Biotechnology as a Social Issue. Eating Agendas: Food and Nutrition and Social Problems. In: Maurer, D., Sobal, J. (Eds.). Walter de Gruyter Inc, New York.
Howarth, R., 2000. The CAP: history and attempts at reform. Econ. Aff. 20 (2), 4–10.
Huckett, B.I., 2009. A compact history of genetic transformation and its influence on crop development. Sugar Cane Int. 27 (1), 3–5.
IAAH, 2008. Interview with Mr. Kostas Stamoulis, Secretary General of the Committee on World Food Security. Retrieved from: http://www.fao.org/news/audio-video/detail-audio/en/?uid=10765.
IBHI, 2019. Independent Bureau for Humanitarian Issues. Retrieved from: http://ibhi.org/wordpress1/.
IDS, 2019. About the the Institute of Development Studies. Retrieved from: https://www.ids.ac.uk/about/.
IFAD, 2019. About the International Fund for Agricultural Development. Retrieved from: https://www.ifad.org/en/about.
IFPRI, 2010. About the International Food Policy Research Institute. Retrieved from: https://www.irri.org/about-us.
IGC, 2009a. 60 Years of Successive Agreements: Before 1949: The Early Years. International Grains Council, London, p. 4.
IGC, 2009b. Grains Trade and Food Security Cooperation: The International Grains Agreement. Retrieved from: http://www.igc.int/en/about/aboutus.aspx.
IMF, 2010. The End of the Bretton Woods System (1972–81). 2010.
IRRI, 2019. What Is IRRI? Retrieved from: https://www.irri.org/.
Jackson, P., 2007. From Stockholm to Kyoto: a brief history of climate change. U. N. Chron. XLIV (2).
Krebs, C.J., 1972. The Experimental Analysis of Distribution and Abundance. Harper and Row, New York.
Kruif, P.d., 1926. Microbe Hunters. Harcourt, Brace, Jovanovich, New York.
Lawson, E.H., Bertucci, M.L., 1996. Encyclopedia of Human Rights. Routledge, London.
Lewin, R., 1992. Complexity: Life at the Edge of Chaos. Macmillan, New York.
Lipson, C., 1981. The international organization of Third world debt. J. Int. Organ. 35 (4 (Autumn)), 603–631.
Macy, I.G., Williams, H.H., 1945. Hidden Hunger. Jaques Cattell Press, Lancaster.
Maxwell, S., Frankenberger, T., 1992. Household Food Security: Concepts, Indicators and Measurements: A Technical Review. UNICEF and IFAD, New York, Rome.
McGovern, S., 1969. Are we well fed?[horizontal ellipsis] the search for the answer. Nutr. Today 4 (1), 10–11.
McMichael, P., 1994. The Global Restructuring of Agro-Food Systems. Cornell University Press, Ithaca, NY.
Meerhaeghe, M.A.G.V., 1998. International Economic Institutions. Kluwer Academic Publishers, Massachusetts.
MercyCorps, 2019. Who We Are: Mercy Corps Timeline: Our History. Retrieved from: http://www.mercycorps.org/10638.
Mitchell, D.O., et al., 1997. The World Food Outlook. Cambridge University Press, Cambridge.

Morrison, A.V., 1986. GATT's Seven Rounds of Trade Talks Span More than Thirty Years - General Agreement on Tariffs and Trade. Business America, Swanton USA, pp. 8–10.

MSF, 2010. Doctors without Borders: History and Principles. Retrieved from: https://www.doctorswithoutborders.org/who-we-are/history.

Munasinghe, M., 1998. Special topic I: structural adjustment policies and the environment: introduction. Environ. Dev. Econ. 4 (9–18).

Najafpour, G.D., 2007. Biochemical Engineering and Biotechnology. Elsevier, Netherlands/Oxford.

Nash, A.E., Humphrey, J.P., 1987. Proceedings of a Conference Held in Montreal: Causes of Refugee Problems and the International Response. Human Rights and the Protection of Refugees under International Law. Canadian Human Rights Foundation and the Institute for Research on Public Policy, Montreal.

Nature, 2007. Milestones in DNA tecnologies. Nat. Rev. Genet. 8 (Suppl. ment).

Nobel Prize, 2010a. Amartya Sen: The Sveriges Riks bank Prize in Economic Sciences in Memory of Alfred Nobel 1998. Retrieved from: http://nobelprize.org/nobel_prizes/economics/laureates/1998/.

Nobel Prize, 2010b. Paul Berg: Asilomar and Recombinant DNA. Retrieved from: http://nobelprize.org/nobel_prizes/chemistry/laureates/1980/berg-article.html.

Ó Gráda, C., 2009. Famine: A Short History. Princeton, University Presses of California, Columbia and Princeton.

O'Neil, M., 1992. Geneticists' Latest Discovery: Public Fear of 'Franken food'. New York Times, New York.

OCHA, 2011. Website of the UN Office for the Coordination of Humanitarian Affairs. Retrieved from: http://ochaonline.un.org/AboutOCHA/.

ODI, 1997. Global Hunger and Food Security after the World Food Summit. Overseas Development Institute, London.

ODI, 2019. About ODI: Our Mission, People and Organisation. Retrieved from: http://www.odi.org.uk/about/default.asp.

OECD, 2019. About the Organisation for Economic Development. Retrieved from: http://www.oecd.org/about/history/#d.en.194377.

Ortiz, R., et al., 2008. Wheat genetic resources enhancement by the international maize and wheat improvement center (CIMMYT). Genet. Resour. Crop Evol. 55 (7), 1095–1140.

Osmanczyk, E.J., 2002. Encyclopedia of the United Nations and International Agreements G-M. Taylor & Francis, London.

Osmanczyk, E.J., 2003. Encyclopedia of the United Nations and International Agreements N-S. Routledge, New York.

Oxfam, 2018. Oxfam International. Retrieved from: https://www.oxfam.org/en.

Perry, M.J., 1998. The Idea of Human Rights: Four Inquiries. Oxford University Press, New York.

Peskin, H.M., 1976. A national accounting framework for environmental assets. J. Environ. Econ. Manag. 2 (4), 255–262.

Peters, P., 1993. Biotechnology: A Guide to Genetic Engineering. Wm. C. Brown Publishers, Inc, Dubuque, Iowa.

Phillips, R.W., 1981. FAO: Its Origins, Formation and Evolution 1945–1981. Food and Agriculture Organisation, Rome.

POPIN, 1994. Background Document on the Population Programme of the UN. United Nations Population Fund, Geneva.

POPLINE, 2010. Food and Nutrition Technical Assistance Project Assessment. POPulation Information onLINE. Johns Hopkins/USAID, Baltimore, MD.

Ray, S., 2004. Roles of Agriculture Project: Socio-Economic Analysis and Policy Implications of the Roles of Agriculture National Synthesis Report India. Roles of Agriculture Project (ROA). FAO, Rome.

Rees, W.E., 1992. Ecological footprints and appropriated carrying capacity: what urban economics leaves out. Environ. Urban. 4 (2), 121–130.

ReliefWeb, 2019. About the ReliefWeb Project: Transparency in Partnerships Fuels ReliefWeb Success. Retrieved from: https://reliefweb.int/.

Roberts, P., 2008. The End of Food. Houghton Mifflin Harcourt, New York.

Rucker, A., 2007. Key players in food aid. eJ. U.S.A.

SAC, 1967. The World Food Problem: A Report of the President's Science Advisory Committee, Volume Lll Report of the Panel on the World Food Supply. The White House: The President's Science Advisory Committee (PSAC), Washington, p. 1.

SADC, 2018. Southern Africa Development Community. Retrieved from: https://www.sadc.int/.

Schneider, K.R., Schneider, R.G., 2010. Genetically Modified Food: What Are GM Foods? Florida. University of Florida: The Institute of Food and Agricultural Sciences (IFAS).
SCN, 2009. The United Nations Standing Committee on Nutrition (SCN): Who We Are. Retrieved from: http://www.unscn.org/en/home/who-we-are.php.
Scrimshaw, N., 2003. Historical concepts of interactions, synergism and antagonism between nutrition and infection. J. Nutr. 133 (316S-321S), 367–403.
Sen, A., 1981a. Ingredients of famine analysis: availability and entitlements. Q. J. Econ. XCVI (3), 433–464.
Sen, A., 1981b. Poverty and Famines: An Essay on Entitlement and Deprivation. Clarendon Press, Oxford.
Sen, A., 1997. Lectures Given at WFP/UNU Seminar- Rome, 31 May 1997: Entitlement Perspective of Hunger. Ending the Inheritance of Hunger. WFP/UNU, Rome.
Slusser, S., 2006. The World Bank, structural adjustment programs and developing countries: a review using resource dependency theory. In: Annual Meeting of the American Sociological Association. Montreal (Paper presented at the).
Sommer, A., et al., 1986. Impact of vitamin A supplementation on childhood mortality. The Aceh study group. The Lancet 327 (8491), 1169–1173.
Stearns, F., Montag, T., 1974. The Urban Ecosystem: A Holistic Approach. Stroudsburg, Penn., Dowden, Hutchinson and Ross.
Styhre, A., 2009. The production of informational objects in innovation work: pharmaceutical reason and the individuation of illnesses. Ephemera Theory Polit. Organ. 9 (1), 26–43.
Subramanian, S., 2003. An analysis of the relationship between the agricultural policies of the European Union and the United States, and their implications on the world food economy. J. Stanf. Course Engr 297b (Controlling World Trade): 32.
Swaminathan, M.S., 1990. Sir John Crawford Memorial Lecture: Changing Nature of the Food Security Challenge: Implications for Agricultural Research and Policy. Consultative Group on International Agricultural Research, Washington.
TCP, 2009. The Technical Cooperation Programme. Retrieved from: http://www.fao.org/tc/tcp/.
Ugaldea, U.O., Castrillob, J.I., 2002. Single cell proteins from fungi and yeasts. Appl. Mycol. Biotechnol. 2, 123–149.
UN, 1948. Universal Declaration of Human Rights. U.N., United Nations. A/RES/217 A (III).
UN, 1966a. International Covenant on Civil and Political Rights and [First] Optional Protocol. U.N., United Nations. A/RES/2200 A (XXI).
UN, 1966b. International Covenant on Economic, Social and Cultural Rights. U.N., United Nations. A/RES/2200 A (XXI).
UN, 1974. Report of the World Food Conference, 5-16 November 1974. World Food Conference. United Nations, Rome.
UN, 1975. Report of the World Food Conference 1974. United Nations, New York.
UN, 1992. Agenda 21. United Nations Conference on Environment and Development (UNCED). United Nations, Rio de Janeiro.
UN, 1993. Institutional Arrangements to Follow up the United Nations Conference on Environment and Development. U. Nations. United nations. A/RES/47/191.
UN, 1997a. Integrated and Coordinated Follow-Up and Implementation of the Major International United Nations Conferences and Summits: Outcome of the World Food Summit. United nations, New York.
UN, 1997b. United Nations Conference on Environment and Development. Retrieved from: https://sustainabledevelopment.un.org/content/documents/Agenda21.pdf.
UN, 2010. World Summit on the Information Society: Regional Dimensions. United Nations, New York, p. 4.
UN, 2019. Universal Declaration of Human Rights. Retrieved from: https://www.un.org/en/universal-declaration-human-rights/.
UN ESCWA, 2009. UN Economic and Social Commission for Western Asia at a Glance. Retrieved from: http://www.escwa.un.org/about/main.asp.
UN ISDR, 2006. ISDR Secretariat Biennial Work Plan 2006 - 2007. United Nations. International Strategy for Disaster Reduction, Geneva, Switzerland, p. 20.
UNCTAD, 2002. About the United Nations Conference on Trade and Development. Retrieved from: http://www.unctad.org/Templates/Page.asp?intItemID=1530&lang=1.
UNDG, 2009. The UN Development Group. Retrieved from: http://www.undg.org/index.cfm?P=2.

UNDP, 2010. United Nations Development Fund: Frequently Asked Questions. Retrieved from: http://www.undp.org.

UNEP, 2010a. The Intergovernmental Panel on Climate Change (IPCC). Retrieved from: https://www.unenvironment.org/resources/report/climate-change-1995-report-intergovernmental-panel-climate-change.

UNEP, 2010b. United Nations Environment Program. Retrieved from: https://www.unenvironment.org/.

UNEP GRID, 2008. The UNEP, Division of Early Warning and Assessment, Global Resource Information Database: About Us. Retrieved from:

UNFPA, 2004. State of World Population 2004: The Cairo Consensus at Ten: Population, Reproductive Health and the Global Effort to End Poverty. UN Population Fund, New York.

UNICEF, 2004. Eliminating iodine deficiency disorders through universal salt iodization in central eastern Europe. In: Commonwealth Independent States and the Baltic States (CEE/CIS/BS). UNICEF, Geneva, p. 18.

UNICEF, 2005. United Nations Childrens Fund. Retrieved from: http://www.unicef.org/about/who/index_history.html.

UNRISD, 2010. What Is the United Nations Research Institute for Social Development. Retrieved from: http://www.unrisd.org/80256B3C005BF3C2/(httpPages)/1889BA294D2950E08025791F005CD710?OpenDocument.

UNU, 2009. United Nations University: Towards Sustainable Solutions for Global Problems. Retrieved from: http://unu.edu/about/.

UOR, 2009. University of reading: agriculture, policy and development. Hist. Agric. Retrieved from: http://www.ecifm.rdg.ac.uk/history.htm.

USAID, 2010. What Is the Famine Early Warning Systems Network (FEWS NET)? Retrieved from: http://fews.net/about-us.

Vlasák, J., Ondřej, M., 1992. Construction and use of Agrobacterium tumefaciens binary vectors with A. tumefaciens C58 T-DNA genes. Folia Microbiol. 37 (3), 227–230.

Wackernagel, M., Rees, W.E., 1996. Our Ecological Footprint: Reducing Human Impact on the Earth. New Society Publishers, Gabriola Island B.C.

Wahlberg, K., 2008. Food Aid for the Hungry?.

Wakeman, R., 2003. Themes in Modern European History since 1945. Routledge, London.

Warnock, K., 1997. Third world: feast or famine? Food security in the new millennium (commentary). J. Race Class 38 (3), 63–72.

Watts, M., 1991. Entitlements or empowerment? Famine and starvation in africa. Rev. Afr. Polit. Econ. 51, 9–26. The Struggle for Resources in Africa.

WHO/FAO, 2006. Understanding the Codex Alimentarius. Joint FAO/WHO Food Standards Programme, Rome, p. 40.

Winderl, T., 2006. UNDP for Beginners: A Beginners Guide to the United Nations Development Fund. UNDP JPO Service Centre, Copenhagen, Denmark, p. 25.

World Bank, 1986. Poverty and Hunger: Issues and Options for Food Security in Developing Countries. World Bank Group, Washington DC.

Worldwatch, 2018. About Worldwatch. Retrieved from: http://www.worldwatch.org.

WRI, 2010. A Brief History of World Resources Institute. Retrieved from: http://www.wri.org/about/wri-history.

WTO, 2019. Understanding the WTO: Basics - the GATT Years: From Havana to Marrakesh. Retrieved from: http://www.wto.org/english/thewto_e/whatis_e/tif_e/fact4_e.htm.

Wu, J.M., 2001. Pangaea central as the coming global access to legal, scientific and technical information through the resource networks of intergovernmental organisations. Int. J. Spec. Libr. 35 (2), 94–112.

CHAPTER

5

The 21st century: ideological convergence

OUTLINE

We now arrive at the 21st century; it was a new decade, more than that it was new millennium, and with it came a renewed sense of belonging, an increased awareness, and a renewed interest in issues of social welfare. Indeed, the new millennium began with a seminal summit (the Millennium Summit) that aimed to tackle these very issues. On top of this, there was also a rallying call to arms against all things antisocialist and an open backlash against genetically modified organisms (GMOs) as people fought toe to toe whether in the courts, on the streets, or at the front line. This decade also had to deal with a global financial crisis, the likes of which had rarely been seen before, and it came hot on the heels of a food price bubble, which gave governments, organizations, and humanity in general a very timely wake-up call.

5.1 The much-heralded Millennium Summit

As mentioned, the new millennium kicked off with the Millennium Summit, which lived up to the occasion with the sheer volume and number of world leaders, dignitaries, and

Food and Society
https://doi.org/10.1016/B978-0-12-811808-5.00005-2

officials in attendance. In its sights the agenda comprised a range of sweeping social objectives representing many of the pressing issues of the day. Setting the mood for much of the Summit was a growing sense of social responsibility to present day and future citizens of the Earth. Goals ranged from reducing poverty, tackling malnourishment, a more equitable sharing of resources, and the natural environment which played a large part in the momentum for change. Gone were the days when the environment could be plundered without consequence, instead a new system of ecological accounting made it easier to promote accountability vis-à-vis the resource base. Among other considerations too was the role and direction of the United Nations in the new millennium. In the end, the Summit came to a close with its principal delegates ratifying the United Nations Millennium Declaration emphasizing that

> ... only through broad and sustained efforts to create a shared future, based upon our common humanity in all its diversity, can globalization be made fully inclusive and equitable. *UN (2000c).*

In achieving such an important set of aims required a fundamental and ideological shift in values to include (and not just for the developed nations) the notion of equality, tolerance, freedom, solidarity, and respect. Furthermore, the Summit agreed a set of eight time-bound measurable goals in achieving the end of poverty; universal education for all; equality including gender equality; improved child health care; maternal health care; and combating HIV/AIDS, focusing on environmental sustainability and finally global partnership. Collectively, these goals are known as the Millennium Development Goals (MDGs), which further broke down into 21 quantifiable targets, fully measured by 60 indicators (appendix Table A5.1) (UN, 2000a; UN, 2006; UNDP/MDG, 2010).

Incidentally to the United Nations definition of food security was also the focus of modification and in 2001 the Committee on Food Security included a social component to the concept when it suggested

> Food security [is] a situation that exists when all people, all times, have physical, social and economic access to sufficient, safe and nutritious food that meets their dietary needs and food preferences for an active and healthy life. *FAO (2001).*

This new focal point or other points talking of safe and nutritious foods, as well as social aspects of food preferences and access brought into sharp focus and ever-growing concept that ultimately became interpretable to the point that it can be made to fit specific needs and agendas. Other events around this time are highlighted in the table below (Table 5.1).

Hot on the heels of the Millennium Summit, the Food and Agricultural Organization (FAO) of the United Nations held another food summit 5 years after the first World Food Summit (WFS).

5.2 World food summit: 5 years later

As the Millennium Summit was still being set up, in 2002, the FAO convened another WFS 5 years after the original. This was put together to assess progress made in the promise toward eliminating or more precisely halving hunger to 400 million by 2015. The Summit met amid disappointing data that offered an annual reduction of only 6 million well below the proposed yearly target of 22 million. Indeed, considering only 32 of the 99 developing

TABLE 5.1 Key dates of the period: 1996 to 99.

2000 International Strategy for Disaster Reduction **(ISDR)**	In response to the rise of increasing natural disasters around the world, the UN leaned upon the (UN/ISDR) to build upon the work of the International Strategy for Disaster Reduction (ISDR) in response to the International Decade for Natural Disaster Reduction 1990–99 (UN, 2000b).
2000 The International Food Relief Partnership Act	The International Food Relief Partnership Act was a US initiative to increase international food-aid programs through the use of grants to nonprofit organizations that would then stockpile food; and further allowing the authorization of the USAID to procure and store commodities overseas in expectation of future emergencies (CBO, 2000).
2000 EC-FAO Food Security Information for Action	The Information for Action Program provided countries undergoing crisis or those transitioning economically, the assistance to formulate more effective policies with the aid of among other things, early vulnerability mapping, warning systems, nutritional surveys, needs assessments; policy analysis and statistical databases et cetera.
2000 UN Millennium Summit	The Millennium Summit was held at UN's New York. The summit looked at the multi-various challenges facing humanity in the 21st century. At the meeting, world leaders ratified the UN Millennium Declaration. The Declaration's eight goals were outlined as a way for achieving the stated aims (UN, 2000c).
2000 The United Nations Millennium Development Goals (**MDG**)	Extensive consultation of the challenges facing humanity was established by the many Millennium Development Goals (MDGs). These were enshrined in a set of quantifiable time-bound (2015) targets addressing such challenges (UN, 2000a; UNDG, 2009).The Millennium Declaration was the working document which set out a framework in tackling these challenges.
2001 Chief Executives Board (**CEB**) for Coordination	In 2001, as a result of a 2 year review the previous Administrative Committee on Coordination (ACC) became the United Nations System Chief Executives Board (CEB) for Coordination (Mezzalama et al., 1999; CEB, 2010).
2001 The Earth Policy Institute (**EPI**)	The Earth Policy Institute (EPI) was founded by former president of the Worldwatch Institute - Lester Brown who worked tirelessly toward sustainable environmental policy solutions (EPI, 2010).
2001 Global Environmental Change and Food Systems (**GECAFS**)	GECAFS launched a 10-year program in 2001 of multi-faceted research focusing on understanding the links between food security and environmental change (GECAFS, 2010).
2001 EuropeAid	In 2001 EuropeAid's goal was to combine the EU Commission's external aid into one centralized body. Consequently, as a result the EU, through its combined bi- and multi-lateral donations effectively became the world's biggest donor of humanitarian aid (ODI, 2000; Mousseau, 2005; EuropeAid, 2009).
2002 World Summit on Sustainable Development (**WSSD**)—Earth Summit_2002	Known as the Johannesburg Earth Summit, the meeting aimed to continue and where possible build on the efforts of the 1992 Earth Summit and to adopt concrete steps and identify quantifiable targets for implementing Rio's 'Agenda 21' (UN DESA, 2006).

(Continued)

TABLE 5.1 Key dates of the period: 1996 to 99.—cont'd

2002 World Food Summit +5	Amid calls for the renewal of the World Food Summit the working group also proposed the acceleration of effort to reduce world hunger (FAO, 2002b).
2002/4 Water Footprint Network (**WFN**)	The water footprint is a form of Environmental Accounting popularized by Arjen Hoekstra which uses the idea of virtual water (the water footprint) to calculate global water use on a per item or per person basis (WFN, 2009).
2003 The Global Footprint Network (**GFN**)	GFN is a kind of think tank that promotes and develops instruments of environmental impacts on various issues based on the concept of Ecological or Environmental Accounting (GFN, 2009).
2003 International Alliance Against Hunger (**IAAH**)	Established on World Food Day 2003, the IAAH in the fight hunger and poverty helps facilitate action through partnerships of civil society and government (FAO, 2003; IAAH, 2009).

Sources: Compiled From Multiple Sources Mezzalama et al. (1999); CBO (2000); ODI (2000); UN (2000a,b,c); FAO (2002a,b); FAO (2003); Mousseau (2005); UN DESA (2006); EuropeAid (2009); GFN (2009); IAAH (2009); UNDG (2009); WFN (2009); CEB (2010); GECAFS (2010).

countries showed a reduction in numbers of undernourished people, disappointing results led to the consideration of other ways to accelerate global efforts and once again delegates vowed to renew their commitment to the goals and resolved to meet the stated 1996 target by 2015 (FAO, 2002b).

In parallel to the WFS, there was a separate conference, a non-governmental NGO/CSO Forum for Food Sovereignty[1]. In attendance were NGOs, civil society groups, and farmers' organizations. They were collectively scornful and united in their condemnation of inaction to date. They had previously pointed out many concerns pertaining to the 1996 WFS and the proposed action plan. After the details mentioned previously, the collection of stakeholders was feeling vindicated. However, rather than being smug, they were more interested in finding out what the problem was. For the parallel conference, it had been suggested that the output goals were fundamentally flawed as, instead of mutual cooperation between the worst affected areas of food insecurity, decisions seemed to be made unilaterally and it was questioned that such policies were forced onto aid packages in the developing countries while the developed countries insisted they maintained the status quo of high support policies themselves. The NGO and CSO also strongly wondered whether the WFS+5 was in fact watering down pledges of the proposed Code of Conduct on the Right to Food to a simple set of voluntary guidelines. Further criticisms focused at the promotion of genetically modified (GM) foods as a panacea by certain governments and transnational corporations (NGO/CSO Forum, 2002). In the WFS+5s view, there were more than enough resources as well as a growing political will to tackle the problem. Yet they appeared united in the causes of widespread hunger and malnutrition and suggested that "… international trade led hegemonic economic model …." This, it was claimed, took away a country or

[1]Food sovereignty is a country's right to define its own policies vis-a-vis agricultural, pastoral, fisheries and food according to their own socioeconomic and cultural ideologies. The notion promotes the rights of food to everyone through small and medium-sized agricultural enterprises that respect different cultures and diversity. Decentralization and democratization are key principles of this movement (NGO/CSO Forum 2002).

communities' control over their own destiny inter alia economic and natural resources. In response, the NGO/CSO Forum offered that Food Sovereignty as one method of socioeconomic development tool over other more familiar alternatives (NGO/CSO Forum, 2002).

Seemingly, on the surface the 21st century was little different from previous decades as the specter of growing population was never far from people's thoughts. Once again, the notion of a carrying capacity of the Earth populated the minds of many. One particular study is worthy of note.

5.3 The Earth to support close to 8 billion

In 2004, likely the most important study on the Earth's ability to sustainably provide for a growing population, i.e., the carrying capacity, was undertaken. The study by Van Den Bergh and Rietveld (2004) looked at 69 historic studies—many previously scrutinized by Cohen (1995) in which estimated population figures ranged between 0.5 billion and 1021 billion. One important difference in Van Den Bergh and Rietveld's study compared with Cohen's in 1995 was the use of metaanalysis in the methodology. Indeed, benefits of metaanalysis over primary research per se are particularly useful when comparing a range of similar studies. Furthermore, metaanalysis gave Van Den Bergh and Rietveld the opportunity to look beyond the output alone instead allowing it to examine the mechanics of the studies (Van Den Bergh and Rietveld, 2004). By doing so, researchers Van Den Bergh and Rietveld asked some important questions about the comparative similarities and differences within the studies that might shed light on any new or overlooked consensus. In this respect, the researchers were fruitful in their analysis discovering that sustainable population limitations were several and included among other things limitations of energy, water, natural resources, and land among other things. However, the two most often quoted limitations were land and limited food resources, which accounted for 71 of the 94 collected estimates. This, however, cautioned Van Den Bergh and Rietveld and was perhaps subject to change as the Earth carried ever-increasing population figures. In this way, truer and more reliable resource limitations other than the ones currently stated could possibly emerge as more important variables in the future (Van Den Bergh and Rietveld, 2004). In the end, discounting erroneous; speculative; or insufficiently rigorous studies by Van Den Bergh and Rietveld concluded that a best estimate of a sustainable carrying capacity of the Earth, based on a median value, was 7.7 billion.

In other news around this time, the World Bank too came under fire for its less than stellar performance.

5.4 The World Bank and nutrition

The World Bank's (WB) admission into the field of nutrition during the early 1970s was not without scornful criticism from the likes of the WHO, FAO, and UNICEF. The main defense of the bank was that issues of nutrition were such that only large-scale interventionist policies would or could make any real difference. And the World Bank (in its view) was uniquely positioned to offer such scaled-up assistance. However, in a timely review of the Banks' progress, Richard Heaver consultant to the Health, Nutrition, and Population Family (HNP) of the

World Bank's Human Development Network paradoxically summed up the work of the WB to date when it suggested:

> ... the Bank has developed the economic justification for large-scale investment in nutrition, and has the experience needed to scale up, it has failed to do so. ***Heaver (2006)***.

Furthermore, Heaver also noted that the WB spent a trifling 2.5% of its human development lending budget on nutrition. This was in blunt contradiction of the initial calls made by the organization in the early 70s. An inconsistency which incidentally defied the WB's own research whereby it had noted that nutrition alone was perhaps "one of," or even "the," best economic investment that could be made in a country's program of development. The review further questioned the incongruent manner in which the Bank's different departments worked in a vacuum where the left hand did not know what the right hand was doing. However, while the review was scathing of existing practices, it did offer a reprieve in the form of future direction.

With improved cohesive policies and the timely dissemination of information to the public as well as within its own corporate body, Heaver also suggested that if the Bank wanted to place more importance on nutritional programs, a branch-wide reevaluation of performance within the specific nutritional departments would need to be accepted (Heaver, 2006).

At about the same time, the GMO backlash was beginning to find common ground.

5.5 The GMO backlash

Regarding the "de facto" European Union (EU) ban on GM foods during the 1990s, the US along with several other countries made an official complaint to the World Trade Organization (WTO) in 2003. The complaint was regarding the EUs de facto violation of international trade agreements. At first the EU was going to fight the accusation; however, political diplomacy quenched a potentially difficult situation by ratifying a UN biosafety protocol that essentially regulated all international trade in genetically modified foods. The protocol, however, still allowed countries to act by banning imports if they felt there was insufficient scientific evidence regarding a particular products safety. By July of the same year, two further European instruments were put in place that between them essentially ended the controversial moratorium. The first was legislation requiring the labeling of food containing more than 0.9% of GMOs which applied to both animal and human food (although this did not apply to animals fed on GM crops). The second legislation ensured the mandatory labeling of any food containing (contaminated) by nonauthorized GMOs of over 0.5%. This was enacted solely for a period of 3 years only after which all nonauthorized GMO foods would be banned. Together both laws sought to safeguard the traceability of GMO products. While such laws essentially ended the embargo in 2004, the WTO ruling of 2006 was found in favor of the United States. The ban had indeed violated international trade agreements. Having said that though, the ruling did not query the future rights of EU member states of any future bans if there was enough evidence to warrant such a measure. While the ruling of the WTO essentially took the middle ground, this left both sides claiming a win while a few environmental groups suggested it in fact changed very little it achieved little or no real change (EC, 2003; Reinhardt and Ganzel, 2003; FOE, 2006).

5.6 2010 UN summit on MDGs

A full decade following the ground-breaking Millennium Summit and its stated goals and objectives (MDGs), a Summit was convened to explore the progress of the target goals. Pre-Summit reports, however, indicated mixed outcomes in terms of successes, failures, and progress, which were ultimately discussed at length in the Summit plus in many of the parallel events. In fact, it was not looking good for the goals as with only 5 years left to go, many observers and officials alike felt that the goals were drifting away, slipping further out of reach. In one particular side event entitled "Raiding the Public till," general consensus was reached in reference to the resources required for achieving the MDGs by 2015. In the parallel event, it was indeed understood that such resources would not be out of reach if

> ... governments were willing to reverse the process of socializing losses and privatizing gains – a process that characterizes the current model of globalization. *Caliari (2010).*

Finally, the outcome document of the main conference, called "Keeping the Promise - United to Achieve the Millennium Development Goals," outlined yet another Action Plan for the achievement of the MDGs by 2015. It was also further accepted that the MDGs were intertwined and mutually reinforcing. In sum, the Action Plan adopted the view that without a single panacea, member countries were encouraged to implement specific country or regional development strategies personalized to their own situations. Even President Obama got into the nitty-gritty when emphasizing a departure from established development strategies. Obama suggested refocusing the way in which the MDGs were measured, i.e., to pull back from long-held metrics of resource and financial inputs to be replaced by a more result-oriented output. This, in many peoples view, would effectively take out, in theory at least, the politics from the stated targets leaving the results to speak for themselves (UNDP, 2011). The MDGs were replaced in 2015 by the Sustainable Development Goals (SDGs).

5.7 Food crisis and financial chaos

During the period between 2006 and 2008, a great upsurge in global food prices took the world by surprise. Triggered by several coincidental factors arising from around the world resulted in price rises of 150% for rice, 75% for corn, and 35% for wheat. This state of affairs came about as a result of multiple factors that began with a drought in Australia—the worst in a century, this effectively halved Australia's wheat production for 2007. Adding to this was a decline in demand with increasing economic growth in countries such as Brazil, Russia, China, and India, in which individuals started to eat less cereals and more meat and dairy-based foods. Furthermore, as oil prices increased and ended up spiking at $60 a barrel, biofuels[2] became more competitive. This put pressure on food versus fuel. Take the

[2]Biofuels derive from corn, sugar cane or sugar beet and cassava among others. These are then converted into ethanol as a greener alternative to fossil-based fuels. The industry is massive and leading the vanguard was Brazil who currently has the longest-running commercial biofuel programme - since the 1970's. Indeed, in 2007/8 the Committee for Food Security (CFS) estimated that a significant proportion (4.7%) of Brazil'ss annual national cereal production was diverted into bio-fuel production (CFS 2008).

United States for instance—in an effort to reduce dependability on oil producing exporting countries (OPEC), the United States had for years now increased the volume of their corn and other crops into ethanol and biodiesel. However, rather than importing the crops for biofuels, the United States instead attempted to be self-sufficient and diverted about 30% of its corn crop for biofuel production. On top of this, a relatively recent phenomenon of food commodity speculation was contributing to the rising prices in both food and fuel. Indeed, it was noted at the time when Frederick Kaufman, contributing editor of Harper's Magazine, wrote an article entitled "How Wall Street starved millions and got away with it" in 2010 in which he concisely summed up the large speculative banks' role in artificially inflating food prices (Phillips, 2008; Vallely, 2009; Kaufman, 2010).

Collectively, the above factors ended squeezing the price of wheat and other cereals considerably upward. Simultaneously, especially in the low-income countries, increased import bills in turn could cause inflationary pressures on the domestic. Moreover, as prices rose, political instability took hold while other governments were forced to take control of some of the prices of basic staples. In Russia, for example, companies were readily forced to arrest and freeze bread and milk prices, while in Argentina people shunned fresh tomatoes when they eventually became costlier than meat. In Italy too, the home of pasta embarked on a 1-day boycott of pasta which was organized in objection to continued rising prices. In other areas, food riots and the threat of food riots occurred in such places as India, Yemen, Burkina Faso, Mexico, and others as shortages and food price inflation were almost out of control (Vidal, 2007).

In the midst of all of this, a global financial crisis exacerbated by unchecked banking operations filtered through eventually affecting ordinary people and coupled with the previously mentioned food price spikes they collectively exacerbated conditions of poverty and hunger.

5.7.1 Financial crisis

The financial crisis of 2007 and 2008 began with, among other financial institutions, high street banks who, in pursuit of ever-increasing revenue, moved into investment banking and began trading risk. Simultaneously, investment banks are also not happy with their lot moved into the home loan market. Book-held risks were increasing, while multifaceted, cleverly created, and vastly complex financial instruments effectually encountered a massive crisis of confidence. Trust in the system began to wane and the whole house of cards began to fall apart. Not surprisingly, the US subprime mortgage market collapsed, in turn creating a chain reaction of no confidence on a global scale. All this came at a time previous to an extended period of economic growth. Moreover, the investment banks that entered into the mortgage arena and that created some of the most complex financial instruments of the time had a knock-on effect in the housing market, which in many industrialized economies simply collapsed. The effect was not localized either, instead the cause and effect seemingly rippled throughout the world. Furthermore, lending slowed to the point where it almost dried. As all of this was happening, financial lenders, fearing their exposure to the problem, began calling in loans. This in turn further affected financial confidence whereby runs on many financial institutions were swift and brutal. In turn, some institutions, lacking sufficient financial support, or reserves quickly went bust while others turned to governments for financial aid. The injection of new government capital and promises of guaranteed capital security did little to assuage the frontal assault. As the crisis began to slow down and subside, banks and other financial institutions went to the other extreme with lending policies becoming so strict; they were counterproductive. The culmination of the above events tightened the liquidity of global financial markets and led to widespread financial and social disorder that had abrupt and overwhelming impacts on the global state of food, hunger, and malnutrition (Rudd, 2009).

As mentioned, the financial crisis profoundly affected many developing countries ability to feed the poor or the hungry and malnourished people. On top of this, as was illustrated in the oil crisis, provision of energy relied too heavily on the OPEC countries and the diversion of crops for bioenergy became questionable. Furthermore, the challenges of climate change and other social ills as outlined in the Millennium Summit became more and more pressing in light of the recent financial crisis. Consequently, the UN called for a High-Level Conference on World Food Security and the Challenges of Climate Change and Bioenergy in 2008.

Before the proposed conference, however, a meeting chaired by the then UN Secretary-General Ban Ki-moon along with other UN agency heads strategized a multitude of ideas to redress the global food crisis. The result was to put together a High-Level Task Force (HLTF) with the sole remit of understanding and advising on the Global Food Security Crisis. The HLTF was chaired by Ban Ki-moon and also included the heads of the IMF, the World Bank, the World Food Program, the IFAD, the FAO, and the WTO, who between them jointly and cooperatively established an action plan for discussion at the forthcoming conference.

The conference was a timely meeting which emphasized concerns over the recent food price debacle, the implications of a runaway biofuel industry, and its effects on global prices of feed fodder and food crops (Campbell-Platt, 2008).

It was also noted that the effects of climate change on the design and supply of global agricultural production might well turn out to be one of the biggest challenges that the world would face over the coming decades. Challenges in the face of changing weather patterns could also uproot certain traditional crops replacing them with new and unfamiliar harvests. This could also mean a radical change in traditional trading partners too. In sum, the conference, recognized the fragility of the world's food supply system and its vulnerability to both known and unforeseen shocks and challenges. In this, questions would need to be answered on how to improve the resilience of worldwide food production in the face of adversity and a lack of biodiversity, etc. Indeed, recommendations of investment, technological development, financial instruments, and dissemination of knowledge should all come together to support adaptation and mitigation of an ongoing change of climate (Campbell-Platt, 2008). Despite such commitments from this and from numerous other conferences, high-level talks, ministerial meetings, etc., the situation was still dismal a year or so later when the report the "State of Food Insecurity" was published announcing that over 1 billion around the world were still undernourished. Alarm bells rang from the rooftops of governmental, humanitarian, and other institutions alike. It seemed that the situation was rapidly failing and in 2009 the Third WFS was held.

Once again heads of state met to reaffirm their commitment and to work on ways to mitigate the worst effects of the financial crises vis-à-vis food and food security. Also, on the agenda was the reformation of global governance. Not helping the cause was the absence of Jacques Diouf—the then Director-General of the United Nations' Food and Agriculture Organization plus other key leaders in the field. This was attributed to a general lack of interest in the global issues of hunger and malnutrition problems. However, more than this, of greater importance appeared to be one important viewpoint, never previously expressed at such a high-level meeting. This was the notion of lost confidence in a global market trading system as a means of ensuring food security. While Venezuela put this down to fundamental problems of the capitalist system, it seemed others downplayed the crisis suggesting that it was just in fact a one-off glitch precipitated by food shortages. Whatever the reason, these jitter. Such feelings and comments put the issue of food self-sufficiency back on the table, which was both welcomed and opposed in equal measure (Christiaensen, 2009; IISD, 2009).

The end of the first decade of the 21st century once again brought with it reflection (Table 5.2) and a renewed promise of achieving the MDGs by 2015.

TABLE 5.2 Key dates of the period: 2000 to 03.

2004 UNU Institute for the Environment Human Security **(UNU-EHS)**	The UNU-EHS is the UN University Institute for the Environment Human Security (UNU-EHS). Programs focus on all things—environment whether environmental hazards, degradation, and vulnerability, as well as societies coping capacities to such disasters (UNU-EHS, 2005).
2004 Millennium Challenge Corporation (**MCC**)	The MCC is an independent US foreign aid agency offering grants to poor countries to help in the fight against poverty and economic development (MCC, 2010).
2004 The Integrated Food Security Phase Classification (**ICP**)	Calls for improved analysis, increased rigor, better comparability, and improved transparency of evidence in food security analysis led to the creation of the IPC. Developed as an add-on to existing frameworks, it helps in the classification of the severity of food security situations (IPC, 2007; FSAU, 2010).
2005 The Central Emergency Response Fund (**CERF**)	There are numerous shortcomings that exist in the UN emergency relief model. Responding to such problems, the UN in 2005 set up CERF, the Central Emergency Response Fund replacing the older Central Emergency Revolving Fund with a remit to speed up the process and reliability of humanitarian aid (CERF, 2007).
2005 UN World Summit	A summit discussing among other things progress to do with the Millennium Development Goals and the world's recommitment to them (UN, 2005).
2008–2013 Food and Nutrition Technical Assistance II Project (**FANTA-2**)	The second FANTA, FANTA-2, extended the work of the original project of refining nutrition, policies, strategies, and programs of food security (FANTA2 2010).
2008 The Special Unit on Commodities	Under the Aegis of UNCTAD, the Special Unit on Commodities helps developing countries to respond to the challenges of commodity markets (UNCTAD, 2009).
2009 FAO World Summit on Food Security	The seemingly deteriorating global food security situation and fear that high food prices would push the undernourished over the one billion mark led to an emergency food summit. The Summit once again reenforced the idea that poor countries drastically needed economic development to boost agricultural production. The Summit recognized too the continual lack of international coherence in global governance (FAO, 2009a,b).

2010 UN Summit on **MDGs**	United Nations Secretary-General Ban Ki-moon hosted the High-level Plenary Meeting of the MDG Summit with the goal of accelerating progress toward all the MDGs by 2015.
2010 FAO Working Group on Animal Genetic Resources for Food and Agriculture	FAO held a meeting to discuss and assess the importance of genetic diversity of livestock at the regional and global levels (FAO, 2010).
2011 FAO and German Federal Ministry of Food, Agriculture and Consumer Protection **IMCF** Conference	An FAO consultation was held on improving the dietary intakes and nutritional status of infants and young children in Asia. This was to be achieved through improved food security including complementary feeding in young children combining food security and nutrition interventions.

Sources: Compiled From Multiple Sources: UN (2005); UNU-EHS (2005); CERF (2007); IPC (2007); FAO (2009a,b); FAO (2010); FSAU (2010); MCC (2010).

References

Caliari, A., 2010. MDGs Still Achievable, but Role of Public Sector is Key. Retrieved 21st Feb 2018, 21st Feb 2011, from. http://www.socialwatch.org/node/12245.

Campbell-Platt, G., 2008. Report on high-level conference on world food security: the challenges of climate change and bioenergy. In: High-Level Conference on World Food Security. FAO, Rome.

CBO, 2000. H.R. 5224: International Food Relief Partnership Act of 2000 Congressional Budget Office, US Congress.

CEB, 2010. The Chief Executives Board (CEB): Overview. Retrieved 23rd March 2018, 2018, from. http://www.socialwatch.org/node/12245.

CFS, 2008. Agenda Item Ii: Assessment of the World Food Security and Nutrition Situation. Committee On World Food Security: Thirty-fourth Session, Rome, Food and Agriculture Organisation.

CERF, 2007. Central Emergency Response Fund: What is the CERF? Retrieved 12th June 2019, 2019, from. https://cerf.un.org/event/high-level-conference/cerf-high-level-pledging-event.

Christiaensen, L., 2009. 2009 World Summit on Food Security. Our World.

Cohen, J.E., 1995. How Many People Can the Earth Support? W.W. Norton, New York and London.

EC, 2003. Press Release: European Commission Regrets US Decision to File WTO Case on GMO's as Unnecessary. Brussels, European Commission.

EPI, 2010. The Earth Policy Institute: About. Retrieved 21st May 2018, 2018, from. http://www.earth-policy.org//index.php?/about_epi/.

EuropeAid, 2009. About EuropeAid. Retrieved 13th Feb 2018, 2018, from. https://ec.europa.eu/europeaid/historical-overview-eu-cooperation-and-aid_en.

FANTA2, 2010. About Food and Nutrition Technical Assistance II Project (FANTA-2). Retrieved 15th July 2019, 2019, from. https://www.fantaproject.org/.

FAO, 2001. The State of Food Insecurity in the World 2001. Food and Agriculture Organisation, Rome.

FAO, 2002a. Report of the world food summit: five years later. In: World Food Summit: Five Years Later. Food and Agriculture Organisation, Rome.

FAO, 2002b. World Food Summit: Five Years Later Reaffirms Pledge to Reduce Hunger. Summit News.

FAO, 2003. FAO urges international alliance against hunger on World Food Day. FAO Newsroom.

FAO, 2009a. Declaration of the world summit on food security. World Summit Food Secur. Food Agri. Organ.

FAO, 2009b. Secretariat Contribution to Defining the Objectives and Possible Decisions of the World. Summit on Food Security on 16, 17 and 18 November. Food and Agriculture Organisation, Rome, p. 10.

FAO, 2010. Status and Trends of Animal Genetic Resources — 2010. Intergovernmental Technical Working Group on Animal Genetic Resources for Food and Agriculture. FAO, Rome, p. 89.

FOE, 2006. Looking behind the US spin: WTO ruling does not prevent countries from restricting or banning GMOs. Briefing Paper. Amsterdam Friends Earth Int.

FSAU, 2010. Food Security Analysis Unit for Somalia: FAQ. Retrieved 8th May 2018, 2018, from. http://www.fsnau.org/.

GECAFS, 2010. Global Environmental Change and Food Systems. Retrieved 23rd June 2018, 2018, from. http://portal.unesco.org/en/ev.php-URL_ID=15559&URL_DO=DO_TOPIC&URL_SECTION=201.html.

GFN, 2009. Global Footprint Network: At A Glance. Retrieved 20th June 2018, 2018, from. http://www.footprintnetwork.org/en/index.php/GFN/page/at_a_glance/.
Heaver, R., 2006. Good Work—But Not Enough of It: A Review of the World Bank's Experience in Nutrition. World Bank: Health, Nutrition and Population Department, Washington, D.C., p. 89
IAAH, 2009. The International Alliance Against Hunger: History and Vision. Retrieved 6th June 2018, 2018, from. http://www.fao.org/english/newsroom/news/2003/20846-en.html.
IISD, 2009. Summary of the world summit on food security 16—18 November 2009. World Summit Food Secur. Bull. 150 (7).
IPC, 2007. The Integrated Food Security and Humanitarian Phase Clasification System: Technical Manual Version 1. Rome/Somalia. FAO/FSUA, p. 47.
Kaufman, F., 2010. The Food Bubble: How Wall Street Starved Millions and Got Away with it. Harpers Magazine. The Harper's Magazine Foundation, New York, 8 pp.
MCC, 2010. The Millennium Challenge Corporation. Retrieved 23rd March 2018, 2018, from. https://www.mcc.gov/.
Mezzalama, F., et al., 1999. Review of the Administrative Committee on Coordination and its Machinery. UN Joint Inspection Unit, Geneva, p. 31.
Mousseau, F., 2005. Food Aid or Food Sovereignty? Ending World Hunger In Our Time. Oakland Institute, USA.
NGO/CSO Forum, 2002. An Action Plan of the NGO/CSO Forum for Food Sovereignty. 2002 Rome NGO/CSO Forum for Food Sovereignty. International NGO/CSO Planning Committee, Rome.
ODI, 2000. Reforming Food Aid: Time to Grasp the Nettle. Briefing Paper. Overseas Development Institute, London.
Phillips, L., 2008. Senators Hear Testimony on Legislation to Reduce Food and Energy Prices by Curbing Excessive Speculation in the Commodity Markets. Senate Committee on Homeland Security and Governmental Affairs, Washington.
Reinhardt, C., Ganzel, B., 2003. A History of Farming. Living History Farm.
Rudd, K., February 2009. The global financial crisis. Monthly 20—29.
UN, 2000a. Millennium Summit: The UN Millennium Declaration. The Millennium Assembly of the United Nations, New York, United Nations.
UN, 2000b. United Nations International Strategy for Disaster Reduction Secretariat (UNISDR). Retrieved 21st Jan 2018, 2018, from. http://www.unisdr.org/.
UN, 2000c. United Nations Millennium Declaration. United Nations, UN. A/RES/55/2.
UN, 2005. 2005 World Summit Outcome. Fact Sheet. United Nations Department of Public Information, New York.
UN, 2006. The UN Millennium Project. Retrieved 12th Feb 2018, 12th FEb 2018, from. https://www.unsystem.org/content/2005-world-summit-outcome-factsheet.
UN DESA, 2006. What is Johannesburg Summit 2002? Retrieved 23rd July 2010, 2010, from. http://www.un.org/jsummit/html/basic_info/basicinfo.html.
UNCTAD, 2009. The Special Unit on Commodities. Retrieved 21st June 2018, 2018, from. http://www.unctad.info/en/Special-Unit-on-Commodities/About-the-Special-Unit-on-Commodities/Who-we-are/.
UNDG, 2009. The Millennium Declaration and the MDGs. Retrieved 19th June 2018, 2018, from. http://www.undg.org/index.cfm?P=70.
UNDP, 2011. UNDP Millennium Development Goals. Retrieved 12th Jan 2018, 2018, from. https://www.undp.org/content/undp/en/home/sustainable-development-goals.html.
UNDP/MDG, 2010. Millennium Development Goals (MDGs). Retrieved 24th June 2018, 2018, from. https://www.un.org/millenniumgoals/.
UNU-EHS, 2005. Human Security in a Chamging World. UNU-Institute for Environment and Human Security, Bonn, Germany, p. 33.
Vallely, P., 2009. The Big Question: Why is so Much of the World Still Hungry, and What Can We Do About it? the Independent. The Independent, London.
Van Den Bergh, J.C.J.M., Rietveld, P., 2004. Reconsidering the limits to world population: meta-analysis and meta-prediction. Bioscience 54 (3), 195—204.
Vidal, J., 2007. Global Food Crisis Looms as Climate Change and Fuel Shortages Bite: Soaring Crop Prices and Demand for Biofuels Raise Fears of Political Instability. Guardian online. London, Guardian News and Media Limited, 3 November 2007.
WFN, 2009. Water Footprint Network: Why a Water Footprint Network? Retrieved 13th July 2018, 2018, from. https://waterfootprint.org.

PART TWO

Food and…

How little we know of what there is to know
Ernest Hemingway
For Whom the Bell Tolls

CHAPTER

6

The beginnings of modern development theory

OUTLINE

Development paradigms or development economics (used interchangeably) have very few commonly agreed upon principles or guidelines. However, even before any cursory explorations of the issues are involved, one must first comprehend what development is and how it is acted upon. In this context, the word "development" is a complex word acting as a descriptor of growth in a variety of sectors such as economic, agriculture, and education, etc. Growth, however, is also a challenging concept in that it is also viewed differently among economists and observers alike. Sometimes the idea refers to goals and is economic in nature, while other times it may represent citizens' social well-being. Of agreement among theorists, though, is the general notion that "development," historically speaking, was firmly bound in the downfall of feudalism and the evolution of capitalism.

In this way, development can be considered as composing of three separate but interlinked aspects. First comes survival; this feature of any development paradigm makes sure the fundamentals of safety, shelter, food, and good health are provided. The second stage is concerned with raising the basic standards of living by creating more job opportunities, better education and awareness, and increased personal and by extension national income. The last stage is social in nature and is engaged with increasing personal and collective social choices. In summary, this skeleton view abridges current perceived wisdom of development

Food and Society
https://doi.org/10.1016/B978-0-12-811808-5.00006-4

and further naturally hints at a continuum from underdevelopment to development. In the first aspect of the continuum, underdevelopment is used to describe the poorly educated or illiterate people of a nation who are essentially poorer and underresourced with limited options and choices. Such people can often be seen living subsistently in terms of basic food and shelter as well as financial and physical resources. This view can and has been extrapolated to the larger population, that is to say groups of underdeveloped people living in underdeveloped regions which in turn characterized by overpopulation, lack of productivity, poor technologies, and lack of equality in terms of general education and training, not to mention a comprehensive health system. Thus, it can be seen that the classic view of developing such countries means increasing gross domestic product (GDP) and structural transformation of the economy, which often entails reducing the reliance on the agricultural sector. This kind of development also entails strengthening socioeconomic diversification and differentiation and further stimulating wider social sophistication taking into account choice and equality. Other measures also include increasing national and global trade and fostering geopolitical relationships while addressing the above issues through good macroeconomic management. Further, healthy fiscal policies, and improved health care education, among numerous other variables are in play—all of which is predicated on a stable political government.

As mentioned previously, there are very few development guidelines; in this way many observers interpret what is needed differently. Some see the failure of underdeveloped countries to improve their economic and social standing as consequences of poor economic efficiency, lack of a stable government, market failure, tyranny, and corruption. Others see such failures as one of dependence. This latter view attempts to explain failure as a result of residual colonialism or external exploitation by wealthy foreign nations. This viewpoint also suggests an ongoing inequitable distribution of wealth through the inappropriate disparity of international terms of trade between the developed and underdeveloped nations, thereby taking advantage or dependency of the imbalance between the rich and poor nations.

Indeed, Contreras (1999) in his paper explored the gaps between the rich and poor nations during the period of the 1940s and suggested that by the 1950s the world would be divided into two camps: (1) the rich and (2) the poor. While the richer nations comprised, many Western countries including several from Europe, the United States, and Canada, etc., it was also suggested that the poorer nations were those in Latin America, Asia, and Africa. This duality of rich and poor nations was already known by many economists and social observers of the time and was discussed at great length and often times with great passion. Many too looked in earnest for ways of closing the gap between the richer and poorer countries (Contreras, 1999). Consequently, a branch of economic theory called development economics ended the default model or prevailing paradigm that aimed to address this imbalance. Essentially development economics fundamental principles aimed to address the poorer nations' economic imbalance with a view to increasing not only the prosperity of the country and its people but also the efficient allocation or reallocation of the country's resources (Mikesell, 1968). This involved creating or introducing new or improved policies that supported wide-ranging initiatives including cultivating educational expansion as well as better working conditions and health initiatives among a plethora of other sweeping changes in all sectors in an effort to create a platform for economic and social growth.

Development theory is firmly rooted in mercantile ideology and espouses nationalist economics, i.e., the control of the nation's resources at the national level. Indeed, Mikesell (1968), an ardent observer, suggested that many preconditions would need to be in place before economic and social growth could take place. These included mobility in terms of the national workforce, increased and ongoing education, political stability, fair terms of trade, including economic freedom, and well-developed capital and financial markets (Mikesell, 1968). In Mikesell's view, he also saw humanitarian aid as an almost necessary means of nations gaining critical momentum to develop self-sustaining economic development. Of course, he suggested because of the inherent nature of many underdeveloped countries, any capital would indeed come from external sources (Mikesell, 1968). In fact, this is indeed what happened immediately preceding the aftermath of World War II. To a great extent, one could confidently argue that modern development economics was raised among a growing need for humanitarianism. In point of fact, many perceived challenges of the post–World War II Eastern European reconstruction programs suggested numerous authors such as Rosenstein-Rodan, Ragnar Nurkse, Kurt Mandelbaum, and Sir Hans Wolfgang Singer, to name a few, who recommended solutions of aid and structural transformation in unison to address the requirements of healthy economic and social development (Pleskovič and Stern, 2001; Meier and Stiglitz, 2002; Riddell, 2007).

Modern aid development theory and ideology emerged from such small beginnings and indeed several notable development models, as cited by Contreras (1999), have been used in various parts of the world at various times over the preceding years. Just a few of the many models that exist include structural change theory; linear stages of growth model; international dependence theory; and neoclassical theory (Contreras, 1999). It is beyond the purview of this book to go into any great detail of these models, suffice to suggest that development theory is a complex concept that acts as a focal point of growth within individual sectors such as economic, agriculture, and education, etc., or in macroroles as a growth model for whole countries or regions. However, having said that, the concept of increased development or growth in and of itself is also problematic to some in that it is viewed through various divergent lenses. By way of example some say developmental growth refers purely to economic goals, while for others it might also incorporate an important social aspect or advancement of the well-being of a population (Table 6.1).

From whichever lens one views economic and social growth, development can be thought of as three separate but interlinked stages. Firstly, one needs to consider survival; this entails obtaining sufficient essentials of food, shelter, good health, and protection against threats. The second stage looks to raise these basic standards and is concerned with creating more and better paid jobs and on top of this better education and increasing personal incomes. The third stage is occupied with increasing personal and social choices. This sums up current perceived wisdom of development and its attainment which suggests a natural continuum ranging from underdevelopment to development.

With this in mind, a so-called "developed" country or region is considered to be one, which usually (but not always) comprises a stable government; good macroeconomic management with healthy trade and fiscal policies; adequate infrastructure; increased education and literacy; good basic health; and personal security among other things. The notion of underdevelopment, by contrast, can be used to describe the poor, ill-educated, and underresourced people of a region or nation. In such countries, one often sees people living subsistently with basic essentials of food and shelter taking up most of their financial and physical

TABLE 6.1 A selection of development models.

Structural-change theory	This particular theory came out of Latin America around the time of the 1940s. It focused on emphasizing the importance of opting out of the agricultural sector and moving employment into the industrial and service sectors. Expanding into these sectors would diversify their local structural economy making them more competitive and more attractive to foreign investment and by extension increase the local economy.
Linear stages of growth model	The linear stages of growth model was first articulated by W. W. Rostow in the 1950s; his ideas tended to focus on accelerated build-up of individual per capita income, which in turn will promote local and international economic growth.
International dependency theory	This theory is more of a reflection than a real development opportunity. It was a controversial contribution to development economics during the 1970s, which almost suggested that weaker economies would piggyback off their more developed counterparts. Fostering a sort of dependence, this theory postulates that such relationships are predicated on maintaining the status quo whereby more developed partners benefit more from their lesser developed counterparts by siphoning off economic benefits in a one-way direction.
Neoclassical theory	Neoclassical theory within developmental economics first gained distinction during the 1980s. It completely dismisses previous neo-Marxist (international dependence) theory as flawed and unrealistic. Instead, the theory promotes less interventionism favoring instead a truly free market uninhibited by excessive government regulation. This has quite possibly been the more influential of the development theories and has, in due course, perhaps become the more prevailing paradigm since the 80s (Contreras, 1999).

Compiled from the works of Contreras, R., 1999. Competing theories of economic development. Transnatl. Law Contemp. Probl. 9, 93–108.

resources of individuals. Moreover, underdeveloped areas/regions are all too often characterized as being overpopulated with low output or productivity; a lack of training and technology; and a general sense of inequality. In such situations, the classic view of development theory looks to achieve structural transformation of the economy through increased GDP; enhancing socioeconomic diversity and differentiation; increasing trade and geopolitical elationships; and engendering wider social sophistication encompassing integration, equality, and choice. Within this model though, as is quite often the case, the resulting structural transformation of the economy is typically coupled by a general decline in reliance on agriculture. While these are generalizations, it should also be noted that there are no real inside tracks to good development guidelines. In fact, development theory is, when one looks closely at the concept, at odds with each other in the sense that when dealing with underdevelopment, the debate takes several approaches. But in doing so, two common themes tend to resurface; these are the inherent and dependence of underdeveloped nations. When talking of inherent failures, it is easy to suggest that underdevelopment is a consequence of poor economic efficiencies. This suggests that resource allocation together with market failure is caused by, or at least exacerbated by, governmental policy failures and/or corruption and tyranny. The other increasingly common failing of underdevelopment is the notion of dependence. This view suggests that present inequality seen between the developed and underdeveloped nations can be traced back to external exploitation or colonization by wealthy foreign nations. This would have been further compounded by the inequitable distribution of wealth through an imbalance of the terms of trade between the rich and the poor.

While as a rule development theory advocates diversifying out of agriculture, the agricultural industry, nevertheless, remained a major sector in many economies. As a result, maximizing this sector became an important mechanism for achieving desired development objectives.

6.1 The growth of the agricultural economic development paradigm

So how has the twin track (mentioned previously) approach come about? Well, for decades now, science and technology have been at the forefront of the agricultural development paradigm. Yet, increasingly over this period, more and more investment for the research and development that lies behind such scientific and technological advances has come from the private sector (Pardey et al., 2006). Not so surprisingly, this investment has been concentrated in the richer developing nations. As more and more investment was channeled into research and development, new technologies were eventually rolled out; and naturally these new technologies often carried with them high research and development costs. As such, and with the need to provide a reasonable return on investments, as is the capitalist way, this meant passing on the costs to the customers and ensured the continuing high cost to consumers of such technologies.

Another outcome of this flood of private investment witnessed was more and more genetic modification techniques and outcomes being granted patents. This inexorable trend of bio-patents saw, and continues to see, an increasing amount of the biological stock being privatized. Once again, this ensured the exclusivity of intellectual property rights and increasing costs to consumers. As a result of the relentless march of the investment–commercialization–reward cycle, many were left unable to compete or pay; and many of these resided in the lesser developed world.

Clearly this places much strain on many small-scale or smallholder farmers in both developed and developing countries who increasingly find themselves in direct competition with capital intensive and frequently highly subsidized production systems that can produce vastly more commodities that are often sold more cheaply on the open market (IAASTD, 2010). Furthermore, the vicious cycle continues as these smallholders are unable to benefit from the increased global traffic brought about by the Uruguay GATT trade rounds. Added to this, the high capital entry barriers of some industrialized farming sectors effectively result in a de facto form of international trade protectionism that precludes any democratic involvement from the small players.

This ensures the brutal and steady decline of local food production systems (Gibson, 2016). It is also a situation that is unlikely to change in the short term either, that is, unless a more significant reduction in trade barriers (subsidies and import tariffs) is vigorously pursued and the negotiating position of the developing countries is properly strengthened.

Unfortunately, while there are many proponents of a true economic level playing field, to rectify the situation, the author feels, is a mountainous challenge and one that will take years to address. This state of affairs ensures that for many, if not most countries in the world, agriculture has become an essential and inextricable component of economic growth. Incidentally, this is a situation that is also tied to the food security of many developing countries (more on this later). Having said that, though, it is wise to note at this juncture that while for many, the current agricultural paradigm is one of the industrialization and concentration there is growing backlash at a production system that is increasingly seen as socially and environmentally destructive. The difficulty here is one of consensus, with alternative models provoking as much disagreement, and with many loud and vociferous advocates of each, the debate is set to continue into the foreseeable future. Unfortunately, this fractious and dogmatic posturing by many opposing camps can be seen at best as unhelpful and at worst damaging to any real structural consensus. Alongside, the agricultural paradigm is not the only model that is being challenged, indeed there are many other competing economic and developmental models to those currently in service but that is another matter.

6.2 Development in practice

Whether centrally planned economies responsible for controlling the production of food or whether supply is met via the global free market system (globalization) or whether in combination, a country's food supply is no accidental undertaking. It is the government and its respective views on such matters that shapes the developmental landscape. With policies of international or national responsibilities looking to promote free trade, self-sufficiency or other instruments of business defines the way a country's or region's food supply is shaped.

Regarding agricultural development paradigms there is far from agreement on just how agriculture fits in with a country's or nation's overall development agenda. Indeed, agricultural paradigms have shifted, sometimes quite radically over time and spatial boundaries. Following post–World War II for instance, structural rebuilding, the prevailing agricultural development paradigm of the time, was to increase productivity—allowing for the security of food as well as any associated benefits of economic growth to naturally filter through the economy and back to the ordinary person (Goodrich, 1947). This, it was thought, would be achieved through a combination of suitable infrastructure; technological innovation; improved institutional frameworks; a good information network; and the right incentives.

The idea of enlarging a country's international agricultural trade as a tool of development at this point had not yet fully emerged. Instead, protectionism and quotas were still seen as legitimate policy instruments with imports supplying any food shortfalls (Goodrich, 1947; Warnock, 1997). This view forced inward facing national policy objectives and a paradigm of self-sufficiency toward food production (Warnock, 1997; McCalla, 2007, pg16). However, change brought about by the Bretton Woods Institutions aided in the formation of international frameworks, which eventually helped open global markets for new and existing training partners in terms of goods, services, and capital. On top of this, the same institutions aimed to create stabilization through outward looking program of advice and monetary assistance. This modernized Western agricultural development practice helped shift insular looking nations to face a more global perspective with the resultant notion of international agricultural trade as a growth development tool gaining traction (Belshaw, 1947). This introduced fundamental radical changes in the way farming was seen; self-sufficiency was slowly being replaced with cheaper imports where comparative advantage became the underlying agricultural philosophy. Developing countries too were encouraged to follow the Western model, and while some adopted the new philosophy (to lesser and greater extents), the now familiar industrialized agricultural model was firmly cemented in the mix of other development strategies. Through this growth and the subsequent occupational redistribution of farm workers, it was hoped developing countries would be able to help themselves in food security while also reducing wealth gaps throughout the world (Belshaw, 1947). This model became very successful in the West; however, it had mixed successes in the developing regions with many failings to properly adopt the full ideology. Despite some stragglers, by the late 1980s, the global free market had firmly taken hold and countries, building on the model of comparative advantage, were encouraged to export whatever they could produce most efficiently. In turn this allowed them to earn the foreign exchange needed to import any required food shortfalls. This became the standard development model with the agricultural sector finally seen as having a valuable role to play in the overall strategy of both the economic and social development of a country or region. In this scenario, McCalla (2007) suggests the agricultural sector:

> ...becomes much more complex, more interdependent with the rest of the economy and charged with meeting multiple goals. [Where] Increasing food production is no longer a goal in itself. ***(McCalla, 2007, pg16).***

However, while in many economies' agriculture has been liberated and aligned with the ideology of global free trade, in others a protectionist approach continues to prevail.

6.3 Globalization

The march of globalization has been relentless. For two, arguably three centuries, advances in media, travel, and technology facilitated by conducive politics has resulted in the relative ease of movement of goods and services across the globe (Gibson, 2016). However, within this context food is a relative newcomer. In fact despite the overt political will that has been bandied about since the 1940s or so it was only by the 1990s that the General Agreements on Trade and Tariffs (GATT) talks was food was finally admitted into the global free trading arena. In theory this effectively allowed countries to buy the food they needed at the best price on the international market unimpeded by economic or political barriers (caveats aside). That said, globalization has brought about economic and cultural liberalization as never before—a new orthodoxy (Vaidya, 2006). Much of this, according to Kennedy et al. (2004), is being spurred on by urbanization and as far as the food is concerned is effectively transforming the whole food supply chain from production, processing, retailing, and marketing through to consumers all around the world (Kennedy et al., 2004).

There are many other benefits of this new global market too. One such benefit has to do with livestock and quality control standards. In dealing on the international market, for instance, bound up within increased global trade is a trend for better quality control standards within the meat and livestock sector. Inherent in this upward trend also exists the possibility of translating such coveted standards into better animal welfare values and by extension the potential of reducing disease and increasing food safety. There are also economic and social benefits of globalization too—as societies are becoming increasingly politically and culturally integrated, so the increased flow of commodities services, labor, and capital means more opportunities for many more individuals and countries. Consequently, greater international trade often raises domestic incomes and standards of living, helping both the financially poor and with improved food security (Parfitt et al., 2010).

Global trade in food and agriculture also brings with it its own challenges. When it comes to animals, for instance, the same strict quality controls that look to raise standards might also act as barriers. Also, when it comes to a finite food supply and with so many countries now looking to be fed from the same trough, the many global inequalities in terms of trade are potentially set to increase with certain established interests looking to benefit considerably more than others. Furthermore, on a political dimension, the FAO, Parfitt, and others also warn that with increased integration and a potential converging global agrarian policy, the rewards might come at a high price as reflected in the increased vulnerability of global production, the lack of development of internal markets, and subsequent price volatilities (McMichael, 1994; FAO, 2003; Parfitt et al., 2010).

Another cautionary note too, says Kennedy (2004), concerns the vast amounts of direct foreign investment by large multinational food companies as well as retailers. While no doubt this investment is resulting in cheaper food, greater availability, and more diversity—such "advances" are inducing fundamental changes in traditional production, procurement, and distribution systems often, suggests Kennedy, at the expense of smaller local agents and long-established food outlets. Moreover, as international trade tends to favor big business and centralized procurement systems, it further acts as an exclusionary hurdle to smallholders or small producers (FAO, 2005).

Lastly, there also appears to be evidence that globalization is likewise bringing about a gradual shift—a convergence—toward a more universal food culture (Kennedy et al., 2004; Dimitri et al., 2005; Scheuerman, 2008; Walker, 2008).

Another important driver of change within the food supply dynamic is increasing urbanization—a trend that is only going to continue.

6.4 Right to food

There are many whose misconception of the morality of the right to food is enshrined in the declaration of human rights back in 1948. Yet, the concept, despite what others would have us believe, had its inception scattered throughout history (Mettrick, 1929; UN, 1948; FAO, 2006; Gibson, 2016). In point of fact, even back in the day of the Babylonian empire, even slaves had the inalienable right to food despite undertaking "service without pay" (Johns, 1904). Another example of this apparently relatively new social construct saw the Rev. Thomas Sherlock, Bishop of London, back in the early 18th century, incorporating the English philosopher John Locke's notions (1714) when he opined that:

> "There is not, I presume, a stronger natural right, than the *right to food* and raiment; this is founded in the common necessity of nature; and 'tis not to be thought that God sent men into the world merely to starve, without giving them a right to use in common so much of it as their necessities require." ***(Sherlock 1718, pg25)***.

Expanding on this idea was Edmund Burke's book in which he comments on the role of governance when reflecting on the French Revolution in 1790 when he wrote:

> Government is not made in virtue of natural rights, which may and do exist in much greater clearness, and in a much greater degree of abstract perfection: but their abstract perfection is their practical defect. By having a right to everything they want everything. Government is a contrivance of human wisdom to provide human wants. Men have a right that these wants should be provided for by this wisdom ... What is the use of discussing a man's abstract right to food or to medicine? The question is upon the method of procuring and administering them. In that deliberation I shall always advise to call in the aid of the farmer and the physician, rather than the professor of metaphysics. ***(Burke 1790, pg88/9).***

Of note here is Burke's separation of the moral right to food to one of the adaptations and practicalities of attaining a person's right to nourishment, it marked an important milestone in regard to the departure of contemporary doctrine.

Separately, in a relatively similar vein was the influential secularist of the 19th century—Charles Cockbill Cattell who also strengthened the idea by unequivocally stating that:

> Every industrious community has the right to food, clothing, shelter, and such social arrangements as will enable it to enjoy an average share of life. ***(Cattell 1874, pg10).***

Last but not least, in the writings of Boyd Orr, one can witness a further drive in the direction of the right to food; in 1939, talking of the virtues of the comparatively new science of nutrition he suggested that

> It is the right of every citizen ... to enjoy the benefits ... of nutrition so that the health of every one ... will be up to the level we now know is possible. ***(Orr 1939, pg80, Orr 1940).***

With the moral momentum providing impetus from these men of foresight it is undoubtedly a feather in the United Nations Human Rights Commission, driven by Eleanor Roosevelt when the Universal Declaration of Human Rights was drafted in 1948. Indeed, it was the first time throughout history that an international body actually recognized food as a right. Yet, despite these promising beginnings, it was not until the World Food Summit 1996 that the right to food was officially preserved in legislation (World Food Summit, 1996). Even this had its limitations, though, as it was not till a decade later that the moral and ethical human rights dimension of the ideology was finally realized (FAO, 2006). Yet all was still not quite fully in place, the reason being, despite the fact that the right to food was now fully accepted in over 40 countries worldwide, hunger and malnutrition was still rife and rising, even before the 2007/9 economic (and food) crisis highlighted the problem that for many, the right to food was at best inadequate (FAO, 2006; SOFI, 2009) and at worst:

> ... a cruel mockery to tell someone they have the right to food when there is nobody with the duty to provide them with food. That is the risk with the rights rhetoric. ***(O'Niel 2002).***

It has been further suggested that despite the guidelines on how best to anchor a rights-based ideology in the fight against hunger and malnutrition, a fundamental lack of cohesion betwixt and between policies ensures that any real coordinated change or progress is difficult to achieve considering that such frameworks are based on voluntary rather than mandatory codes (Hartmannshenn, 2004; UN, 2004).

6.5 Questioning agricultural free trade

In playing devil's advocate, there are some important caveats that have been expressed in the current model of global free trade vis-à-vis food and food security. These concerns surround the fact that common in the free trade paradigm and contrary to much of the postwar government-managed market economy is the notion that global trade would naturally grease the wheels of development. However, critics have argued, scathingly in some instances, that this belief is wholly and inappropriately misplaced. Instead, some have suggested relying on the private sector to replace those important functions that had been previously provided by governments, which resulted in more food insecurity rather than less (Murphy, 2010). This view gained further credence in the 2007 World Development Report that essentially acknowledged this very shortcoming and sure enough; following on the heels of these concerns was the 2007–08 food price crises which saw widespread civil unrest. These developments firmed up the realization in policy-maker's minds of a basic need to manage the inherent volatility in global free trade markets and to further protect people from volatile food costs. This view marks a shift from the laissez-faire, free market mechanism of free trade and a return to a more cautious approach with some promoting increased government managed intervention and social safety nets (Murphy, 2010). From such developments, another emerging paradigm emerged to challenge prevailing trends in development economics—that of food sovereignty. The idea of food sovereignty marks a departure from the hitherto large-scale intensification of agriculture and a liberalized trade-based food security model toward more localized, small-scale sustainable solutions (see Politics of Food) (Windfuhr and Jonsén, 2005).

References

Belshaw, H., 1947. The food and agriculture organization of the United nations. Int. Organ. 1 (2), 291–306.

Burke, E., 1790. Reflections on the Revolution in France, and on the Proceedings in Certain Societies in London Relative to that Event. In a Letter Intended to Have Been Sent to a Gentleman in Paris. Printed for J Dodsley, London. Pall Mall.

Cattell, C.C., 1874. Co-operative Production. G. H. Reddalls, Birmingham,.

Contreras, R., 1999. Competing theories of economic development. Transnatl. Law Contemp. Probl. 9, 93–108.

Dimitri, C., et al., 2005. Economic Information Bulletin: Number 3: The 20th CenturyTransformation of U.S. Agriculture and Farm Policy. United States Department of Agriculture, USDA.

FAO, 2003. Trade Reforms and Food Security: Conceptualizing the Linkages. Food and Agriculture of the United Nations, Rome.

FAO, 2005. The Globalizing Livestock Sector: Impact of Changing Markets. Committee on Agriculture Report. Food and Agriculture Organisation, Rome.

FAO, 2006. Policy brief: food security. Rome Food Agric. Organ. 4.

Gibson, M., 2016. The Feeding of Nations: Re-defining Food Security for the 21st Century. CRC Press, Boca Raton, Florida.

Goodrich, L.M., 1947. From league of nations to United nations. Int. Organ. 1 (1), 3–21.

Hartmannshenn, T., 2004. FOOD SECURITY: Guidelines for the Promotion and Execution of Food Security Projects by German Agro Action. German Agro Action, Bonn.

IAASTD, 2010. Agriculture and Development: A Summary of the International Assessment on Agricultural Science and Technology for Development. IAASTD Synthesis Report. S. A. T. F. D. I. International Assessment of Agricultural Knowledge. GreenFacts, Johannesburg, South Africa, p. 6.

Johns, C.H.W., 1904. Library of Ancient Inscriptions: Babylonian and Assyrian Laws, Contracts and Letters. Charles Scribner's Sons, New York.

Kennedy, G., Nantel, G., Shetty, P., 2004. Globalization of Food Systems in Developing Countries: Impact on Food Security and Nutrition: A Synthesis of Country Case Studies. FAO, Rome.

Locke, J., 1714. The Works of John Locke Esq: In Three Volumes. printed for John Churchill at the Black Swan, London. Paternoster.

McCalla, A.F., 2007. FAO in the Changing Global Landscape. University College Dublin, Dublin. UCD. ARE Working Papers, vol. Paper 07-006.

McMichael, P., 1994. The Global Restructuring of Agro-Food Systems. Cornell University Press, Ithaca, NY.

Meier, G.M., Stiglitz, J.E., 2002. Frontiers of Development Economics: the Future in Perspective. Oxford University Press, New York.

Mettrick, E.F., 1929. Population, poverty, and ethical competence. Int. J. Ethics 39 (4), 445–455.

Mikesell, R.F., 1968. The Economics of Foreign Aid. Aldine Publishing Company, Chicago.

Murphy, S., 2010. Trade and Food Reserves: What Role Does the WTO Play? Institute for Agriculture and Trade Policy, Minneapolis, Minnesota.

O'Niel, O., 2002. Second Reith Lecture. BBC Radio, London, p. 4, 10th April.

Orr, J.B., 1939. Diet and nutrition: nutrition problems, dietary requirements for health. Can. Med. Assoc. J. 41 (1), 78–80.

Orr, J.B., 1940. Diet and nutrition: nutrition problems, dietary requirements for health. Br. Med. J. 1 (4146).

Pardey, P.G., et al., 2006. Agricultural Research: A Growing Global Divide. D.C, IFPRI, Washington,.

Parfitt, J., et al., 2010. Food waste within food supply chains: quantification and potential for change to 2050. Phil. Trans. R. Soc. 365 (1554), 3065–3081.

Pleskovič, B., Stern, N., 2001. Annual World Bank Conference on Development Economics. World Bank, Washington D.C. Annual World Conference on Development Economics.

Riddell, R.C., 2007. Does Foreign Aid Really Work? Oxford University Press, Oxford.

Scheuerman, W., 2008. Globalization. The Stanford Encyclopedia of Philosophy. E. N. Zalta. Stanford University, Stanford, CA.

Sherlock, T., 1718. A Vindication of the Corporation and Test Acts: in Answer to the Bishop of Bangor's Reasons for the Repeal of Them. To Which is Added: A Second Part, Concerning the Religion of Oaths. London Printed for J Pemberton at the Buck and Sun.

SOFI, 2009. The State of Food Insecurity in the World 2009. Food and Agriculture Organisation, Rome, p. 59.

UN, 1948. Universal Declaration of Human Rights. U.N., United Nations. A/RES/217 A (III).

UN, 2004. Voluntary Guidelines to Support the Progressive Realization of the Right to Adequate Food in the Context of National Food Security: Adopted by the 127th Session of the FAO Council November 2004. Food and Agriculture Organisation, Rome.

Vaidya, A.K. (Ed.), 2006. Globalization: Encyclopedia of Trade, Labor, and Politics. ABC-CLIO, Santa Barbara.

Walker, P., 2008. The Origins, Development and Future of the International Humanitarian System: Containment, Compassion and Crusades. ISA's 49th Annual Convention, Bridging Multiple Divides. International Studies Association (ISA), San Fransisco, CA.

Warnock, K., 1997. Third world: feast or famine? Food security in the new millennium (commentary). J. Race Class 38 (3), 63–72.

Windfuhr, M., Jonsén, J., 2005. Food Sovereignty: Towards Democracy in Localized Food Systems. Warwickshire, ITDG/FIAN, p. 70.

World Food Summit, 1996. World Food Summit Plan of Action, paragraph 1. In: Rome Declaration on World Food Security and World Food Summit Plan of Action. World Food Summit, Rome, Italy, pp. 13–17.

CHAPTER

7

Agriculture, forestry, and fisheries

OUTLINE

Food or more specifically agriculture in this sense is also responsible for a large proportion of global employment.

7.1 The agricultural landscape: a global structure

By 2010, there were approximately 525-570 million farms worldwide (Lowder et al., 2016). Collectively, these farms employed over 1 billion people, equivalent to just over one-third (35%) of the global employable population. Furthermore, on top of this, another 2.6 billion men, women, and children directly relied on such agricultural systems for their livelihoods (GHI, 2010).

Food and Society
https://doi.org/10.1016/B978-0-12-811808-5.00007-6

However, these figures can be misleading, for on the face of it these numbers might suggest a healthy and thriving global agricultural sector, yet the actual underlying trend is one of consolidation and industrialization. And this is having a downward impact on the employment numbers and a concomitant increase in the number of larger farm sizes. Moreover, the debate is not clear-cut; for instance, mechanization allows fewer people to achieve greater production which is not necessarily a bad thing in and of itself. However, on the flipside, this industrialization often results in benefits from economies of scale and labor-saving costs, which often combine to undermine their smallholder competitors.

What we end up with is a global agricultural production system on a two-track footing: the global and local. The global track can be considered as comprising an agricultural market driven and catered for on an industrial scale; the local track on the other hand can be considered as comprising predominantly small-scale or subsistence farming catering to the individuals or limited local markets. This is perhaps an unforgivable generalization, yet it suffices to highlight the inherent dichotomy within the system.

7.1.1 A global track

In terms of the local track, of the previously mentioned 525 million farms, there are between 400 and 450 million strong smallholder farms comprising 2 hectares or less. These collectively employ about 1 billion people, approximately one-third (35%) of the entire global population. Furthermore, the 525 million farm lands are equivalent to about 60% of worldwide arable land and provide the bulk of many developing countries food requirements, or from another perspective about half the world's total food supplies (Hartmannshenn, 2004; Båge, 2008; IAASTD, 2009). The global track is increasingly characterized by larger farm sizes and the ever-growing mechanization and industrialization of the industry. While in the developed world this trend has been ongoing for the last 60 years plus, it was the advent of the Uruguay GATT trade rounds in 1992, with the inclusion of agriculture in the barrier reducing targets that really defined the beginning of the complex global food chain that is familiar to us today. In fact, in the last 20 years or so, global agricultural trade has increased more than threefold, although having said that this still only represents a small part of total overall global production. This industrialization of the agricultural product has its focus on both the domestic and international markets. This is increasingly reflected in the intensification of farming and the overall increase in farm sizes particularly in the industrialized world (Table 7.1). As a result, today's complex global food system remains

TABLE 7.1 Approximate farm size by region.

Region	Average farm size (ha)
Africa	1.6
Asia	1.6
Latin America and Caribbean	67.0
Western Europe	27.0
North America	121.0

Source: Compiled from UNCTAD report, Technology and Innovation for Sustainable Agriculture. (Geneva, Switzerland, UNCTAD: Commission on Science and Technology for Development, 2010).

populated with larger farm sizes, which are increasingly owned by or under the management of multinational corporations who continue to combine both sovereignty and leverage over unwitting end users such as consumers and agricultural producers alike.

7.1.2 The local track

The size of the small-scale farm holdings identified previously, the local track comprises numerous other pastoralists or herders; hunters and gatherers; forest dwellers; fisherfolk; gardeners; and the landless and other rural workers. While the benefits of local and smallholder farming have seen a renaissance of late, there is, however, one obvious disadvantage. That having such a large proportion of a country's domestic agricultural production supply in the hands of such small-scale farmers and the like, then this predominantly, subsistence-oriented approach becomes extremely vulnerable particularly from direct drivers of change such as climate change, pests, disease, natural disasters and economic instability, etc.

Moreover, particularly within developing countries with little political or economic power, the small farmers have not readily benefited from the globalization and increased traffic within the sector. Barriers such as international protectionism coupled with the high capital entry deterrents exclude any real democratic participation of smallholders. Sadly, changes are not on the near horizon either, not until that is, trade reforms aiming to reduce protectionism in the form of subsidies and import tariffs are lowered or removed completely, will a more level playing field be created. In fact, until the negotiating position of small-holders is strengthened in the developing countries, the status quo is likely to remain.

So, it can be seen that agriculture can be viewed as a critical point of economic growth for the poorest countries; however, their exclusion from the global platform ensures the exclusion of many, potentially strong, players. Crucially, there are several economic and agricultural development paradigms with as many believers in each camp. Thus, back-and-forth posturing does little to progress the situation for all parties.

7.2 Increased food production?

By 2050, the population of the Earth is anticipated to tip over the 9 billion mark. Between now, then, and beyond, agriculture will be required to not only provide food, feed fodder, and fiber to meet our existing populations' needs but also over and above that it would have to meet the needs of the projected increase. This equates to an increase of 0.35% year-on-year food production if one is to generate the 70% extra food essential to meet the expected population numbers. Achieving this though is only a small part of the challenge, the problem or challenge is one of balance. Are we collectively able to produce this amount of extra food without seriously affecting the environment or unduly raising the price of food remains to be determined (GHI, 2010). This challenge also comes during a period which sees agriculture sector continue to decline in popularity. In the United States and Western Europe, for example, the past two generations or so have collectively seen their share of global agricultural output fall to around 1%–2% of GDP. This is in comparison to fewer developing countries whereby the agricultural sector still represents (albeit declining) an average of 20%–40% of overall GDP and which continues to employ 60% or so of the labor force.

This is a very real challenge, food chain waste savings, economies of scale, and eating lower on the food chain as well as other initiatives aside, one can legitimately ask where is this 70% increase in production supposed to come from? Ignoring for now any off-farm food generation, the core challenge of increasing the quantity of food must come about one of the two ways; the first is increasing yields and the second is increasing land under production (possibly from the regeneration of depleted land). Yet the Earth's budgeted land is a finite resource. Furthermore, when it comes to yields, there is also a natural limit which is proportional to the quantity and even the quality of any agricultural inputs. Once again, the outcomes are financially driven whereby economic trade-off is one of the inputs versus outputs. This creates what is known as a yield gap, i.e., the difference pertaining to a crop's maximum potential yield and to what is actually grown. This in turn depends to a large extent on the prevailing economic or policy environment of the day. Consequently, the future needs of the next generations rely to a large extent on conducive political and economic agendas (DFID, 2004).

These agendas might promote policy changes including the direction, for instance, of agricultural development paradigms with possible changes inter alia market-oriented over subsistence farming, smallholdings versus large farms, and self-sufficiency among others. However, these policy changes need to be considered carefully. Improving the lot of smallholder farms might increase rural employment, productivity, which on a national level improves overall agricultural output and which in turn can effectively lower food commodity prices. Having said that, an all-too-familiar response from farms in these circumstances is to increase production, further depressing prices creating a seesaw effect of over and under production (Zhang et al., 2007). There are indeed ways to account for these kinds of ups and downs and it becomes necessary to find a suitable balanced solution or trade-off to maximize potential food output while satisfying as many stakeholders as possible.

Such considerations are predicated on sustainable agricultural models. More on this is discussed in the next section.

7.3 Modern agricultural practices

Boserup (1965) in his study specifically targeted lesser developed nations in which he identified five unique agricultural systems based on the time and length of fallow between periods of productivity. These included (1) multicropping where no fallow was given, (2) annual cropping where fallow only lasted a few short months, (3) short fallow, which allow periods of fallow to last between 1 and 2 years, (4) bush fallow, which lasted a little longer at between 6 and 10 years, and (5) slash and burn forest fallow lasting between 15 and 20 years (Boserup, 1965).

This is neither a definitive list nor does it take into account the various other farming practices such as fisheries and livestock production. Interestingly too, while all of the above methods are still practiced to some extent in various parts of the world, today there is a growing propensity (driven by need and profit) for shorter or no-fallow periods of agriculture. Of course, this trend aside, the underlying consideration over which farming method is chosen still remains to a large extent dependent on available resources and constraints such as government policy, climate change, and numerous other economic and social or political pressures. The essential thing to take from this is that different practices have different impacts vis-à-vis soil quality, sustainability, and ultimately future productivity. Take for instance fallow-based farming—this method allows sufficient time between cropping to allow the soil to naturally replenish its stock of nutrients. The trend for shorter or no-fallow period farming asks much more of the land and requires the land to be micromanaged. Nutrients, given time to naturally replenish in the fallow farming process, now, had to be added manually. This intensification also requires that every aspect or every process is now considered at the micromanagement level; these include biodiversity, land degradation, irrigation, pest control, fertilizer, and even agricultural effluent. Farming at this level is as much a science as it is an art, and getting the right balance between productivity and sustainability is as much a political hot potato as it is a way of life.

In summary, it can be seen that while both monocultures[1] and polycultures[2] (multiple cropping and intercropping) are in common use, monocropping has undoubtedly become the dominant form of agriculture these days.

Some of these will be looked at in the following text.

7.3.1 Slash and burn

Slash and burn, or more commonly shifting cultivation, is a term used to describe land that is temporarily cultivated and then left to fallow for certain periods of time. If the slash and burn practice takes place in the forest or woodland, then it usually involves the burning of the shrub land or trees releasing much needed nutrients back into the soil. After cultivating the prepared nutrient-rich land crops are resown until such time as yields are no longer productive. The area is then abandoned or left to fallow while surrounding trees and shrubland as well as the forests natural flora can regrow and become fertile again. A process which could take years to achieve, although this is not necessarily a bad thing as the land, is still used for firewood, fetching, timber, berries, and other cultivatables, after which the process is repeated again and again. Slash and burn farming requires large amounts of land at any one point, most of which are out of use. In terms of sustainability of the slash and burn method, the skill lies in understanding how long to cultivate and how long to fallow each piece of land: the aim of course is to not allow the net nutrient loss or degradation of a particular parcel of land. Other questions of sustainability are determined by many factors such as time, availability of land, the intensity of cultivation, and the cultural appropriateness of the slash and burn practice (Brush, 1987).

7.3.2 Annual, monocultivation, and polycultivation

Annual cultivation drastically reduces periods of fallow or time when the land is allowed to rest between cropping seasons. This method of farming is used a lot in the intensification of

[1]**Monoculture** – is the practice of growing a single crop over a wide area. It is widely used in modern agriculture and its characterized by its large harvests with minimal labor UOR (2009). "University of Reading: Agriculture, Policy and Development, History of Agriculture." Retrieved 4 June 2019, from http://www.ecifm.rdg.ac.uk/history.htm.

[2]**Polyculture** includes multi-cropping - the practice of planting several crops sequentially in any 1 year is known as multiple cropping; and intercropping - is when several crops are grown at the same time on the same land.

modern agriculture. Consequently, the land has to be managed properly to ensure that nutrients are not leached from the soil and that pests are kept under control. This usually means greater financial cost and issues of sustainability, biodiversity, and soil degradation (Boserup, 1965).

It's difficult to say what came first, whether industrial advances or agricultural paradigms; however, one of the other led to the practices known as mono- and polyculture. Although both are in use today, monocropping is without doubt the most predominant type of agricultural practice today.

Monoculture is the growing of a single crop. It is in wide use around the world and is considered the modern method of agricultural cropping. It is characterized by the minimal labor input and large agricultural output (UOR, 2009).

Polyculture is the planting of multiple crops consecutively in a single year called as multiple cropping; on the other hand, intercropping is the term used when several crops are sown together within the same area.

Inherent advantages exist for each method whether monocultures or polycultures practice. Monoculture, for instance, largely benefits from lack of competition for certain nutrients; furthermore, advantageous cost savings can be made to large-scale industrial output. Lastly, although it receives a lot of bad press, monoculture is not by itself a bad agricultural practice; in other words, it is not inherently unsound when it comes to the environment. More and more studies show that the problems attributed to monocultural practices are due in large part to poor management and then to the practice itself. Polycultures tend to benefit from year-round production and provide diversity for the ecosystem, as well as reach less nutrients from the ecosystem.

7.3.3 Pastoralism or pastoral farming (animal husbandry)

Pastoralism, or animal husbandry, is that part of agriculture that deals with animal livestock such as goats, chickens, yaks, camels, sheep, and bovine, etc. Not only are they great sources of proteinaceous meat, but also many provide milk, eggs, leather, and fiber too. There are several types of pastoralism—the first is nomadic whereby humans move along with their herds in search of grasslands to grade; then there are the herders who migrate seasonally also in search of pastures new; and lastly there is the branch of pastoralism called transhumance, which is similar to the herders in which they move seasonally between higher and lower pastures. Being an important part of the agricultural sector, pastoralism faces challenges and is especially vulnerable to natural and man-made risks/disasters. One such risk involves the decreasing availability of pastures. Moreover, apart from inclement weather, other challenges encountered in pastoral communities may include competition for land in such things as expansion of sedentary agriculture or the expansion of other agricultural projects like the conversion of lands to things such as animal sanctuaries, game reserves, and wildlife parks, thus creating competition for valuable natural resources. One aspect of which type of pastoralism is considered over another depends partly by the climatic conditions. Tropical areas, for instance, allow for all types of pastoral practices, while subtropical and arid environs are limiting because of sunshine, rainfall, and temperature.

7.3.4 Permaculture

Permaculture is a relative newcomer to the more traditional farming models. It is a holistic approach to farming that attempts to mimic relationships found in nature. By mimicking nature, permaculture establishes relationships and aims to sit side by side with the natural ecosystem. It embodies a set of goals and values that is almost a way of life rather than a way of agriculture. In these values, permaculture espouses an environmentally beneficial paradigm of education, population numbers, and a light touch when it comes to the utilization of Earth's natural resources.

Some of the more central principles laid out in this method includes the following:

- interacting with nature for the more efficient use of resources;
- utilizing better the abundant natural energy provided by the sun, wind, and water;
- encouraging self-reliance through the incorporation of marginal lands, urban gardening such as window boxes, and even replacing ornamental gardens with dual-purpose multifunctioning esthetic and edible plants;
- producing little or no waste by connecting input and output elements so that they meet each other's needs.

All in all permaculture aims to be fruitful by integrating agricultural systems within the ecosystem rather than outside of it (Permaculture Ass, 2011).

7.3.5 Agroforestry

In terms of agriculture, forests provide both produce and employment either through wood and wood products or specialist food commodities. Agroforestry or forest farming is the practice of combining trees and shrubs with crops and/or livestock. These can either be integrated in a spatial sense or temporally (over time) but it is not uncommon to find agroforestry specializing in high-value crops. By mixing forestry and agriculture—in this way biodiversity and ecological systems remain healthy and varied as well as being an added environmental benefit which also allows for valuable water retention in the soil while further protection is offered against wind and rain erosion. In general, the successful practice of agroforestry relies on knowledgeable and careful selection of trees and crops to optimize production and minimize negative competitive effects. Several categories of crops are particularly suited to this type of farming, including mushrooms; nuts; vegetables such as radish and beetroot; honey; herbs; fruits such as blueberries; elderberries; blackberries; raspberries, and so on; edible flowers; and sap products such as maple syrup, birch sap wine, etc.

7.4 Fisheries and aquaculture

Fish is a highly nutritious animal protein that contributes significantly to the diet of around 1 billion people worldwide—in fact, it has been suggested that globally as much as 15%–20% of all animal proteins. It is also understood that much of the world's oceans, rivers, and lakes are overwhelmingly fished by small-scale artisanal fishers contributing vital food sources to the overall dietary makeup of many poor localized communities. Not surprising then that in

1996 fish and fisheries accounted for the income of an estimated 30 million people worldwide of which 95% reside in the developing countries (FAO, 2011).

Of note too is the growing tendency for traditional small fisheries to divert traditional foods away from the local community, instead choosing to supply the international market in exchange for valuable incomes and alternative food sources. This comes about largely through the growing demand of industrialized nations. This is coupled with advances in processing, packing, and transporting perishable goods over the last two to three decades. While this decreases the available fish for local consumption, there are those that argue that the increased processing and fisheries employment opportunities create more jobs and income. Moreover, some argue that by selling premium fish products on the international market, cheaper imported food substitutes effectively create a net inflow of foreign earnings. With regard to traditional capture fishing, it can also be seen that the industry is receiving competition from a growing aquaculture[3] (or aquafarming) industry. This according to the Food and Agriculture Organization (FAO) is the fastest growing food-producing sector now accounting for nearly 50% of the world's food fish (FAO, 2011). Systems like these and the more efficient processing entailing less spoilage and wastage together with a reduction of discards at sea have the potential to considerably increase the productivity of the sector. However, poor fishery resource management has been seen to have an adverse impact on the aquatic environment and resources, which in turn create enormous pressures in terms of wastage and the sustainability of the sector (Kurien, 2004).

Another important characteristic of the modern agricultural landscape is the industrialization or more specifically the intensification, extensification, concentration, and specialization of agriculture.

7.4.1 The green and evergreen revolutions

When people talk of the green revolution, they generally focus around the 1960s, as new technologies became widely available and were taken by many farmers across the world. However, in point of fact, foundations for the green revolution were actually sown long before when agricultural yields really began to take off. This so happened around the agricultural industrial and chemical revolutions of the 18th century. New advances in science and technology are all about the industrialization of the agricultural sector. However, this was fairly short-lived as yields once again began to stagnate around the 1850s (Johnson, 1997, 2000; Borlaug, 2000; McCalla and Revoredo, 2001; Webb et al., 2008). However, all that changed once again from the 1940s onward, which is characterized by scientific and technical progress in terms of genetic breeding, creating hybrid crops and the industrialization and increasing mechanization of agriculture (Brewster, 1945; Barton and Cooper, 1948; Borlaug, 2000; Gardner, 2002; UOR, 2009). Of course, as mentioned previously, much of the progress of increasing food yields and the mechanization of agriculture came about toward the end of the war as the allies needed to maximize food production and do so quickly. Such urgency necessitated the multilateral pooling of resources through technical assistance programs led by the likes of the FAO, the United States, and also through the dissemination of information

[3]The cultivation of aquatic organisms whether fresh or saltwater under controlled or farm conditions.

by the Cooperative Mexican Government-Rockefeller Foundation agricultural program of 1943.

Such resources were developed in no small measure by Nobel Laureate Norman Borlaug and others which brought about hybrid crop varieties, crops more resistant to disease and more pest-resistant plants and crops. On top of this, the now widely available synthetic fertilizers brought about increasing yields per hectare (Sharma and Gill, 1983; Perkins, 1990; Herdt, 1998; Borlaug, 2000; Troyer, 2004; Wu and Butz, 2004; USAID, 2009). The problem with the notion of the green revolution was that these technologies and advances in science were not particularly widely used or disseminated outside of the industrial nations. Instead this happened around the 1960s and hence the reason why people believed that the green revolution began about this time. Indeed, the phrase "green revolution" was kneeling coined until 1968 in an address to the Society for International Development by the then United States Administrator of the Agency for International Development (USAID) William S. Gaud, who offered

> These and other developments in the field of agriculture contain the makings of a new revolution. It is not a Violet Red Revolution like that of the Soviets, nor is it a White Revolution like that of the Shah of Iran. I call it the Green Revolution. *Gaud (1968).*

By the 1980s, the so-called green revolution had been adopted by many developing countries excluding sub-Saharan Africa. The results were tangible with large 70% plus crop yield averages between 1961 and 1999. This led to the conclusion by many that research and development would play a pivotal role in securing future global needs (Nellemann, 2009).

Nowadays, however, there is much blame aimed at the green revolution especially in connection with the industrialized nations. This is due in no small part to criticisms about the heavy use of fossil fuels, inorganic fertilizers, and the environmental consequences of both the scientific and industrialization practices in general (Muir, 2010). One of the more frequently repeated issues in modern agriculture is the use of the inorganic or mineral fertilizers as they are sometimes called. As both can be made from natural and synthetic sources in recent years, close to 97% is now regularly produced from synthetically created ammonia. The major problem is to do with the huge amount of energy required to produce this ammonia and some have suggested that the process consumes up to on average 5% of global natural gas supplies, which equates to just over 1.5% of the entire global energy production (McLaughlin et al., 2000; GEO, 2007). To place this in perspective to produce one ton of nitrogen (chemically fixed) in the form of it takes around 30 to 40,000 ft^3 of feedstock in the form of natural gas and a further one ton of oil to drive the process. Furthermore, to produce this one ton it produces on average 1.8 tons of CO_2 (Sundquist and Broecker, 1985).

This is not the end—every year between 30% and 40% of global food production is damaged due to birds, mammals, insects, viruses, bacteria, weeds, and fungus. Indeed, insects alone are responsible for 14% of such losses. Therefore, the need for pesticides cannot be underestimated. This is not a problem it in and of itself; however, the term pesticides incorporate insecticides, herbicides, fungicides affecting any chemical, biological, or physical agent that has the ability to destroy or control such pests (Muir, 2010). While we have recently learned a lot about the use of such agents, first- and second-generation compounds including dichlorodiphenyltrichloroethane, cyanide, and arsenic were found to be highly toxic.

By contrast, modern-day pesticides are rigorously tested for potential damage to the soil, the crops, and the potential for toxicity in humans through excessive use, so lessons have been learned yet the fear among many doubters remains (Muir, 2010).

The second important charge delivered at the feet of the green revolution is those of environmental consequences. For instance, unsuitable use of fertilizers, pesticides, and any number of persistent organic pollutants has been linked to the long-term degradation of the soil itself. Most damage comes from poor farming practices where buildup of excessive nitrate, heavy fertilizer run-off, eutrophication, salinization, and the chemical "burning" of the soil affecting much needed trace elements and microbial species has taken place (Eifert et al., 2002). People are also aware and offer considerable concern about how the green revolution affected the biodiversity of the plants and crops we grow. Many crop hybrids have been harvested over the years that it has resulted in some cases to 90% reduction in crop varieties grown. Such reliance on so fewer crops brings with it its own set of problems to deal with increased vulnerability, fragility, and effect of unknown influences, not to mention a loss of dietary diversity and related nutritional concerns (Darmawan et al., 2006).

Finally, the green revolution was party to the huge structural changes within the industry. Essentially, farming now encouraged larger, more industrial-focused farms at the cost of their smallholder brethren. The larger farms meant ever-increasing agricultural inputs due to continual depletion of base soil fertility. This resulted, according to Pimentel et al. (1998), in a cost–benefit trade-off between productivity and the local ecosystems so as to avoid damaging our future generation's environmental habitat. This message took on increasing potency as the world observed a not so slow and steady decline in the quality of local ecosystems. This view quickly became mainstream affecting policies of energy, environmental protection, and agriculture, etc.

7.4.2 Evergreen revolution

There is no doubting the effect the green revolution had on the world and on global production. Indeed, it can be seen that over a comparatively short period of time, 8% growth came from increasing cultivatable land, whereas a whopping 92% increase was a direct result of the green revolution technologies. Furthermore, despite a continuing shocking number of people who remain hungry and malnourished, there are still more people in the world, to date, who have achieved increased nutrition at less cost than any other time in history. Increased yields further developed peoples and countries' economies, population levels, and a further distribution of equity. As an aside, the specter of the Malthusian prophecy was indeed kept at bay (Brown, 1981).

Moreover, the question of sustainability in the face of intensive agricultural productivity cannot be brushed aside although the yield gap is decreasing, and as the revolution continues, more and more people are looking for another more fully sustainable "evergreen" revolution devoid of expensive monetary and environmentally costly practices. More favorably, one can see that research spending governed by a more knowledgeable political and social awareness has emerged fully supportive of the postgreen or evergreen technologies. By fully maintaining a balance of equilibrium, modern-day evolution within the field is sympathetically maintained with the resultant sustainable, natural resource base.

Indeed, more focus on the environment, from a holistic perspective, ensures the overall efficiency of inputs while effectively controlling for pests through a sweeping array of complimentary practices. Through the introduction of conservation tillage and other environmentally friendly practices of drip irrigation and multiple cropping practices among others, it is hoped the adoption of a new agricultural, economic, and social paradigm is a snug fit for the modern ideology of developmental and economic and social growth (GHI, 2010).

7.5 The intensification, concentration, and specialization of agriculture

By the middle of the 20th century, fears over population growth spurred a dramatic acceleration in global food production. This has been achieved through improved technologies including agrochemicals, mechanization, and plant breeding, "progress," which has ultimately been responsible for the intensification, extensification, concentration, and specialization of agriculture.

In intensive agricultural systems, large amounts of capital relative to the area of land cultivated as well as large quantities of fertilizer, insecticides, fungicides, and herbicides are employed to improve crop yields and manage pests. Furthermore, intensification also sees farming becoming increasingly mechanized with capital intensive high-efficiency machinery being used for planting, irrigation, cultivating, and harvesting. This kind of agriculture is often employed where population densities are high and/or where land values are also high; such situations also often occur near primary markets.

Unlike the large inputs of intensive agriculture, extensive agricultural practices rely heavily on the natural fertility and productivity of the soil. As this type of agriculture usually provides lower yields, large tracts of land are needed. This type of farming can usually be seen in areas of low population density where land is plentiful and cheap in relation to labor and capital. It also means this sort of practice is oftentimes far from primary markets (Britannica, 2009; UOR, 2009).

This above gives a brief snapshot of the many styles of farming and whether working the land, with livestock, farming the seas or the forests, each has its benefits and each faces a myriad of challenges. All are important too when it comes to the challenges of food waste. One concept closely associated with the current drive to reduce food wastage is that of sustainability; it is a concept that runs parallel to both the idea of efficiency within the food supply chain and the good stewardship required to successfully manage the natural resource base. As with food wastage, sustainability seeks to ensure the ongoing viability of a supply stream hitherto seen as limitless.

7.6 The impact of food consumption on the agroecological resource base

Presently, global agriculture production is already an extremely resource-intensive sector. On top of this, pressure on worldwide resources is being further compounded by unrelenting demand from growing populations with ever greater food production needs (UNEP, 2012).

Currently, global food production sector utilizes about one quarter of all habitable land on Earth and uses more than 70% of total freshwater consumption. As a result, the world's food production sector has become the largest single cause of biodiversity loss in the world; it produces over 30% of global greenhouse gas emissions; it is responsible for as much as 80% of total annual deforestation contributing to its roles as the single greatest cause of land-use change throughout the world. The following gives a small glimpse of the effects our food production sector is having on our planet's land and water resources.

Only one quarter of the Earth's land surface is suitable for cultivation and within these confines' cattle occupy about 25% of this land with feed required to sustain these and other livestock requiring another 25%, also recalling that livestock is an inefficient use of resources. In US style feedlots, animals are reared intensively and fed almost exclusively on grain. Furthermore, as a result of this, the conversion factor of meat versus wheat varies from animal to animal and from the intensity and frequency with which they are fed, as well as efficiency in which various animals convert grain into protein. All this varies widely. And by way of example, we take approximately 7 kg of grain (once again in American style feedlots) to produce 1 kg of live weight beef. In other animals, pork, for instance, it takes 4 kg to make just 1 kg of live weight pork. And for poultry the ratio is 2:1, and lastly for farmed fish it takes less than 2 kg of wheat to produce approximately 1 kg of fish protein (Worstall, 2012).

With this in mind it comes as no surprise that 35% of the total global grain harvest is ultimately used in the production of animal protein. Indeed, in many regions around the world, the majority of new agricultural land brought into play is employed simply for this purpose of supplying the growing market for animal protein, whether directly in the production of animals or for animal feed. This is perhaps not more evident than in the high-profile deforestation patterns as seen in Brazil's national treasure—the Amazon. In 2007, for instance, ranches in the Amazon covered about 8.4 million ha[4], which, according to Nellemann, is suggestive of simple market responses to global demand (Nellemann, 2009). Furthermore, if consumption trends continue as predicted, animal protein consumption is set to increase by a factor of 4 by 2050, placing further pressure on existing land resources with this in mind (UNEP, 2012).

Available land is only one part of the equation; maintaining adequate food production, whether land- or water-based, relies on well-functioning ecosystems, which in turn rely on healthy and sustainable practices with the long term in mind. Unfortunately, continued misuse and abuse of the land has led to widespread degradation the full scale of which is only recently coming to light.

By way of example, intensive agricultural practices during the second half of the 20th century directly led to the degradation of 2 billion hectares of arable land (about a quarter of all arable land). And it does not stop there, it is estimated a further 2–5 million hectares are added to this figure annually (Nellemann, 2009). Overall, this has lead to very real losses; according to the International Food Policy Research Institute (IFPRI), such degraded lands lead to nonoptimized yields amounting to the loss of about 20 million tons of grain or 1% of global annual grain production each year (IFPRI, 2012).

In terms of freshwater resources, agriculture, as mentioned, utilizes about 70% of global annual supply. Continued agricultural pollution from intensification is threatening the

[4]In Brazil in 2007 ranches alone accounted for an estimated 70% of deforestation in that year.

current sustainability of the system, resulting in unnecessary acidification and dead zones within our river and aquifer ecosystems, etc. Apart from implications of land-based production, this also comes at a time when fish consumption is at an all-time high with fish contributing to 15% of animal protein intake for over 3 billion people (or equivalent to about 17 kg per global capita) (UNEP, 2012). Yet amid these figures, the industry is in perilous condition as approximately 75% of the world's major marine fish stocks are either depleted or overexploited.

Furthermore, while aquaculture supplies a good proportion of global fish protein needs—growing in excess of 60% between 2000 and 2008 (from 32.4 million tons to 52.5 million tons), it is not always the panacea that it is sometimes portrayed. For example, each 1 kg of farmed tuna requires 20 kg of wild-caught fish feed during the course of its life.

References

Båge, L., July 7, 2008. G8 Summit Special Report: Supporting Smallholders is Crucial to Food Security. Financial Times, London.

Barton, G.T., Cooper, M.R., 1948. Relation of agricultural production to inputs. Rev. Econ. Stat. 30 (2), 117–126.

Borlaug, N.E., 2000. The green revolution revisited and the road ahead. In: 30th Anniversary. Lecture. The Norwegian Nobel Institute, Oslo.

Boserup, E., 1965. The Conditions of Agricultural Growth: The Economics of Agrarian Change under Population Pressure. G. Allen and Unwin, London.

Brewster, J.M., 1945. Farm technological advance and total population growth. J. Farm Econ. 27 (3), 509–525.

Britannica, 2009. Encyclopædia Britannica Online.

Brown, L.R., 1981. World population growth, soil erosion, and food security. Science 214, 995–1002.

Brush, S.B., 1987. Comparative Farming Systems. Guilford Press, New York.

Darmawan, et al., 2006. Effect of green revolution technology from 1970 to 2003 on sawah soil properties in Java, Indonesia: I. Carbon and nitrogen distribution under different land management and soil types. Soil Sci. Plant Nutr. 52 (5), 634–644.

DFID, 2004. Agriculture, Hunger and Food Security. Department for International Development (DFID), UK.

Dyer, G., 2006. Feeding the World: How long can the world feed itself? Retrieved 6 July 2019. http://gwynnedyer.com/2006/feeding-the-world/.

Eifert, B., et al., 2002. The World Grain Economy to 2050: A Dynamic General Equilibrium, Two Sector Approach to Long-Term World-Level Macroeconomic Forecasting. University Avenue Undergraduate Journal of Economics (Online).

FAO, 2011. Fisheries and Food Security. Retrieved 5 Jan 2019, from http://www.fao.org/FOCUS/E/fisheries/intro.htm.

Gardner, B.L., 2002. American Agriculture in the Twentieth Century: How it Flourished and What it Cost. Harvard University Press, Cambridge, Massachusetts.

Gaud, W.S., 1968. The Green Revolution: Accomplishments and Apprehensions. The Society for International Development, Shorehan Hotel, Washington, DC. US Agency for International Development.

GEO, 2007. Planet's Tougher Problems Persist, UN Report Warns. Global Environment Outlook 4.

GHI, 2010. GAP Report: The Global Harvest Initiative: Measuring Global Agricultural Productivity. The Global Harvest Initiative, Washington, D.C.

Hartmannshenn, T., 2004. Food Security: Guidelines for the Promotion and Execution of Food Security Projects by German Agro Action. German Agro Action, Bonn,.

Herdt, R.W., 1998. The Rockefeller Foundation: The Life and Work of Norman Borlaug, Nobel Laureate. The Rockefeller Foundation, New York.

IAASTD, 2009. IAASTD Global Report: Agriculture at a Crossroads. IAASTD Synthesis Report. Johannesburg, South Africa, International Assessment of Agricultural Knowledge. Science and Technology for Development (IAASTD), p. 6.

IFPRI, 2012. IFPRI books, International Food Policy Research Institute (IFPRI), number 2011 Global Food Policy report. Washington, D.C.

Johnson, D.G., 1997. Agriculture and the wealth of. Nations. Am. Econ. Rev. 87 (2), 1–12.

Johnson, D.G., 2000. Population, food, and knowledge. Am. Econ. Rev. 90 (1), 1–14.

Kurien, J., 2004. Responsible Fish Trade. Towards Understanding the Relationship Between International Fish Trade and Food Security. FAO and the Ministry of Foreign Affairs of Norway (MFA), Rome.

Lowder, S.K., Skoet, J., Raney, T., 2016. World Development: The Number, Size, and Distribution of Farms, Smallholder Farms, and Family Farms Worldwide, 87. FAO, Rome, Italy, pp. 16–29.

McCalla, A.F., Revoredo, C.L., 2001. Prospects for Global Food Security: A Critical Appraisal of Past Projections and Predictions. International Food Policy Research Institute, Washington, DC.

McLaughlin, A., et al., 2000. Comparison of energy inputs for inorganic fertilizer and manure based corn production. Can. Agric. Eng. 42 (1).

Muir, P., 2010. BI301: Human Impacts on Ecosystems. Oregon State University.

Nellemann, C., 2009. The Environmental Food Crisis: The Environment's Role in Averting Future Food Crises. United Nations Environment Programme (UNEP) GRID-Arendal, Geneva.

Perkins, J.H., 1990. The Rockefeller Foundation and the green revolution, 1941–1956. Agric. Hum. Val. 7 (3/4), 6–18.

Permaculture Ass, 2011. Website of the Permaculture Organisation. Retrieved 12th Feb 2019, from http://www.permaculture.org.uk/.

Pimentel, D., et al., 1998. Energy use in agriculture: an overview. Agric. Eng. Int. CIGR J. 1.

Sharma, H.C., Gill, B.S., 1983. Current status of wide hybridization in wheat. Euphytica 32 (1), 17–31.

Sundquist, E., Broecker, W., 1985. The Carbon Cycle and Atmospheric CO_2: Natural Variations Archean to Present. American Geophysical Union, Washington.

Troyer, A.F., 2004. Breeding widely adapted, popular maize hybrids. Euphytica 92 (1–2), 163–174.

UNEP, 2012. The Critical Role of Global Food Consumption Patterns in Achieving Sustainable Food Systems and Food for All. UNEP Discussion Paper, Nairobi, Kenya. United Nations Environment Programme: 43.

UOR, 2009. University of Reading: Agriculture, Policy and Development, History of Agriculture. Retrieved 4 June 2019, from http://www.ecifm.rdg.ac.uk/history.htm.

USAID, 2009. FrontLines: Borlaug, Father of Green Revolution, Dies. Retrieved 5th March 2019, from https://pdf.usaid.gov/pdf_docs/PNADQ606.pdf.

Webb, P., et al., 2008. More Food, but Not yet Enough: 20th Century Successes in Agriculture Growth and 21st Century Challenges. Food Policy and Applied Nutrition Program. School of Nutrition Science and Policy, Tufts University, Boston, Massachusetts.

Worstall, T., 2012. It Does Not Take 7 kg of Grain to Make 1 kg of Beef: Be Very Careful with Your Statistics. Forbes, Jersey City, USA, New York.

Wu, F., Butz, W.P., 2004. The Future of Genetically Modified Crops: Lessons from the Green Revolution. US, Rand Corporation.

Zhang, X., et al., 2007. A Typology for Vulnerability and Agriculture in Sub-saharan Africa. IFPRI, Washington, DC, 2010.

CHAPTER

8

Organic food and agriculture

John Paull

School of Land & Food, University of Tasmania, Australia

OUTLINE

Food and Society
https://doi.org/10.1016/B978-0-12-811808-5.00008-8

8.1 What is it?

The premise of organic agriculture is that we cannot poison our way to prosperity. Organic food is produced without the use of synthetic fertilizers and pesticides. The German supermarket chain, Aldi, describe organic food as: "grown as nature intended with no chemicals or additives, altogether a better way to eat" (Aldi, 2007).

As new intrusive chemical food technologies have developed so too has the definition of organic grown. Organic food production now excludes not only synthetic fertilizers and pesticides, but also food irradiation, genetically modified organisms (GMOs), and engineered nanoparticles.

The word "organic" has multiple meanings. The Oxford English Dictionary lists six meanings for "organic' as an adjective: "**1** relating to or derived from living matter"; "**2** not involving or produced with chemical fertilizers or other artificial chemicals"; "**3** Chemistry relating to or denoting compounds containing carbon and chiefly or ultimately of biological origin"; "**4** relating to or affecting a bodily organ or organs"; "**5** (of the elements of a whole) harmoniously related; "**6** characterized by natural development" (COED, 2007).

When the terms "organic agriculture," "organic farming," and "organic food" are used, it is the second meaning of "organic" that applies: "not involving or produced with chemical fertilizers or other artificial chemicals." This usage is a twentieth century development and nineteenth century dictionaries do not mention it. With this meaning it was first used in the 1940 book 'Look to the Land' by Lord Northbourne in which he wrote of the contest of "organic versus chemical farming" (1940).

Organic agriculture is now recognized internationally. The United Nations' Food and Agriculture Organization's definition of "organic agriculture" in *Codex Alimentarius* is "Organic agriculture is a holistic production management system which promotes and enhances agro-ecosystem health, including biodiversity, biological cycles and soil biological activity. It emphasizes the use of management practices in preference to the use of off-farm inputs, taking into account that regional conditions require locally adapted systems. This is accomplished by using, where possible, agronomic, biological and mechanical methods, as opposed to using synthetic materials, to fulfill any specific function within the system" (FAO, 2001, p.1).

8.1.1 Who cares?

Around the world, there are lots of consumers who care. A global survey of more than 23,000 consumers in 17 countries revealed that "organic" was important to many in making their food choices (GfK, 2017).

Respondents were asked "When deciding which food or beverage product to eat or drink, how important are the following in making your decision?" Globally, 44% of respondents rated "It is organic or made from organic ingredients" as "very important" or "extremely important." "Organic" was rated as an important selection criterion by 22% in the Netherlands and 24% in the United Kingdom and rising to 52% in Russia and 58% in China (Fig. 8.1) (GfK, 2017).

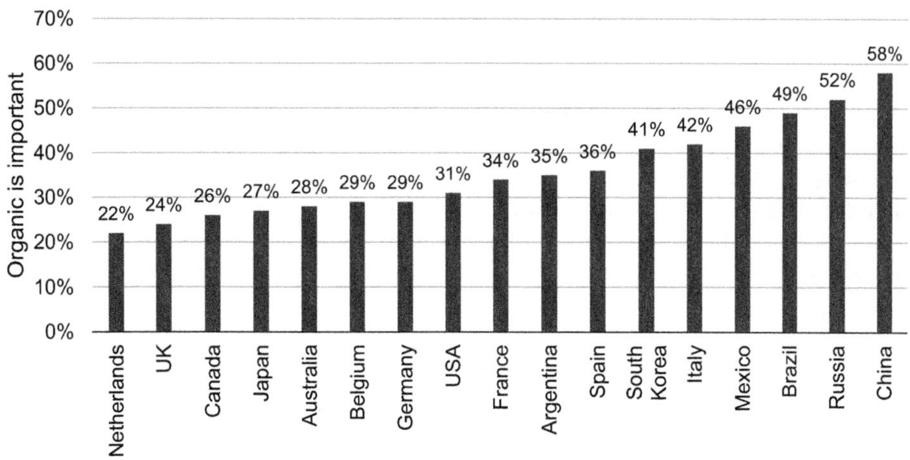

FIGURE 8.1 Percentage of consumers who rated "organic" as important for their food choices. *Source: John Paull.*

Ten decision factors were presented to consumers. For respondents in China, the two leading decision factors nominated were GM-free ("It is free from GMO (genetically modified) ingredients) and organic (Figs. 8.1 and 8.2). There were some variations and trends across demographic factors but also strong consistencies. Organic was equally important for men and women (45% and 44% respectively). Importance of organic peaked for the 30–39 age group (at 50%). The importance of organic ramped up from 41% to 50% with rising income bands.

Consumers purchase organic food for reasons of health, environment and animal welfare (Nielsen, 2005). A study of 21,261 consumers in 38 countries, reported that 72% of consumers purchase organic food, either "regularly" or "sometimes" (Nielsen, 2005). The main reasons consumers nominated for purchasing organic food were: "healthier for me" (51%); "healthier for my children" (17%); "better for the environment" (15%); "kinder to animals" (7%); and "other" (10%) (Nielsen, 2005).

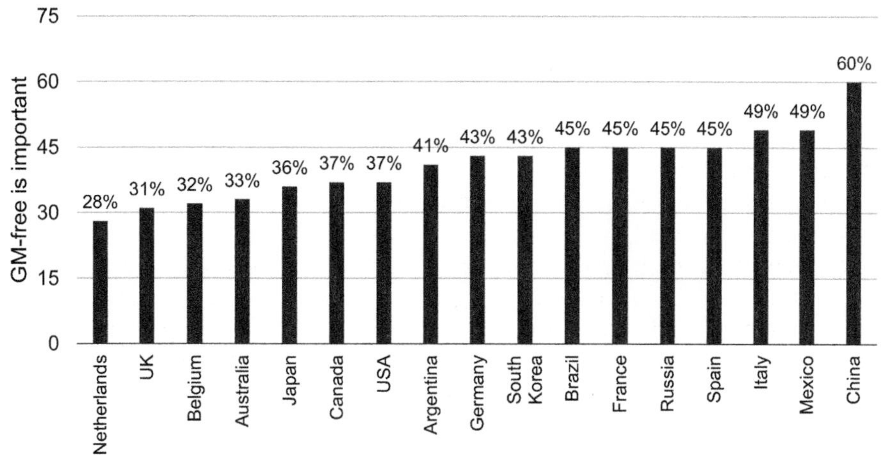

FIGURE 8.2 Percentage of consumers who rated "GM-free" as important for their food choices. *Source: John Paull.*

Farmers care for organics because (i) it is healthier for them and their family, (ii) it is better for the farm, it does not poison their soil, and (iii) they receive a premium price for their produce. Premiums for certified organic most typically range between 0% and 100%, but for a variety of reasons they may be outside this range. Price premiums depend on availability, seasonality, provenance and what the market will bear. The premium paid by the consumer may be quite different to the premium achieved by the producer who may be a price-taker rather than a price-maker. Some examples of price premiums are: 15% premium for bananas, 29% for raisins, 40% for apples, 106% for pears, and 139% for carrots (PANUK, 2017).

8.2 Organics by the numbers

Organics has been described as the fastest growing food sector (e.g., Brumby, 2007, p.1) The global organic sector has a market value of €75 billion (US$82 billion) with the United States as the largest single market, followed by Germany and France.

Farmers all over the world are growing organic produce (Fig. 8.3). One hundred and seventy-two countries report statistics of organic production. There are 51 million hectares of certified agricultural land globally, with Australia as the largest host of such land, followed by Argentina and United States. In addition, there are 40 million hectares of certified wild culture land, with Finland as the largest host of such land, followed by Zambia and India. There are 2.4 million organic producers, led by India and followed by Ethiopia and Mexico. Each year the numbers creep up a little or a lot; however, certified organic agriculture accounts for just 1.1% of the world's agricultural land so there is no cause for complacency.

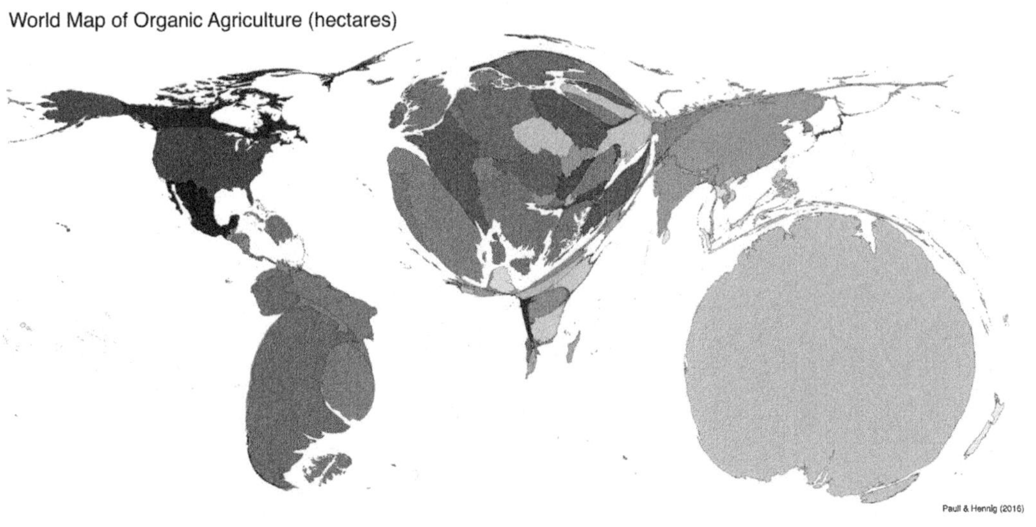

FIGURE 8.3 World map of organic agriculture: a density-equalizing map (a cartogram) with countries sized according to their reported organic agriculture hectares. *Source: John Paull.*

8.3 Organic pioneers

Organics advocacy has been the work and the passion of many. From the outset, the development of organics has been an international affair. There was US Professor Franklin King reporting favorably on the agriculture of China, Japan and Korea (in 1911). There was the New Age Austrian philosopher, Dr. Rudolf Steiner, lecturing, in what is now Poland, to an audience from six countries (Germany, Poland, Austria, Switzerland, France and Sweden) and calling for an agriculture differentiated from chemical agriculture (in 1924).

A German chemist, Dr. Ehrenfried Pfeiffer, was coordinating research in Switzerland and managing a farm in Holland (in the 1920 and 1930s). There was the Oxford lecturer in agriculture, Lord Northbourne, hosting a conference of continental advocates of biodynamics at his estate in Kent (in 1939). When WW2 broke out Northbourne published his manifesto of organic farming pitting organics in contest with chemical farming (in 1940).

There was US entrepreneur, Jerome Rodale, who took up the ideas and terminology of Northbourne and published his periodical "Organic Farming and Gardening" (in 1942). There were the Australians, including the pastoralist Harold White, who founded the world's first association devoted to the promulgation of organic agriculture (in 1944).

There was US biologist, Rachel Carson, who harnessed the work of two New York biodynamic farmers, along with advice from Ehrenfried Pfeiffer, to write "Silent Spring" giving worldwide impetus to the fledgling organics movement (in 1962). Later came Frenchman, Roland Chevriot, President of the French national farmer organization, Nature et Progres, who brought together delegates from France, South Africa, Sweden, United Kingdom, and United States, to form an international organics advocacy group (in 1972).

8.3.1 Franklin King, China, Japan, Korea and "permanent agriculture"

An American professor of agriculture and United States Department of Agriculture (USDA) chief, Franklin Hiram King (1848–1911) was disillusioned with the direction that US agriculture was taking. He took off, in 1909, for an extended trip to study the time-proven farming practices of Asia where soil fertility had been maintained over centuries.

King's account, "Farmers of 40 Centuries or Permanent Agriculture in China, Korea and Japan" was published posthumously (1911). King reported favorably of the agricultural practices that he witnessed, and which he termed collectively "permanent agriculture." His book was cited by the earliest advocates of organic agriculture in continental Europe, United Kingdom, United States, and Australia (Paull, 2011c).

8.3.2 Rudolf Steiner at Koberwitz, and anthroposophic farming

An Austrian mystic and New Age philosopher, Dr. Rudolf Steiner (1861–1925), delivered a 10-day course on agriculture to farmers at Koberwitz (now Kobierzyce in Poland) in 1924. Steiner taught that "the farm is an organism" and a decade and a half later this characterization spawned the term "organic farming." Steiner taught that synthetic fertilizers were unnecessary. He urged attendees to put his ideas to the test, and to extend by experiment, the ideas espoused in his course (Steiner, 1924).

During Steiner's agricultural course at Koberwitz, he founded the world's first organic agriculture research entity, the Experimental Circle of Anthroposophical Farmers and Gardeners (ECAFG). Anthroposophist's, followers of Rudolf Steiner, around the world including Europe, United Kingdom, United States, Australia, New Zealand, India, South Africa, took up the challenge and signed on to the ECAFG.

Just a few months after the Agriculture Course, Steiner retired from all public life due to ill health. It was said that he had been poisoned by adversaries. Steiner died in early 1925. He had never repeated the Agriculture Course and he stepped aside from the Experimental Circle at the outset. Over the next decade and a half, the work of the ECAFG led to the formulation of "biodynamic farming." The work of advancing Steiner's ideas of agriculture fell to the Experimental Circle coordinated in Dornach by Ehrenfried Pfeiffer.

8.3.3 Ehrenfried Pfeiffer in Switzerland, and biodynamic farming

Dr. Ehrenfried Pfeiffer (1899–1961) was the leading exponent of biodynamic agriculture until his death. Pfeiffer had studied under the tutelage of Steiner, although he did not attend the Koberwitz course. In the 1920 and 1930s, Pfeiffer was the director of the Bio-chemical Research Laboratory of the Natural Science Section at the Goetheanum in Dornach, Switzerland. This was at the headquarters of Steiner's Anthroposophical Society.

Pfeiffer developed a biodynamic farm in Holland to test, refine and progress the ideas of Steiner. Pfeiffer's experimental farm was also used to showcase these new ideas in agriculture. Visitors from Britain included Viscount Lymington (The Earl of Portsmouth; 1898–1984), the author of "Horn, Hoof and Corn: The Future of British Agriculture" (Lymington, 1932) and Sir Albert Howard (1873–1947), the co-author of "The Waste Products of Agriculture: Their Utilization as Humus."

Steiner set the Experimental Circle the task of putting his ideas from the 1924 Agriculture Course to the test. Then, when the results were in, to publish a coherent account of this no-synthetic-chemicals approach to agriculture. Pfeiffer published his book "Bio-Dynamic Farming and Gardening: Soil Fertility Renewal and Preservation" in 1938. The book appeared simultaneously in five languages: English, French, German, Dutch and Italian (Pfeiffer, 1938a,b,c,d,e). It was Pfeiffer's book that took the terminology "biodynamic" to the world and beyond Anthroposophic circles (Paull, 2011a,b). The vision of Steiner was that this was to be an agriculture for all farmers (Pfeiffer, 1956).

8.3.4 Lord Northbourne and his manifesto of organics "Look to the Land"

A British agriculturist, Lord Northbourne (birth name Walter James; 1896–1982), coined the term "organic farming." His manifesto of organic agriculture "Look to the Land" (1940) argued that: "We must remember that food of better quality is food which has vitality, individuality, freshness; food which is grown right, not only food that looks right; food which is effective as a vehicle of life and is not either mere stimulant or mere filling" (p.129).

Northbourne was an early advocate of localism and local production: "It is ludicrous to cart stuff about all over the world, so someone can make a 'profit' out of doing so, when that stuff could much better be produced where it is wanted" (p.104).

Northbourne spoke out for attention to biodiversity: "Large scale monoculture (the growing of one crop only) upsets the balance of factors in the soil in many ways. There is no give and take between crops. Disease spreads easily. Nature always provides a mixture of plants, and of animals; only so can living matter be kept constantly in circulation without wastage" (p.21).

"Look to the Land" introduced Northbourne's view of the contest of "organic versus chemical farming." He was making the strong case for organic agriculture but he recognized that the triumph of organic farming would be no quick or easy romp: "It is a task for generations of concentrated effort, slow and laborious, needing all available skill and resources ... A combination of cooperation and individual effort ... And those engaged will be fighting a rearguard action for many decades, perhaps for centuries" (p.115).

Northbourne's ideas were foundational in launching the worldwide organics movement and defining the terminology for the movement. He was a very shy man and his advocacy mostly was via his writings and his example. The book was a turning point in his own life. He was a graduate and lecturer in agriculture of the University of Oxford, a talented artist, a capable linguist, a keen sportsman, an Olympic silver medalist (in rowing), a lifelong farmer, he was profoundly spiritual, and an accomplished author (Paull, 2014).

Northbourne's interest in biodynamics led him to visit Switzerland in 1939 to visit the leading advocate of the times, Dr. Ehrenfried Pfeiffer. He invited Pfeiffer to present the first conference on biodynamic farming in Britain, the Betteshanger Conference at his estate in Kent. The following year "Look to the Land" appeared.

Rather than the mechanics or the agronomic practices of organics, Northbourne's book presents the philosophy, the rationale, and the imperative of organic farming. The ideas of his organics manifesto were quickly taken up as they spread globally. Jerome Rodale was an early adopter in the United States (from 1942). The Australian Organic Farming and Gardening Society (founded 1944) was the first association in the world to be founded with the express purpose of proselytizing for organic agriculture.

Northbourne's Betteshanger Conference was held in July 1939. It is the link between biodynamic and organic farming. The week-long biodynamic conference was held at Betteshanger on the estate of Lord Northbourne. Lectures were presented by Ehrenfried Pfeiffer and other biodynamics instructors from continental Europe. Talks included Pfeiffer's accounts of "The Soil as a Living Organism" and "The Farm as a Biological Organism" (Paull, 2011a,b). Shortly after the Betteshanger Conference, World War II broke out. Interactions with Pfeiffer were then limited during the course of the conflict and the appetite in Britain for Germanic-originating ideas evaporated. Northbourne took Steiner's idea of "the farm as organism" and crafted the term "organic farming" from it. He secularized the ideas of Steiner and Pfeiffer and presented a clear philosophy of "organic farming" in well-crafted language in "Look to the Land."

8.3.5 Jerome Rodale in the United States and "Organic Farming and Gardening"

American entrepreneur Jerome Irving Rodale (1898–1971) was the first to realize the commercial opportunity, and perhaps also the ecological imperative, presented by Northbourne's "organic farming." Rodale brought a decade of successful entrepreneurial publishing experience, and his considerable marketing verve, to the task.

Rodale had no farming experience but he was by then well experience with mining, "uplifting" and repurposing ideas, jokes, and stories from Britain. Rodale transformed the idea of "organic farming" into a publishing empire which is still with us as Rodale Inc. and the Rodale Institute based in Pennsylvania, USA.

Rodale's first move was to launch a new magazine "Organic Farming and Gardening" (in May 1942). It was the world's first organics periodical. He at once sought to broaden the audience appeal by extending "organic farming" to "organic farming and gardening." It was to prove a masterful stroke of marketing.

Rodale promptly found that US farmers were slow on the uptake of his ideas (and his magazine). In contrast, there were many more US gardeners (than farmers), and he discovered that they were a receptive audience. He quickly switched the magazine name around to read "Organic Gardening and Farming." For some of its publishing history the "farming" was dumped altogether from the title and the magazine became simply "Organic Gardening."

Rodale spent the remaining three decades of his life as the most prominent North American advocate of organics. He has been described as: "one of the earliest and loudest advocates of organic farming," and as one who: "enjoyed a reputation as both a genius and a crackpot" (Feldman, 2004, p.1060). He built the world's largest organics publishing house, Rodale Press.

The "New York Times Sunday Magazine" ran a story titled: "Guru of the Organic Food Cult" (Smil, 2001). Rodale was the subject of that story. According to his son Robert, he was: "particularly pleased by the cover story" and he was: "at last riding the crest of a growing wave of acceptance" (Rodale, 1971, p.37).

Rodale's pleasure, at being mainstreamed after 30 years of intrepid advocacy for the organics cause, was short lived. Just the day following the publication of the "New York Times" account that had hailed him as a guru of organic food, he died (Toma, 1967).

Jerome Rodale had founded the world's first "organic" periodical: "Organic Farming and Gardening." It had the longest run of any organics periodical (from 1942 to 2017) and it has now spawned "Rodale's Organic Life."

Rodale had written presciently in that first edition in May 1942 that: "One of these fine days the public is going to wake up and will pay for eggs, meats, vegetables, etc., according to how they were produced. A substantial premium will be paid for high quality products such as those raised by organic methods" (Rodale, 1942a,b, p.4).

In his "Autobiography," Rodale (1965) recalled the founding of&G: "What did I know about farming and gardening - practically nothing. But I began to read and I began to practice" (p.42–43). Inspired by his vision for organics, he purchased a 63 acre farm for US$7000 (Gross, 2008).

Rodale took a commercial view from the outset, and additionally he was an accomplished and prolific wordsmith, as well as a successful publisher and entrepreneur. He recalled: "Little did I realise what I was touching off - that I would be the one to introduce this great movement into the United States "(Rodale, 1965, 42).

8.3.6 Harold White and the AOFGS, Australia

The Australian Organic Farming and Gardening Society (AOFGS; 1944–55) was the world's first association to style itself "organic." It was founded 2 years before the UK Soil Association. The AOFGS was the first organization in the world to develop a set of organic farming principles. The first of their 10 principles of organic farming stated: "The Society holds that overstocking, over cropping, monoculture and the excessive use of chemical fertilizers, poisonous sprays and dusts have resulted in erosion and serious loss of soil fertility in Australia. The Society associates with these unwise land practices, the destruction of bird life, the denuding of forests by over-exploitation and bushfires, and the pollution of rivers and streams" (Paull, 2008a, p.155).

The AOFGS was founded during WW2. At the time there were unprecedented constraints and war-time controls on resources by the government. The AOFGS could not get a release of paper to publish its proposed journal. They had to wait 2 years until peace once again prevailed before their quarterly journal appeared. The "Organic Farming Digest" (1946–54), was the first organics periodical to be published by an association. Two leading contributors to this Australian venture were Ehrenfried Pfeiffer and Jerome Rodale.

New South Wales (NSW) grazier, Colonel Harold White, was a leading contributor to the Organic Farming Digest. He wrote 20 articles, beginning with two articles in the first edition (1946), (White, 1946a,b) and finishing with the 29th and final issue.

White's article "The Why of it" appeared in the first issue of "Organic Farming Digest." He stated that: "We have been lured from the sound practices of our ancestors and are wandering in a maze; we must get back to the narrow path laid down by Nature for our safeguard and well-being." He added: "We appeal to all land holders that they hold the health of the community in their hands, for the way the food is grown makes all the difference to the health of the persons eating it" (White, 1946b, p.6).

White was an experienced NSW farmer and he predicted that: "the time is not far distant when England will demand that imported foods be grown by organic methods" (White, 1946a, p.6). That has not yet come to pass but he was right in that exports have come under closer and closer scrutiny for pesticide residues and GM-free status.

The Living Soil Association of Tasmania (LSAT; 1946–60) adopted the AOFGS's quarterly periodical, the *Organic Farming Digest,* as its official journal and distributed it to its members. The LSAT adopted a social inclusion policy of recruiting government agencies, farming bodies, and civil society organizations onto its council. It was the first organics advocacy group to allow for junior members and groups (Paull, 2009).

Both societies, the AOFGS and the LSAT, ultimately folded. They failed to get themselves onto a sustainable financial footing. These were the days before organic certification (that came decades later) and so before certification was available as a money-stream. Unlike Jerome Rodale in the United States, the early proponents of organics in Australia were enthusiasts rather than business entrepreneurs. And unlike the Soil Association in the United Kingdom, they did not achieve charity status.

In the end, the AOFGS acknowledged that: "The Society has always operated under a financial handicap, and for this reason the Digest fell short in some respects. However, the principles of organic farming have been sufficiently publicised for the work to continue, and the supporters of the organic movement can best promote it by their own example of wise land use" (The Executive Officers, 1954, p.1).

At the close-down of the AOFGS, its officers were correct in stating that that: "Although the termination of this magazine will be regretted by many, there is solace in the fact that it has performed a service in publicising organic farming principles in Australia" (The Executive Officers, 1954, p.1). The AOFGS had folded after more than a decade of advocacy. They left footprints of their organic's advocacy in the printed media of the period. Coinciding with the life and fortunes of the AOFGS, mentions in the Australian printed media of "organic farming" were zero in the 4 years before the AOFGS. Then mentions ramped up quickly from 1944 and further in 1945. Media mentions of "organic farming" peaked in 1946 and subsequently trailed down to just a sole mention in 1955 (Paull, 2017a,b).

This first chapter of Australian organics advocacy was demonstrably successful in promoting their agricultural ideas. They were not successful in finding a sustainable business model, and not successful in passing the baton on to a next group of advocates. Theirs's was an early start, and now Australia leads the world in certified organic hectares, accounting for 22.69 million hectares which is 45% of the global total.

8.3.7 Eve Balfour and the Soil Association, United Kingdom

Eve Balfour (1899–1990) published her compendium "The Living Soil" (1943) in the United Kingdom. Her book extensively quoted contemporary agricultural authors, mostly British. "The Living Soil" borrowed a large slab of text, over four pages, of Northbourne's book "Look to the Land" (1940). Despite her clear familiarity with Northbourne, Balfour's book lacked any mention of the term "organic."

Balfour had studied agriculture at the University of Reading, graduating just after WWI. Her family put in the start-up capital in 1919 for the New Bells Farm, in Suffolk. She then struggled to make a go of farming during the Great Depression at this farm. Balfour was a founding member of the UK Soil Association (1946 to present).

Balfour proposed the so-called "Haughley Experiment" to compare various farming styles (1940). This enterprise was supported by donations raised by the Soil Association which was registered as a charity. The exercise was ultimately wound up with little to show for the substantial expenditure of time, money, and the goodwill of donors (Hall, 1962).

Both Balfour and the Soil Association were very slow to take up Northbourne's term "organic farming." The foundational documents of the Soil Association make no reference to "organic" farming, agriculture or gardening. The objects of the Soil Association were stated as: "1. To bring together all those working for a fuller understanding of the vital relationships between soil, plant, animal and man; 2. To initiate, co-ordinate and assist research in this field; and 3. To collect and distribute the knowledge gained so as to create a body of informed public opinion" (Soil Association, 1947, p.1).

Once the flagship project of the Soil Association, the Haughley Experiment, folded, not to say failed, the raison d'être and the finances of the Soil Association were uncertain.

The publication of Rachel Carson successful blockbuster book "Silent Spring" (1962) generated fresh interest as well as controversy. The appearance of Carson's book served to breathe new life into the Soil Association (Reed, 2010). The formalization of organic certification then created a funding stream for the Soil Association which continues to this day. The Soil Association is the largest organics certifier in the United Kingdom.

8.3.8 Marjorie Spock, Mary Richards, Rachel Carson, and "Silent Spring"

In a US government campaign to eradicate an insect, the gypsy moth, three million hectares of north east United States were aerially sprayed in the late 1950s with DDT. "Aircraft pilots were paid according to the number of gallons sprayed." So there was little or no incentive for restraint nor "from spraying the same areas more than once" (Kögel and Prskawetz, 2001, p.257).

Two biodynamic farmers with a farm on Long Island took exception to having their farm and their produce (and themselves) sprayed with DDT in this ill-considered government program. Marjorie Spock (1904–2008) and Mary T. Richards (1908–90) had both studied at Rudolf Steiner's Goetheanum at Dornach, Switzerland. They were passionate about Anthroposophy in general and biodynamics in particular. They were taking guidance from Ehrenfried Pfeiffer who by this time had moved to the United States (Paull, 2013a).

Spock and Richards were about to be impacted by: "one case that borders on the surreal, the New York state and federal departments, citing an implausible threat from the gypsy moth to New York City and environs, announced plans to spray densely populated Nassau County, Long Island, with DDT in fuel oil" (Kögel and Prskawetz, 2001, p.257).

A group of six residents, and swelling to 13, of Long Island, took legal action against the government (Murphy v. Benson, 1957; Murphy v. Benson, 1959). The prime movers of the group were Spock and Richards (Brooks, 1972). The group included organic gardener and past president of the National Audubon Society, Richard Murphy (Ladejinsky, 1943, p.43). From the United Kingdom, the Soil Association sent a $100 cheque to Spock in support of the campaign (Spock, 1960) but most of the funds in the fighting kitty were from Richards.

The Long Island Spray Trial ran from 1957 to 1960. Spock and Williams corresponded with experts around the world and collated a massive file of data and contacts. First, they applied for an injunction to stop the spraying. That was denied. The case went to trial in the District Court. It ran for 22 days. They lost. They appealed. The Court of Appeals upheld the original decision. They then took the case to the Supreme Court, again with no success. Spock and Williams had generated over 2000 pages of scientific documentation on pesticides and toxic spray programs and spent US$100,000 in the process (Paull, 2013b).

What was a notable failure in the courts of the land was about to be retested in the court of public opinion? And with a different outcome. Although, it must be admitted that Long Island and the biodynamic farm of these two warriors was sprayed with DDT. They sold up and purchased a farm elsewhere.

The trial documentation was shared with biologist and author Rachel Carson. She had contacted Spock and Richards after the initial injunction to stop the spraying was lost (Murphy v. Benson, 1957) and while the District Court case (Murphy v. Benson, 1958) was in preparation (Carson, 1958a,b).

Carson wrote to Spock and Richards: "I am most grateful for all the material you have sent me … I am delighted there is so much sound material" (1958). She wrote "I am undertaking to do a small quick book … You have been so enormously helpful to me, and apparently are so familiar with a vast amount of material on the subject" (1958).

The challenge as Carson saw it was how to get traction with this story which had failed so miserably in the courts. Carson wrote: "It is a great problem to know how to penetrate the barrier of public indifference and unwillingness to look at unpleasant facts that might have to be dealt with if one recognised their existence. I have no idea whether I shall be able to do so or not, but knowing what I do, I have no choice but to set it down … I guess my own principal reliance is in marshalling all the facts and letting them largely speak for themselves" (1960).

Against the Spock-Williams-Carson view was the countervailing view expressed in "Pesticides: the Price of Progress" (Time, 1962). Readers of Time magazine were assured that: "many insecticides are roughly as harmless as DDT" (p.46). A dash of DDT should be of no concern to citizens:

> "Dr Wayland J. Hayes, chief of the toxicology section of the U.S. Public Health Service in Atlanta says that every meal served in the U.S. probably contains a trace of DDT, but that is nothing to worry about" *Time (1962), p.45.*

Carson's book "Silent Spring" appeared in 1962 to acclaim and ridicule. "Chemical Week" (1962, p.5) declared that:

> "Her technique in developing this theme is more reminiscent of a lawyer preparing a brief, however, than a scientist conducting an investigation … the industry is facing a hostile and to some extent uninformed prosecuting attorney. Her facts are correct, her conclusions less certain, and her innuendoes misleading." They were unaware of her informants.

The "Conservation News" hit out: "In any large-scale pest control program … we are immediately confronted with the objection of a vociferous, misinformed group of nature-balancing, organic gardening, bird-loving, unreasonable citizenry that has not been convinced of the important place of agricultural chemicals in our economy" as quoted by Soraci in 1962 (Jezer, 1987, p.169).

The journal "Science" informed readers that: "Mankind has been engaged in the process of upsetting the balance of nature since the dawn of civilization … modern agriculture and modern public health, indeed modern civilization, could not exist without an unrelenting war against a return to a true balance of nature" (Baldwin, 1962, p.1043).

Nevertheless, despite (or was it because of) a well-orchestrated campaign against Carson and her book, it was the book that prevailed. Carson died of cancer in 1964, living long enough to see her book send great waves of disquiet through governments and populations. Governments had been in thrall of chemical companies, and populations were largely in the dark about chemical harm and contaminations.

The US Environmental Protection Agency (EPA) was founded in December 1970. It canceled all Federal registrations of DDT products on 14 June 1972, and from 31 December 1972 usage of DDT was banned in the United States. The EPA has largely adopted the approach of examining pesticides on a case-by-case basis. So, the list of available pesticides operates as a revolving door, and as old chemicals exit, new ones enter the arsenal of registered toxins. In August 1972, the "Australian Agricultural Council recommended that all existing registrations for DDT should be reviewed as a matter of urgency, with the view to withdrawing all uses for which acceptable substitutes exist" (Harrison, 1997).

What had failed in the courts worked as a powerful and effective narrative in the hands of Rachel Carson. At no time did she publicly acknowledge the organics and biodynamics movements, she did not align herself with them, and nor did she give any public credit to her sources, Spock and Williams, nor to Pfeiffer. She recorded her debt to Spock and Williams in her private correspondence (Paull, 2013a,b). She was very particular to let her book stand alone as a testament against pesticides in general and DDT in particular.

Carson fueled a debate that organics advocates had been trying to kindle for two decades. The Chairman of the House of US Representatives' Appropriations Subcommittee for Agriculture, US Congressman and lawyer, Jamie Whitten was alerted: "I became aware that there was a sizable movement at work aimed at severely curtailing or even eliminating the use of pesticides … This made me afraid that an aroused public opinion might stop the use of materials that I had become convinced are absolutely essential to our health and prosperity. And so I began to speak out in defense of the role of pesticides" (1966, p.vi).

As Reed has observed: "After Silent Spring, no public discussion of pesticides could be held without her critique being invoked" (Reed, 2003, p.167). Carson's book gave fresh impetus to the organic's movement worldwide. Carson herself always publicly stood at a distance from the organic's movement. Her message was not news to the organics sector, it was their story after all. What was new is the publicity, the dispersion of the narrative, and the public discourse generated, and that Carson mainstreamed the issues.

8.3.9 Roland Chevriot in France, and IFOAM

Parisian engineer, Roland Chevriot, was the President of the French national farmer organization, Nature et Progres, when he called a meeting of likeminded organizations. His vision was of a "federation respecting all particularities and individualities" (Chevriot, 1972). He sent out over 50 invitations. A meeting was convened at Versailles, France, in 1972. Five organizations from five countries attended: his own Nature et Progres, the Soil Association (United Kingdom), the Soil Association of South Africa, the Swedish Biodynamic Association, and Rodale Press (USA) (Paull, 2010).

As Denis Bourgeois of Nature et Progres states: "We created a federation of 'movements' at a time when organic agriculture was weakly organised and needed activists and volunteers to push the idea" (1997). The name of the new organization, the "International Federation of Organic Agriculture Movements," appeared in the initial invitation to meet. The group of five founding members adopted the proposed name (in English) with the acronym IFOAM. None of these five organizations bore the term "organic" in their title. Nevertheless, the choice of name acknowledged "organic" as the term to signify their common cause. It secured "organic" as the core narrative term and as the international descriptor of what is now a clearly identifiable and differentiated segment of the global food and farming sector.

The formation of IFOAM created an entity which united the aspirations, the philosophies and the hopes of disparate groups around the world. From the outset "biodynamic" was accepted as a special case of "organic." The headquarters of IFOAM was relocated to Bonn, Germany. IFOAM remains the global umbrella advocacy group for the organics sector.

There are now 848 "affiliates" of the federation from 121 countries (IFOAM, 2017). The name "IFOAM" has always been somewhat ungainly. In 2015 the name was changed to "IFOAM - Organics International" which some will consider even more cumbersome. The organization hosts an International Congress every 3 years in a location that is announced 3 years in advance.

8.3.10 Pawan Chamling in Sikkim, India

The Chief Minister of Sikkim, Pawan Chamling, declared, in 2003, the "goal of making Sikkim entirely organic" (Chief Minister's Office, 2016, p.23). This trail-blazing goal was achieved in 2016 and the northern Indian state of Sikkim is now 100% organic (Paull, 2017a,b).

Chamling's ambitious objective was for a "Total Organic State." The vision was backed up with policy interventions. An action plan "Going for Organic Farming in Sikkim" was prepared. Four points of the action plan were "(1) promotion of on-farm production of organic manures; (2) capacity building; (3) establishment of bio-fertilizer production units; and (4) establishment of soil testing laboratories" (Chief Minister's Office, 2016, p.392).

The Sikkim State Organic Board (SSOB) was established to oversee the implementation of the plan. The implementation began with 100 villages which were declared as "bio-villages" and farmers were trained in organic practices. The success of the inaugural bio-villages was a proof of concept for Sikkim villages and it attracted other farmers. By October 2009 there were 396 bio villages.

Under the Sikkim Organic Mission of 2010 farmers were provided with seeds, manure and training. In 2014 the Sikkim Agricultural, Horticultural Input and Livestock Feed Regulatory Act banned the use of synthetic fertilizers and pesticides. By 31 December 2015, 75,000 ha of agricultural land were certified organic (Chief Minister's Office, 2016, p.23).

These innovations culminated in 2016 with acclaim from the Prime Minister of India, Nahendra Modi, who declared that: "Sikkim has paved its way into history and has set an example for the entire world that nature needs care and protection" (Chief Minister's Office, 2016, p.387).

Multiple states of India have expressed aspirations of replicating Sikkim's success. They include: Goa, Karnataka, Kerala, Madhya Pradesh, Maharashtra, Meghalaya, Mizoram, Rajasthan, Tamil Nadu, Uttar Pradesh, and Uttarakhand (Paull, 2017a,b).

If Sikkim can do it, others can do it. The message from Sikkim is: "Let us work together to recreate a world that is Totally Organic" (Chief Minister's Office, 2016, p.636). As a mission statement this injunction carries the authority of one who has actually achieved it for his own state of Sikkim, India.

8.3.11 Vladimir Putin in Russia

Under President Vladimir Putin, Russia has acted to close down GMO farming and ramp up organic farming. The share of GMO in Russian food reportedly declined from 12% to 0.01% in the decade top to 2014. At that time there were only 57 registered food products containing GMO ingredients. Agriculture Committee of the State Duma decided that Russia needs a ban on the registration and sale of GMOs The Ministry of Agriculture supported "a complete ban on growing and using genetically altered organisms in the country, with the exception of those used in scientific research" (RT, 2014).

Hand in glove with the close down of GMOs in Russia is the ramping up of organics. Russia has aspirations of becoming a major exporter of organic food. The government news agency has reported that: "Putin vows to make Russia a major supplier of organic food to Asia–Pacific Region" (RT, 2017). Putin stated that "Russia is one of the world's leaders in exports of grain, vegetable oils, fish, and a number of other foods. We expect to become the leading supplier of ecologically clean food to our neighbors in the Asia–Pacific region" (Putin, 2017, p.3).

Russia's Deputy Agriculture Minister, Aleksandr Petrikov, is reported as stating that "the ban" could afford Russia large economic gains if the country chooses to become a major global producer of GMO-free products. With anti-GMO sentiments sweeping multiple continents, Russia could be looking to position itself to meet international demands as a major supplier of organic foods and ingredients" (Coatney, 2014).

There are several prongs to Russia's rejection of GMOs. There are the uncertain long-term health impacts of GM food. Public sentiment is against GM food with 80% of Russians reportedly against GM food (Voice of Russia, 2014). There is the not unimportant issue that profits from GM seed are mostly repatriated to the United States. Petrikov makes an economic point that the "introduction of GMO cultures carries economic risks – Russia cannot compete with foreign producers when it comes to costs, but still can position itself as a producer of high-quality, GMO-free agricultural goods. Thus, any use of GMO cultures would harm the national export potential" (RT, 2014).

President Putin is reported as stating that "Russia should become the world's largest supplier of organic food" (Interfax, 2015). Russia already exports organic buckwheat (grechka), wheat millet, alfalfa, flax, berries, mushrooms, cedar nuts, and herbs.

In a message to the Federal Assembly, Putin states that "We not only can feed ourselves but taking into account our land and water resources - which is especially important - Russia can become the world's largest supplier of healthy, environmentally friendly, high-quality food that has long been lost to some Western producers. Especially as the demand in the global market for such products is constantly growing" (Interfax, 2015).

Putin is spearheading a bold vision for Russia as a GM-free zone for food production and a major producer of organic food. It is much easier for organics to prosper in a GM-free zone. This avoids the problems of contamination such has been experienced in Australia (Paull, 2015). It avoids the costs of segregation and the consequences of segregation failures. It is also reassuring for export markets and consumers, and it positions Russia as a high-quality premium quality food producer.

For the moment, the organics sector in Russia is modest and the certified organics sector is in its infancy. Russia made a serious bid to host the triennial Organic World Congress for 2020, unfortunately without success. The future for organics looks bright in Russia with public support, ministry support, and presidential support.

8.4 Pesticides and externalities

The great uncoated externality of chemical farming is the pollution-load on the environment, producers, and consumers. This pollution is a cost of production which in chemical agriculture is thus externalized and is not reflected in the market price. The cost is instead dispersed over time and space, over the whole community and environment.

Some pesticides are foliar (they are applied to the foliage) and some are systemic (they are taken up by the plant and incorporated into the cells). In any event they are designed not to wash off but rather to bind to the plant. Agricultural chemicals cannot be confined to the farm. They percolate through the supply chain and some quantities reach the consumer.

A study of fruit and vegetables supplied to British school children under a government scheme, the School Fruit and Vegetable Scheme (SFVS), reported the presence of 123 different pesticides. They reported the presence of 62 different insecticides, 50 different fungicides, five herbicides, as well as insect growth regulators, plant growth regulators, and microbicides. These contaminants include carcinogens, endocrine disruptors, neurotoxins and developmental toxins (PANUK, 2017).

Of the fruit supplied into the Department of Health's £40 million school children program, one sample of apples contained 11 different pesticides. One sample of pears contained nine different pesticides. Overall, 84% of the produce tested positive for at least one residue, and 66% tested positive to multiple residues. Almost all the apples (97.5%) contained at least one pesticide residue, and 67.7% carried residues of multiple pesticides. There were 64 different pesticides identified in the apples sampled. Of the children's raisins tested, 100% tested positive for multiple pesticides. None of the produce was organic (PANUK, 2017).

Converting to an organic SFVS was estimated to add 14% to the cost of the program (PANUK, 2017). Organic produce generally is sold at a premium of somewhere between 0% and 100%. Converting to an organic diet is an effective way of reducing children's exposure to pesticides(Göen 2016).

8.5 Certification and the organics hinterland

Statistics of organic agriculture do not tell the whole story. For starters, they underestimate the size and achievements of the organic's movement, simply because they report certified organic agriculture (rather than both certified and uncertified organic). The statistics are seductively precise but they are merely the countable manifestation of a larger phenomenon, perhaps much larger.

Just how large is the uncounted "world of organic agriculture," as compared to the counted world of certified organic agriculture? It is a matter of speculation. A study in India compared the experience of organic farmers (N = 350) and chemical farmers (N = 200). All of the organic farmers lacked certification (Sudheer, 2013). Reasons given for the absence of certification were the cost of certification, the lack of information to achieve certification, and the size and scale of the operation (Sudheer, 2013). It may be that there are more uncertified organic producers than certified ones. There may be more uncertified organic agriculture hectares than certified hectares. While these hypotheses seem likely, the relevant statistics (or estimates) are not available.

There are reasons to be organic and there are reasons to be certified, and they are different reasons. Early pioneers such as Rudolf Steiner and Lord Northbourne make no mention of certification. Organics certification generally postdates the 1972 founding of IFOAM. In Australia, active advocacy of organic agriculture dates from the 1944 founding of the Australian Organic Farming and Gardening Society (AOFGS) but organics certification in Australia dates only from the founding of organics certifiers beginning in 1987 (Paull, 2013a). Certified organic food and agriculture is a subset of organic food and agriculture.

The viability of organics certification relies on the availability of a premium price for the certified produce. If a premium price cannot be achieved in a particular market then it is difficult to justify the cost of certification. There is a multitude of other reasons why organic producers are not certified, including: they cannot afford it, the scale of their operation does not warrant it, there is no suitable certification scheme available, they agree with the philosophy of organics but not of certification, they want to avoid the intrusion into their affairs.

The hinterland of Certified Organic includes not only noncertified organic, but also like-minded production methods. In the United States, there is "Certified Naturally Grown" founded in 2002 by organic farmers in reaction to the USDA taking control of US organic standards and certification. The coupling of USDA with certified organic has been a cause of disquiet considering the track record of the USDA in championing pesticides and GMOs and its long prior history of unfriending the organics sector. As a response to the disenfranchisement of the grassroots organics movement there has been "a backlash against the federal takeover of the organic program ... Certified Naturally Grown has expanded over the past decade to include more than 700 farms in 47 states" (Reighart, 2013).

China has developed a certification scheme, Green Food, which is a step-down from organic but a step-up from chemical agriculture. Green Food production is a reduced input agriculture opting for a balance between health and environment, and economics and productivity. Green Food products are sample tested for pesticide residues, and annual inspections are conducted (Giovannucci, 2005, p.12).

Green Food is a Ministry of Agriculture initiative: "In 1991, the MOA began to implement the strategy of 'Green Food' in agro-food production in order to improve the health and environmental safety and market competitiveness of Chinese food products, as well as agricultural environmental sustainability" (Lu, 2005, p.17).

For Green Food certification the area should conform to four criteria: the area meets the highest grade of air standards in China; heavy metal residues are restricted in irrigation, water and soil (with tests for mercury, cadmium, arsenic, lead, chrome, etc.); processing water meets the National Drinking Water Standard; and chemical applications are restricted and regulated, with some of the most poisonous pesticides and herbicides banned (Giovannucci, 2005, p.12).

Of the certified food eco-labeling schemes in China, certified organic is reported as accounting for 9%, Green Food accounts for 29%, and Hazard-Free accounts for 62% (based on hectares certified). These eco-certifications account in total for 28% of China's cultivated land (Paull, 2008b). These certification schemes provide levels of reassurance to consumers.

The organics sector has experienced steady growth over the past two decades (there are no global statistics prior to the year 2000). The annual growth rate of 11.9% per annum is cause for celebration. But any congratulations must be tempered by the realization that the growth is coming off a low base, and organic agriculture remains a niche agriculture that aspires to be the global agriculture.

8.6 Conclusion

Organic agriculture and food meet the expectations of consumers for clean and healthy food. Organic production methods protect the health and welfare of the environment, farmers, farm workers, consumers, and animals.

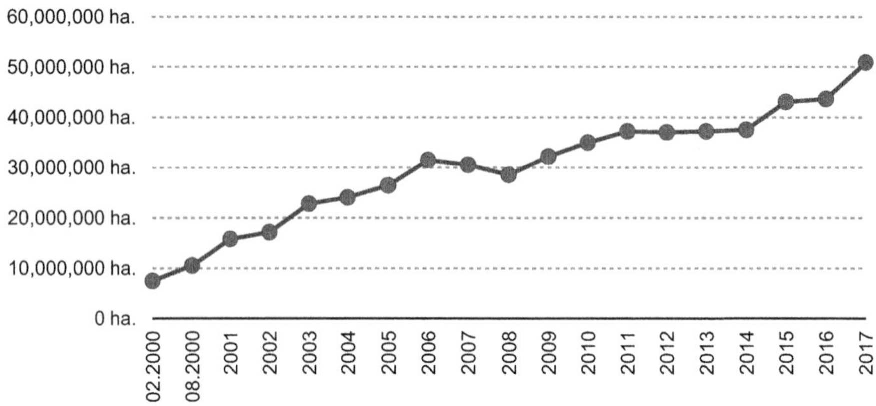

FIGURE 8.4 Global certified organic hectares show steady growth over the past two decades, year reported Growth. *Source: John Paull.*

Organics is a global enterprise—however, certified organic agriculture land accounts for just 1.1% of global agriculture land. Uncertified organic agriculture may account for some (unknown) multiple of certified organics and thereby increase that percentage. Nevertheless, for most of the time, most of the world's population is not eating an organic diet.

The state of Sikkim is the standout example of a success story of conversion to organic production. Pawan Chamling has shown what can be done when there is the will. Sikkim has achieved a 100% organic state in a short period (2003–16). Chamling's vision of a world that is "Totally Organic" (Chief Minister's Office, 2016, p.636) is the vision of the pioneers of organics, and it remains the challenge for every advocate of organics. Only 11 countries report more than 10% of their agriculture land as organic, while 111 countries report less than 1% of their land as certified organic (Paull, 2017a,b).

Some satisfaction can be drawn from the global annual growth rate of organic agriculture of 11.9% compounding over the past two decades (Fig. 8.4). This, however, must be tempered with the disappointment that global certified organic agriculture still only accounts for 1.1% of global agriculture hectares.

References

Aldi, 2007. Weekly News. Aldi Stores, Australia.

Baldwin, I.L., 1962. Review: chemicals and pests. Sci. New Series 137 (3535), 1042–1043.

Balfour, E.B., 1940. The Haughley Research Farms: Proposal for an Experiment in Soil Fertility in its Relationship to Health. East Suffolk County Council. The Haughley Research Farms Trustees, Ipswich.

Balfour, E.B., 1943. The Living Soil: Evidence of the Importance to Human Health of Soil Vitality, with Special Reference to Post-war Planning. Faber and Faber, London.

Bourgeois, D., 1997. How it all began. Ecol. Farming (17), 12–14.

Brooks, P., 1972. The House of Life: Rachel Carson at Work. Houghton Mifflin, Boston, MA.

Brumby, J., 2007. Minister's Forward in Victorian Organic Food Products Directory 2007. Regional Development Victoria, Victorian Government, Melbourne, Australia.

Carson, R., 1958a. MS Typewritten Letter: Dear Mrs. Spock; 14/3/1958. Yale Collection of American Literature, Beinecke Rare Book and Manuscript Library, New Haven, CT.

Carson, R., 1958b. MS Typewritten Letter: Dear Mrs. Spock; 26/3/1958. Yale Collection of American Literature, Beinecke Rare Book and Manuscript Library, New Haven, CT.

Chemical Week, 1962. Viewpoint: The Chemicals Around Us. Chemical Week, 14 July: 5.

Chevriot, R., 1972. Subject: Creation of an International Federation. Typescript Letter, 1 Page, Undated. Nature et Progrès, Paris.

Chief Minister's Office, 2016. Sikkim : Under the Leadership of India's Greenest Chief Minister Shri Pawan Chamling - Sustainable Developmenet through Greening, Organic Farming and Unique Social Enginering. Chief Minister's Office, Government of Sikkim, Gangtok, India.

Coatney, C., 2014. Why Does Russia Plan to Stop GMO Cultivation and Imports? 6 June. Biology Fortified, Stockton, CA.

COED, 2007. Compact Oxford English Dictionary. Retrieved 3 August 2009, from: www.askoxford.com.

FAO, 2001. Codex Alimentarius - Organically Produced Food. FAO/WHO Food Standards Programme Codex Alimentarius Commission, Food and Agriculture Organization of the United Nations, Rome, Rome.

Feldman, J.W., 2004. Rodale, J. I. (1898-1971) Organic farming advocate. In: Krech, S., McNeill, J.R., Merchant, C. (Eds.), Encyclopedia of World Environmental History, vol. 3. Routledge, London, p. 1060.

GfK, 2017. Decision Factors on what to Eat or Drink: Global GfK Survey (October 2017). GfK (Growth from Knowledge), London.

Giovannucci, D., 2005. Organic Agriculture and Poverty Reduction in Asia: China and India Focus. International Fund fot Agricultural Development (IFAD), Rome.

Göen, T., 2016. Greenpeace-Japan-Study of the Effect of Nutrition Change on the Pesticide Exposure of Consumers: Final Report. Erlangen, Germany, Institute and Outpatient Clinic of Occupational, Social and Environmental Medicine. Friedrich-Alexander Universität Erlangen-Nürnberg.

Gross, D., 2008. Our Roots Grow Deep: The Story of Rodale. Rodale Inc, Emmaus, PA.

Hall, J.R.M., 1962. Comment on Haughley. Mother Earth 12 (October), 419.

Harrison, S., 1997. Organochlorides in Australia. In: Proceedings, UNEP Workshop on Persistent Organic Pollutants (POPs). Bangkok.

IFOAM, 2017. Membership E-Directory 2017. IFOAM - Organics International (IFOAM), Bonn.

Interfax, 2015. Џутин Назвал РФ Џотенциально КруЏнейшим В Мире Џоставщиком ЛколоҐически цистыч Џродуктовћ Ь Гдвдлбдр. interfax.ru, Moscow.

Jezer, M., 1987. The Dark Ages, Life in the United States, 1945-1960. South End Press, Cambridge, MA.

King, F.H., 1911. Farmers of Forty Centuries, or Permanent Agriculture in China, Korea and Japan. Mrs. F. H. King, Madison, Wisconsin.

Kögel, T., Prskawetz, A., 2001. Agricultural productivity growth and escape from the malthusian trap. J. Econ. Growth 6 (4), 337–357.

Ladejinsky, W.I., 1943. The food situation in a world at war: the food situation in Asia. Ann. Am. Acad. Pol. Soc. Sci. 225, 91–93.

Lu, W., 2005. Trade and Environment Dimensions in the Food and Food Processing Industries in Asia and the Pacific, A Country Case Study of China. Department of Agricultural Economics, Zhejiang University, Hangzhou.

Lymington, V., 1932. Horn, Hoof and Corn: The Future of British Agriculture. Faber & Faber, London.

Murphy v. Benson, 1957. Robert Cushman MURPHY, Mary T. Richards, Marjorie Spock, John C. Homer, Gladys Weeks, and David Kennedy, Plaintiffs, V. Ezra Taft BENSON, Secretary of Agriculture of the United States, Lloyd Butler, Area Supervisor of Plant and Pest Control Division, Daniel Carey, Commissioner of Agriculture and Markets of the State of New York, Defendants. US District Court for the Eastern District of New York, Murphy v. Benson, 151 F. Supp. 786 (D.N.Y. 1957).

Robert Cushman MURPHY et al., and Archibald B. Roosevelt et al., Plaintiffs, v. Ezra Taft BENSON, etc., et al Murphy v. Benson, 1958. Defendants. US District Court for the Eastern District of New York. Murphy v. Benson, 164 F. Supp. 120 (D.N.Y. 1958).

Murphy v. Benson, 1959. Petition for Rehearing: Robert Cushman Murphy et al., Plantiffs-Appellants v. Ezra Taft Benson, Secretary of Agricultutre of the United States, Defendant, and Lloyd Butler, Area Supervisor, Plant and Pest Control Division of the United States Department of Agriculture et al., Defendants- Appellees. US Court of Appeals, Murphy v. Benson, 270, Docket No: 25448 (2d Cir. N.Y. 1959).

Nielsen, 2005. Consumer Attitudes towards Organic Foods, A Global Consumer Survey. ACNielsen.

Northbourne, L., 1940. Look to the Land. Dent, London.

PANUK, 2017. Food for Thought: Pesticide Residues in the School Fruit and Vegetable Scheme (SFVS). Pesticide Action Network UK (PANUK), Brighton, UK.
Paull, J., 2008a. The greening of China's food - green food, organic food and eco-labelling. In: Sustainable Consumption and Alternative Agri-Food Systems Conference (SUSCONS). Liege University, Arlon, Belgium.
Paull, J., 2008b. The lost history of organic farming in Australia. J. Org. Syst. 3 (2), 2–17.
Paull, J., 2009. The Living Soil Association: pioneering organic farming and innovating social inclusion. J. Org. Sys. 4 (1), 15–33.
Paull, J., 2010. From France to the world: the international federation of organic agriculture movements (IFOAM). J. Soc. Res. Policy 1 (2), 93–102.
Paull, J., 2011a. The Betteshanger Summer School: missing link between biodynamic agriculture and organic farming. J. Org. Syst. 6 (2), 13–26.
Paull, J., 2011b. Biodynamic agriculture: the journey from Koberwitz to the world, 1924-1938. J. Org. Syst. 6 (1), 27–41.
Paull, J., 2011c. The making of an agricultural classic: farmers of forty centuries or permanent agriculture in China, Korea and Japan, 1911-2011. Agric. Sci. 2 (3), 175–180.
Paull, J., 2013a. A history of the organic agriculture movement in Australia. In: Mascitelli, B., Lobo Ballarat, A. (Eds.), Organics in the Global Food Chain. Connor Court Publishing, pp. 37–60.
Paull, J., 2013b. The Rachel Carson letters and the making of silent spring. Sage Open 3 (July-September), 1–12.
Paull, J., 2014. Lord Northbourne, the man who invented organic farming, a biography. J. Org. Syst. 9 (1), 31–53.
Paull, J., 2015. The threat of genetically modified organisms (GMOs) to organic agriculture: a case study update. Agri. Food 3, 56–63.
Paull, J., 2017a. Four new strategies to grow the organic agriculture sector. Agrofor Intl. J. 2 (3), 61–70.
Paull, J., 2017b. Organic farming: the arrival and uptake of the dissident agriculture meme in Australia. In: Etinghoff, K. (Ed.), Sustainable Development of Organic Agriculture: Historical Perspectives. Apple Academic Press, Toronto, pp. 31–52.
Pfeiffer, E., 1938a. Bio-Dynamic Farming and Gardening: Soil Fertility Renewal and Preservation. Anthroposophic Press, New York.
Pfeiffer, E., 1938b. De Vruchtbaardheid der Aarde: haar Behoud en Haar Vernieuwing het Biologisch-Dynamische Principe in de Natuur. Uitgevers-Maatschappij & E. Kluwer, Deventer, Netherlands, N.V.
Pfeiffer, E., 1938c. Die Fruchtbarkeit der Erde Ihre Erhaltung and Erneuerung: Das Biologisch-Dynamische Prinzip in der Natur. Basle. Verlag Zbinden & Hugin, Switzerland.
Pfeiffer, E., 1938d. Fécondité de la Terre, Méthode pour conserver ou rétablir la fertilité du sol: Le principe biodynamique dans la nature. Editions de La Science Spirituelle, Paris.
Pfeiffer, E., 1938e. La Fertilita della Terra come Restaurarla e come Conservarla. La Prora, Milano, Italy.
Pfeiffer, E., 1956. Bio-dynamics: a short, practical introduction. Bio-Dynamics (41), 2–9.
Putin, V., 2017. Article by Vladimir Putin, the 25th APEC Economic Leaders' Meeting in Danang: Together towards Prosperity and Harmonious Development. 8 November. Kremlin, Moscow.
Reed, M.J., 2003. Rebels for the Soil: The Lonely Furrow of the Soil Association 1943-2000. University of West England. PhD, Bristol.
Reed, M.J., 2010. Rebels for the Soil: The Rise of the Global Food and Farming Movement. Earthscan, London.
Reighart, S., 2013. York County Farms Opt Out of Organic Certification. York Daily Record 8 September: http://cumberlink.com/news/.
Rodale, J.I., 1942a. Introduction to organic farming. Org. Farm. Garden. 1 (1), 3–5.
Rodale, J.I. (Ed.), 1942b. Organic Farming and Gardening. Emmaus, vol. 1. Rodale Press, Pennsylvania, pp. 1–16 (1).
Rodale, J.I., 1965. Autobiography. Emmaus. Rodale Press, MA.
Rodale, R., 1971. J. I. Rodale, in memoriam. Org. Gard. Farming 18 (8), 36–41.
RT, 2014. Duma Seeks Moratorium on GMO Production in Russia. 26 February. rt.com, Moscow.
RT, 2017. Putin Vows to Make Russia Major Supplier of Organic Food to Asia-Pacific Region. 11 November. rt.com, Moscow.
Smil, V., 2001. Enriching the Earth: Fritz Haber, Carl Bosch, and the Transformation of the World Food Production. Massachusetts Institute od Technology, Massachusetts.
Soil Association, 1947. Objects. Mother Earth 1 (2), 1.
Spock, M., 1960. Correspondence: the long Island gypsy moth case. Mother Earth 11 (January), 103.

Steiner, R., 1924. Agriculture Course. Goetheanum, Dornach, Switzerland.
Sudheer, P.S.K., 2013. Economics of organic versus chemical farming for three crops in Andhra Pradesh, India. J. Org. Syst. 8 (2) (XXXX-XXXX).
The Executive Officers, 1954. Farewell. Farm Garden Dig. 3 (5), 1–3.
Time, 1962. Pesticides: the price for progress. Time 45–48, 28 September.
Toma, P.A., 1967. The Politics of Food for Peace. University of Arizona Press, Tucson.
Voice of Russia, 2014. 80% of Russians Feel Negative towards GMOs. 13 February. Voice of Russia, Moscow. sputniknews.com.
White, H., 1946a. Organic farming. Org. Farm. Dig. 1 (1), 15–17.
White, H., 1946b. The why of it. Org. Farm. Dig. 1 (1), 4–6.
Whitten, J.L., 1966. That We May Live. D. Van Nostrand Company, Princeton, NJ.

CHAPTER

9

From agriculture to the global food chain/system

OUTLINE

Food waste occurs all along the supply chain; so, understanding the global food system and food supply chain (FSC) in particular as well as their intrinsic nuances is key to better comprehending the complexities of the wastage problem. In this endeavor, the following takes a snapshot of the agricultural, livestock, and fisheries primary farming landscapes to better set the backdrop of the supply chain characteristics of the proceeding sections. However, with regard to the food supply and the FSC, it must be noted at the outset that the two are two very different things; the food chain on the one hand is the chain that sees the bird eating the bug, the animal eating the bird, and humans eating the animal, and so on;

Food and Society
https://doi.org/10.1016/B978-0-12-811808-5.00009-X

the FSC on the other hand is just a supply chain—a sum of all the processes described in the next sections responsible for bringing food to the table.

9.1 Agriculture, forestry, and fisheries

Global food production in general (with some small caveats) has for a long time been a somewhat nationally protected sector. In fact, since the first General Agreement on Trade and Tariffs (GATT) back in 1947 (WTO, 2010), the industry was vociferously protected by producers and governments alike. However, after several decades and amid growing pressure from the World Trade Organization (WTO), as well as from a critical mass of capitalist momentum, all that changed in 1995 as agriculture was finally placed on GATT discussion table. Subsequent GATT rounds have further reduced trade barriers across nations resulting in a hefty increase in global agricultural trade (WTO, 2010).

As a result, for many if not most countries in the world now, agriculture has become an essential and inextricable component of economic growth. It comes as no surprise then that such developments have altered the traditional agricultural landscape. Incidentally, this is a situation that is also tied to the food security of many developing countries (more on this later). A closer look at the current structure of the global agricultural landscape and its impact on the global FSC helps to elucidate this point.

9.2 What is the food supply chain?

So, what about FSC themselves? In its simplest form, an FSC is the process that describes how food from the farm reaches our tables. Adding a layer of complexity, the FSC can further be described as the production, the processing, the distribution and consumption and to some extent the disposal of foods. With this in mind, FSC can be characterized by the flow of food in one direction and the flow of money in the opposite direction as shown in Fig. 9.1.

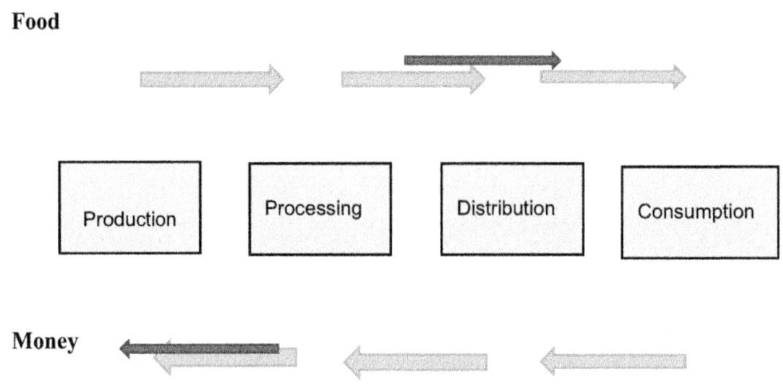

FIGURE 9.1 A simple food supply chain. *Courtesy of Pat Newsham*

From the perspective of food, producers, processors, and retailers push (supply) foods as consumers pull (demand) food. Conversely from the viewpoint of money, consumers push money as retailers, processors and producers pull money along the chain. In this way it can be seen that when one aspect of the FSC is affected, the whole chain is affected, thus representing two-sided causality or co-dependence. It has to be noted too that even in spite of this interdependence, the ability of the chain to function well is not a given. On the contrary, inefficiencies in early supply chains were noted among developing and developed countries alike and part of early developed countries strategic economic development policy gave high priority to the improvement of such systems (Gra, 1990). As a result, during the early part of the 20th century, technical and economic improvements facilitated by political goals, many countries were able to develop their FSC systems to meet the needs of the majority of their populations. Initially this meant increasing national production as well as building and improving storage and transportation facilities; later, however, it was also recognized that improvements along the whole of the chain were required if supply and demand were to be in equilibrium (Gra, 1990). For many this has ultimately resulted in systematic and continuous improvements in the production and distribution of food; for others—particularly in the developing economies, many barriers still remain.

9.2.1 Food supply chain characteristics

Food systems are predicated on the previously mentioned well-functioning primary farming models and as a whole the global food system can be thought of as a collection of FSC that make up the total supply and demand around the world. Over the recent past, the supply chain itself has witnessed many profound changes evolving through technological advances, globalization and politicization. More importantly, however, every aspect of the FSC is becoming increasingly accountable—particularly at the social level. Although thus far, this accountability has largely been confined to moral and ethical pressures and as is so often the case, such influences can be limited. That said this marks the culmination of pressure both at public and institutional levels. This trend is not seen in isolation either, accountability of the FSC is just one of many such social goals including inter alia: sustainability, natural resource allocation, an increased sense of fair trade and of course food wastage—and that's just the beginning; the pressures for change are mounting (SOFI, 2005; Ericksen, 2008; Gibson, 2016).

Social pressures aside, over the past few decades trends in global production have been characterized by Erickson and others as the intensification of agriculture accompanied by a corresponding trend for larger farm sizes and the increasing fragmentation of marginalized smallholders (Ericksen, 2008). On top of this, these last few decades have also seen increased "value-added" foods in the processing sectors and a concentration of corporate businesses up and down the supply chain (vertical integration) as opposed to the hitherto dominant traditional horizontal model. This can be seen in many levels of the food chain and it effectively allows increasing corporate control by large multinationals over large sections of the food system. Some of these changes were highlighted by Polly Ericksen in a recent paper contrasting food systems and their effects on societal outcomes (Ericksen, 2008) In a similar vein both Maxwell and Slater's work build on similar ideas and draws attention to some of the

TABLE 9.1 Features of "traditional" and "modern" food systems.

Food system feature	Traditional food systems	Modern food systems
Main employment in food sector	Predominantly food production	In food preparation and processing, packaging production and retail
Supply chain	Short and local food-chains	Long-chains with increasing numbers of food miles
Food production system type	Diversified, varied productivity	Only a few crops dominate; which require intensive, high input requirements
Typical farm size	Smallish family owned business	Industrial, large scale
Typical food consumed in each chain	Basic local staples	Processed foods with brand names; more animal products
Purchased food bought often from	Small-scale, local shops or local market	Large supermarket chains
Nutritional worries	Undernutrition, malnutrition	Long-asting dietary diseases
Main areas of national food shocks	Production shocks, for poor practices, rain, and harvesting	Price and trade problems
Main areas of household food shocks	Production shocks, for poor practices, rain, and harvesting	Income shocks and food poverty
Important environmental concerns	Degradation of the Soil	Nutrient loading, agricultural and industrial chemical runoff, poor water usage, and greenhouse gas emissions
Characteristics of Food Chain	Local to national	National to global

Based on *Ericksen, P.J., 2008. Conceptualizing food systems for global environmental change research. Glob. Environ. Chang. 18 (1), 234–245.*

fundamental shifts occurring within the food system as a whole. Combine these trends with the more recent Foresight program's findings, the UK government's think-tank (Foresight, 2010) of traditional and urbanized systems and we get a clear indication of the way modern developed food systems are behaving (see Table 9.1).

Food, along with many other natural resources these days is increasingly being seen in a holistic fashion, both as an independent system as well as in its symbiotic relationship with the environment. Indeed by offering the notion that food systems as well as other social goals of sustainability etc. (Table 9.2) in turn are both affected by, and influenced by, outside considerations of policy, economics and the environment, this explicitly closes the circle that sees the global food system increasingly at the heart of political and economic as well as social considerations (Ericksen, 2008; Gibson, 2016). Furthermore, with on-going environmental concerns of water availability; pollution; energy use; land degradation and biodiversity among others, this ensures the continuation of a holistic concept that becomes increasingly difficult to separate from wider societal issues. In sum modern food chain systems are

TABLE 9.2 Food systems from traditional to industrialized.

Traditional systems	Intermediate systems	Industrialized systems
• Simple technological tools, customs and practices • Very labor-intensive • Traditional harvesting and basic storage techniques • Lack of proper integration with local markets • Growers do not particularly understand urban needs • Very limited access to international markets • Unstable prices • Poor market information	• Facilities co-exist alongside traditional systems such as refrigeration and storage • Produce quantity and quality varies • Local and export markets • Requires closer integration of the supply chain • Consumer demands, in distant markets not fully cohesive or understood	• Access to sophisticated technologies • Harvesting is highly mechanized • Food processing/manufacturing sector are more sophisticated • Medium- and large-scale farms are increasingly more prevalent • Quality and cosmetic produce increasingly demanded by consumers • Achieves many deadlines of quality, safety, and volume of retailers and consumers • High volume, low cost foods produce more wastage at both retailer and consumer levels

Based on observations from *Foresight, 2010. Expert Forum on Reduction of Food Waste, UK Embassies Science and Innovation Network and Foresight, London.*

unrecognizable from those of just 2 or 3 decades ago. Importantly too, while back then good governance was exercised evenly from one end the chain to the other, subsequent integration and concentration both vertically and horizontally is proving to be a meaty challenge. As a result, and as mentioned, this hitherto socially unregulated aspect of modern food culture is attracting increasing calls for public accountability in terms of fair play, good governance and open transparency.

9.2.2 Multi-disciplinary FSC's

While the above represents a typical FSC in its simplest form, in reality between the production and consumer stages there are many considerations of economics, politics, sociology, science and technology, health, agronomy entomology, pathology and others that must be deliberated (Gra, 1990). Consequently, within each of these stages there is a diverse and disparate collection of participants all with varying roles and motivations.

From this it can be seen the influences or the "pulls" and "pushes" along the food chain are numerous. Factors such as geography, population demographics, the structure and health of the financial sector, technology, the health of the economy, inward investment, the political environment, not to mention geographic benefits and limitations and climate change as well as consumers ever demanding requirements—all these factors collude to shape and determine the efficacy of all FSC's chains across countries and products (Minten et al., 2009).

As mentioned earlier, any changes along one part of the chain have a domino effect along the entire supply chain ultimately affecting both producers and consumers alike. Taking a small family farm producer for the moment, changes in trade policy might mean local wheat prices become too expensive compared to cheaper imports. If this situation continues for several seasons with no matching subsidies from the government or trade associations, then the producer will have to adapt either by dropping prices to compete (affecting income), or perhaps adapt by changing crops which would mean a complete new set of agricultural inputs and associated costs. Moreover, increased or decreased labor, altered land management systems, extra investment in new technologies and a whole host of other considerations are dealt with for what may or may not turn out to be better or worse farmer incomes. This example reflects considerations at the individual level. In the wider picture, as has been mentioned another important driver of change in many developed economies is the trend of vertical integration that sees companies merging and buying up and down the chain. That is retailers buying processors, or farmers and manufacturers merging in alliances etc. This concentration of large parts of the chain has a profound effect on all other actors along the chain. The farming sector too, as has been mentioned earlier is also trending toward consolidation with increasingly larger farm sizes (Minten et al., 2009).

The trends and speed of change are different for different countries and regions.

9.2.2.1 Drivers of FSC change

As can be seen there are numerous components of a typical FSC from the wealth of individual's/countries; speed of growth; as well as the economic systems in operation (including transitional countries) among others. Add to these, however, social factors inter-alia the propensity toward globalization; urbanization; consumer preferences (i.e., more locally grown produce; changing dietary habits and more sustainable products etc. and the FSC ultimately becomes very complex with many competing and conflicting goals etc.

As for change, well it is not hard to see drivers of change in all supply chains are any one or combination of factors already described here and in the previous chapter. Yet, understanding such drivers in this context is paramount if policy and economic goals of food security, efficiency and wastage limitation are to stand any chance of success.

Placing all of this in context is incumbent on the big picture, so taking the global FSC as a whole the following chapter looks at the current food production capacity, usage and future needs.

9.3 Size of the global food sector

The food market is a difficult sector to pin down in terms of breadth and size. The reason being is that food is a confusing sector; it is many things—it traverses many industries, it is both a commodity and an ingredient, and of course a meal. As such quantifying the food industry is beset with difficulties as its value can be measured at every stage along the food chain (Murray, 2007). Take the output from the farm sector as an example, it is sold and processed and sold again—and potentially processed and sold again—each time adding value. Then there is the different categorization of foods—primary, processed, animal-human-biofuel uses, and these are just a few examples. On top of this there are also the huge amounts of food being bought and sold informally, on the gray and black markets, in fact, all in all the potential for miscalculation, double counting and omission is great. Yet despite these challenges for this book we have used the raw production values of 2009 to serve as an example because the full spectrum of figures are readily available. Using this data, it can be seen that the global industry was valued or estimated to be worth in the region of USD 4.8 trillion annually (UNEP, 2012). While, with growth rates of about 4.4% annually between 2009 and 14, the food industry result in a global market worth in excess of USD5.3 trillion annually (Murray, 2007; Alpen Capital, 2011). This growth was largely underpinned by continued rises in per capita income (next section), especially in emerging markets, which was expanding faster than the developed world (Murray, 2007; Alpen Capital, 2011).

9.3.1 Global food production

Of the 8.9 billion tons produced in 2009, stripping out food grown for alcoholic beverages this left 8.6 billion tons, from which when we factor in imports, exports and stock holding variations we arrive at 8.5 billion tons. Now set aside food produced for seed, for animal feed and industrial uses such as biofuels or the pharma industries and the like, we are left with approximately 4.2 billion tons (excluding alcoholic beverages)—less than half of what we originally started with. This is the average food produced annually for human consumption. Table 9.3 shows these figures in a little more detail.

TABLE 9.3 Global food production commodities by Volume (2009) (million tons).

Item	M. Tons
Cereals, excluding beer	2251
Sugar crops	1890
Vegetables	1008
Starchy roots	720
Milk, excluding butter	697
Fruits, excluding wine	593
Oil crops	490
Alcoholic beverages	310
Meat	285
Sugar and sweeteners	189
Vegetable oils	144
Fish, seafood	143
Eggs	68
Pulses	63
Animal fats	35
Offal's	17
Stimulants	17
Treenuts	14
Spices	7.5
Total	8945

Based on FAO's database - FAOSTAT, 2013. Food and Agriculture Statistics, Food and Agriculture Organisation.

A little more detail helps with perspective here. By re-arranging the above table, we can better visualize the main components we produce in order of volume as classified by similar commodity groups.

It can be seen too from that by far the biggest food commodity produced was cereal grains at between 2.2 and fast approaching 2.4 billion tons annually. This growth was pretty much universal although perhaps Africa and Latin America trailed the rest of the world in actual volumes.

When it comes to meat and, as mentioned previously, of the global total production in 2009 of about 285 million tons. We consume about 210 million cattle, 418 million sheep or goats, 1.1 billion pigs and 55 billion chickens annually—or about 40 kg per person per year (approx. 10, 2, 15 and 13 kg of beef, lamb, pork and poultry respectively). Production wise, breaking this figure down regionally we learn from FAOSTAT that predominant livestock in South and Southeast Asia mainly comprises pig (7 million ton) and chicken (9 million ton). In Europe pig production was dominant at approx. 27 million tons while in North America and Oceania it was more diversified with chicken (18 million ton), cattle (16 million ton) and pig (12 million ton). In Latin America, meat production was largely dominated by cattle (around 15 million ton) and chicken (around 17 million tons) while lastly, in both sub-Saharan Africa and the North African, Western and Central Asian regions it was mostly chicken (around four million tons each).

The fish and seafood production industry as a whole cannot be underestimated—indeed it reportedly supports the livelihoods and incomes of an estimated 540 million people—or about 8% of the world's population (FAO, 2012; UNEP, 2012). We can also see from the FAO statistics of 2011 there was an estimated 154 million tons available (a little increase on the 142 million tons of 2009) of which 80%–85% or 131 million tons was available for human consumption[1] (FAO, 2010b; FAO, 2012; FAO, 2013a; FAO, 2013b). Total production comprise about 90 million tons of captured seafood and 64 million tons of aquaculture giving a total per capita food fish supply of about 18.8 kg. On top of this there is also the aquatic plants component made up of 89 1000 tons of capture and 19 million tons of aquaculture for a total of about 20 million tons (FAO, 2013b).

China remains by far the largest fish-producing country with production of 52 million tons in 2010 (37 and 16 million tons from aquaculture and capture fisheries, respectively) with India being the next largest producer at just over nine million tons (FAO, 2012).

This means, looking at the figure almost as much fish, seafood and aquatic plants were supplied through farming (aquaculture) as was captured from the sea. Yet this is not the whole picture for while aquaculture or aqua-farming is indeed a growth industry it is not the overarching panacea to the problem of a declining fisheries industry that many had hoped for. The reason is simple—aquaculture, although it implies the farming of fish independent of the marine and freshwater ecosystems the truth is that for every 1 kg of farmed fish we produce we still require a significant amount of caught fish to feed them. In this regard therefore, in order to maintain current per capita fish consumption levels to 2050 it will require approximately 56% growth in the aquaculture industry. In turn this would require

[1]Based on extrapolated trends from FAO yearly statistics in the "Disposition of world fishery production" report 2009 FAO (2013b). "Yearbook of Fishery Statistics Summary tables." Retrieved 5 Jan 2019, 2019, from http://www.fao.org//fishery/publications/yearbooks/en.

about 23% increased fish landings as feed to support this production and any further collapse in marine ecosystems would consequently have a major detrimental effect on future production sustainability not to mention prices (Gustavsson et al., 2012).

9.3.2 Current food usage

As we have already highlighted, we utilize less than half of the food we grow, catch or rear for human consumption. The following highlights some of the key food usage statistics:

- Total food supply, approximately 8.9 billion tons
- Feed 1.133 billion tons
- Seed 142 million tons
- Processing 2.426 billion tons
- Other Utilization 743 million tons
- Food 4.2 billion tons (excluding alcoholic beverages)

Factoring in these deductions we can calculate that which is left for human consumption equates to about 1.8 kg of food or approximately 2828 kcal for every person on the planet every day. One of the important concerns relates to the way we utilize our food especially when we are dealing with the efficient and equitable allocation of this precious resource. One of the key considerations relates to the competition our food resources face from other uses such as animal feed or those used as industrial substrates (inputs) such as in the biofuel and pharma industries (UNEP, 2012). Indeed, many argue that such competition only adds to food insecurity around the world and suggest alternative uses other than food inputs in such industries bringing about a more equitable sharing of the food supply. The argument, however, is never that simple.

So, going back to the equitable use of the food supply, we can examine just how the levels of global food production are translated into nutritional status of individuals around the world. This can be seen in the prevalence of over and undernutrition, or more overarchingly malnutrition.

9.4 Employment in agriculture

According to the International Labor Organization (ILO) in 2011 about 1billion people around the world were employed in the agricultural sector. This represents about 35% (36.2% women, 32.8% men) of the total employed global workforce and the second greatest source of employment after services (43.8%)—with the industrial sector trailing at about 22.1%. In many countries too, the agricultural sector is also the most important sector for female employment especially in Africa and Asia (ILO, 2012). illustrates the global workforce by region in terms of gender and industry (ILO, 2012).

Taken together these figures while representative of total employment and considering global working population in 2008 was in the region of three billion (male 1.8 billion, women 1.2 billion), seen in isolation these figures can be somewhat confusing. The following graph combines this data and presents the figures as a proportion of the three billion globally employed.

Furthermore, the trend for employment in agriculture is a downwards one and one set to continue. As the literature highlights, of interest is that over the 10 years to 2008 in every region in the world without exception agricultural employment numbers have been on this marked downward trend. The 10-year trend also shows the rapid rises in the industrial and service sectors in almost all regions.

In the past the role of women in agriculture has not generally received broad-based recognition. However, trends in globalization and more acknowledgment of the valuable contribution that women play in both industry and the service sectors are redefining the relationships of women in economic and rural development paradigms (Hall, 1998). Sadly though, while progress is being made it has been a little slow and women continue to be plagued by low incomes and limited access to things like education, health care and equitable access to land and natural resources.

When it comes to food and improvements vis-à-vis employment in agriculture, the food sector and general food security—a strong case has been made regarding the combined importance of education and employment. Education it seems is at the heart of many attempts to improve human development although an important barrier to adult education remains the relatively high illiteracy rate (Mukudi, 2003).

9.4.1 Education and employment

Over the last few decades there have been great improvements in global literacy. However, in 2016 although high on the development agenda, literacy shows that 750 million adults (of which two-thirds are women) still lack basic skills in reading and writing. Accordingly, the latest data for 2016 tells us that 102 million of the illiterate population were between the ages of 15- and 24-year olds. Conversely, the global adult (25+) literacy rate was in the region of 86% in 2016, while the youth literacy rate was 91%. Once again gender disparity is an important consideration within these figures and it could be seen that in Southern Asia 73% of men compared to just 51% of women could adequately read and write resulting in a gender parity index of just 0.70. As a consequence, general low levels of education combined with high levels of illiteracy can drastically hinder the economic development of individual's and their countries (Gibson, 2016). It also has an important bearing too on the nutritional status of children as parental education has been found to be an influential factor in theirs and their children's nutritional status, so much so in-fact that children of illiterate parents are consistently seen to score more poorly on nutritional status indices (Mukudi, 2003).

The rationale for this is widely agreed, the better educated a person is, the more empowered they are with improved social skill sets which might also help in which more reasoned choices can be made regarding nutrition and health and of course in other social ancillary costs of food such as wastage and the environment (Dollahite et al., 2003; Rosegrant and Cline, 2003; CFS, 2007).

9.5 Commercial control over the supply chain

Everyone is involved in food; from production to processing to retail to consumption, everyone is affected. Yet increasingly of concern within the Food Supply Chain (FSC) dynamic are the fundamental relationships along the chain from farmer to consumer. This becomes especially important as the trend toward a more integrated global food chain increasingly sees fewer and fewer Small to Medium Enterprises (SMEs) in favor of a small band of large multi- or transnational corporations concentrating the lion's share of the market. A lot has been written about corporate power, even the United Nations Food and Agriculture Organization acknowledges that large transnational corporations have come to increasingly dominate world agricultural commodity markets (FAO, 2004; Mousseau, 2005). This allows such enterprises to wield direct and increasing influence on what is produced and distributed. Put in perspective, the top 500 businesses by market value in 2010 included in their ranks—23 food producers or retailers; 20 biotechnology companies; and 13 oil and gas giants. The top four food companies were: Wal-Mart-US, Nestle-Switzerland, Unilever-Netherlands/UK and Tesco-UK with a combined market value of $533.2 billion and a turnover in 2010 $637.1 billion with Wal-Mart alone accounting for $405 billion of this. Interestingly while Wal-Mart was only seventh in the rankings by market value it had by far the biggest turnover during that year, beating its nearest rival, the oil giant Exxon Mobil by over $100 billion. In-fact, so great are the food giants' economic power that in turnover alone the big four: Wal-Mart, Nestle, Unilever and Tesco's individual 2010 turnovers each out-earned the 2009 Gross Domestic Product (GDP) earnings of 171, 137, 127 and 133 countries respectively (Table 9.4). In-fact the combined turnover of the 23 top food companies of that year almost equaled that of the Russian Federation earnings of 2009 (Financial Times, 2010; World Bank, 2010).

From another perspective we can see that the top five corporations' share of the global food processing sector represents 18% of the entire global market. Adding in the next five companies and this figure rises to 28% (Mulle and Ruppanner, 2010).

Bringing these figures up to date (2017), according to Forbes one can see the top five food and beverage companies by market value are many of the same as in previous years:

1. Nestle—headquarters: Switzerland, market value $229.5 billion
2. PepsiCo—headquarters: US, market value: $159.4 billion
3. Coca-Cola—headquarters: US, market value: $182.9 billion

TABLE 9.4 Top corporations' share of the global food processing market 2009.

Share of the global market	%
Nestlé (Switzerland)	26
PepsiCo Inc. (USA)	12
Kraft foods (USA)	12
The Coca-Cola company (USA)	9
Tyson foods (USA)	8
Mars (USA)	7
Archer Daniels Midland company (USA)	7
Cargill (USA)	7
Danone (France)	6
Unilever (The Netherlands)	6

Inspired by Mulle, E.D., Ruppanner, V., 2010. Exploring the Global Food Supply Chain Markets, Companies, Systems. THREAD Backgrounder No 2, May 2010. V. Ruppanner. Online, 3D: 35.

4. Kraft Heinz Company—headquarters: US, market value: $110.4 billion
5. Anheuser-Busch InBev—headquarters: Belgium, market value: $213.1 billion

Such concentrations in the food retail sector can have negative consequences on the supply chain too. It has been noted that with the likes of Wal-Mart for instance, the sheer size of the company threatens market competition. Indeed, with over $400 billion in annual sales, the company is able to impose strict rules on quality control and enforce its own price targets on suppliers. Such is the bargaining position of these corporations that consumers can continue to pay relatively higher prices for goods on the shelf as retailers force price reductions on farmers and suppliers simultaneously threatening revenues of smaller individuals and companies further along the supply chain (Mulle and Ruppanner, 2010). Moreover, it has also been said that in pursuit of profit, larger corporation's practices also tend to reduce wages and working standards—although to be fair this is not confined to multinational agri-food retail businesses either. Such practices do not significantly differ by sector; however, one thing that is different is that other sectors by comparison rarely share the same inelastic captive markets. Furthermore, while globalization of trade opens up opportunities for agricultural exports it represents a clear potential threat in the development of internal markets through displaced competition—the cheap imports of commodities of a higher quality than can be found locally (Parfitt J, Barthel M et al., 2010).

Further elaborating on the charge that corporate profits are made all too often at the expense of the primary producers is the example of the genetics industry and the growing proclivity of biologically based intellectual property rights (IPR) ownership. Combined, genetic engineering and IPR's are charged with promoting monopolistic privilege over material that many believe should be common property. Effectively what is happening is the growing trend for proprietary ownership of seed and livestock at the genetic level thus ensuring farmers' continuing servitude. Interestingly in this point Windfuhr and Jonsén observe that:

> Whereas more than 90% of genetic resources for food and agriculture are from biotopes in the South, corporations in developed countries claim 98% of the patents on genes and living organisms. ***Windfuhr and Jonsén (2005).***

Furthermore, IPR's are also said to be an obstacle of technological transfer too—a barrier of common social development. This is especially so if you factor in the reality that many developing countries often lack the sophisticated patenting infrastructure enjoyed by many industrialized countries.

This is just one example and there are many more along the full length of the food chain. In fact, as more of our food is being innovated, transacted, processed and retailed by an ever-decreasing number of transnational or multinational companies, so many are beginning to question the sustainability of continuing trends. In this sense, one observer, Olivier De Schutter, the UN's special rapporteur on the right to food, at the 17th Session of the UN Commission on Sustainable Development commented:

> Trade is mostly done not between States but between transnational corporations … The expansion of global supply chains only shall work in favor of human development if this does not pressure States to lower their social and environment standards in order to become 'competitive States', attractive to foreign investors and buyers." ***CSD (2009).***

As a result, there has been an increasingly vocal backlash against what is seen as the commodification of the food chain. At the heart of the argument is the fundamental dichotomy inherent in both food-as-profit and food-as-a-right ideologies (Actionaid, 2010; La Via Campesina, 2011). The argument unfolds like this, while corporations are small to medium-sized, the many and varied stakeholders enable a more equitable division of power between producers, processors, retailers and finally consumers—and consequently, any benefits received are more evenly shared. By contrast, as the concentration of the agri-business sector (both in traditional horizontal[2] and vertical[3] integration strategies) sees the democratic free market being replaced by a de facto oligopoly.[4] From this point it does not take an economic mastermind to understand that a market concentrated with fewer companies effectively increases a company's economic power base. This argument incidentally, also underpins many anti-globalization and anti-capitalist movements; moreover, the juxtaposition of the two arguments is increasingly being polarized. That is to say, on the one side, as relentless capitalist economies demand ever growing profits, so companies respond with increasing innovation, ingenuity and novelty—further consolidating and concentrating their competitive advantage. On the other hand, traditionalists and anti-capitalist movements dismiss what some see as "progress" out-of-hand descending instead on G8 Summits and the like.

9.6 Food price volatility

One disturbing trend over the past few years has been the increased frequency of wild fluctuations in basic food commodity prices. As a result, in 2007/8 alone, it had been estimated that increased prices pushed a whopping 200 million people (or 133 or 115 million) depending on who you read into food poverty (SOFI, 2008; World Vision, 2019; GFMG, 2010).

This escalation in volatility has been attributed to many causes, not least of which is a direct consequence of globalization—or more precisely, a flawed global development model based on competition, self-interest and speculation. Others speculate that the rapid expansion of the developing economies, especially BRIC countries (Brazil, Russia, India and China) coupled with a growing change in dietary habits as being one important root cause. Others still cite increased general demand owing to increased wealth coupled with higher fuel costs (Shah, 2007; Watts, 2007; FAO, 2008; Smith and Edwards, 2008).

Indeed, regarding this last point, it serves to highlight the fact that there is an apparent strong correlation between cereal grains and oil prices. It is not the first time this has been suggested either with some even suggesting that were this convergence to continue in the same vein then the possibility of oil prices acting as a de-facto agricultural price index is certainly conceivable (von Braun, 1995; Schmidhuber, 2003; Brown, 2008; SOFI, 2008).

[2]Horizontal integration - is a microeconomic principle that sees expansion of the firms at the same level in the value chain as in retailers buying up other retailers.

[3]Vertical integration - is a microeconomic principle that sees expansion of the firms into other areas of the supply chain as in retailers buying up or merging with food processing industries or investing in primary producing activities.

[4]It takes approximately 6.5–7 kg of wheat grain to produce 1 kg of beef.

Notes: indices set at 2005 = 100.

In reality, however, price fluctuations that were seen over recent years was the result of a combination of factors beginning with a drought in Australian—the worst for a century, which halved farmer's wheat production volumes in 2007. Then there was the weight of increased demand from such growth markets as mentioned earlier (BRIC countries etc.) in which individuals are trending toward more meat-based diets.[4] Couple this with the fact that biofuel[5] production was seen at that time as something of a benevolent industry[6] and as oil prices spiked reaching $60 a barrel, so biofuels become much more competitive, further incentivizing the diversion of crops away from food. Moreover, consideration too needs to be made of a relatively new phenomenon of food commodity speculation. A phenomenon cited to be just as responsible for artificially inflating food commodity prices as these other drivers and one which, according to Frederick Kaufman (of Harper's Magazine) is unconscionable (Phillips, 2008; Vallely, 2009; Kaufman, 2010).

Ultimately the confluence of these four factors: shortages, increased demand, speculation, and competition from nonfood industries provided almost unprecedented volatility within the commodity markets (OECD-FAO, 2007; Shah, 2007; FAO, 2008; UNEP, 2012).

There are many widely accepted consequences too of these erratic movements in price. On a very basic level the demand for food is price inelastic, that is to say—overall, the quantity of food people require or consume varies little with changes in prices (Lee, 1993; UOVS, 2008). That said, people still exercise food choices; in poorer low-income countries for instance, people are very responsive to price and as prices rise so they may end up reducing demand for meat, dairy and vegetable products and increasing staples like bread and cereals. More affluent countries are not immune either and equally, rising prices too might force some to substitute cheaper food in place of expensive items rather than reducing the quantity consumed. However, while both communities suffer, it is generally agreed that it will the poorer sections of the community that will suffer the most in such situations. A worrying statistic in developing countries which places this notion in perspective was made by Regmi in 2001 when he suggested that for every 1% increase in food prices, food consumption expenditure decreased by about 0.75% and caloric intake reduced by 0.5% (Regmi, Deepak et al., 2001; Brown, 2008).

Not surprisingly then, frustration at rising food prices over the last two decades and more recently also translated into social unrest. For example, there were tortilla demonstrations in Mexico, pasta protests in Italy, maize disputes in Kenya, and even food riots in some areas, these were just a few of the many issues around the world which eventually led some governments to intervene (Lee, 1993; von Braun, 1995; King and Elliott, 1996; Regmi, Deepak et al., 2001; Eifert et al., 2002; Watts, 2007; BBC, 2008; Brown, 2008; Delva, 2008; UOVS, 2008; Clapp, 2009; FAO, 2009b, SOFI, 2018). Some of these interventions included the removal

[5]Biofuels are made from corn, sugar-cane/beet and/or cassava among others, crops are converted into ethanol as a greener alternative to fossil fuels. The industry is vast, led in spirit by Brazil. In 2007/8 the Committee for Food Security (CFS) estimated that the biofuel industry utilized a whopping 4.7% of annual cereal production in the production of alternative fuels sources CFS (2008). Agenda Item II: Assessment of the World Food Security and Nutrition Situation. Committee On World Food Security: 34th Session, Rome, Food and Agriculture Organization.

[6]*The US, in a bid to reduce reliability on oil exporting countries increased the use of corn and other crops in the production of ethanol and bio-diesel, however, rather than importing these crops the US attempted to be self-sufficient and diverted approximately a third of its corn crop for biofuel production.*

of quotas and tariffs on imports; the banning or introduction of export duties; buying food at preferential rates; increasing grain stockpiling and the increasing of food subsidy interventions—all in an effort to control the prices of certain basic food staples. However, it was recognized that this was largely a short-term fix and one which was not sustainable in the longer term (Vidal, 2007; SOFI, 2008; Dupont and Thirlwell, 2009).

However, while such inflationary pressures are on course to increase it must be noted that high food commodity prices in and of itself, is not necessarily a bad thing. There are inherent winners and losers in food price increases—higher prices, for instance, are good for sellers or farmers who are net sellers of food. This can have a trickle-down effect too, translating into increased agricultural employment, better wages and potentially increased agricultural investment (SOFI, 2008).

9.7 Current global food situation: production, usage and needs: food balance sheets

The food balance sheets (FBS) illustrate the structure of a country's food supply during a specified reference period. They were first introduced during the First World War, used extensively after the war and were improved by the Second World War. By the inception of the United Nation's Food and Agriculture Organization great emphasis was placed on both the preparation and utilization of the FBS. Since then the food balance sheets have been prepared and published by FAO on a regular basis. The use of this data allows governments, organizations and individuals to analyses trends over time in the overall national food supply and to determine whether supply, as a whole, is sufficient in relation to its nutritional requirements. Moreover, this data can also be used to determine changes in production and consumption patterns of its population, better enabling good stewardship and governance in matters of food policy (FAO, 2002).

Specifically, the FBS disseminates data on both supply and utilization of each food item.[7] In detail the FBS comprises data on: production; imports and exports; that used for feed and seed; some wastage figures; and other utilization data to arrive at total food supply figures for human consumption (FAO, 2002).

Production and Trade: Production and trade data are usually part of regular national official statistical analysis within most countries and are either based on direct survey's and enquiries or in its absence, estimated by Government bodies or agencies. Any changes in stock holdings are derived from information from any combination of marketing authorities, factories or farmer stock surveys. With regard to trade statistics, these data usually obtained from industrial and manufacturing censuses or surveys and these figures can be used for instance, as an indication of a country's dependency on its imports (import dependency ratio) to feed itself among other things.

Feed and Seed: The FBS collects data on that part of total food production that is used for animal feed and for future seeding. These figures are generally obtained from cost of production surveys or can be estimated by the relevant government bodies and agencies concerned.

[7]Food items consist of a basket of primary and a number of processed commodities which are potentially available for human consumption.

Once again both indices can be used to garnish a picture of the degree to which primary food resources are used in the production of both seed and animal feed.

Processing and other utilization: As we shall see in later sections not all food that is grown is used for food. Other industries such as the biofuel, pharma industries vie for the same finite resources and as such valuable data on nonfood uses are collected from industrial or manufacturing censuses and surveys.

Stock changes and waste: Stock changes are calculations which aim to reflect actual changes in what is produced, what has been imported/exported as well as that part of a country's stock holdings which has been utilized or added to. Wastage covers are estimates, themselves are often based on assumptions are based on expert opinion obtained in a country:

Supply: What is left after the above is taken into consideration is total proportion of food remaining for human consumption. This is an important data stream for analyzing trends in food demand and consumption expenditure as well as things like income elasticity coefficients etc (FAO, 2002). This data is often represented in several forms either as total volumetric production figures in tons or in terms of per capita supply as in kg, grams or calorie equivalents per capita per.

The proceeding sections look at what and how all the food that is produced year on year is utilized. Yet before any of this can be discussed it is incumbent to understand something of the difficulties inherent in statistics such as these and those described above.

9.8 Nutritional status: over- and undernutrition

According to both the UN Standing Committee on Nutrition (SCN) and the World Health Organization (WHO) and others, malnutrition today is one of the most striking public health problems and a major contributor to the total annual global disease burden (DFID, 2009; WHO/EMRO, 2010; WFP, 2019).

Malnutrition by itself refers to the overarching problems of "mal" or bad-nutrition; in other words, it describes all deviations from adequate or optimal nutritional status. This is an all-encompassing definition which includes not only deficiencies as is common but also inappropriate nutritional combinations too. Moreover, this view of malnutrition also incorporates poor nutrient absorption and/or the poor biological uptake of nutrients (bioavailability[8]) and lastly, as is becoming increasingly common, the excessive intake of nutrients too (Thomas, 2007; DFID, 2009).

Undernourishment/undernutrition results from the insufficient intake or the inadequate utilization of ingested nutrients (Shetty, 2006; Thomas, 2007; DFID, 2009). Historically, this was associated with overall caloric or energy intake, but strictly speaking it can and does include insufficient micronutrient intake too. This is because when calories are deficient,

[8]Bioavailability is the availability, irrespective of the quantity consumed, of useable nutrients to the body. That is, not all ingested nutrients are necessarily available or properly metabolized owing to perhaps such barriers as infection and disease Shetty (2002). Measurement and Assessment of Food Deprivation and Undernutrition: Keynote Paper: Measures of nutritional status from anthropometric survey data. International Scientific Symposium, Rome.

the likelihood is that micronutrients (vitamins and minerals) are also likely to be deficient too (Thomas, 2007).

Overnutrition—whether overweight or obesity on the other hand, are far from being the once labeled "diseases of affluence." Instead both are becoming increasingly prevalent in developing countries (Cleaver et al., 2006; WHO, 2010). Sweeping dietary changes at the global level see largely hitherto plant-based diets being quickly replaced by energy-dense, high-fat, high-sugar processed diets (Muller and Jahn, 2009; WHO, 2010). The health implications of over eating are also in turn being further compounded by physical inactivity as the propensity for more sedentary lifestyle is becoming the norm in both developing and developed regions alike. In fact, so prevalent is the problem that the World Health fully expect to see chronic disease such as Cardiovascular Diseases (CVD) (heart disease, hypertension, stroke) and its attendant comorbidities such as diabetes to be the leading cause of death in all developing countries in the very near future (WHO, 2010).

However, for the full picture, irrespective of whether under- or overnutrition malnutrition can be thought of, broadly speaking, as taking two forms—chronic and acute (Cleaver et al., 2006).

9.8.1 Chronic

In cases of chronic malnutrition usually occurs in prolonged or protracted crisis where people continually fall short of adequate and/or appropriate nutritional intake. Without regular or sufficient key vitamins such as iron, zinc, vitamin A, and iodine individuals, especially children, development is severely hindered. For this reason, chronic malnutrition is often seen to result in stunting (short stature for age), and it is this form of malnutrition that is ultimately responsible for most of the hunger-related deaths every year (USAID, 2007; World Vision, 2019). In addition, as chronic undernutrition occurs in children early in life (between 6 and 24 months) this particularly affects their mental development as well as their physical ability to cope with similar circumstances when adults, making them more vulnerable to chronic illnesses throughout the course of their lives (Checchi et al., 2007; USAID, 2007).

9.8.2 Acute

Acute malnutrition by contrast is often brought on quickly and not always with sufficient warning. Sudden food shortages whether from poor harvests or conflict mean that food, perhaps once freely available is no longer obtainable in significant amounts. This can manifest in rapid weight loss and is often associated with wasting (thin stature for height). Because of the sudden nature of acute malnutrition, as opposed to chronic or slow-onset malnutrition, acute malnutrition generally has a more immediate and profound effect on an individual's immune system. This is particularly so in children under-5. As a result, there is a strong correlation between acute malnutrition and mortality and in such emergencies, it is the acutely malnourished therefore who are to mostly dealt with first (Checchi et al., 2007; USAID, 2007; World Vision, 2019).

However, in terms of equitable food distribution there is more focus on undernutrition than there is with overnutrition. This is simply because the focal point of the social lens over the last few years has been firmly focused on issues of food security, and while there are no extant measures of food insecurity—undernourishment or measures of underweight serves as effective proxies.

One of the most widely used measures of undernourishment is the FAO's "Prevalence of Undernourishment" figures. It is calculated on the basis of three parameters: average food consumption (availability); inequality of access; and the minimum calorie requirement for a "typical" average person weighted by the population demographic (MDG, 2010) highlights the regional prevalence of undernourishment as of 2010/12. In this year there were an estimated 868 million undernourished people in the world—or about 12.5% of the population. It must be said too that although these figures were way down on 2009 figures of 1.02 billion, the downturn has been attributed more to improved economic conditions than any great progress in hunger alleviation policies (FAO, 2010a). Despite this caveat this figure of 868 million people is still large by any standards (CIA Factbook, 2010; FAO, 2010a; SOFI, 2012).

As for trends over time, it would seem that little progress has been made over the last 40 years or so suggestive, even by FAO's understanding:

> ... that present solutions are insufficient ... ***SOFI (2009).***

On the face of it, it can be seen that in the 3 decades to the late 1990's there was a slow but definite decline in both absolute numbers as well as in the prevalence or percentage of globally undernourished people in the world. Over the last 10 years or so, however, this trend seems to be reversing with prevalence of undernourishment figures over the last 5 years or so hovering at a low double digit percentage of the total world population (CFS, 2008; SOFI, 2008; FAO, 2009a; SOFI, 2009; WFP, 2019).

While this gives an overall trend, such figures do little to pinpoint regional trends. Much literature attests to the fact that food insecurity is predominantly a phenomenon of lesser developed countries. And so it is that while these developing countries account for approximately 70% of global population—19% of their numbers are considered to be suffering from undernutrition (SOFI, 2009). Furthermore, year on year the developing world continue to collectively represent in excess of 95% of the global prevalence of undernutrition numbers.

On a positive note and for the sake of the big picture if we look at the figures from another perspective, we can see that today's 19% is actually way down on the near 35% figures of 1969. In-fact the intervening period has witnessed some great gains. Take for instance, the near 35% figure of the developing world during the 1969–71 period and project that figure on 2010's population figures—this would, (based on a population of 5.67 for the region) represent undernourished figures of nearly 1.99 billion (DESA/POP, 2010). Compare these to actual figures of 852 million for the region (2010/12 figure's) and the reduction of well over one billion undernourished and progress can definitely be seen (SOFI, 2012).

Yet even this does not adequately portray the full picture of regional trends. For that we can combine the previous two figures into a picture that represents both the numbers and prevalence (or percentage of populations) of undernourishment of individual regions. In doing so the following uses the fully disaggregated regional figures from the rolling period 2010—12 (FAO, 2011).

As can be noted from the literature one can see the full grim reality of regional inequalities. Of the total 815 million undernourished, the vast majority were seen to come from the developing countries. Notably, most of these numbers comprised Asia and sub-Saharan Africa, representing the bulk of undernourished people in the region. Further, from these figures it can be seen that within the sub-Saharan African region, Ethiopia is the worst affected followed by Kenya Sudan, Nigeria and Tanzania. In Asia too, widespread undernourishment was found in Southern Asia India followed by Pakistan and Bangladesh while in East Asia, China. In South-eastern Asia the story continues with Indonesia and Philippines respectively. In the America's, particularly in Latin America the worst affected are Brazil and Columbia; while in Haiti in the Caribbean people were also underfed.

Note: The FAO are not the sole publishers of undernourishment metrics—both the United States Department of Agriculture's Economic Research Service (USDA/ERS) as well as the World Health Organization (WHO) also collect and collate indices of nourishment. In the case of the ERS their indices differ from the FAO in two main areas; firstly, the ERS use a different methodology relying on a universal cut-off point of 2100 kcals per person to determine undernourishment; and secondly the data is restricted to 70 developing countries. The WHO by contrast concentrate on anthropometric indices of both adults using a measure of body mass index (BMI) and in children under five using three measures namely: a low height-for-age—**stunting** metric; a low weight-for-height—**wasting** metric; and a low weight-for-age—**underweight** metric. Moreover, unlike the measures of other institutions which are subject to complex statistical manipulation, the WHO's figures are largely unadulterated.

9.9 Sufficient food

It can be seen clearly from the above statistics that the vast majority of malnutrition fall in the regions of Africa, Asia the Caribbean and parts of South America. This inequality is not a recent phenomenon and marks a longtime trend spanning decades if not centuries. This regional inequality in availability and food consumption patterns is further reflected in the food security status of these regions. Indeed with many underdeveloped countries receiving a less than equal share of world food availability it is perhaps not surprising that perceived worldwide food shortages continue to persist (Charles, 2008; Sachs, 2008).

The reality, however is very different and there is not, nor have there been for many years now, global food shortages; in fact, the opposite is true. This is evidenced in which shows clearly that the global food supply continues to provide an adequate diet consisting of over 2800 kcals for every single man, woman and child on the planet (SOFI, 2012).

Put simply, thanks in no small part to the green revolution, global food production continues to produce enough food for the entire world population. From this it can be seen that while food shortages, hunger and starvation are still prevalent in many regions of the world, insufficient food production is not the cause (DeRose et al., 1998; Freedom 21, 2008). This state of affairs serves to highlight a fundamental and underlying truth that while there is no global food shortage as such, there does in-fact exist vast regional disparities in availability, access, and ultimately consumption patterns.

References

Actionaid, 2010. Actionaid. Retrieved 23rd Jan 2019, 2019, from: https://www.actionaid.org.uk/about-us.

Alpen Capital, 2011. GCC Food Industry. Bahrain, Alpen Capital: 71.

BBC, 2008. Clashes at Nairobi Food Protest. BBC Online.

Brown, L., 2008. Why Ethanol Production Will Drive World Food Prices Even Higher in 2008. Retrieved 7 June 2019, 2019, from: https://www.issuelab.org/resource/why-ethanol-production-will-drive-world-food-prices-even-higher-in-2008.html.

CFS, 2007. Assessment of the World Food Security Situation. Committee on World Food Security: Thirty-Third Session. Food and Agriculture Organisation, Rome.

CFS, 2008. Agenda Item II: Assessment of the World Food Security and Nutrition Situation. Committee on World Food Security: Thirty-Fourth Session. Food and Agriculture Organisation, Rome.

Charles, C., 2008. Food Shortages: How Will We Feed the World? Telegraph online).

Checchi, F., et al., 2007. Public Health in Crisis Affected Populations A Practical Guide for Decision-Makers. Humanitarian Practice Network (HPN) at ODI, London.

CIA Factbook, 2010. The CIA World Factbook. Retrieved 12th Oct 2019, 12th Oct 2019, from: https://www.cia.gov/library/publications/the-world-factbook/geos/xx.html.

Clapp, J., 2009. The global food crisis and international agricultural policy: which way forward? Global Governance 15 (2), 299–312.

Cleaver, K., et al., 2006. Agriculture and Rural Development: Hunger and Malnutrition. World Bank Seminar Series: Global Issues Facing Humanity. The World Bank, Washington DC.

CSD, 2009. Contribution of Mr. Olivier De Schutter Special Rapporteur on the Right to Food; 17th Session of the UN Commission on Sustainable Development. UN Commission on Sustainable Development, New York.

Delva, J.G., 2008. Uneasy Calm in Haiti after Food Price Protests. Reuters.

DeRose, L., et al., 1998. Case study: the importance of non-market entitlements. In: DeRose, L., Messer, E., Millman, S. (Eds.), Who's Hungry? and How Do We Know? Food Shortage, Poverty, and Deprivation. United Nations University Press, Tokyo, New York, Paris.

DESA/POP, 2010. World Population Prospects. UN Department of Economic and Social Affairs.

DFID, 2009. Draft Terms of Reference Tackling the Neglected Crisis of Undernutrition Research Programme Consortia. Department for International Development (DFID), London.

Dollahite, J., et al., 2003. The impact of nutrition education on food insecurity among low-income participants in EFNEP. Fam. Consum. Sci. Res. J. 32, 127–139.

Dupont, A., Thirlwell, M., 2009. A new era of food insecurity? Survival 51 (3), 71–98.

Eifert, B., et al., 2002. The World Grain Economy to 2050: a dynamic general equilibrium, two sector approach to long-term world-level macroeconomic forecasting. University Avenue Undergraduate Journal of Economics (Online).

Ericksen, P.J., 2008. Conceptualizing food systems for global environmental change research. Glob. Environ. Chang. 18 (1), 234–245.

FAO, 2002. Training in the Preparation of Food Balance Sheets. Food Balance Sheets: History, Sources, Concepts and Definitions. FAO, Italy, Rome, p. 6. Paper no. 5.

FAO, 2004. The State of Agricultural Commodity Markets 2004. U.N. Food and Agriculture Organisation, Rome, p. 30.

FAO, 2008. Crop Prospects and Food Situation No.1 2008: Global Cereal Supply and Demand Brief. U.N. Food and Agriculture Organisation, Rome.

FAO, 2009a. Address by Jacques Diouf Director-General of the Food and Agriculture Organization of the United Nations (FAO). Food Security for All. Food and Agriculture Organization, Madrid.

FAO, 2009b. Food Outlook. Rome, Food and Agricultural Organisation.

FAO, 2010a. Global Hunger Declining, but Still Unacceptably High: International Hunger Targets Difficult to Reach. News Brief. Rome, Food and Agriculture Organisation.

FAO, 2010b. The State of World Fisheries and Aquaculture 2010. Food and Agriculture Organisation, Rome, Italy, p. 197.

FAO, 2011. Food Security Statistics: Country Profiles. Food and Agriculture Organisation.

FAO, 2012. The State of World Fisheries and Aquaculture 2012. Food and Agriculture Organisation, Rome, Italy, p. 209.

FAO, 2013a. Fisheries and Aquaculture Department. Retrieved 5 June 2019, 2019, from: http://www.fao.org/fishery/statistics/en.

FAO, 2013b. Yearbook of Fishery Statistics Summary Tables. Retrieved 5 Jan 2019, 2019, from: http://www.fao.org/fishery/publications/yearbooks/en.

FAOSTAT, 2013. Food and Agriculture Statistics. Food and Agriculture Organisation.

Financial Times, 2010. World top 500 companies by market value. Financ. Times. Retrieved 21st March 2019, 2019, from: https://www.ft.com/ft500.

Foresight, 2010. Expert Forum on Reduction of Food Waste. UK Embassies Science and Innovation Network and Foresight, London.

Freedom 21, 2008. Meeting essential human needs. In: 9th Annual National Conference. Freedom, Dallas, p. 21.

GFMG, 2010. The 2007/08 Agricultural Price Spikes: Causes and Policy Implications. DEFRA/The Global Food Markets Group, London.

Gibson, M., 2016. The Feeding of Nations: Re-defining Food Security for the 21st Century. CRC Press, Boca Raton, Florida.

Gra, J.l., 1990. A Commodity Systems Assessment Methodology for Problem and Project Identification. University of Idaho, Moscow, Idaho.

Gustavsson, J., et al., 2012. Global Food Losses and Food Waste. Study Conducted for the International Congress SAVE FOOD! at Interpack2011 Düsseldorf, German. Swedish Institute for Food and Biotechnology (SIK) and FAO: 38, Rome, Italy.

Hall, D.O., 1998. Food security: what have sciences to offer? Paris, France. The International Council for Science.

ILO, 2012. Global Employment Trends 2012. International Labor Organisation, Geneva, Switzerland, 121 pp.

Kaufman, F., 2010. The Food Bubble: How Wall Street Starved Millions and Got Away with it. Harpers Magazine. The Harper's Magazine Foundation, New York, 8 pp.

King, M., Elliott, C., 1996. Education and debate: averting a world food shortage: tighten your belts for CAIRO II. Br. Med. J. 313, 995–997.

La Via Campesina, 2011. La via Campesina: The International Peasant's Voice. News & Views. Retrieved 21st Feb 2019, 21st Feb 2019, from: http://www.viacampesina.org/en/index.php?option=com_content&view=frontpage&Itemid=1.

Lee, J., 1993. Observations on Agricultural Policy, Policy Reform and Public Policy Education. 43rd Annual Conference of the National Public Policy Education Committee, Florida.

MDG, 2010. Millennium Development Goals Indicators, United Nations Development Group. Retrieved 14th June 2019, from: http://unstats.un.org/unsd/mdg/.

Minten, B., et al., 2009. Linking Urban Consumers and Rural Farmers in India: A Comparison of Traditional and Modern Food Supply Chains. The International Food Policy Research Institute (IFPRI), New Delhi.

Mousseau, F., 2005. Food Aid or Food Sovereignty? Ending World Hunger in Our Time. Oakland Institute, USA.

Mukudi, E., 2003. Education and nutrition linkages in Africa: evidence from national level analysis. Int. J. Educ. Dev. 23, 245–256.

Mulle, E.D., Ruppanner, V., 2010. Exploring the Global Food Supply Chain Markets, Companies, Systems. THREAD Backgrounder No 2, May 2010. V. Ruppanner. Online, 3D: 35.

Muller, O., Jahn, A., 2009. Malnutrition and maternal and child care. Matern. child health 287–310.

Murray, S., 2007. Food: The World's Biggest Industry. Forbes, Online.

OECD-FAO, 2007. Agricultural Outlook 2007–2016. OECD-FAO, Paris/Rome.

Parfitt, J., et al., 2010. Food waste within food supply chains: quantification and potential for change to 2050. Philosophical Transactions of the Royal Society 365 (1554), 3065–3081.

Phillips, L., 2008. Senators Hear Testimony on Legislation to Reduce Food and Energy Prices by Curbing Excessive Speculation in the Commodity Markets. Senate Committee on Homeland Security and Governmental Affairs, Washington.

Regmi, A., et al., 2001. Cross-country analysis of food consumption patterns. In: Regmi, A. (Ed.), Changing Structure of Global Food Consumption and Trade. United States Department ofAgriculture Economic Research Service, Washington, D.C.

Rosegrant, M.W., Cline, S.A., 2003. Global food security: challenges and policies. Science 302 (5652), 1917–1919. Global food security will remain a worldwide concern for the next 50 years and beyond. Recently, crop yield has fallen in many areas because of declining investments in research and infrastructure, as well as increasing water scarcity. Climate change and HIV/AIDS are also crucial factors affecting food security in many regions.

Although agroecological approaches offer some promise for improving yields, food security in developing countries could be substantially improved by increased investment and policy reforms.
Sachs, J., 2008. The Power of One: How to End the Global Food Shortage. Time Magazine Online.
Schmidhuber, J., 2003. The Outlook for Long-Term Changes in Food Consumption patterns: Concerns and Policy Options. FAO Scientific Workshop on Globalization of the Food System. FAO, Rome.
Shah, A., 2007. Food Aid.
Shetty, P., 2002. Measurement and assessment of food deprivation and undernutrition: Keynote paper: measures of nutritional status from anthropometric survey data. In: International Scientific Symposium, Rome.
Shetty, P., 2006. Malnutrition and undernutrition. Medicine 31 (4), 18–22.
Smith, K., Edwards, R., 2008. The year of global food crisis. Sunday Herald Online.
SOFI, 2005. State of Food Insecurity. Food and Agricultural Organization, Rome, p. 40.
SOFI, FAO, IFAD, UNICEF, WFP, WHO, 2018. The State of Food Security and Nutrition in the World 2018, Building climate resilience for food security and nutrition. FAO, Rome.
SOFI, 2008. The State of Food Insecurity in the World 2008. Food and Agriculture Organisation, Rome, p. 59.
SOFI, 2009. The State of Food Insecurity in the World 2009. Food and Agriculture Organisation, Rome, p. 59.
SOFI, 2012. The State of Food Insecurity in the World 2012. SOFI. Rome, Food and Agriculture Organisation, p. 65.
Thomas, D.R., 2007. Disorders of Nutrition and Metabolism: Undernutrition. The Merck Manual Online Medical Library. R. S. Porter and J. L. Kaplan. Merck Sharp & Dohme Corperation, New Jersey.
UNEP, 2012. The Critical Role of Global Food Consumption Patterns in Achieving Sustainable Food Systems and Food for All. United Nations Environment Programme, Nairobi, Kenya, p. 43. UNEP Discussion Paper.
UOVS, 2008. Demand and Supply Theory.
USAID, 2007. Food for Peace: FY 2008 P.L. 480 Title II Program Policies and Proposal Guidelines. United States Agency for International Development, Washington, D.C.
Vallely, P., 2009. The Big Question: Why Is So Much of the World Still Hungry, and what Can We Do about it? the Independent. The Independent, London.
Vidal, J., 2007. Global Food Crisis Looms as Climate Change and Fuel Shortages Bite: Soaring Crop Prices and Demand for Biofuels Raise Fears of Political Instability. Guardian online. London, Guardian News and Media Limited, 3 November 2007.
von Braun, J., 1995. A New World Food Situation: New Driving Forces and Required Actions. International Food Policy Research Institute, Washington, D.C (The International Food Policy Research Institute).
Watts, J., 2007. Riots and hunger feared as demand for grain sends food costs soaring. Guardian Online.
WFP, 2019. Website of the World Food Program. Retrieved 25 Feb 2019, 2019, from: http://www.wfp.org/.
WHO, 2019. Website of the World Health Organisation. Retrieved 2nd April 2019, 2019, from: http://www.who.int/.
WHO/EMRO, 2010. Regional Strategy on Nutrition 2019–2010: You Are what You Eat. Regional Strategy on Nutrition WHO/EMRO. World Health Organisation: Eastern Mediterranean Regional Office, Cairo, p. 56.
Windfuhr, M., Jonsén, J., 2005. Food Sovereignty: Towards Democracy in Localized Food Systems, vol. 70. ITDG/FIAN, Warwickshire.
World Bank, 2010. WorldBank: World Development Indices. World Bank Group.
World Vision, 2019. Website of World Vision. Retrieved 23rd May 2019, 2019, from: http://www.worldvision.org.uk/.
WTO, 2010a. International Trade Statistics 2010. World Trade Organisation, Geneva.
WTO, 2010b. Understanding the WTO: Basics - the GATT Years: From Havana to Marrakesh. Retrieved 30 March 2019, 2019, from: http://www.wto.org/english/thewto_e/whatis_e/tif_e/fact4_e.htm.

CHAPTER

10

Global food waste

OUTLINE

Most people would not deny that food wastage is a problem; it is also in direct conflict with the fundamentals of good housekeeping and compromises the integrity of the notion of sustainable living. Interestingly too, in large part from the perspective of societal or social responsibility, food issues till now have often been framed within contexts of food and human rights; of food security; food safety; health, and nutrition; and of animal welfare among others. Of course, on occasion, when these issues are broadened, they might further include employment matters, fair trade/fair wages, sustainability, and natural resource limitations.

Food and Society
https://doi.org/10.1016/B978-0-12-811808-5.00010-6

Coming late to the party, however, are issues of food waste; albeit part of the broader spectrums of natural resources and sustainability, till now the role of food waste has generally been that of an endnote—a side line. The good news though is this is changing. Thankfully, more and more people are beginning to get the message that all the world's people, without exception, have to be supported, clothed, housed, and fed solely from the earth's finite natural resources (EC, 2011a,b).

Broken down, the essence of the issue is the fact that a shamefully large part of food is wasted compared with what is being produced. Indeed, food waste, which also has other political, economic, environmental, and social implications, is of increasing public concern worldwide. In the *political* arena, issues of governance and responsibility are at the forefront of the problem; waste-less targets for instance have to be met and regulation, policy and change have all to be addressed. In *economic* terms, food waste is wasted money; money lost to the consumers, to the suppliers, lost in the upkeep and maintenance of landfill sites and other associated expenses of waste management such as transport and operational costs, etc (Monier et al., 2010). The *environmental* costs of wasted food too, produced at different stages of the food supply chain (production, transport, manufacturing, distribution and consumption) must also be well-thought-out. Considering the food sector represents a significant impact on Global Warming Potential (GWP) (a tool used to compare the global warming impacts of different gases.) (Pachauri and Meyer, 2014). So it can be seen that there are considerable environmental impacts from wasted food within the food sector itself (Monier et al., 2010). Wasting food also raises many uncomfortable *social* questions. Given the fact that there are approximately 815 million (a staggering 10.7% of the population) undernourished people around the world; add this to the fact that obesity is a fast growing global problem affecting 1.9 billion people (WHO, 2018), and furthermore, given the fact that every year approximately nine million people die of hunger and hunger-related diseases (MercyCorps, 2018) while current food production is sufficient to feed the world's entire population; in that context and taking into account periodic financial crisis and rising food prices - the fact that estimates put global wastage quantities at between 30% and 50% of production (depends on who one reads - see below). Consequently, some believe that food waste must be seen for the shameful scourge it surely is—simply criminal!

Of the total world food production, approximately four billion tons per year, food waste occurs at every level of the food supply chain from agricultural production through to the point of consumption. Whether from unavoidable losses of weather and pests through to avoidable waste from poor practices, food waste is composed of raw and cooked foods such as vegetable peelings, meat trimmings, bones, carcasses and organs as well as food which is simply out of date or spoiled or that which is simply left on our plate after we have eaten (Europa, 2012; IMechE, 2013b).

A few years ago, an important, albeit under-publicized report, was authored by Gustavsson, Cederberg and Sonesson of the Swedish Institute for Food and Biotechnology (SIK) on the problem of food wastage. It was written for the International Congress "SAVE FOOD!" in Düsseldorf, Germany in 2011 and highlighted the sheer volume of lost or wasted food that occurs every year on a global scale (Gustavsson et al., 2012). The report, compiled at the behest of both the Food and Agriculture Organization of the United Nations (FAO) and Interpack 2011, was a brief summary based on considerable evidence. While estimates of global losses vary considerably, in this report the major finding suggested that annually, roughly one-third of all global food produced for human consumption was lost or wasted. The reports' estimates, amounted to a whopping 1.3 billion tons (approximately) of food per year. By contrast another report published by the Institution of Mechanical Engineers in January 2013 suggested a little more waste than this—in fact as much as between 30% and 50%, or between 1.2 and 2bn tons (IMechE, 2013a,b). Indeed, other estimates from reports around the world suggest food waste ranges from between one quarter to one third to 50 percent, depending on who one reads (Stuart, 2009; Parfitt et al., 2010; Parfitt and Barthel, 2010; Gustavsson et al., 2012; Kummu et al., 2012). Despite these wildly varying figures the one unavoidable truth is the fact that they are all excessively high. Another common finding too is that in low-income countries, much food is wasted mostly at the early to middle stages of the food supply chain; while among the medium- and high-income countries food was largely wasted at the consumption stage (Gustavsson et al., 2012).

10.1 Food waste: historical perspectives

Food waste was not always the side issue it has become over the last few decades. In the 18th century for example, Robert Heron writing for the Board of Agriculture in 1794 commented on the wasteful mill practices within the Isles of Great Britain and suggested change for the sake of progress (Heron 1794). In another example in a 19th article in the Popular Science magazine, Frederic Mather (1894) talks of the potential for waste to sustain mankind:

> It has been stated that if the waste products of the world had been saved they would sustain the present population for more than a hundred years. ***Mather (1894).***

Furthermore, citing the benefits of converting waste products into tradable commodities Frederic Mather looks at the once wasted by-product from cotton seed oil production as an example of ingenuity and technology.

> What was garbage in 1860 was fertilizer in 1870, cattle food in 1880, and table food and many things else in 1890. ***Mather (1894).***

Sometime later, and in a different setting the horrors of food waste were forefront in many people's minds when in 1917 it was estimated that food waste in Chicago alone amounted to around 1,250,000 pounds per day while total waste from the entire US amounted to $700 million per day/or year. A staggering amount which was enough, as experts at the time agreed, to sustain every man, woman and child in the 13 largest cities in the US (Windsor, 1917).

Meanwhile, in Britain during the war with up to two thirds of its food requirements eventually being imported, an odd situation arose where both food surpluses and food waste existed alongside each other. During the same period tackling the issues of food waste was one of the initial targets of the Women's Institute (WI). Set up in Britain in1915,[1] the WI initially set out to encourage women to get involved in food production during the war

[1]The WI initially began in Canada in 1897 in Ontario and soon spread to Belgium, Poland and the USA.

(WI, 2013). The government too realized that they needed to find ways of producing more food at home. This was tackled on many fronts but perhaps one of the most well-known ways in which the food situation was confronted involved giving the women's institute a grant whereby they would travel the country giving talks and demonstrations about methods of intensive cultivation, using subsidized fertilizers so farmers and countryfolk could grow plants more cheaply and quickly. However, the WI is better remembered for their workshops on preservation methods of surplus fruit and vegetables through jam making, bottling; pickling; and canning techniques, etc.

10.2 Attitudes to food: personal and cultural

Not surprisingly, within the cultural fabric of society, food takes in many roles from general sustenance, to celebrations, weddings, Christmas, and of course cultural references. Indeed, of the many faces of food, wastage is noticeable in many levels not least of which can be described as the individual level as well as within the community as a whole. This relationship, however, develops and changes over time and circumstance. As certain foods become cheaper, for example, perceptions of quality, desirability and ultimately value changes accordingly. Inadvertently adding to the problem, market signals can add to the problem too, with ever increasingly cheap—or the buy one get one free deals, the perception of food's value is being interpreted as an indicator of low-value, disposable food (WRAP, 2009). Furthermore in the cultural stakes, it is not uncommon for household guests to be offered more food than they can consume (WRAP, 2009).

An interesting but rarely considered problem of attitudinal behavior, concerns the idea that consumers might be overwhelmed by their ever-increasing environmental obligations (Monier et al., 2010). The difficulty here, suggests the EU Preparatory study, is that with so much food waste being generated on a personal/private level, the actual extent of the problem is difficult for the individual to see at the individual level and is therefore easier to avoid or ignore.

10.3 Terminology and typology

before you any further, there is some confusion over the terminology used in describing food waste. Indeed, the default position for describing any food that is thrown away, discarded for one reason or another, lost or otherwise inedible is simply referred to as "food waste". However, this does not reflect the true reality of the phenomena.

In European legislation, in particular Resolution 2011/2175(INI), this very confusion is recognized, and by way of example it is suggested that people believe that "food waste" is generally used to describe all discarded foodstuffs from along the food supply chain owing to several reasons including economic, aesthetic or simply owing to the nearness of the products "use by" date, but which may still be fit for human consumption. Furthermore, in the absence of any alternative useful application of food passed its "use by" date (as opposed to its "sell by" date) is, in the end disposed of (Caronna, 2011). So back to the definition of food waste, in the absence of any harmonized definition the European Commission was called upon to legislate specific definitions for food waste, food loss, biofuels and biowaste among others.

Understanding the differences between losses and wastage is an important exercise. Not surprisingly, the notion of food loss and wastage means different things to different people. In the Swedish Institute for Food and Biotechnology report for example, the report distinguished food waste and loss in the following ways:

Food loss refers to the decrease in edible food mass at the production, postharvest and processing stages of the food supply chain.
Food waste on the other hand referred wastage the profligacy of retailers' and consumers' behavior, i.e., at the retail and final consumption stages of the chain.

In both cases, losses and wastage referred only to that part of the chain that was originally primarily planned for use by human consumption. All else, i.e., food that was intended for use as animal feed and production such as biofuels, etc., were not included (Pachauri and Meyer, 2014).

While this is a good place to start, and indeed proved very useful in the report, it is this author's view that such treatment of food loss and waste in this way might be somewhat limiting. Semantically, the idea of loss and waste intrinsically contain within both words, different measures of culpability and as such are suggestive of varying degrees of blame. In light of this a better definition might separate loss to mean that part of the equation that we have little or no control over (i.e., adverse weather conditions such as flood or drought or perhaps conflict or pest infestation), while waste is more suggestive of culpability through such things as lack of knowledge, poor management or simple profligacy.

It is in this author's opinion, and solely for the purpose of this book, the terms "food waste" and "food losses" will be used sparingly, concentrating more on the term "food waste." While strictly speaking, this is incorrect, for the sake of clarity this chapter will not dig deeply into the full terminology of food waste. Instead, this author aims to get the message across irrespective of the correct terminology.

10.4 Why now?

While large figures of food waste are currently being bandied around which, in and of themselves are staggering, what is perhaps not so obvious are the many associated wasted resources that are used to produce the food in the first place. These include things such as soil inputs, manpower and land usage, etc., or indeed the many incurred costs such as emitted greenhouse gases, soil degradation, fertility, or pollution, etc. From this it can be seen that by reigning in our wasteful habits, by promoting more efficiency, through technological innovation, and through a sea-change of habit, humanity has the ability to make considerable inroads, not only on the headline wastage figures but also on the many ancillary costs too. Ultimately then, if we are to reduce this burden of waste while achieving the goals of food for all, sustainable living, and generally reduce the size of our environmental footprint, then it is incumbent on us to understand the issues involved.

To supplement the fragmentary picture and to flesh out the big picture of wasted food this chapter aims to highlight the many considerations, arguments and areas of acrimony that fall between what we as a species produce in terms of sustenance and what we consume, waste and lose through both fault and non-fault. A tall order without a doubt and one in which the chapter tackles in a very specific way. Where much of the data is empirical in nature and where supporting data cannot be found, alternative acceptable estimates, guestimates and considered opinions will be sought where suitable. On the note of quantification of the problem, this chapter importantly, does not seek to add to the guestimates of global waste but rather, to bring together the varying figures for consideration as a whole. Inevitably, this means that there will be gaps, and but by highlighting such gaps it is hoped that future research will ultimately close these with hard evidence.

By highlighting the significance of the problem; identifying the worst culprits; by investigating the scope for improvement within the food supply change; looking at system efficiencies and the challenges of affecting behavioral change to reduce postconsumer waste; as well as other initiatives for waste reduction, this work looks to bring together, to assimilate if you like, into one chapter as much of the debate as possible.

10.5 Key causes of food waste

There are many causes of waste along all aspects of the food chain. That said, food waste can also be described in various ways, not just according to which part of the food chain it falls on. The following takes a liberal approach using many different ways of describing the key causes of waste. It does so too not in any particular order but rather as the narrative flows.

10.5.1 Economic development

One way of looking at food waste is through the lens of developing and developed nations. Characteristically, in the developing world, wastage usually occurs through a lack of infrastructure, managerial skills in food production as well as postharvest processing and other technical difficulties. Thus, it can be said that much of the developing nations waste is preconsumer oriented. As suggested too, perhaps one of the biggest barriers is that of knowledge transfer, especially from developed nations (Parfitt et al., 2010) (http://www.ncbi.nlm.nih.gov/pmc/articles/PMC2935112/WRAP, 2009). This is in stark contrast with that in the developed countries where the majority of food waste continues to occur at the consumer and postconsumer stages. One of the key findings of many reports point out cost of food, being low compared to disposable incomes. Furthermore, consumers have high expectations when it comes to the cosmetics of food (more on this later). All in all, it is suggested there is an almost tangible disconnect with how food is grown as well as the true value of food in terms of production, employment and the general reality of the back end of the food chain (Parfitt et al., 2010). In a similar vein, many transitioning countries follow the same economic development path, this involves a general shift away from raw ruralism to urbanization; and as above, this will likely generate the same kind of ignorance or disconnect from the way food is being grown. Of course, this will take a generation or two, but the danger is still there. Lastly, as food begins to reach its peak output levels, and while climate change alters the traditional growing patterns of food, irrespective of the arresting variable, food values will begin to rise and perhaps potentially reflect the true worth of the product in question. This would encourage less wastage and a more sensible relationship with the food we eat (Parfitt et al., 2010).

10.5.2 Personal preferences

As has been mentioned, as food becomes relatively cheaper compared with personal or household incomes so food choices take on more and more influence (WRAP, 2009). Dichotomously, this means more food choice decisions are based on whim rather than on need. So prevalent is personal preference approximately 1.5 million tons per year in the UK alone. Demographic trends, also play a part in the wastage profile of people. Take for instance single occupant households who, because of their lack of sharing food, will tend to generate more waste than two or more people would (Monier et al., 2010). On top of this personal preference, increased disposable income and changing cultural values are all colluding to increase wastage at all levels. This brings us nicely to wholesale retail sector in general.

10.5.3 Wholesale/retail sector

Within the wholesale and retail sector wastage occurs at many levels and for many different reasons. These are summarized as: management issues; technical problems; and consumer demands. In management issues the problem lies in the communication inefficiencies that commonly exist between retailers, distributors, wholesalers as well as manufacturers; poor stock management especially in the difficult areas of forecasting demand and the like. When it comes to technical issues that generate waste within these sectors is often related to the sensitivity of food produced that require certain temperatures for operation and storage. Poor conditions within the areas of storage, distribution and transportation can all lead to premature spoiling. Perhaps the most disturbing or at least most worrying trend over recent decades, and one that is increasingly recognized these days for the frivolity that it is, is the trend of aesthetics' (Monier et al., 2010). This has been mentioned before and concerns the perceived quality improvements of having all the same shape and same sized carrots, straighter cucumbers or oranges all of the same hue for example. There are many areas of food waste common to both the wholesale/retail sector and the manufacturing/processing sectors too. In particular food waste can be found in the management of stock and forecasting issues as well as packaging problems. Areas of interest here include the carrying of too much stock and relying on such schemes including the generous "take-back" schemes; the often-contractual obligations of suppliers to accept the return of unsold goods with up to 75% residual shelf life from retailers. While it is great for retailers it is not good for the suppliers who then have to quickly offload deteriorating food before it spoils. In this way stock holding is a managerial problem in which inaccurate ordering and demand forecasting is prevalent.

10.6 Undervaluation of food

In circumstances where food is not considered of value the potential for waste is ever-present. We mentioned previously that as food becomes increasingly cheaper in relation to incomes so consumers relative spending power increases. This leads to a situation where perhaps the proper value of food is undermined; that is to say the intrinsic value of food in this situation is clearly not just to be found in the cost of purchase but rather in the aggregate of its economic plus associative costs of environmental, political and social tags too. In such situations, such as those of free school lunches for instance, the perception among children is one of plenty; ergo children are potentially open to the temptation of taking more than is needed (Monier et al., 2010).

10.6.1 Poor food planning

In an era of low cost of food in relation to increasing household incomes, throughout much of the developed world, poor food planning ranks high within the household food wastage profile. Simply popping down the local supermarket and buying a week's supply of groceries, especially within the EU, without too much thought or consideration of what is going to be eaten on a day-to-day basis sees too much food being discarded (Monier et al., 2010). Whether from the mismatching of food items (in terms of balanced meals); or from over production for dinner guests; food that is not wanted by other household members; or simply having too much food in the trolley which could not be eaten before being spoiled are all too common excuses (Monier et al., 2010).

10.6.2 Portion sizes

When it comes to portion sizes, both at retail but more importantly within the foodservice sector there has rarely been anything other than a one-size-fits-all approach; and one that is, according to the 2010 EU preparatory report on food wastage, a major cause of concern (Monier et al., 2010). This is particularly prevalent in canteen style operations, particularly in schools, army messes' and the like, where people are given a fixed-cost, standard portion size irrespective of the individual's needs or wants. Not surprising then, that some people will eat all of the food on their plate while others might waste a considerable proportion. However, the story is different in situations where customers serve themselves. The report tells us that in self-serve style operations, whether buffet, vending machine or similar as much as 92% of the food chosen is actually consumed. While this is good news for the wastage figures, there is a major trade-off in such situations. This trade-off relates to the success or otherwise of the self-serve operation in question; i.e., whether or not it is well attended and all the food has been sold or is in a position to be reused (Monier et al., 2010).

Another interesting study along these lines was conducted in 2006 by Wansink et al. whose experiments showed that given larger bowls, people actually eat on average 31% more food and while given larger spoons the increase was about 14.5% (Wansink and Cheney, 2005; Wansink et al., 2006).

This occurred even for nutrition experts who took part in the study. This taps into study by Monier et al. in that as people generally eat most of what they serve themselves so the extra portion sizes of people serving themselves with larger utensils results in overeating. This is what Wansink and colleagues alluded to as the Ice Cream Illusion and one important counter measure of this would be to consider carefully the many contextual cues which are commonly used in foodservice operations. This, according to the study can also apply to the home situation or simply anywhere where food is being served (Wansink and Cheney, 2005).

These considerations also tap into single serving items like those served in hotels and many catering facilities such as jams, cereals, juice and milk cartons for example where waste could be avoided simply by allowing customers to serve themselves from central containers. Once again there is an important caveat in this scenario and that concerns the trade-off between longevity and shelf life versus holding bulk food items in large storage containers. Take bulk containerized cornflakes in a hotel in Macau or Hong Kong for example; during the spring and summer month's humidity can reach as high as 100%, not the best conditions for serving the hydrophilic breakfast cereal, especially if lax procedures see the containers being improperly sealed and stored. This is just one example; another might include the cross-contamination of foodstuffs through careless or plain thoughtless customers (or poorly trained staff) who might use the wet prune spoon in the rice crispies bowl. In both cases the whole stock might have to be discarded and waste arising out of such practices will offset any savings of the centralized serving operation. Such practices, and potential gains and losses need to be carefully considered (Wansink and Cheney, 2005; Wansink et al., 2006).

In sum serving styles, preportioned and canteen/buffet style foodservice operations carry with them, inherent pluses and minuses when it comes to potential food waste. Both service styles need to be looked at for innovative preventative measures. In restaurants or non-buffet style operations where set portion sizes are served for instance, one solution already introduced by the TGI Friday's chain in the United States is the offering of smaller portion sizes of existing dishes. In canteen or buffet style service operations on the other hand, due diligence, training and measures to increase public awareness will only benefit the cause in the long run. In both cases though, consideration of contextual cues such as bowls, plates and serving spoon sizes could act as powerful subconscious messages of moderation (Wansink and Cheney, 2005; Monier et al., 2010).

10.6.3 Consumer demands

An increasingly worrying trend over recent decades is the trend that the "customer is always right." This oft quoted mantra pampers the customer to the tune that what the customer wants the customer gets. I realize raising such issues is treading on thin ice but if we are to tackle the problem head on then some home truths might need to be explored. The idea that the customer is always right recognizes the power of a consumer led society; in this scenario - money talks; the problem here is that sometimes money talks a language all its own. To elaborate on this, a crude parallel can be drawn between a petulant or spoilt child and a person who spends without regard for the consequences. Buying food that is deemed aesthetically pleasing at the expense of the wasted produce that did not make the grade (and in the face of undernourishment around the world) smacks of the spoilt child within. This of course would be a different story were there options (graded options) on the shelves, then customers, if they so choose could pay a premium for the grade A carrots, etc. On the flipside, however, consider this—if carrots grew like gnarled root ginger; would we then be willing to spend that extra time trying to peel such an awkward vegetable? Undoubtedly some would, yes, especially if there was a cost saving involved (Monier et al., 2010). This begs the question then, who exactly decided we all wanted uniformly straight vegetables? Market surveys? Wholesalers? Retailers? Is this a product of industrial needs, requirements? Or are we being led?

The problem of conformity is not limited to the straightness of vegetables either, oranges, or tomatoes for instance with a less than uniform hue might be seen to be of an inferior quality; misshapen or non-uniform sized and shaped fruit and vegetables too also fall within this category (EC, 2011a). In this sense, aesthetics' is an ugly word, mere shameful examples of style over substance; a frivolous distraction. There is light, however, at the end of the EU tunnel, for as of 2009, new rules were introduced relaxing regulations on the uniformity of fruit and vegetables vis-à-vis size and shape. This goes some way to addressing the legal requirements of heterogeneous foods but a more fundamental shift in the quality-perception dynamic of foods might be needed in the longer term.

10.6.4 Stock management

All across the supply chain, inefficiencies in business operations, especially in the area of stock management have been highlighted as an area for concern in the food wastage issue. Of particular note concerns the area of forecasting which is constantly blighted with the difficulty in predicting market demand and by extension-stock holding. Also, of interest here include the carrying of too much stock as a result of such "take-back" schemes; the often-contractual obligations of suppliers to accept the return of unsold goods with up to 75% residual shelf life from retailers. This adds to stock holding and premature spoilage as well as the difficulties previously mentioned of inaccurate ordering and demand forecasting. Anticipating demand and the resultant issue of overstocking is prevalent across most product groups and can be exacerbated by seasonal fluctuations (OECD, 2002; Wansink and Cheney, 2005; Monier et al., 2010). Take Christmas cakes or Easter eggs for example whose demand is notoriously difficult to predict and which are inherently sensitive to spoilage because of their short shelf lives.

10.6.5 Overordering

Being habitual and intuitive, household food behaviors are a little more involved and causes of waste can be attributed to a wide variety of actions that the consumer may not even think about.

Across the nations of Europe, a large portion of food (37 million tons) is thrown out by households every year (Monier et al., 2010). Furthermore, based on evidence from the UK it has been said that as much as 60% of this could be avoided, translating to an average per-household saving of about €565 each year (Monier et al., 2010). Complicating the picture are notions of the undervaluation of food; the lack of necessity in utilizing food efficiently (overordering, taking notice of sell-by dates, etc.) or using up leftovers as well as a general lack of knowledge in such things as the environmental problems that food waste presents among other things. Add to this the widely recognized problem of food labeling vis-à-vis misinterpretation and confusion over sold-by dates, use-by dates and display-until dates, etc., and it's not difficult to see just how the problem of waste grows. Poor food planning too plays a major role in the food waste profile of the EU at the household level. This, according to Monier et al. tracks back to the relative abundance and low cost of food in relation to household income, throughout the member states. Arising from this then is the practice that sees far too many people continuing in the bad habit of poor food shopping planning or simply buying too much food. Food purchased that perhaps does not combine well with other unplanned food purchases in the household meal; food that was not wanted by the other members of the household; or food that just could not be eaten before being spoiled are all too common excuses of food waste (Monier et al., 2010). Retailers and wholesalers too must take some responsibility for compounding this problem with on-going promotional sales of two-for-one deals, etc.

There are two other important areas recognized within this report and this concerns demographics and preferences (Monier et al., 2010). It would seem that single person households as well as the younger generation are in the habit of wasting more but more widely, food preferences across the whole socio-dynamic cause for concern. Within the more developed EU for instance, people are nowadays commonly discarding often nutritious parts of food due to personal preferences, such things as bread crusts, apple skins even potato skins are often discarded because people simply don't like them or are fussy.

10.7 Foodservice styles

The various foodservice outlets have to be considered too whether food is served at hotels, restaurants, hospitals, army mess halls, corporate canteens, airlines, cruise ships. In fact, everywhere food is served to the public, employees or guests, etc., and each has its challenges in terms of wastage. To help understand the different challenges consider the various foodservice styles from a la carte (cooked to order); table d'hôtel (fixed or menu of the day/week); or buffet style, whether served or self-served. It is worth noting too that the majority of outlets can employ one or several of these service styles based on a number of considerations including manpower, economic costs, perceptions of quality and speed or convenience among others. So, whether serving buffet style lunch at your place of work or a la carte menus at your local restaurant the challenges are many and varied (Monier et al., 2010). A quick note on a face style service appears that customers often expect that nothing will run out. This is especially so in the luxury market which encourages businesses to prepare more, sometimes substantially more than is likely to be consumed Within a la carte style foodservice food is cooked according to the "just in time" principle. The difficulty here is when food is overcooked or not cooked in time with the other ordered dishes, the food can, in far too many instances be discarded and the process is restarted (Monier et al., 2010).

In all cases there are common difficulties shared across foodservice styles. The challenge of not knowing how many people dining might be being a constant source of frustration. To reduce this unknown variable, encouraging customers to make better use of reservations perhaps with some form of incentivized deal might place less pressure on stock holding and other logistics. Also encouraging the practice of taking home restaurant leftovers could substantial reduce the prevalence of restaurant food waste; a practice unfortunately which is frowned upon frowned upon in some parts of the world, particularly in Europe.

10.7.1 Staff feeding

Staff feeding at foodservice businesses can be a sensitive issue. At the point of production, whether processers or foodservice retailers, food waste is generated for a variety of reasons ranging from overstocking, difficulties in forecasting demand, canceled orders, overcooking, and poor management among others. Whatever the reason though, while some establishments allow employees to eat leftovers or to take food home with them at the end of their shift, others ban the practice outright. The problem is a complex one. The difficulty here is that there are two sides to the equation. In restaurants, or factories where too much food has been produced, or is coming close to the end of its useful shelf life the issue is what to do with the leftover food. On the one hand if the staff are allowed to eat it for free or take it home, they are happy and fed, after-all the food was only going to be thrown away wasn't it? It's a win-win situation the restaurant was going to lose valuable stock, irrespective of what was going to happen to the food postdiscard; staff morale is raised and the company's standing among its employee's increases. On the other hand, consider this, waste food in this sector comes from numerous sources; overordering, poor management, difficult forecasting, and mistakes among others. With this in mind, if the staff are to be fed from the over stock or leftovers, or canceled orders what is to stop the chef or food manager from overordering in the first place? By ordering or cooking too much food, the chef might be erring on the side of caution but he also might be under peer pressure to feed a potentially otherwise disgruntled workforce. That is to say, if there are no leftovers, to feed the staff or to supplement existing staff feeding then unhappy staff can be difficult staff. The easy way out is purposely over order, over stock, and over cook. It is a culture that this author has witnessed first-hand over two decades, in many foodservice style operations and across many continents.

That said, employers are not blind to the practice and the resulting difficulty is two-fold. On the one hand a lot of these practices are gray areas and employers have to rely on the trust and professionalism of their workforce. On the other hand, many establishments, for fear of abuse, simply ban the practice; that is staff are not allowed to eat any of the establishment's food under any circumstances other than as customers. In this case any leftovers generated are just binned; discarded (Monier et al., 2010).

Retailers and other food stockists too are fearful of allowing or encouraging this practice and far too often the default line is to bin-it. Take into account too that the majority of the developed world operates within a market economy where capitalist forces prevail and, in many instances, the surest way to avoid abuse is to ban the practice in the first place. As a counter measure, the alternative to binning this waste has to be a priority; whether; allowing more flexible staff purchasing schemes, animal feed or redistribution; this waste stream is one that requires considerable business cultural modification. However, in this last regard, much more work needs to be done as Veronique et al. comments; the lack of clearly defined channels in which excess food from retailers and the hospitality industry in general can be redirected toward charitable organizations is sorely missing.

10.7.2 Lunch timings

One consideration that is given little thought is that of timings of institutional meal breaks. Particularly in this category are schools; according to Jonathon Bloom, studies in the USA have found that scheduling lunch after break-time when children have exercised, creates an appetite that reduces the occurrence of waste. Also, it has been suggested that if break-times came after lunch then there is the possibility that children might be a little preoccupied with the opportunity of play and socializing which might interfere or distract them from the task of eating (Bloom, 2010).

Mirroring this issue of meal times, again within institutional environments such as hospitals, prisons and the like is the lack of individual's autonomy over meal times, choice or even the size of the meal. This could result in patients or inmates, etc., not being hungry at the allotted time eating less than they might otherwise.

10.7.3 Transportation/distribution

Related to the above, transportation, is another area of concern that can lead to both packaging and storage problems. Poor packaging standards and poor packaging performance can lead to damage during transportation and the ultimate discarding of the product; often too, irrespective of whether the food itself is damaged or not. On top of this, any drastic changes in temperature during shipment can also lead to the premature shortening of shelf-lives, or worse actual spoilage during the journey. This is an area in which meat and fish are particularly susceptible. In fact, current research has just been carried out that successfully tracked temperature changes in food products along the supply chain. The idea of these Time/Temperature Indicators (TTI) was to introduce an active "smart label" that could record measurable, time-temperature dependent changes reflecting the full or partial temperature history of a food product to which it was attached (FRESHLABEL, 2008). The technology has been tested and is now ready for industry roll-out. Identifying these breaks, which by themselves or cumulatively lead to spoilage has the potential to drastically reduce spoilage due to temperature fluctuations and perhaps at the same time alter certain characteristics of the food chain itself along the way.

10.7.4 Storage, handling and packaging

Inappropriate or poor storage and packaging practices contribute to the premature spoilage of food. Of course, conditions of climate and household temperatures vary with region but in the UK, WRAP has indicated that as much as two million tons of food is being stored incorrectly contributing enormously to the wastage problem (WRAP, 2008). It would seem too that this is not just an issue at the household level alone either. Poor storage, distribution, and transportation (next section) practices at all stages from processing to wholesale distribution and retailing can and do lead to premature spoiling—issues of light (overexposure), temperature, improper, or inadequate packaging materials.

The difficulty here is the trade-off between storage and packaging for while some products, say cucumbers might have their shelf life extended fivefold by simply wrapping them in plastic film, the wastage of the packaging material must also be considered. In many instances, the use of reusable plastic containers for fresh fruits, pasta, and grains, for instance, might markedly lengthen shelf life at minimal cost to the environment. In sum improved knowledge of preservation or storage techniques, must prevail.

10.8 Data limitations

Significant gaps exist in the overall understanding of the food waste problem. From quantification to identifying poor food practices to technological challenges through to policy and social and economic dimensions, food waste knowledge is still in its infancy. That said, there is mounting evidence that agrees that losses are much higher at the harvest and immediate postharvest stages in developing countries while being higher in affluent economies at the postconsumer levels. However, on the other hand, the conventional view that postconsumer food waste is the sole preserve of industrialized countries alone is challenged with some countries and regions, especially the BRIC countries (Brazil, Russia, China and India), experiencing similarly high levels of food wastage (Parfitt and Barthel, 2010).

The lack of reliable data on food waste at every level seems to be a recurring obstacle in bottom line figures as well as in the comparability of studies. Difficulties continue in terms of defining waste: what is diverted to animal feed, what is unavoidable waste (bones seeds and skins) as well as the unavailability of time series data.

This presents major difficulties when attempting to accurately identify trends, in addition investment in data collection and analysis will go a long way to improving recognition and reporting accuracy. Indeed, one only need take a leaf out of the Danish Environment Agency which conducts a national food waste study every 10 year at a cost of several hundred 1000 Euros.

10.9 The challenge ahead

With regards to being fed, the difficulty inherent in this notion is, that on the surface there appears to be a shortfall of food; that is presently, and for many years now, there still exists a large cohort of the population (approximately 815 million people) who are, and have been year-on-year persistently undernourished; are not being sufficiently fed. This situation taps into the issue of food security that is ensuring everyone has access and available food supplies. Further compounding this problem is the expected population growth that will see close on 9 billion people living off the planet by 2050. Couple this with improved nutrition standards and shifting dietary p and it is not hard to understand that the global food supply chain is coming under further increased pressure.

The challenge then is that any increase in the food supply has to meet not only present shortfalls but also future projected needs of an extra 2 billion people by 2050. With this in mind, the global community is without doubt in its agreement on the need to increase the global food supply; and such a rise has been estimated to be in the region of around 70% (Caronna, 2011; IMechE, 2013a,b). Of course this is perhaps an unforgiveable oversimplification and without getting into too much detail for the moment, the elephant in the room here is there currently does exist sufficient production to feed the existing population on a daily basis (Gibson, 2016). The problem is one of food security, discussed in more detail later in the book.

Adding to these notions is the fact that food production and consumption patterns have changed drastically over the last three decades. Changes in production have resulted from shifting agricultural and retail paradigms that witness more and more of the vertical integration of an increasingly larger part of the supply chain as well as the propensity for larger farm sizes. Technological innovation too has further increased the availability and variety of foodstuffs in ever more variations of packaging, shelf-lives and states (cured, smoked, freeze dried, etc.). Changes in consumption patterns too have largely been the result of rising per capita incomes, lifestyle changes and demographic shifts (rural-urban shift as well as the increase in single person household (Gibson, 2016). Within the context of these changes then, and despite the advances in food waste management systems, with many innovations in waste treatment including the increased practice of home-composting, waste volumes continue to grow (Monier et al., 2010).

Such difficulties present mankind with unprecedented and far-reaching social, economic, environmental as well as political challenges, which need to be addressed if any semblance of a sustainable future for all is to be achieved. Yet while political will is on the increase and while engineers, scientists and agriculturalists have the know-how and the tools that can help achieve the required productivity increases for instance, pressure is still set to grow on the finite resources of land, energy, and water. The key issue here is just how to produce more food in a world of finite resources while at the same time reducing the significant environmental, economic, and social impacts of food waste (IMechE, 2013a,b).

Indeed, for a while there was much talk by the United Nations and others of the need for a 70% increase in food production - only now though are people beginning to accept alternative views that rely less on big production increases but more on a suite of initiatives that promises to reduce food waste and the concomitant strain on the natural resource base. In fact, by simply eliminating losses, the potential exists to provide between 60% and 100% more food while simultaneously freeing up land, energy and water resources for other uses (Monier et al., 2010; Kummu et al., 2012; IMechE, 2013a,b).

10.10 What can be done: waste prevention initiatives?

As has been mentioned food is wasted all along the food supply chains (FSCs). On top of this there is much anecdotal evidence suggesting strong links between food wastage patterns or profiles with the economic development of countries and regions. From this it can be determined that under these different conditions, food supply chains (FSCs) are organized and work completely differently; they are malleable and heterogeneous. FSCs then, reflect both individuals and countries wealth status, their attitudes and their respective relationships with food. There is no doubt where the blame lays according to the UK's Institution of Mechanical Engineers (IMechE). Comprising several areas of culpability, the first firmly resides with the market mechanism of the unnecessarily strict sell-by and buy-one-get-one free dates. This only encourages UK households to buy more than they actually need. In 1-year UK households can throw away up to 20% by weight of the food purchased for that year equating to over 7 million tons. Of this figure more than 60%, about 4.4 million tons, was still edible at the point of discard (WRAP, 2013a). It has been argued too, that by preventing such waste the average UK family could save over £680, while simultaneously adding to the environmental benefits of less landfill, transport costs as well as storage among others, thus mitigating some of the worst proponents of climate change. Further still, in postharvest losses, especially along the transport, distribution and retail lines there is the potential to substantially increase, through good practice, available foods for the consumer while encouraging more efficient use of the resource base (WRAP, 2013a,b). While despite several decades of awareness, food waste is still regarded as being in its infancy, furthermore, as awareness of the problem is disseminated so waste prevention is rapidly gaining momentum.

10.10.1 Production

One of the many ways we can help reduce the wastage in the areas of pre- and postharvest production is to become less wasteful. While this might sound like an oversimplification, it is also the basic truth. The obsession of standardization as well as the perception of quality of uniformly straight cucumbers, odd or misshapen fruit is frankly absurd (WRAP, 2013a,b). Indeed, following the lead of the European Union's hierarchical Waste Framework Directive, in which priority is given to waste reduction at source while observing a preferential list including reusing, recycling and waste management, the US Environmental Protection Agency adopts a similar practice. However, EPA waste treatment options does not differentiate between anaerobic digestion, incineration and landfill. Of course, most environmentally friendly of these options is the anaerobic digestion (EPA, 2018). Having said that, these are not the only options. As far as the European Union is concerned one third of all food waste is in edible so alternative disposal methods must be found. Anaerobic digestion is one method; however, options such as diversion to animal feed, energy capture, industrial uses, especially of recycled cooking oils and composting are perhaps the more environmentally friendly options of choice (WRAP, 2009; Caronna, 2011; WRAP, 2013a,b; EPA, 2018).

10.10.2 Targeted awareness campaigns

As far as information dissemination goes, without a doubt, the last three decades have seen a steady increase in general environmental awareness for all. Having said that, however, food waste has not been at the forefront of policy priority since the First World War. This, according to Monier et al. (2010) can be laid at the feet of complacency in the face of abundance. Historically, as food production increased and became cheaper so, not surprisingly, not everyone thinks about what they are throwing away. Indeed, wasteful behavior or profligacy might for some actually be an entirely unconscious way of living (Monier et al., 2010). Consequently, and as has been mentioned, despite a growing awareness of the issues it would seem that food waste prevention is not yet well-established. Thus, if this turns out to be the truth so the early stages must deal with increasing awareness to initiate change. In doing so, this will be an important first step on the road to the new food revolution.

Stakeholders tend largely to agree that targeted awareness campaigns are essential if humanity's profligate behavior is to be adjusted. The direct benefits of such strategies too are thought to be closely linked to the amount of money invested in such strategies with expected potential savings, at the household level at least, might just turn out to be considerable. Indeed, there is much evidence pointing to the fact that the largest proportion of avoidable food waste is being generated at the personal or household level. Moreover, while individuals and households have been identified as the principle generators of avoidable food waste, many of today's awareness campaigns are principally targeting this group. There are many good examples of effective approaches that aim to tackle this dilemma which include engaging schools and institutional feeding in places like hospitals and the like, as well as privately owned restaurants too. In awareness campaigns then, the is huge, acting as a first line of defense, not just in the examples given but rather in engaging all sectors of industry in the fight against food waste (Monier et al., 2010; Caronna, 2011; WRAP, 2013a,b).

In addressing this worrisome state of affairs, two potential targeted streams are identified that may together initiate significant change. In the first instance, strategies are aimed at producers and retailers to help prevent or at least reduce individual and household food waste. Of course, the good news is that by simply drawing attention to the extent of the wastage problem with awareness campaigns and the like, the cumulative effect can be disproportionately effective. As an example of just how successful targeted awareness campaigns can be in initiating behavioral change just take a look at the UK's "Love Food Hate Waste" campaign which, in a single year, managed to prevent an amazing 137,000 tons of food being wasted—a near 3% reduction in avoidable individual and household food waste; equal to about 1.8% of overall total food waste (WRAP, 2009).

In practical terms, taking a leaf out of our grandparent's book can shed light on many lost and forgotten practices. Take leftover items such as meat, bread, rice or pasta, for instance, these were historically reemployed by our forefathers in many classic dishes throughout the European Union; the same items which are now being discarded on a regular basis. Even the much heralded classic regional bread dish, great British bread pudding, Panzanella in Italy, Pain Perdu in France, where commonly made with old stale bread (Monier et al., 2010).

Affecting long-term behavioral change is a key in such strategies and is also the motive behind encouraging households to separate out food waste from other rubbish discarded within households, restaurants, and cafeterias. In this simple action, the awareness generated (as touched upon earlier) has the potential simply to draw much needed attention to the waste individuals are responsible for (Monier et al., 2010).

References

Bloom, J., 2010. American Wasteland: How America Throws Away Nearly Half of its Food (And what We Can Do about it). Da Capo Press, Cambridge, Massachusettes.

Caronna, S., 2011. REPORT on How to Avoid Food Wastage: Strategies for a More Efficient Food Chain in the EU. EUROPEAN PARLIAMENT, Brussels, p. 23. <NoDocSe>A7-0430/2011.

EC, 2011a. Commission Implementing Regulation (EU) No 543/2011 of 7 June 2011 Laying Down Detailed Rules for the Application of Council Regulation (EC) No 1234/2007 in Respect of the Fruit and Vegetables and Processed Fruit and Vegetables Sectors. E. Commission, Strasbourg, 543/2011/EU.

EC, 2011b. A sectoral approach to CSR to tackle societal issues in the food supply chain. In: EP2.008 Final. Brussels, High Level Forum for a Better Functioning Food Supply Chain: Expert Platform on the Competitiveness of the Agro-Food Industry, p. 7.

EPA, 2018. Sustainable Management of Food. Retrieved 15 January, 2019, from: https://www.epa.gov/sustainable-management-food.

Europa, 2012. What Can I Do in My Daily Life to Limit Food Waste from: http://ec.europa.eu/food/food/sustainability/index_en.htm.

FRESHLABEL, 2008. Integrated Approach to Enable Traceability of the Cooling Chain of Fresh and Frozen Meat and Fish Products by Means of Taylor-made Time/Temperature Indicators. Bremerhaven, Germany.

Gibson, M., 2016. The Feeding of Nations: Re-defining Food Security for the 21st Century. CRC Press, Boca Raton, Florida.

Gustavsson, J., et al., 2012. Global Food Losses and Food Waste. Study Conducted for the International Congress SAVE FOOD! at Interpack2011 Düsseldorf. Swedish Institute for Food and Biotechnology (SIK) and FAO, German. Rome, Italy, p. 38.

Heron, R., 1794. General View of the Natural Circumstances of Those Isles, Adjacent to the North-west Coast of Scotland, Which Are Distinguished by the Common Name of Hebudæ or Hebrides: –of the Various Means Which Have Been Employed to Cultivate and Improve Them—and of Some Other Means, Which Are Humbly Proposed, as Likely to Contribute to Their Farther Improvement. Drawn up for the Consideration of the Board of Agriculture and Internal Improvement. Board of Agriculture, Great Britain.

IMechE, 2013a. Global Food Waste Not, Want Not: Feeding the 9 Billion: The Tragedy of Waste. Retrieved 30 January, 2019, from: http://www.imeche.org/knowledge/themes/environment/global-food.

IMechE, 2013b. Global Food: Waste Not, Want Not. The Institution of Mechanical Engineers, Westminster, London, p. 35, 2013.

Kummu, M, d., et al., 2012. Lost food, wasted resources: global food supply chain losses and their impacts on freshwater, cropland, and fertiliser use. Sci. Total Environ. 1 (438), 13.

Mather, F., 1894. Waste Products: Cotton Seed Oil. Popular Science. Bonnier Corporation, USA, p. 146. May 45.

MercyCorps, 2018. Quick Facts: What You Need to Know about Global Hunger. Retrieved 25th March 2018, 2018, from: https://www.mercycorps.org/articles/quick-facts-what-you-need-know-about-global-hunger.

Monier, V., et al., 2010. PREPARATORY STUDY on FOOD WASTE across EU 27. Technical Report. European Commission, Brussels, p. 213.

OECD, 2002. Towards Sustainable Household Consumption? Trends and Policies in OECD Countries. The OECD Policy Briefs. The UN Organisation for Economic Cooperation and Development, p. 12.

Pachauri, R.K., Meyer, L.A., 2014 [core writing team. Climate change. In: Pachauri, R.K., Meyer, L.A. (Eds.), Climate change 2014: Synthesis report. Contribution of working groups I, II and III to the Fifth assessment report of the intergovernmental Panel on climate change. Intergovernmental Panel on Climate Change, Geneva, Switzerland, p. 151.

Parfitt, J., et al., 2010. Food waste within food supply chains: quantification and potential for change to 2050. Philos. Trans. Royal Soc. A 365 (1554), 3065—3081.

Parfitt, J., Barthel, M., 2010. Global food waste reduction: priorities for a world in transition. In: Foresight Project on Global Food and Farming Future: Science Review: SR56. Foresight Project, Banbury, UK, p. 44.

Stuart, T., 2009. Waste: Uncovering the Global Food Scandal. W. W. Norton, New York.

Wansink, B., Cheney, M., 2005. Super bowls: serving bowl size and food consumption. Am. J. Prevent. Med. 31 (3), 240—243.

Wansink, B., et al., 2006. Ice Cream illusions: bowls, spoons, and self-served portion sizes. J. Am. Med. Soc. 293 (14), 1727—1728.

WHO, 2018. Website of the World Health Organisation: Obesity and Overweight. Retrieved 2nd April 2018, 2018, from: http://www.who.int/en/news-room/fact-sheets/detail/obesity-and-overweight.

WI, 2013. Becoming Active Citizens — National Federation of Women's Institutes in England and Wales - 1915—1925. T. N. F. O. W. S. Institutes. The National Federation of Women's Institutes, United Kingdom, p. 15.

Windsor, H.H., 1917. Popular Mechanics Co. Popular Mechanics. Hearst Magazines, Chicago, United States, p. 290.

WRAP, 2008. Research into Consumer Behaviour in Relation to Food Dates and Portion Sizes. Waste & Resources Action Programme, p. 77.

WRAP, 2009. Household Food and Drink Waste in the UK. Waste & resources action programme.

WRAP, 2013a. Solutions to prevent household food waste. Love Food Hate Waste. Retrieved 2nd Feb 2019, 2013, from; http://www.wrap.org.uk/content/love-food-hate-waste.

WRAP, 2013b. Waste & resources action programme. Love Food Hate Waste. Retrieved 22nd Feb 2019, 2013, from: http://www.wrap.org.uk.

CHAPTER

11

The Hunger files: food security

OUTLINE

When considering food and its various roles through the lens of social, economic, and political perspectives, there is perhaps no single issue more forefront in the political agenda than that of food security. When considering that every year, more than 9 to 10 million people die of hunger and hunger-related diseases, with approximately 6 million of these being children under the age of 5. Understanding just how this situation can still occur under the auspices of a vigilant global developmental community and amid the ever-increasing social engagement of the global population is most definitely one of the key challenges of our time.

This is not a new phenomenon either; over the last 40 years or so (since records have been kept), the numbers of hungry and malnourished around the globe have fluctuated between 800 million and 1.2 billion people (SOFI, 2011). Food security then can be thought of as understanding just how and why this phenomenon exists and continues to exist to the extent that it does, as well as taking action into ways of mitigating the worst effects of food insecurity. Thus, in achieving food security, global multilateral institutions and government aim to ensure people regularly have enough food to eat—not just for today or tomorrow, but for next month and next year too.

However, in achieving this, it is only by understanding the complexity and multidimensional nature of the phenomenon that practitioners of global multilateral hegemony can hope to shape policy in tackling such issues. The difficulty, however, is that food security is increasingly difficult to separate from an ever-growing bundle of societal aspirations including interalia undernutrition, poverty, sustainability, free trade, national self-sufficiency, and so on.

Food and Society
https://doi.org/10.1016/B978-0-12-811808-5.00011-8

Any solutions, therefore, involve fully comprehending that what is indeed included, implied, understood, or excluded within the food security catchall. Yet despite the global reach of the phenomenon and this deceptively simple definition, food security still engenders widespread misconception and misunderstanding (SOFI, 2011). One frequently perpetuated misconception is that food insecurity exists solely within the confines of developing countries (Gibson, 2016). Once again, however, the reality is much different—on the contrary in fact as it can be seen that food insecurity exists in many developed economies too, although granted largely to a lesser extent.

11.1 Food security: what is it?

When considering food and its various roles through the lens of social, economic, and political perspectives, there is perhaps no single issue more forefront in the political agenda than that of food security. Every year, more than 10 million people die of hunger and hunger-related diseases, with 6 million of these being children under the age of 5. Understanding just how this situation can still occur under the auspices of a vigilant global developmental community and amid the ever-increasing social engagement of the population is most definitely one of the key challenges of our time. The following figure (Fig. 11.1) gives visual insight into the phenomenon while also providing background knowledge as to cause and effect.

This is not a new phenomenon either, over the last 40 years or so (since records have been kept), the numbers of hungry and malnourished around the globe have fluctuated between 800 million and 1.2 billion people. Food security then can be thought of as understanding just how and why this phenomenon exists and continues to exist to the extent that it does, as well taking action into ways of mitigating the worst effects of food insecurity. Thus, in achieving food security global multilateral institutions and government aim to ensure people regularly have enough food to eat; not just today or tomorrow, but also for next month and next year too.

However, in achieving this it is only by understanding the complexity and multidimensional nature of the phenomenon that practitioners of global multilateral hegemony can hope to shape policy in tackling such issues. The difficulty however is that food security is increasingly difficult to separate from an ever-growing bundle of societal aspirations including inter alia undernutrition, poverty, sustainability, free trade, national self-sufficiency, and so on. Any solutions therefore involve fully comprehending that what is indeed included, implied, understood, or excluded within the food security catchall. Yet despite the global reach of the phenomenon and this deceptively simple definition, food security still engenders widespread misconception and misunderstanding (SOFI, 2011). One frequently perpetuated misconception is that food insecurity exists solely within the confines of developing countries (Gibson, 2016). Once again, however, the reality is much different—on the contrary in fact as it can be seen that food insecurity exists in many developed economies too, although granted largely to a lesser extent.

By the Food and Agriculture Organizations reckoning food security is the product of food access, food availability, stability of supplies, and biological utilization. The following elaborates.

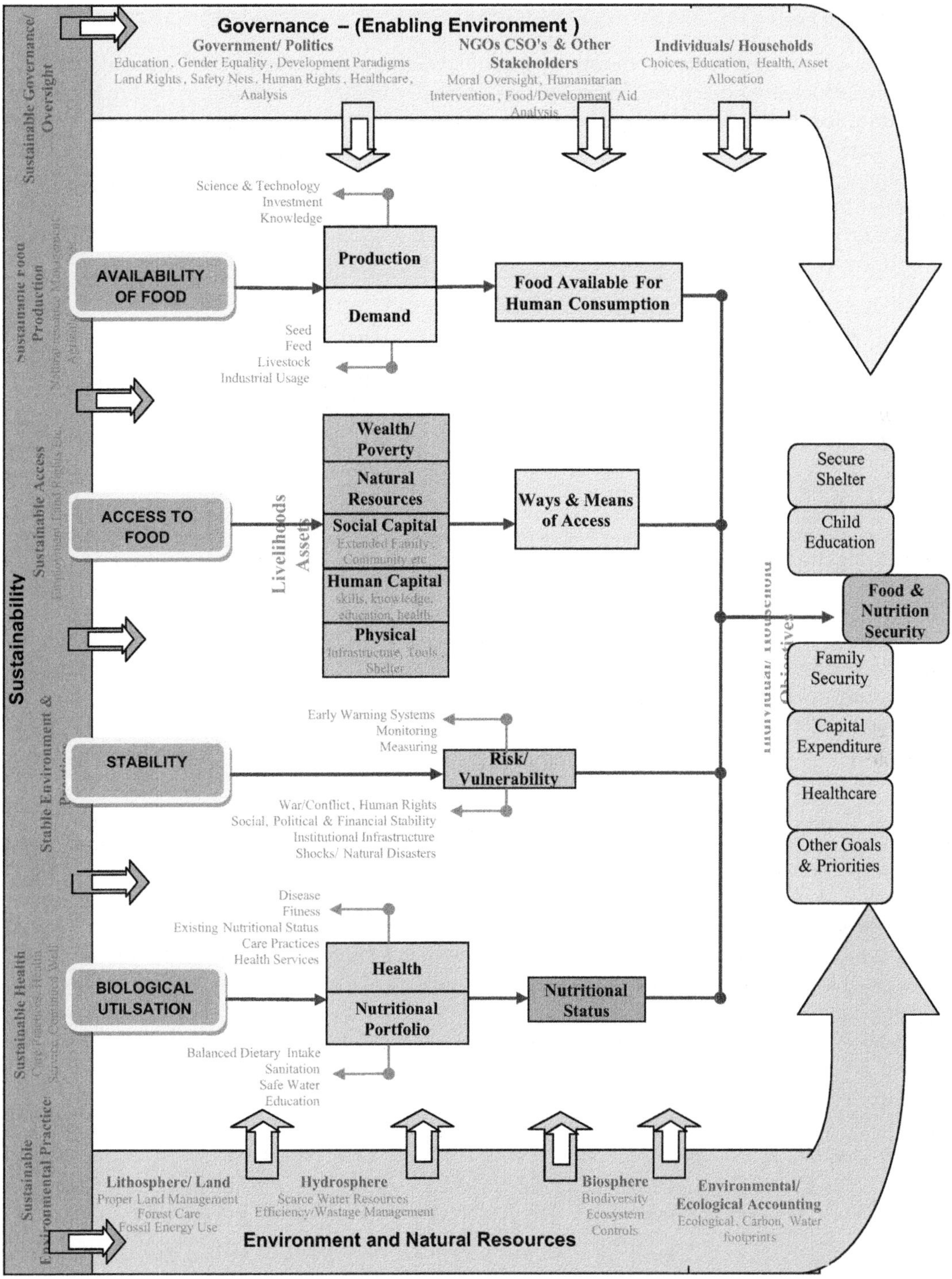

FIGURE 11.1 Conceptual framework of food security. *Source: From Gibson, The use of Conceptual Frameworks in Understanding Food Security. (© Mark Gibson).*

11.1.1 Access

The first dimension entails ensuring people have adequate access, both physical and economic to sufficient and safe food through growing it; purchasing it; being gifted it; bartering or trading for it, etc. (Sen, 1981; SOFI, 2011). This can be thought of as a package of entitlements that allows individuals to acquire and maintain sufficient foods for an adequate and nutritional diet. This can be obtained directly as mentioned through own-grown produce, sufficient incomes, barter, or exchange or indirectly through social arrangements at the community or national levels such as family, welfare systems, traditional rights, access to common resources, and of course emergency food aid (SOFI, 2011; EuropeAid, 2012).

11.1.2 Availability

When it comes to availability, food is provided through two streams—domestic production and/or imports. In food security thought and consideration needs to be given to the physical availability of food at farms and in local markets. In turn this is predicated on well-functioning market infrastructures with adequate road and rail networks as well as ensuring adequate storage and processing technologies (SOFI, 2011).

11.1.3 Stability

There is an understanding that food security can be lost as well as gained and as such understanding and managing risk is gaining much credibility as a tool in the fight against hunger. Risk management involves in-depth understanding of issues of stability and vulnerability. This might include understanding of the wider economy in general, of livelihoods in particular, of incomes, or even of food supplies and how they are affected by such things as shocks like floods, droughts, or pests, etc.

11.1.4 Biological utilization

Lastly, the concept of biological utilization is simply the ability of a person to properly absorb the food they eat. In turn this ability is closely related to a person's health status which itself is also predicated not only on food but also on important nonfood inputs. As such it can be seen that optimum biological utilization necessitates the need for proper health and child care; clean water and sanitation services; adequate knowledge of nutritional and physiological needs; and the proper application of this knowledge.

The UN's definition has had many incarnations over the years; however, the most often quoted is that which was defined in the report The State of Food Insecurity 2001, in which it suggested

> Food security [is] a situation that exists when all people, at all times, have physical, social and economic access to sufficient, safe and nutritious food that meets their dietary needs and food preferences for an active and healthy life *(SOFI, 2001).*

The United States on the other hand employs several loosely aligned definitions depending on need and the institutional body in question. Of these bodies, the USDA usually focuses on national hunger issues while USAID mainly operates with an international remit.

In the first, the US Department of Agriculture (USDA) utilizes just one definition and it defines food security as

> Access by all people at all times to enough food for an active, healthy life. Food security includes at a minimum: (1) the ready availability of nutritionally adequate and safe foods, and (2) an assured ability to acquire acceptable foods in socially acceptable ways (e.g., without resorting to emergency food supplies, scavenging, stealing, or other coping strategies) ***(USDA/ERS, 2013).***

The second agency, the US Agency for International Development (USAID), uses multiple definitions dependent on purpose. USAID's current general classification though is based on the USAID Policy Determination #19 from 1992, which states that food security exists:

> When all people at all times have both physical and economic access to sufficient food to meet their dietary needs in order to lead a healthy and productive life. ***(USAID, 1992).***

The United States also utilizes the Agricultural Trade Development and Assistance Act or more commonly the Public Law 480 (PL 480) program. In this program they offer a more flexible definition allowing for a broader range of possible interventions. Again, based on Policy Determination #19, food security is seen as

> Access by all people at all times to sufficient food and nutrition for a healthy and productive life. ***(USAID, 2007).***

Furthermore, on top of these definitions, the USDA Economic Research Service (ERS) also defines the food insecure as those consuming less than 2100 kcal per day (USDA/ERS, 2013).

There are of course many other definitions from many other institutional bodies, governments, and individuals but suffice to say there are many competing definitions of food security. Looking at the above definitions, perhaps the first thing of note here is that collectively they represent the more convergent of those on offer.

Yet even in spite of certain striking similarities in overall underlying philosophies, in detail there are also many key differences. This is where notions of food security begin to take on a complexity of its own and it throws up numerous difficult questions. For instance, there is little ambiguity in both the FAO and US PL480 when they include "all people at all times" and "safe" foods, yet what of "sufficient, nutritious" (FAO and USAID) and "... nutritionally adequate" (USDA) foods? What are and who decides the nutritional requirements of humankind? Is there a one-size-fits-all or are there different nutritional needs for different people? Furthermore, even when this can be agreed upon, who sets or determines the standards? For that matter too, what is the difference—if one exists—between sufficient and adequate foods?

It does not stop there either, what of the FAO's dimensions of physical, social, and economic access? And, does the availability of food in sufficient quantity take into consideration the quality of the food—and if so, again by what standards? Also, what of the market mechanism itself, is it to be governed by free trade or through government manipulation of agricultural subsidies, quota's, and import tariffs, etc.? Similarly, who sets food safety standard policies? Are there sufficient market infrastructures? Are they well maintained?

In fact, the questions are ad infinitum and while for some, such differences are merely a matter of semantics, for others—policy-makers, statisticians, and the like—especially if policy is to be predicated on such concepts, such nuances are important and need to be clarified to the nth degree. Such ideas are also further complicated by the realities of perspective. That is, food security affects the individual, the household, and at national or regional levels, and the concepts are very different within each foci.

Food security is a political and social behemoth and affects policy at the most fundamental level; it affects agricultural decisions, subsidies, trade, employment, population growth, dietary habits, health care, and of course governance among numerous other considerations. All in all, the concept of affordable and equitable access to food for all is the goal of food security, yet it is a goal replete with complexity and difficulty both at the conceptual and practical levels.

11.2 Pandora's box

The above descriptions, although quite explicit, still only scratch the surface of the concept of food security. Yet even in this incarnation, hints as to the depth, the breadth, and the true extent of the multidimensionality of food security issues start to emerge. And so, it is with such a complex phenomenon many have attempted to define food security with mixed success. Indeed, it is worth noting that even back in 1992 a study by Maxwell and Frankenberger had already by that time identified close to 200 separate definitions. However, while many of these captured some or all of the essence of the issues at hand, definitions from two bodies in particular are among those most widely quoted; these are the United Nations and various bodies of the United States.

References

EuropeAid, 2012. About EuropeAid. Retrieved 13th Feb 2019, from https://ec.europa.eu/europeaid/general_en.

Gibson, M., 2016. The Feeding of Nations: Re-defining Food Security for the 21st Century. CRC Press, Boca Raton, Florida.

Sen, A., 1981. Poverty and Famines: An Essay on Entitlement and Deprivation. Clarendon Press, Oxford.

SOFI, 2001. The State of Food Insecurity in the World 2001. Food and Agriculture Organisation, Rome.

SOFI, 2011. The State of Food Insecurity in the World 2011. SOFI. Food and Agriculture Organisation, Rome.

USAID, 1992. Policy Determination: Definition of Food Security. USAID, US Agency for International Development.

USAID, 2007. USAID P.L. 480 Title II Food Aid Programs and the President's Emergency Plan for Aids Relief: HIV and Food Security Conceptual Framework. USAID Bureau for Democracy, Conflict & Humanitarian Assistance, Office of Food for Peace and the U.S. President's Emergency Plan for AIDS Relief, Washington.

USDA/ERS, 2013. Website of the Economic Research Service. Food Security Assessment. Retrieved 20th April 2019, from http://www.ers.usda.gov/topics/food-nutrition-assistance/food-security-in-the-us/measurement.aspx.

CHAPTER

12

Food: natural and environmental considerations

OUTLINE

12.1 Sustainable natural resource limitations and resource wastage

Environmentalism and sustainability are two buzzwords that have come to represent an awakening of the people's collective conscience over the last two decades or so. Pedantically, the two words have slightly different meanings, yet there is sufficient overlap that they are commonly used interchangeably. Environmentalism dwells on the larger picture, whereby the earth's physical surroundings as well as the flora and forna are the focal points. All in all, the environmentalist movement takes a balanced overarching view to earth's continuing conservation and maintenance the big picture, so to speak. On the other hand, sustainability takes a slightly different view, looking at the environment from usage and finite resource perspective. In this way, sustainability is a term promoting a sensible way of utilizing natural resources that neither overstresses their existing or future potential nor degrades resources to the point of collapse.

Food and Society
https://doi.org/10.1016/B978-0-12-811808-5.00012-X

Over the past 50 years or so, there has been an increasing awareness of the finite nature of the Earth's natural resources. More than this, environmentalism, although covering the full gamut of humanity's needs, is increasingly being used in the same breath as food and food production (Gibson, 2016). That aside, an increasing reliance on the natural resources from an equally increasing population places greater stress on resources, which in turn jeopardizes the sustainability of such resources. As a consequence, there has been growth. Consequently, over the last few decades there has been continuing growth not only in environmental organizations, programs, and charities but also a concomitant fundamental shift in the way we, as a race, view ourselves, particularly when it comes to our relationship with Mother Nature. This is especially prevalent in the way that some people have chosen to coexist with Mother Nature. They no longer see themselves as the sole end users of the Earth's products; instead, many see themselves as custodians of a finite resource base: "a mutually beneficial symbiotic relationship" (Gibson, 2016). What was once a marginal movement had now become mainstream; environmentalism and sustainability were here to stay. This will have ramifications that continues to touch upon all aspects of our lives not only today but for future generations to come (Steck, 2014).

The second half of the 20th century witnessed increasing global population numbers, which point that food production had almost doubled. Yet such an achievement was without much consideration to the environment or the sustainability of the Earth's natural supplies production. With overreliance on chemical inputs; excessive use of fossil fuels; and from inappropriate water management, among others, agricultural production left rather a large environmental footprint on an already stressed ecosystem (Khan and Hanjra, 2009). It is not as if we did not see it coming, as such issues were tentatively pointed out in the 1920s with Rudolph Steiner's organic biodynamics and in the 1960s with Rachel Carson's views on questionable agricultural practices of the day. Yet even in spite of this, it was not until perhaps 1972 with the United Nations Environment Protection conference in Stockholm that such issues were received by a more aware and understanding audience. The message was well received and by the 1980s the notion of the need for an almost ever-expanding food supply and the potential negative consequences for the environment was drastically reshaping the agricultural production industry. More sustainable agricultural practices, conservation programs, and an all-round modified attitude toward sustainability was an easy sell firmly grounded in common sense. Consequently, the environmental movement now enjoys widespread priority in policy decision-making that simultaneously serves to protect the environment and to improve standards of living, as well as food security and numerous other goals (Nathoo and Shoveller, 2003; Ericksen, 2008).

However, nowadays, people have gone further by understanding the implications of doing nothing that there has been considerable backlash against agricultures high-energy, high-intensity industrialization of the food chain leading to a sort of modern renaissance represented by a sweeping array of alternative agricultural systems, whether organic, low input biodynamic, traditional, or indigenous or any other number of environmentally friendly systems (Chappell and LaValle, 2011).

12.2 Sustainable agriculture

In large part thanks to the green revolution, more food is grown today than at any other time in history, yet there is much discussion over whether modern agricultural practices are sustainable. Many reasons are cited for this, ranging from the intensive application of fossil-based technologies (fertilizers, mechanization, transport) to the reduction of biodiversity associated with modern farming techniques, as well as general environmental degradation (land, climate) (GFS, 2011). As a result, the past few decades have witnessed the increased implementation of sustainability theories and technologies. Sustainability has as its ultimate focus the ecosystem. Ecosystems are found both at the micro- and macrolevels; they can be small and revolve around small geographic areas such as local rivers or marshlands or they can be thought of in larger scales as in the Sahara or the Arctic, etc. The main point is while the earth can be thought of as one large ecosystem, there are a myriad of smaller sometimes overlapping ecosystems that make up the whole; just how many there are is a matter of debate.

A well-functioning ecosystem provides us not only with the food and other natural raw materials we so rely on but also it ensures management in areas of land and soil erosion or formation control, nutrient cycling, and carbon sequestration, among others. Ecosystems also regulate the hydrological and biological regimes as well as controlling pests and diseases, pollination, detoxification of wastes, and climate. Within this framework, sustainability looks to preserve these ecosystems in perpetuity.

That being said, sustainability is not a new concept—lack of diversity and the fear that the Earth was becoming "man-heavy" led Mehta in 1929 to quote from the expressive language of the Sanskrit's in cautioning against short-termism:

> … the excessive milking of the agricultural cow so that nothing is left for the nourishment of the calf …
> *Mehta (1929), pg84.*

As a result, today's growing ethos is more about producing food and other goods in a responsible, environmentally friendly manner that looks favorably on the inclusivity of the global allotment. Yet sustainability practices, however laudable, are slowly and sometimes reluctantly adopted. This can be for all sorts of reasons from short sightedness to greed, economic considerations, rights-based issues, political support, and persuasion—in fact any number of reasons. Another hindrance, albeit one that is slowly changing as momentum increases, is that sustainability gains in local ecosystems are usually first and foremost felt at the local level through increased robustness of the system, reduced vulnerability, and better overall resilience, etc. While this translates into better productivity, improved longevity of the system and the many esthetic and recreational impacts people are only now fully coming to understand these local ecosystems are in fact interconnected. With this comes the realization that one cannot improve local land potential at the expense of nutrient runoff for example, which is going to affect the fish farm or the water supply in the next village or town. Nor can we fell swathes of forested land through slash and burn practices without consideration of the wider climate change potential.

While sustainability looks to preserve these ecosystems, it also looks to address years of abuse or neglect. Thus, a mixture of preventative and remedial measures such as reversing land degradation, climate change mitigation (reduced emissions and enhanced carbon sequestration), the conservation of biodiversity, and the protection of water resources are all employed to address the issues at hand. In terms of agricultural sustainability, this progress manifests itself in many forms from consideration of chosen methods of production to the more sympathetic use of inputs, etc. One consideration over agriculture for which system is chosen over any other is partly determined by the climatic environments. Tropical environs allow for any or all of the above practices while subtropical and arid environments are limited to such things as sunshine and/or rainfall. There are inherent advantages of each method too, monocultures, for instance, although the practice receives a lot of bad press, in and of itself the monoculture is not intrinsically environmentally "unsound," rather the problems of monoculture reflect the poor execution of this farming method. However, monoculture does benefit from reduced competition for nutrients and the cost reduction bonuses of large-scale industrial production. Polycultures, on the other hand, benefit from year-round production and diversity in the ecosystem as well as less leaching of the nutrient soil-base. Therefore, understanding the geophysical and climatic limitations of a particular region ensure the right practice is chosen for the right set of circumstances. In turn this has the potential to maximize productivity and lessen the risk of waste due to inappropriate, wasteful, or unnecessary inputs.

However, in reality, the best or most appropriate farming practice is not always chosen, perhaps also the right system might be in place but is subject to abuse, or simply a resource or practice is overwhelmed through improper or overuse. The following looks at some of the impacts our current global food supply setup is impacting on our natural resources.

12.3 The impact of food consumption on the agroecological resource base

Amidst these concerns and a growing awareness of the ill effects of humankind's relentless drawings of the earth's natural resource base, more so over the last few decades as many people are now tuned into the understanding that continuing on the present course is simply unsustainable (GEO, 2007; Dyer, 2006a,b). By way of example, in the European Union (EU), food production and consumption alone are responsible for 28% of material resource use in the EU, in turn generating the following:

- 20%–30% of total EU environmental impacts
- 17% of all direct greenhouse gas emissions (Europa, 2012)

Thus, out of necessity and more importantly fundamental belief in the ideology of sustainability, many people began to question the status quo that saw the Earth as little more than a bottomless larder. This view was confirmed back in 1992 in a moment of clarity, which was also reaffirmed at the Rio Earth Summit by the Union of Concerned Scientists when they suggested

> Human beings and the natural world are on a collision course ... The Earth is finite. Its ability to absorb wastes and destructive effluent is finite. Its ability to provide food and energy is finite. Its ability to provide for growing numbers of people is finite. And we are fast approaching many of the Earth's limits. *UCS (1992)*.

Some of the key drivers behind these unsustainable consumption patterns include growing population, increasing urbanization, and rising incomes (GEO, 2007). And these drivers are likely to continue; yet, if current trends of resource-intensive consumption patterns continue, it has been reported that humanity will require the equivalent of two Earths to support the projected population by 2030 (UNEP, 2012). This does not take into account other associated challenges of production such as competition from animal feed, industrial usage, or climate change variability. Indeed, on this last point, it is suggested that the effects of climate change alone are expected to reduce yields by up to 25% in some areas (Nellemann, 2009; UNEP, 2012).

Presently, global agriculture production is already an extremely resource-intensive sector. On top of this, pressure on worldwide resources is being further compounded by unrelenting demand from growing populations with ever greater food production needs section (UNEP, 2012). Currently, global food production sector utilizes about one-quarter of all habitable land on earth and uses more than 70% of total freshwater consumption. As a result, the world's food production sector has become the largest single cause of biodiversity loss in the world; it produces over 30% of global greenhouse gas emissions; it is responsible for as much as 80% of total annual deforestation contributing to its roles as the single greatest cause of land use change throughout the world. However, the following gives a small glimpse of the effects our food production sector is having on our planet's land and water resources.

Only one-quarter of the earth's land surface is suitable for cultivation and within these confines' cattle occupy about 25% of this land with feed required to sustain these and other livestock requiring another 25%. Recalling too that livestock is an inefficient use of resources—requiring 7 kg of grain to produce just 1 kg of animal protein. Therefore, it should come as no surprise that approximately 35% of the total global grain harvest is ultimately used in the production of feed, resulting in animal protein. Indeed, in many regions around the world, the majority of new agricultural land brought into play is employed simply for this purpose of supplying the growing market for animal protein, whether directly in the production of animals or for animal feed. This is perhaps no more evident than in the high-profile deforestation patterns as seen in Brazil's national treasure—the Amazon. In 2007, for instance, ranches in the Amazon covered about 8.4 million ha[1], which, according to Nellemann, is suggestive of simple market responses to global demand (Nellemann, 2009). Furthermore, if consumption trends continue as predicted, animal protein consumption is set to increase by a factor of 4 by 2050 placing further pressure on existing land resources (UNEP, 2012).

Available land is only one part of the equation; maintaining adequate food production, whether land or water based, relies on well-functioning ecosystems, which in turn rely on healthy and sustainable practices with the long term in mind. Unfortunately, continued misuse and abuse of the land has led to widespread degradation the full scale of which is only recently coming to light.

By way of example, intensive agricultural practices during the second half of the 20th century directly led to the degradation of 2 billion hectares of arable land (about one-quarter of all arable land). And it does not stop there—it is estimated a further 2–5 million hectares are added to this figure annually (Nellemann, 2009). Overall, this leads to very real losses; according to the International Food Policy Research Institute (IFPRI) such degraded lands

[1]In Brazil in 2007 ranches alone accounted for an estimated 70% of deforestation in that year.

lead to nonoptimized yields amounting to the loss of about 20 million tons of grain or 1% of global annual grain production each year (IFPRI, 2011). In terms of freshwater resources, agriculture as mentioned utilizes about 70% of global annual supply. Continued agricultural pollution from intensification is threatening the current sustainability of the system resulting in unnecessary acidification and dead zones within our river and aquifer ecosystems, etc. Apart from implications of land-based production, this also comes at a time when fish consumption is at an all-time high with fish contributing to 15% of animal protein intake for over 3 billion people (or equivalent to about 17 kg per global capita) (UNEP, 2012). Yet amid these figures, the industry is in perilous condition as approximately 75% of the world's major marine fish stocks are either depleted or overexploited.

Furthermore, while aquaculture supplies a good proportion of global fish protein needs—growing in excess of 60% between 2000 and 2008 (from 32.4 million tons to 52.5 million tons), it is not always the panacea that it is sometimes portrayed.

12.4 Trophic's in action: wheat versus meat

The trophics describe the level of an organism's place in the food chain, i.e., where it belongs in the pecking order or who eats who if you like (Johnson et al., 2014). By way of example consider the following by Miller in 1971 which suggests

> Three hundred trout are needed to support one man for a year. The trout, in turn, must consume 90,000 frogs, that must consume 27 million grasshoppers that live off of 1,000 tons of grass. ***Miller (1971), pg233.***

Other examples are numerous and often anecdotal, but all aim to highlight a particular food's role and relationship within the food chain. An alternative way of looking at this upward chain relates to the energy and food equivalents of each food type/species as we go up each level food.

The trophics are a good way of ecological accounting, taking stock if you like of what is bioavailable and what is actually being used, and in that usage just how big or small the ecological footprint.

Consider the overall food chain picture and we find there are approximately 10–50,000 edible plants species in the world, yet human kind relies to the largest extent on less than 200 of these to make up the total variety of food we eat. Concentration and mechanization also have seen even this paltry number reduced to just 15 important species on which we rely to provide up to 90% of our food. Furthermore, just three such species, rice, wheat, and corn, provide between them more than 65% of this 90%. In the animal world to the story is similar with just a couple of dozen species currently being used for food. In fact, as much as 99.9% of all animal protein comes from just nine animals: buffalo, cattle, sheep, pigs, chickens, ducks, geese, and turkeys (FAO, 1995; Joseph, 2009). Considering all this, this may not be seen as the best usage of the planets resources (Johnson et al., 2014).

The issues here are manyfold. First and foremost is the concern for the lack of variety within the average diet; this has implications in the overreliance on concentrated food sources, particularly in nutritional considerations. But more than this it has consequences within the agricultural model too—i.e., the intensification of agriculture at the expense of biodiversity in general. Of course, this is a double-edged sword in terms of food wastage as introducing more diversity within the diet will encourage more stockholding and consequentially the potential for more waste. However, with regard to the trophic's, the issue is not really the fact that people are eating more vegetables, on the contrary this is a good thing; the concern for many is the trend for more meat within the average diet. Health issues aside, the reason is straightforward—meat is a resource hungry commodity. Rearing livestock is a very grain-intensive process and this has implications not only for the obvious food security aspect but also for issues surrounding the sustainability and management of resources (Bonhommeau et al., 2013). The main issue as mentioned is the resource intense requirements of meat production. Firstly, to put this in context, globally we annually consume about 210 million cattle, 418 million sheep or goats, 1.1 billion pigs, and 55 billion chickens[2]. That works out at about 10, 12, 15, and 13 kg of beef, lamb, pork, and poultry, respectively, or about 40 kg of meat in total for every person every year. This requires about 25%–30% of total currently available cropland and about one-third of all grain grown in the rearing of these animals, not to mention the associated environmental and energy costs too. The grain drain becomes clearer still if we convert the cost of these inputs on a per animal basis. As a simple guide, it takes approximately 3 kg of grain and 16,000 L of virtual water[3] to produce just 1 kg of meat (Nellemann, 2009). The logistics from this perspective are worrying; meat is clearly an inefficient use of resource inputs that many feel could be put to better use. And of course, the trend is for increased meat consumption, which then places further upward pressures on cereals grains. In addition to these considerations, further increases would also result in corresponding increases in water, crop, and land requirements, which might be difficult to maintain.

To further elaborate on this inefficient resource usage—we consider that nearly all life on Earth is fueled by solar radiation; that is, plants, algae, and photosynthetic bacteria convert solar energy to grow. These in turn are eaten by bacteria, insects, and animals, etc. As this takes place, the transfer of energy up the trophic levels is reduced—it diminishes. This concept refers to the ten percent rule of ecological efficiency, which is a rule of thumb that loosely suggests that energy, after consumption, metabolic processes, and energy expenditure in general, only about 10% of the original energy consumed at one trophic level is converted into stored biomass available to the next trophic level (Russell et al., 2007; Kling, 2010; Johnson et al., 2014).

Thus, from this it can be seen that by eating more meat we are effectively not utilizing valuable resources in the most efficient or optimum way; resources in fact, which could either be used to feed a greater number of people or the total production of meat can be reduced without compromising food security and/or other environmentalist issues. Of course, eating

[2]This is the based on best guestimate. Calculations are derived from FAOs food balance sheets of 2007 as well as industry standard carcass weights. Which equates to about – 63.2 million tonnes (mt) of Bovine Meat with a slaughtered per carcass weight of approx 300 kg, which adds up to 210.6 million equivalent cattle; Mutton & Goat Meat are eaten in large numbers too, to the tune of 12.6 mt with a slaughtered carcass weight of approx 30 kg which aggregates to 418.6 million equivalent sheep; Pig meat collectively equals 99 mt with a slaughtered carcass weight of approx 90 kg. This totals about 1.1 billion pigs; and lastly, Poultry Meat at 83 mt with a slaughtered carcass weight of approx 1.5 kg yields a total of 55.4 billion chicken equivalent.

[3]Virtual water is a nominal amount of water calculated to give an idea on the quantity of water required to produce a good or a service throughout the course of its growth, lifetime or production.

less meat from a finite resource pool makes sense all round, yet the question then arises; with sufficient food being grown (even after the livestock feeding) to meet global requirements, is this not then just a moot point? Of course, it would be yet to cease discussion there would be to ignore the impact of both increased future meat demands and the consequences for both the allocation of resource usage as well as the issues of the environment and energy costs too. As a result of these concerns, there are increasing campaigns aimed at reducing meat consumption in the industrialized world while attempting to slow or restrain the changing dietary patterns in developing countries (Brown, 2005; Dyer, 2006a,b).

By thinking more efficiently and more sustainably simply by changing people's diets, we could in short help support more people and help these people to live healthier lives (Bonhommeau et al., 2013). Even the World Health Organization recommends eating lower on the food chain, although in fairness this is less to do with energy efficiency or the ecological issue than it is the health aspects of eating less animal fats, etc. (WHO, 2017). Yet, for whatever reason, attempting to change dietary habits might not be as difficult as some are suggesting. By way of example just looking at a little-known paper in 1968 which looked at malnutrition and national development, Alan Berg enshrined the cultural diet in a romantic notion of immovable and transcendental importance:

> ... food habits also have deep psychological roots and are associated with love, affection, warmth, self-image and social prestige. As a result, there is perhaps no aspect of personal life less flexible than one's eating pattern. ***Berg (1968).***

Yet, while this plausible and even intuitive view might invoke a sense of truth in us, trends in the interim 40 years have shown just how fickle this notion of a traditional or cultural dietary identity has become.

References

Berg, A.D., 1968. Malnutrition and national development. J. Trop. Pediatr. 14 (Suppl. 3), 116–123.

Bonhommeau, S., et al., 2013. Eating up the world's food web and the human trophic level. Proc. Natl. Acad. Sci. 110 (51), 20617–20620.

Brown, L., 2005. Outgrowing the Earth: The Food Security Challenge in an Age of Falling Water Tables and Rising Temperatures. W.W. Norton and Co, New York.

Chappell, M.J., LaValle, L.A., 2011. Food security and biodiversity: can we have both? An agroecological analysis. Agric. Hum. Val. 28, 3–26.

Dyer, G., 2006a. How long can the world feed itself? from. https://www.resilience.org/stories/2006-10-13/food-agriculture-oct-13/.

Dyer, G., 2006b. Population and human development – the key connections. Retrieved 25th March, 2018, from. https://www.resilience.org/stories/2006-10-13/food-agriculture-oct-13/.

Ericksen, P.J., 2008. Conceptualizing food systems for global environmental change research. Glob. Environ. Chang. 18 (1), 234–245.

Europa, 2012. What can I do in my daily life to limit food waste. from: http://ec.europa.eu/food/food/sustainability/index_en.htm.

FAO, 1995. Dimensions of Need: An Atlas of Food and Agriculture. Food and Agriculture Organisation, Rome.

GEO, 2007. Global Environment Outlook 4: Environment for Development. United Nations Environment Programme, Malta.

GFS, 2011. Global Issues. Retrieved 12th March 2019, from: http://www.foodsecurity.ac.uk/issue/global.html.

Gibson, M., 2016. The Feeding of Nations: Re-defining Food Security for the 21st Century. CRC Press, Boca Raton, Florida.

International Food Policy Research Institute, 2011. IFPRI: Brief - Economics of land degradation, the costs of action versus inaction. Retrieved 12th February, 2019, from: http://ebrary.ifpri.org/utils/getfile/collection/p15738coll2/id/124969/filename/124970.pdf.

Johnson, S., et al., 2014. Trophic coherence determines food-web stability. Proc. Natl. Acad. Sci. 111 (50), 17923–17928.

Joseph, B., 2009. Environmental Studies. Tata McGraw Hill, New Delhi.

Khan, S., Hanjra, M.A., 2009. Footprints of water and energy inputs in food production – global perspectives. Food Policy 34 (2), 130–140.

Kling, G., 2010. The Flow of Energy: Primary Production to Higher Trophic Levels. University of Michigan, Michigan.

Mehta, V.N., 1929. Famines and standards of living. Ann. Am. Acad. Pol. Soc. Sci. 145, 82–89.

Miller, G.T., 1971. Energetics, Kinetics and Life: An Ecological Approach. Wadsworth Publishing Company, Belmont, CA.

Nathoo, T., Shoveller, J., 2003. Do healthy food baskets assess food security? Chronic Dis. Can. 24 (2–3), 65–69.

Nellemann, C., 2009. The Environmental Food Crisis: The Environment's Role in Averting Future Food Crises. United Nations Environment Programme (UNEP) GRID-Arendal, Geneva.

Russell, P.J., et al., 2007. Biology: The Dynamic Science. Thompson Brooks/Cole, Canada.

Steck, T.L., 2014. Human Population Explosion. Retrieved 25th March, 2019, from: https://editors.eol.org/eoearth/wiki/Human_population_explosion.

UCS, 1992. 1992 World Scientists' Warning to Humanity. Retrieved 21st July 2019, 2019, from: http://www.ucsusa.org/about/1992-world-scientists.html.

UNEP, 2012. The Critical Role of Global Food Consumption Patterns in Achieving Sustainable Food Systems and Food for All. UNEP Discussion Paper. United Nations Environment Programme, Nairobi, Kenya, p. 43.

WHO, 2017. Website of the World Health Organisation. Retrieved 2nd April 2019, 2019, from: http://www.who.int/nutrition/topics/5_population_nutrient/en/.

CHAPTER

13

Food: population rise and sustainability

OUTLINE

Fears of the adverse effects of unchecked population growth are not new. In fact, they go back millennia to at least 200 AD Rome and to a particular Quintus Septimius Florence Tertullianus (Freen, 1996). Indeed, in his paper on population growth and agriculture, Johnson (2000) quotes Quintus Septimius Florence Tertullianus as suggesting that populations had in fact (even back then) become "burdensome to the world" (Johnson, 2000, pg1). Not only this, the foresighted Tertullianus also touched on many of the complexities and causes of overpopulation which remain relevant in today's society; these include urbanization, biodiversity, deforestation, and farming marginal and unsuitable lands among others. Such concerns were raised in an era when population was a mere fraction of today's figures. Indeed, during this particular period, global population has been estimated at a mere 200 million. Furthermore, according to Johnson, the growth rate of the population at that time (revised in 2000) stipulated that from about 200 AD forth, the rate of growth—until about 1000 AD—was a steady 0.04% per annum after which rates increased to 0.12% annually until about 1700 AD—after which, for the next 100 years or so, a noticeable annual increase of around 0.41% meant that population levels had reached the world's first "one billion" around 1800 AD (Zhang, 2008). From this point, various regions grew at differing

https://doi.org/10.1016/B978-0-12-811808-5.00013-1

rates with the resultant world population growing to 2 billion over the next 130 years to 1930. From here, each additional one billion people were added to total global population in ever shorter cycles reaching 3 billion in 1960; 4 billion in 1975; 5 billion in 1987; 6 billion in 2000; and 7 billion in 2011, which is 130, 30, 15, 12, 13, and 11 years, respectively (Zhang, 2008). In terms of future extrapolations, Zhang (2008) further suggested that populations would increase to between 7.94 and 8.33 billion by 2030 and 9.5 billion by 2050. This is in line with the United Nations and US Census Bureau's as well as the UN estimates to 2050 (UNDP, 2006; USCB, 2008a).

13.1 Population trends

Johnson (2000), when considering the initial slow population growth before 1800 AD, hypothesized that the slow growth was in large part due to restrictions in available calories per capita of approximately between 1600 and 2000 Kcal per person. These figures were based on two sets of statistics; the first were estimated accounts of calorific availability in England and France before the rise of the industrial revolution—a period purported to be a time of underdevelopment. Secondly, Johnson factored in the figures from developing countries between the years 1934–38 (being the earliest available records); these were used to extrapolate the 1600–2000 Kcal per person mentioned above (Johnson, 2000). Johnson (2000) in his work also noted that contemporary life expectancy in early civilization remained at a continual average of 25–30 years. That was, until after about 1650 AD, at which time longevity noticeably increased, likely, in no small part, to the increased availability of greater per capita nutrition.

As population began its marked increase after about the 17th century (Fig. 13.1), concern mounted over real or imagined threats of excessive population growth and by extension pressures on both the social fabric and the sustainable resources of the day (Malthus, 1798; Chalmers, 1852). Over time two separate but related hypotheses emerged; the first was the theory espoused by Thomas Malthus's that resources of food would ultimately become the limiting factor in population growth; while tentatively similar, a second concept was that of an optimum population—a measurable and sustainable carrying capacity of the Earth. Collectively, both occupied the minds of leading thinkers of the day. In fact, the debate is so important; it has been, and continues to be, central to the Earth and its population capacity debate today. In this regard it has become one of the most prominent and dominant discourses vis-à-vis the Earth's capacity to sustain human life. So, what of current population growth estimates? The next sections take a brief look at the current situation.

So, what of population trends? Well, latest figures (as of 2015) show that the number of total human inhabitants tops out at just over 7.34 billion and is growing at about 80 million people per year. With growth at this rate, the global population is set to increase to about 7.94–8.33 billion by 2030 before trailing off at around 9.2–9.5 billion by 2050/60 (ESA, 2007; GEO, 2007; Zhang, 2008). Within these figures, the demographic makeup will more likely see the lesser developed regions accounting for the bulk of this growth, from 5.4 billion in 2007 to an estimated 7.9 billion in 2050. Compare this to the more developed areas and regions of the world, which are forecast to remain stationary or static at around 1.2 billion during the same period of growth (ESA, 2007). The population profile, more properly the

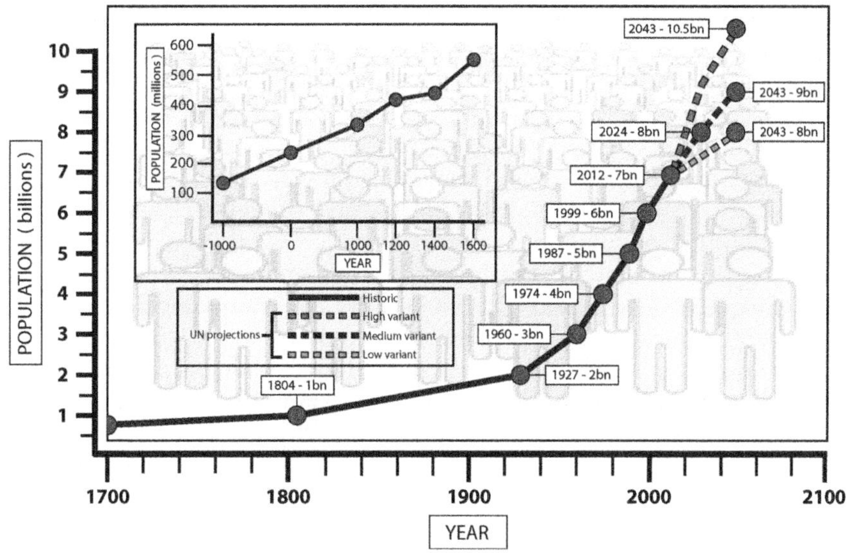

FIGURE 13.1 Population: historical and projected trends. *Source: Courtesy of Pat Newsham based on averages of past trends offered by multiple sources (UN, 1973; McEvedy and Jones, 1974; Tomlinson, 1978; Biraben, 1980; Johnson 2000; Haub, 2002; USCB, 2008a,b; Zhang 2008) and future figure based on UN medium, high and low variant projections (Johnson, 2000; UNPP, 2009).*

demographic, is also likely to be transformed. This is illustrated by the Institution of Mechanical Engineers that, based on characteristics of current and projected economic development stages, identifies three principal emerging groups of societies (IMechE, 2013). These are identified as mature or fully developed; postindustrial societies characterized by stable or declining populations whose age profile would be high; late-stage developing nations (those currently industrializing) as in the likes of China for example, characterized by decelerated population growth rates, increasing affluence, and an increasing age profile; and lastly the newly developing countries in the early stages of industrialization, as in Africa, whose population demographic is likely to be characterized by high growth rates and a predominantly young age profile (IMechE, 2013). For some, this apparent two/three track affair raises significant concerns when it comes to the food supply. Others, like the FAO, however, feel these forecasts are somewhat overstated (FAO, 2011). In the eyes of FAO, the reason is simple and they cite the adequate quantities of present-day global food production as an example of just how efficient and innovative humankind can be when called upon out of necessity. However, how such notions pan out is one of the future challenges that face both the agricultural and biotechnological industries of the present and future.

13.2 The Malthusian hypothesis

As mentioned, the most notable and indeed the most vocal of the Earth's inability to sustain unchecked population growth was the influential 18th/19thth century, political economist and demographer Thomas Robert Malthus. Compellingly, Thomas's work, published in 1798 called "Essay on the Principle of Population," later revised and elaborated on in 1803; he reasoned against the "perfectibility of man" (Malthus, 1798, 1803). Indeed, together the publications combined to formally outline, perhaps for the first time, one of the most quoted and well-debated theories concerning the Earth's ability to provide for ever-increasing population numbers and moreover how it ultimately comes to adversely affect society in the end (Grigg, 1982; Abramitzky and Braggion, 2009). Malthus's central tenet was the notion that population growth and the food supply would increase or grow at different rates. Populations, it was posited, grew exponentially (at ever-increasing percentage intervals), while it was suggested that food supply could only increase arithmetically or by fixed amounts periodically. This meant that in Malthus's prediction, if left unchecked, population growth would eventually outstrip the ability of the food supply to cater for the masses with predictable results, i.e., leaving many hungry and at risk of famine (Malthus, 1798, 1803). This basic notion is illustrated in Fig. 13.2, which shows how the exponential population growth curve (the "s" curve) overwhelms the ability of the food supply (the straight line) to cater for the population at the intersection "A." It is at this cross section, in Malthus' view, that the food supply departs and is unable to keep up with a growing population's food requirement.

While this formed the mainstay of Malthus's central theme, he also understood that since population appeared, for the most part to be in equilibrium at the time and was not running away vis-à-vis the food supply; so, he postulated that certain checks and balances had to be in play aiding or regulating the situation as it were. A few checks in particular, both preventative and positive were in Malthus's view seen to be acting as key gatekeepers of unsustainable population growth. The first couple of preventative checks were attributed to things like sensible population decisions such as moral restraint in the forms of marriage deferment of

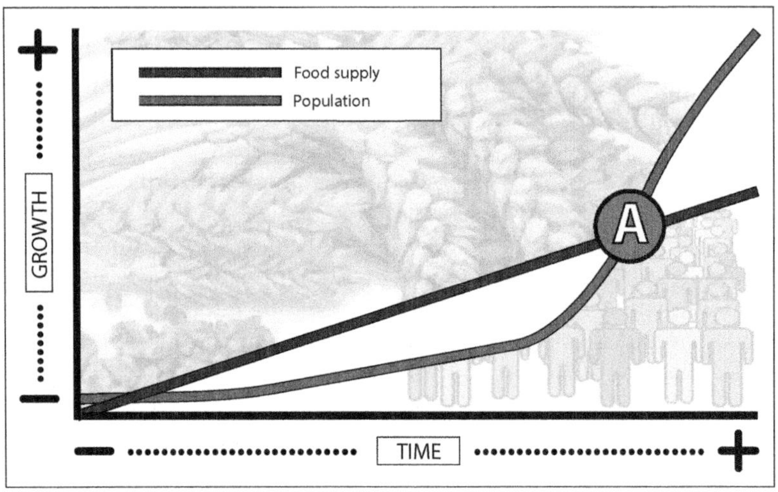

FIGURE 13.2 Predictions of comparative population and food growth rates according to Thomas Malthus. *Based on Thomas Malthus's ideas from "An Essay on the Principle of Population." (London, Printed for J. Johnson, in St. Paul's Church-Yard, 1798). Courtesy of Pat Newsham.*

having kids until such times as food and social structures were in place to feed and look after a growing "needy" population. Secondly, Malthus considered certain events - as positive checks or measures that effectively shortened life expectancy. Principally in Malthus' view these came in the form of war, famine and disease (Malthus, 1798, 1803). Cumulatively, Malthus' theory led to what became known in later times as the Malthusian Cycle or "oscillating" population numbers. Simply put, populations would increase as incomes rose, which in turn allowed for earlier marriages and more children. This growth was then halted or decreased as the food supply was not enough to keep up with demand, leading once again to lower fertility rates and/or deferred marriages (Grigg, 1982). Fundamentally, in Malthus' view, population could or would not proliferate without adequate means of nutritional sustenance and the two forces, he contested, should at all times be kept in equilibrium. Malthus was also vocal on his ideas and presented certain outcomes to his argument. Most notable was the idea that much of the fault of overpopulation lay at the feet of the "reproductive and productive" nature of the less well-off. This, he pondered, was often in excess of their ability to provide for themselves and their families (Vorzimmer, 1969; Ross, 2003). Malthus went further too, controversially regarding famine, mortality, and environmental degradation as natural occurrences of overpopulation. From this perspective, Malthus viewed this as an almost necessary means of keeping the population numbers in check (Malthus, 1798, 1803; Ross, 2003; ÓGráda, 2009).

It did not stop there either; Malthus theories were also populated with more examples of potential runaway population scenarios. In this, it was opined that population could grow out of control by growing grains and legumes to maximize agricultural efficiency; together with the free labor market; the freedom to work wherever it was available; the establishment of workhouses reducing or at least negating the need for public assistance, especially for the poor. This went hand in hand with the idea that seemed to only encourage the poor to conceive children they could not feed. Furthermore, rather than aiding an already provocative problem, Malthus was fervently against the idea of the redistribution of wealth. He did so in the belief that by increasing incomes this would only encourage the poor to have yet more children (Malthus, 1798, 1803).

Even despite this, Malthus was a man of vision and remains a controversial figure of both contemporary and of current times. In this, irrespective of one's own current view, Malthus was no doubt well ahead of his time, bringing to the fore a debate that has lasted, with no clear-cut resolution, to today's debate. Indeed, while there are many advocates of Malthus's theory, there are equally as many critics – some just as prominent. Among early detractors were Karl Marx and Friedrich Engels who were severe critics of what they saw as Malthus's bourgeois attempts to cement the status quo in terms of existing social strata by disregarding the exploitation of the wage-workers (Gimenez, 2008). On top of this Engels was particularly against the notion of the denigration of the poor and hapless, as unavoidable outcomes of universal natural laws. Engels also maintained in 1843/4, that Malthus's promoted more negative ideas concerning population growth than remedial solutions. In response to the poor, Engels cited Malthus in the area of the poor and the food supply in which the poor would:

> … multiply in excess of the available means of subsistence … [as] the root of all misery and all vice. And as a:
> … vile, infamous theory, …[a] hideous blasphemy against nature and mankind ***Engels (1844).***

Others too were just as taken aback, cynical almost of Malthus's ability to make such charges. Patten (1912) for instance, perceived Malthus's as completely lacking of awareness of economic influences on populations choices. Equally, author of "The Dynamics of Agricultural Change: The Historical Experience" Grigg (1982), regarded Malthus's notion that increased populations were restrained by subsequent growth in food supplies; was simply wrong, as Malthus seemingly ignores agricultural production's technical advances that has thus far increased to keep pace with ever-increasing population growth. Johnson (2000) too convincingly puts forth the argument that during the mid-19th century, major population increases in the developed world had come about more so through advances in knowledge and technologies that had successfully reduced mortality rates, rather than increasing fertility (Davis, 1956; Johnson, 2000). It does not stop there, indeed, contrary to Malthus's opinion that food was a major controlling factor of unrestricted population growth Boserup (1965), Woodruff (1909) and Johnson (2000) as well as others suggested that it was in fact the other way around. That is to say, increased population numbers are pivotal in increasing or spurring growth within the agricultural industry as well as in the wider economy in general. This advancement of the agricultural industry, it was considered, was forced on mankind more through necessity than as an act of social responsibility i.e., – supply and demand.

Grigg (1982), on the other hand, had slightly different views vis-a-vis population growth and food supply. Griggs balanced view, while observing many flaws in the Malthusian theory, fell short of accepting this notion, conceding instead that growth in agricultural output was more likely a result of a multitude of factors beyond just population numbers. Nonetheless, the core precept still remains the subject of much conjecture both today and in the past. Indeed, in the words of Weisdorf (2005), he suggests population and agricultural output was in fact a:

> "... 'chicken-and-egg issue' [which] remains unresolved; did human societies domesticate plants and animals as an adaptive response to population pressure or did domestication give rise to a larger population?" *Weisdorf (2005), pg566.*

While the proponents and followers publicly debated the phenomenon, one man, in particular quite possibly Malthus's most vocal critic, was writer and philosopher William Hazlit. In a series of open letters rejected not only much of Malthus's doctrine (particularly in Malthus's view of the poor) but also his fundamental contribution to science in general. In fact, Hazlit, while describing Malthus's views on population, openly offered:

> "... I have no hesitation in saying that his work is the most complete specimen of illogical, crude and contradictory reasoning ... Argument threatens argument, conclusion stands opposed to conclusion ..." *Hazlit (1807), pg15.*

Further adding:

> "There is hardly a single statement in the whole work, in which he seems to have a distinct idea of his own meaning. The principle is neither new, nor does it prove anything new; least of all does it prove what he meant it to prove ... His whole theory is a contradiction; it is a nullity in the science of political philosophy." *Hazlit (1807), pg16.*

As an aside, one interesting observation regarding the Malthusian prophecy and the food supply—at least from the historical perspective—is worthy of note. The Malthusian prophecy essentially stipulated that the population growth and the potential of the food supply were on divergent trajectories. Yet the true nature of attempts over the years to feed our growing populations more likely resembles a sigmoid curve than that of a straight arithmetic line proposed by Malthus. Building on this and in answer to the charge of a limited food supply, Malthus clearly did not factor in man's ability to innovate in times of need. Of course, this does not render the debate moot far from it; as populations grow, the question still remains, just how far can humankind innovate to feed the masses.

Yet for all Malthus's verbiage, another vociferous detractor of his hypotheses, Engels did not completely discount the fact that Malthus's ideas did in fact possess some modicum of truth (Benton, 1996).

13.2.1 A lasting legacy

Despite the above continuous and often vocal opposition, Malthus also enjoyed much commendation among his peers. Indeed, with all the criticisms of the Malthus's concept, one thing cannot be denied; and that is the fact that simply by raising the profile of the debate, Malthus's work has without doubt contributed to the ongoing discourse on population pressures and the food supply. This is best summed up in the commemoration of the centenary of his death in which Bonar et al. (1935) praised the work of Malthus:

> Is he [Malthus] worthy of remembrance?.Yes … If only for the impulse he gave to the serious study of a branch of economics which before him, hardly a branch at all … [Although] I do not say it was universally well received, but it [certainly] excited universal attention. ***Bonar et al. (1935), pg222.***

Furthermore, John Maynard Keynes, economist and co-author of the commemoration, was also mindful of Malthus's indelible contribution to the ongoing debate, when he summed up Malthus's canniness for intellectualizing the population problem.

In sum, at its core, the notion of finding a balance between that of natural resources versus population growth is without doubt a pertinent and valid point in the food population versus agricultural output debate of both then and now. Yet, difficulties arise for others as Malthus attempts to qualify his point of view with seemingly inconsistent and unsupported opinion rather than empirically based evidence. Having said that, however, with all things considered, it cannot be denied that by simply elevating the debate to the level of both the public and his peers, Malthus's work undoubtedly continues to contribute to the ongoing discourse of population growth and food supply challenge. Indeed, such compelling, provocative, and emotive arguments marshaled the seed of a popular and well-researched debate whose dominant dialogue began in the late 19th and early 20th centuries and one which has contributed significantly in the preoccupations of laymen, philosophers, and policymakers alike (Prentice, 2001; Sulistyowati, 2002; Ross, 2003; Brentano, 1910; Patten, 1912; Flugel, 1915; Wolfe, 1928; Hiller, 1930; Spengler, 1949; Meade, 1961; Ehrlich, 1967; Galor and Weil, 1999; Kögel and Prskawetz, 2001; Trewavas, 2002; Lagerlöf, 2003).

13.3 Population sustainability and the carrying capacity of the Earth

Continuing in parallel to Malthus' debate is the notion of the carrying capacity of the Earth. While some might be confused by the similarities in both arguments, it is not surprising, as they have indeed very similar issues. Having said that, they are, however, tackled from opposing perspectives. Also, as with the population debate, the carrying capacity theory is equally controversial. Called the "optimum population capacity" or the "sustainable carrying capacity of Earth," the debate came to the front in the early part of the 17th century but did not really attain traction or critical mass until the early part of the 20th century where, just like the population debate, it began to be seen as a challenging issue.

First off, it must be said that the optimum capacity in terms of population that the Earth can support is not quite the same concept as the maximum number of people that can live on the Earth. Instead, optimum capacity is regulated by the notion of a sustainable "carrying capacity." This idea hinges on the concept that there is an unmistakable relationship between population numbers and Earth's natural resource base, upon which it relies for sustenance, shelter, health, and numerous other resources (Roughgarden, 1979; Daily et al., 1994). The idea proposes a fundamental limit to population numbers, which is oftentimes, but not always, related to limiting factors of natural resources—the keyword here is sustainability (Lidicker, 1962; Grigg, 1982; Van Den Bergh and Rietveld, 2004; Gilland, 2006).

As one can see from this, the concept of a sustainable carrying capacity is in principle fundamentally similar to Malthus' population and food theory. However, by contrast, the carrying capacity argument attempts to analyze the equation from the flipside of the argument—that of a limited or finite natural resource base. And, as has been mentioned, just like Malthus' theory, notions of the Earth's carrying capacity has had a long history. It has often been quoted that one of the first to tackle this idea head on was Anton van Leeuwenhoek, an eminent microbiologist, in the 17th century who, in a letter to the Royal Society of London, made the first real calculation of optimum population numbers in 1679 (Leeuwenhoek, 1686; Neill, 1926; Egerton, 1968; Van Den Bergh and Rietveld, 2004; Gilland, 2006). Leeuwenhoek's simple calculation extrapolated a global maximum population figure of 13,385 million based on the density of the Netherlands at the time (Leeuwenhoek, 1686, pg14–16). In reality, however, this calculation was more an exercise in its metaphoric inference on his groundbreaking spermatozoa studies than it was any real attempt at offering any viable contribution to the carrying capacity deliberation. However, he did whet people's appetite for the debate, and after sporadic interest throughout the 17th and 18th centuries, it was in the late 19th and early 20th centuries that the idea resurfaced with any great fervor. Indeed, by the middle of the 20th century and gaining momentum on the back of the Malthusian debate, population growth, the food supply, and the limiting natural resource debate were already becoming more pertinent to contemporary society. In fact, the awareness of the potential detrimental relationship had, by this time, crept out of the realms of theory and was fast becoming a point of policy for government and institutions alike. Unfortunately though, during this period, few institutions outside of the FAO, the USDA, and the World Bank were conducting globally representative surveys on the state of the food supply and other natural resource supplies in regard to population pressures (Ruxin, 1996; Gibson, 2012). There were several issues at play: the first required knowledge of population numbers, distribution, and trends; the second required thought to be given to the maximum potential resources that the Earth could sustainably use to support its inhabitants; while thirdly, consideration was needed in respect of the potential of the food system to cope with these emerging challenges. As mentioned, however, few surveys were carried out during this period and consequently many relied heavily on the data and statistics contained within just a few limited publications (Ruxin, 1996). The reading was grim, and by the 1950s and 1960s, numerous reports emerged claiming new global Malthusian catastrophes and that the world was ultimately on the brink of mass starvation. Hunger and malnutrition became buzzwords in the popular press, although the specter of doom was not confined to the media either. Indeed, many academic authors, it seemed, were also convinced of humanities inability to feed itself under present population trajectories. Such was the growing fear, and amid the obvious gap in research, that a task force was set up by the White House in 1967 to look into the situation. The task forces report downplayed existing fears of imminent food shortages at the time suggesting instead that the problem's

> ... size and significance tend[ed] to be obscured by rhetorical overkill. ***SAC (1967).***

This downplaying of the situation was in part, according to Thomas Poleman, promoted by the few global reports on the subject and which incidentally were biased toward exaggeration and sensationalism (Poleman, 1972). However, the fears would not subside and during the 1970s and 1980s many more surveys were conducted, reports published, and theories postulated.

During this time, many diverse limiting factors were cited as being the pivotal component in the restraint of population growth. Ravenstein in 1891, for instance, thought that limiting factors of cultivatable land would ultimately constrict population growth, while Swedish economist Knut Wicksell opted for economic and technological resources in defining limits of optimum population figures (Felkin, 1891; Gottlieb, 1945; Uhr, 1951; Spengler, 1983; Ravenstein, 1990). Others, like Woodruff (1909), proposed that:

> "The saturation point for population closely corresponds to the mean annual rainfall ... the more there is the more grass and grain ..." ***Woodruff (1909), pg21.***

These early salvoes were the opening shots in a phenomenon that has paralleled the food-population debate during many decades of the 20th century. Indeed, building on these early proposals, a plethora of studies followed, each convincingly perceived as defining limiting factors backed by much compelling evidence. Many, like the FAO, saw agricultural land as the limiting factor; the FAO, in their calculations based on current and future potentially available agricultural land, determined that the Earth could in fact support nearly 9 billion people. Another, later study maintained that available freshwater would eventually surface as the most likely restrictive force (AEZ, 2000). However, within this debate, one study really stands out from the rest—that of Van Den Bergh and Rietveld (2004). This study is of specific interest as it metaanalyzed 69 historic studies, many also previously explored by Cohen (1995).[1] Based on metaanalysis of studies covering research from 1679 to the present,

[1]Cohen's carrying capacity study of the Earth in 1995 determined a median sustainable value of 12 billion people.

Van Den Bergh and Rietveld (2004) stripped out any wild outlying figures and determined a median value of 7.7 billion—a number incidentally which Van Den Bergh and Rietveld cautioned might itself not even be properly sustainable (Van Den Bergh and Rietveld, 2004). Their studies also showed that the two most often quoted limitations—as indicated by the research studied—were spatial extrapolation (land) and limited resources (food) which collectively accounted for 71 of the 94 collected estimates. However, while these two factors received the most attention, it was cautioned that actual limiting factors of the future might just as likely be those more stringent or inflexible resources such as water, forest, and other nonrenewable resources (Van Den Bergh and Rietveld, 2004). Yet in spite of this "reasonable" estimate and with so many other wildly varying figures (0.5 billion to 1021 billion) based on equally divergent limiting factors inter alia: food; water; natural ecological and geophysical constraints; photosynthesis; carbon cycle, etc., and with no likelihood of convergence in estimates in the foreseeable future, many feel that it is a debate that is set to continue for a long time to come (Constantino and Constantino, 1988; Murai, 1994; AEZ, 2000; Gilland, 2006; Cao et al., 2007; Ferguson, 2008; OPT, 2009).

Another important driver of the food supply chain (FSC) change is the relentless move toward globalization and urbanization.

13.3.1 Urbanization: the Rural–Urban dynamic

In later sections it will be seen that agriculture remains an important sector in terms of global employment; however, the trend is one of contraction. Coinciding with this reduced reliance on an agricultural workforce is a concomitant increase in urbanization (cities with more than 5000 inhabitants). Today, some 50% of the world populations now live in urban environments (Parfitt et al., 2010). This rural–urban shift has only recently become, over the last few centuries, a cultural phenomenon or dynamic. Indeed, urbanization between 1300 and 1800 AD, according to Johnson, remained largely unchanged at about 10%–12% in Europe (excluding Russia), India, China, and much of the developed world (Johnson, 1997, 2000). Incidentally, by 1800 AD, there was only one city in the world with a population of over 1 million inhabitants—Beijing. However, by the end of the 19th century, urbanization was drastically reshaping the world demographic with Europe's urban population then reaching approximately 37.9%. During the same period, there were over 16 cities worldwide that had increased their population figures to over 1 million (Johnson, 1997, 2000). This trend continued with urbanization increasing throughout the world—going from an approximate 67/33% rural/urban split in 1962 to the 50/50% milestone in 2008. This will undoubtedly not stop there either—as the figure is set to increase over the next two decades or so with urbanization ultimately accounting for approximately 4.72–5 billion by 2030 (about 57.8% of projected population numbers). It is also said that much, if not most of this, is set to take place in Asia and sub-Saharan Africa where urban populations are expected to double over the said period. By contrast, the rural population is expected to remain static or perhaps fall slightly, accounting for around 3.12 and 3.41 billions over the same time period (FAO, 1995; von Braun, 1995a; Cohen, 2006; UNFPA, 2007; USDA/ERS, 2008; Zhang, 2008).

It has been suggested that increasing poverty, hunger, and malnutrition were some of the principal accelerants of this migration in developing countries. Exacerbating this drive seems to be several factors. For one, rising agricultural productivity through applied technologies and greater intensification, for example, has led to concomitant decline in demand for rural agricultural labor. Furthermore, strong "pulls" of urbanization include apparent higher standards of living; better social mobility; diversity of employment and to higher wages; improved access to education; and increased opportunities as well as greater inherent choices associated with urban living. Importantly too, urban dwellers typically have improved access to things such as health care, clean water, and sanitation, which all together act as very powerful motivators of urbanization (Food Summit, 1996; Ericksen, 2008; USDA/ERS, 2008).

Urbanization, it must be cautioned, also has a double impact on the food economy; on the one hand, higher wages often mean greater food choices and dietary diversity, often leading to more meat and dairy consumption. While on the other hand, the local agricultural economy is often negatively affected in terms of rural sector employment, as well as questions regarding whether or not the rural landscape has adequate time to develop alternative strategies to cope with such, often speedy, drastic changes (von Braun, 1995b; King and Elliott, 1996; Ravallion et al., 2007). Furthermore, as a larger share of the population comes to live in urban centers, urbanization is creating the need for new or extended existing FSCs to feed this growing segment. For these new and extended FSCs to operate with efficiency, many countries, particularly developing countries, need real improvements in transportation, storage, and marketing infrastructures to keep food affordable for all, including the lower income groups (Parfitt et al., 2010).

That said, while the dynamics of the collective "push and pulls" of urbanization is a complicated matter, the perceived benefits are strongly identified as improvements all round. However, this reality ignores the fact that while, in absolute terms, rural poverty outstrips urban poverty, the urban poor population, as a group, is increasing more rapidly—in some cases by as much as 30%—than those in rural areas. Adding to the fact is that 90% of the world's urban slums are located in developing countries; there is mounting concern that issues of food, particularly food security, in such areas are set to worsen (USDA/ERS, 2008).

References

Abramitzky, R., Braggion, F., 2009. Malthusian and Neo-Malthusian Theories: Malthus' Legacy.

AEZ, 2000. Inventory of Land Resources. Retrieved 2nd June 2019, 2019, from: http://www.fao.org/ag/AGL/agll/gaez/index.htm.

Benton, T. (Ed.), 1996. The Greening of Marxism. The Guilford Press, New York.

Biraben, J.-N., 1980. An essay concerning mankind's evolution. J. Hum. Evol. 9 (8), 655–663.

Bonar, J., et al., 1935. The commemoration of Thomas Robert Malthus. Econ. J. 45 (178), 221–234.

Boserup, E., 1965. The Conditions of Agricultural Growth: The Economics of Agrarian Change under Population Pressure. G. Allen and Unwin, London.

Brentano, L., 1910. The doctrine of Malthus and the increase of population during the last decades. Econ. J. 20 (79), 371–393.

Cao, S., et al., 2007. Disharmony between society and environmental carrying capacity: a historical review, with an emphasis on China. Ambio. J. Hum. Environ. 36 (5), 409–415.

Chalmers, C., 1852. Notes, Thoughts, and Inquiries. John Churchill, Princes Street, Soho: London.

Cohen, J.E., 1995. How Many People Can the Earth Support?. W.W. Norton, New York and London.

Cohen, B., 2006. Urbanization in developing countries: current trends, future projections, and key challenges for sustainability. Technol. Soc. 28 (63).

Constantino, R., Constantino, L., 1988. Distorted Priorities, the Politics of Food. Foundation for Nationalist Studies, Quezon City, Philippines.
Daily, G.C., et al., 1994. Optimum human population size. Popul. Environ. 15, 469–475.
Davis, K., 1956. The amazing decline of mortality in underdeveloped areas. Am. Econ. Rev. 46, 305–318, 2 Papers and Proceedings of the Sixty-eighth Annual Meeting of the American Economic Association.
Egerton, F.N., 1968. Leeuwenhoek as a founder of animal demography. J. Hist. Biol. 1 (1), 1–22.
Ehrlich, P., 1967. Paying the piper: there is no longer any hope of feeding the population of the world. New Sci. 36, 652–655.
Engels, F., 1844. Umrisse Zu Einer Kritik der Nationalokonomie. In: Milligan, T.M., Struik, D.J. (Eds.), Outlines of a Critique of Political Economy, vol. 3. Collected Works, London, Lawrence and Wishart, pp. 418–443, 1975.
Ericksen, P.J., 2008. Conceptualizing food systems for global environmental change research. Glob. Environ. Chang. 18 (1), 234–245.
ESA, 2007. World Population Prospects: The 2006 Revision. UN Department of Economic and Social Affairs:Population Division, New York.
FAO, 1995. Report of the First External Programme and Management Review of the International Irrigation Management Institute (IIMI). Food and Agriculture Organisation, Rome.
FAO, 2011. Website of the Food and Agriculture Organisation of the United Nations. Retrieved 15th Feb 2019, from: http://www.fao.org/themes/en/.
Felkin, R.W., 1891. On acclimatisation. Scott. Geogr. J. 7 (12), 647–656.
Ferguson, A.R.B., 2008. Food and Population in 2050. Optimum Population Trust, p. 8.
Flugel, J.C., 1915. Ethics and the struggle for existence. Int. J. Ethics 25 (4), 518–539.
Food Summit, 1996. Rome Declaration on World Food Security World Food Summit. Food and Agriculture Organisation, Rome.
Freen, R.H., 1996. Ancient Greek philosophical concerns with population and environment. Popul. Environ. 17 (6), 447–458.
Galor, O., Weil, D.N., 1999. From Malthusian stagnation to modern growth. In: The American Economic Review, vol. 89, pp. 150–154, 2 Papers and Proceedings of the One Hundred Eleventh Annual Meeting of the American Economic Association.
GEO, 2007. Global Environment Outlook 4: Environment for Development. United Nations Environment Programme, Malta.
Gibson, M., 2012. The Feeding of Nations: Re-defining Food Security for the 21st Century. CRC Press, Boca Raton, Florida.
Gilland, B., 2006. Population, nutrition and agriculture. Popul. Environ. 28 (1), 1–16.
Gimenez, M.E., 2008. The Population Issue: Marx vs Malthus.
Gottlieb, M., 1945. The theory of optimum population for a closed economy. J. Political Econ. 53 (4), 289–316.
Grigg, D., 1982. The Dynamics of Agricultural Change: The Historical Experience. St Martin's Press, New York.
Haub, C., 2002. How many people have ever lived on earth? Popul. Ref. Bur. 30 (8), 3–4.
Hazlit, W., 1807. A reply to the essay on population. In: Malthus, R.T.R. (Ed.), A Series of Letters to Which are Added, Extracts from the Essay; with Notes Paternoster Row. Longman, Hurst, Rees and Orme, London.
Hiller, E.T., 1930. A culture theory of population trends. J. Political Econ. 38 (5), 523–550.
IMechE, 2013. Global Food Waste Not, Want Not: Feeding the 9 Billion: The Tragedy of Waste. Retrieved 30 January, 2019, from: https://www.imeche.org/policy-and-press/environment-theme.
Johnson, D.G., 1997. Agriculture and the wealth of. Nations. Am. Econ. Rev. 87 (2), 1–12.
Johnson, D.G., 2000. Population, food, and knowledge. Am. Econ. Rev. 90 (1), 1–14.
King, M., Elliott, C., 1996. Education and debate: averting a world food shortage: tighten your belts for CAIRO II. Br. Med. J. 313, 995–997.
Kögel, T., Prskawetz, A., 2001. Agricultural productivity growth and escape from the Malthusian trap. J. Econ. Growth 6 (4), 337–357.
Lagerlöf, N.-P., 2003. From Malthus to modern growth: can epidemics explain the three regimes? Int. Econ. Rev. 44 (2), 755–777.
Leeuwenhoek, A.v., 1686. Ontledingen en ontdekkingen van levende dierkens in de teel-deelen van verscheyde dieren, vogelen en visschen; van het hout met derselver menigvuldige vaaten; van hair, vlees en vis; alsmede van de groote menigte der dierkens in de excrementen. Cornelis Boutesteyn, Leydon.

Lidicker Jr., W.Z., 1962. Emigration as a possible mechanism permitting the regulation of population density below carrying capacity. Am. Nat. 96 (886), 29–33.
Malthus, T.R., 1798. An Essay on the Principle of Population. As It Affects the Future Improvement of Society with Remarks on the Speculations of Mr. Godwin, M. Condorcet, and Other Writers. Printed for J. Johnson, in St. Paul's Church-Yard, London.
Malthus, T.R., 1803. An Essay on the Principle of Population. Or, a View of its Past and Present Effects on Human Happiness; with an enquiry into our Prospects respecting the Future Removal or Mitigation of the Evils which it occasions. Printed for J. Johnson, in St. Paul's Church-Yard, London.
McEvedy, C., Jones, R., 1974. Atlas of World Population History. Facts on File, New York.
Meade, J.E., 1961. Mauritius: a case study in Malthusian economics. Econ. J. 71 (283), 521–534.
Murai, S., 1994. Global environment and population carrying capacity. In: Uitto, J.I., Ono, A. (Eds.), Population, Land Management, and Environmental Change. U.N. University, Tokyo.
Neill, R.M., 1926. Microscopy in the Service of Man. Williams & Norgate, London.
Ó Gráda, C., 2009. Famine: A Short History. Princeton, University Presses of California, Columbia and Princeton.
OPT, 2009. Earth Heading for 5 Billion Overpopulation? Optimum Population Trust.
Parfitt, J., et al., 2010. Food waste within food supply chains: quantification and potential for change to 2050. Phil. Trans. R. Soc. 365 (1554), 3065–3081.
Patten, S.N., 1912. Theories of progress. Am. Econ. Rev. 2 (3), 61–68.
Poleman, T.T., 1972. Food, Population, and Employment: Ceylon's Crisis in Global Perspective. Staff Paper. Cornell University: Department of Agricultural Economics, New York, p. 31.
Prentice, A.M., 2001. Fires of life: the struggles of an ancient metabolism in a modern world. Nutr. Bull. 26 (1), 13–27.
Ravallion, M., et al., 2007. New Evidence on the Urbanization of Global Poverty. World Bank, Washington D.C.
Ravenstein, E.G., 1990. Ravenstein on global carrying capacity: originally published in: proceedings of the royal geographical society, new monthly series, Vol. 13, 1891. Popul. Dev. Rev. 16 (1), 153–162.
Ross, E.B., 2003. Malthusianism, capitalist agriculture, and the fate of peasants in the making of the modern world food system. Rev. Radic. Political Econ. 2003 35, 437–461.
Roughgarden, J., 1979. Theory of Population Genetics and Evolutionary Ecology: An Introduction. Macmillan, New York.
Ruxin, J.N., 1996. Hunger, Science, and Politics: FAO, WHO, and Unicef Nutrition Policies, 1945 –1978, Chapter II the Backdrop of UN Nutrition Agencies, by Joshua Nalibow Ruxin. University College London. PhD, London.
SAC, 1967. The World Food Problem: A Report of the President's Science Advisory Committee, Volume lll Report of the Panel on the World Food Supply. The White House: ThePresident's Science Advisory Committee (PSAC), Washington, p. 1.
Spengler, J., 1983. Knut Wicksell, father of the optimum. Atl. Econ. J. 11 (4), 1–5.
Spengler, J.J., 1949. The world's hunger: Malthus, 1948. Proc. Acad. Political Sci. 23 (2), 53–72.
Sulistyowati, C., 2002. Can Gmo ensure food security amid the population growth. In: Institute of Development Studies and Technological Assistance. BIC News.
Tomlinson, H., 1978. 'Not an instrument of punishment': prison diet in the mid-nineteenth century. Int. J. Consum. Stud. 2 (1), 15–26.
Trewavas, A., 2002. Malthus foiled again and again. Nature 418, 668–670.
Uhr, C.G., 1951. Knut Wicksell: a centennial evaluation. Am. Econ. Rev. 41 (5), 829–860.
UN, 1973. The Determinants and Consequences of Population Trends: New Summary of Findings on Interaction of Demographic, Economic and Social Factors, Vol. I, Population Studies, No. 50. United Nations, New York, Sales No. E.71.XIII.5.
UNDP, 2006. Human Development Report 2006 – Beyond Scarcity: Power, Poverty and the Global Water Crisis. United Nations Development Programme, New York.
UNFPA, 2007. State of the World's Population: Unleashing the Potential of Urban Growth. U.N. Population Fund.
UNPP, 2009. World Population Prospects: The 2008 Revision Population Database. United Nations Population Fund.
USCB, 2008a. Historical Estimates of World Population. U.S. Census Bureau, Washington, D.C.
USCB, 2008b. Total Midyear Population for the World: 1950–2050. U.S. Census Bureau.
USDA/ERS, 2008. Food and Consumer Price Index (CPI) and Expenditures. Food Security Assessment. USDA/Economic Research Service, Washington, DC.

Van Den Bergh, J.C.J.M., Rietveld, P., 2004. Reconsidering the limits to world population: meta-analysis and meta-prediction. Bioscience 54 (3), 195–204.
von Braun, J. (Ed.), 1995a. Employment for Poverty Reduction and Food Security. International Food Policy Research Institute, Washington D.C.
von Braun, J., 1995b. A New World Food Situation: New Driving Forces and Required Actions. International Food Policy Research Institute, Washington, D.C. The International Food Policy Research Institute.
Vorzimmer, P., 1969. Darwin, Malthus, and the theory of natural selection. J. Hist. Ideas 30 (4), 527–542.
Weisdorf, J.L., 2005. From foraging to farming: explaining the neolithic revolution. J. Econ. Surv. 19 (4), 561–586.
Wolfe, A.B., 1928. The population problem since the world war: a survey of literature and research. J. Political Econ. 36 (5), 1928.
Woodruff, C.E., 1909. The Expansion of the Races. Rebman Company, New York.
Zhang, W., 2008. A forecast analysis on world population and urbanization process. Environ. Dev. Sustain. 10 (6), 717–730.

PART THREE

Anthropology of Food

Hunger is insolent, and will be fed.
Homer
The Odyssey

CHAPTER

14

Governance and food

Mukta Das

SOAS Food Studies Centre, SOAS University of London

OUTLINE

Historically, governance in matters of food production, trade, and distribution—in fact all along the supply chain—was a general mix of national policies, private and commercial influences, nongovernmental organizations and similar institutional influences and legislation. Within this arena, guidance or stewardship was pretty much a collaboration of these collective interests. However, with the advent of globalization as a new orthodoxy, particularly in the latter half of the 19th and the first half of the 20th centuries, big picture politics took over. The result, power, and influence in these matters gravitated toward the dominant powers—particularly at the global level. A dominant power in this sense often came about either through fear, as in the cold war USSR; economic, as in the United Kingdom, then the United States, and more recently China or through ideology as in the United Nations. This is a loose description and without getting drawn into political ideology for the moment, the point made here is that global power fluctuates. With this power, whether political,

Food and Society
https://doi.org/10.1016/B978-0-12-811808-5.00014-3

economic, or ideological, comes responsibility and in an ideal world a benevolent dominant power facilitates good governance. It is also a circular argument and yet despite this apparent definitive route it is not always easy to navigate. And so, it was with the politics of food. As more and more of the food supply chain became globalized, more and more of it also became politicized. The challenge here was that it was not always easy to discern the ideology or motives of policy and policymakers. In all of this though, especially with the inception of the League of Nations and its predecessor of the United Nations, international hegemony is increasingly being established through multilateralism. Multilateralism and multilateral agreements are an appropriate and cooperative means of achieving common goals to manage coordination and to resolve conflicts of interest.

There are many dimensions to food politics ranging from legislation to terms of trade to ideological paradigms—all of which collectively seek to shape the structure of the food industry. A few of these are discussed in the following sections.

Perhaps the first consideration of the political dimension of food is that of food security, ensuring people have adequate access to the food they need.

Freedom and Democracy

Some argue that you cannot separate good governance from democracy; indeed, some like Lord Acton strongly believed that

> Power tends to corrupt, and absolute power corrupts absolutely. ***Engel-Janosi (1941).***

14.1 From human rights to food rights

The notion of human rights for all—a country's investment in its own people—has gained much traction over the last few decades. Indeed, under the banner of many guises including the "human development" concept, a country's "social capital" and "social justice" for all are all paradigms that aim to improve the "lot" of the common person. Since civilization first congregated, cogitated, and sought council, human rights, in its many incarnations (Table 14.1), focuses, in one way or another, on the collective betterment of individuals and nations.

With "human rights" came human progress, and while the idea of social capital or human development generally means different things to different people, central to all notions is a collection of social goals bound up in a "rights"-based ideology. This includes notions of social, mobility, justice, and the right to food among others which have all been preserved in some form or other—not only in the United Nations Declaration of Human rights of 1948 but also within many other similar multilateral instruments over the years. These ideas built on earlier social, economic, and cultural enlightenments, but specifically flourished during the last half of the 20th century when there was considerable progress in the socioeconomic development in the lives of numerous people throughout the world. One setback though was that despite the many socioeconomic advances afforded to many, there were many more who were, in fact, still lagging behind. Consequently, by the late 1980s and early 1990s, people began to take a closer look at the various aspects of development and what it in fact entailed. The first challenge that came to the fore was just how were the concepts of social development and social capital to be defined, let alone measured? Up to this point the idea of a country's development—its progress if you like—was solely and inextricably tied to its economic

TABLE 14.1 Selected instruments of human rights over the centuries.

1792–1750 BC	The code of hammurabi
559-530 BC	The Persian cyrus cylinder
697 AD	The Irish cáin adomnái
1215	The English magna carta
1216	Great Charter of Ireland
1222	The Hungarian golden bull
1525	Twelve articles
1689	The English bill of rights
1776	The Hindu gentoo codes
1776	The US declaration of independence and the Virginia declaration of rights
1789	The French declaration of the rights of man and of the citizen
1791	The US bill of rights
1856	Ottoman empire Hatt-ı Hümayun
1919	Basic rights and liberties in Finland
1948	United Nations universal declaration of human rights
1949	Fundamental rights and duties of citizens in People's Republic of China
1976	International bill of human rights
2000	United Nations millennium declaration

Source: Halhed (1776); Hurlbut (1847); UN (1948); UN (2000); Ishay (2008).

activity, as measured by its Gross National Product or Gross Domestic Product (GDP). Much thought was given to the idea of defining and measuring alternative social progress and finally by the 1990s, the UN Development Program's (UNDP) Human Development Report defined development in terms of individual's longevity, knowledge, and living standards among other social constructs. Once the methodology was ironed out, these concepts ultimately became the benchmark for indirect measures of social well-being. From such metrics, it was quickly observed that a country's economic growth did not always correspond to measures of social development as described above. That is to say, some countries seemed to achieve better social development with less economic progress. Not surprisingly, divergence of economic and social metrics gave rise to much debate over the concept of economic growth for betterment. This helped focus people's minds and aided in the rebalancing of perspectives—taking the development paradigm and placing human development (or capital) as the central goal of progress. Importantly too, this paradigm shift turned out to be a real eye-opener as human capital was now being seen not only as a goal to be achieved as in the likes of GDP but also as a major input "... previously unaccounted for in neo-classical economic models ..." (Gibson, 2016). This new thinking implicitly tied economic productivity to the education, health, and well-being of the populace—its human "development" or capital (Hasan, 2001; MacAuslan, 2009). Put another way, an "holistic" approach in which food was but one component part became the cornerstone of the concept of food security. Furthermore, it was quickly understood that by employing skilled or educated healthy workers, labor would, more likely, be more productive allowing businesses to capitalize on more efficient economic and social capital. While such laudable progress toward human rights can be seen throughout history, the concept of the human right to food has also been well observed in "moral, philosophical, and ideological terms over a similar period" (Gibson, 2016). Also, while great social strides have been made with the establishment of the inalienable right to food, in reality this "right" is not a new concept. The right to food is peppered throughout history, being invoked on numerous occasions (Mettrick, 1929; UN, 1948; FAO, 2006) from the slaves of ancient Babylonian times (Johns, 1904) to the words of English philosopher John Locke (1714), which are echoed by the Rt. Rev. Thomas Sherlock, Bishop of London when he opines that

> There is not, I presume, a stronger natural right, than the *right to food* and raiment; this is founded in the common necessity of nature; and 'tis not to be thought that God sent men into the world merely to starve, without giving them a right to use in common so much of it as their necessities require. ***Sherlock (1718), pg 25.***

Further expanding on this idea was Edmund Burke's comments on the role of government in his book reflecting on the *French Revolution* in 1790 when he penned:

> What is the use of discussing a man's abstract right to food or to medicine? The question [should focus] upon the method of procuring and administering them. In that deliberation I shall always advise to call in the aid of the farmer and the physician, rather than the professor of metaphysics. ***Burke (1790), pg 88/9.***

Of interest here is the separation, by Burke, of the philosophical or moral "right to food" with the practicalities of acquiring rightful nourishment. At the time, this viewpoint set precedence, marking an important departure from contemporary doctrine.

In another point of view, the influential 19th century secularist Charles Cockbill Cattell also promoted the notion food as a right by stating clearly that

> Every industrious community has the right to food, clothing, shelter, and such social arrangements as will enable it to enjoy an average share of life. ***Cattell (1874), pg 10.***

Lastly, food as a right was given further kudos through the writings of Boyd Orr. In advocating the merits of the "new" science of nutrition in 1939, Boyd Orr observed that

> It is the right of every citizen of our great British Empire to enjoy the benefits of this newer knowledge of nutrition so that the health of every one of His Majesty's subjects will be up to the level which we now know is possible. ***Orr (1939) pg 89.***

And so, it was, in a moment of moral momentum, that the UN Human Rights Commission drafted the Universal Declaration of Human Rights in 1948; and the first time in history that rather than being implicit, the human right to food was made very explicit (albeit nonlegislatively). Legislation came in 1996 in the form of the World Food Summit (Food Summit, 1996). Such slow progress, according to several UN commentators, was the result of the politics of procrastination and the notion that it is all well and good talking about a rights-based ideology but yet another itself to act upon it (FAO, 2006; SOFI, 2009). In fact, such was the hypocrite and distance between ideology and practical solutions led O'Neil back in 2002 to comment

> It is a cruel mockery to tell someone they have the right to food when there is nobody with the duty to provide them with food. That is the risk with the rights rhetoric. ***O'Niel (2002).***

It has further been suggested that, in spite of guidelines on how best to anchor a rights-based approach to food in such, (largely) politically-oriented instruments, a lack of real progress is still being made. This is especially difficult as such frameworks are more often than not based on voluntary rather than mandatory codes (Hartmannshenn, 2004; UN, 2004).

14.2 The food gap

Furthermore, if less meat and other resource-intensive foods were consumed, the food gap between the rich and the poor would narrow. The food gap, which sees 815 million or so food insecure, could also be further narrowed if the necessary agricultural growth develops in a way that provides more jobs and better incomes to supplement the 2 billion plus workers (many of whom are women) in the industry, besides addressing the food/poverty gap.

14.3 Land use and land ownership rights

With regard to food, land use and land ownership rights are of paramount importance. In the more developed countries, land rights are fairly well established and food supply is more stable as a result. Throughout the developing world, however, land is one of the most critical resources that the rural poor can access. Land tenure or the system of rights governs access to land and other resources. Such rights might exist as a rule of law; through customary use, marriage, inheritance, or power, they might also exist under freehold, leasehold, or by mutual agreement with no contractual basis. The difficulty in many of these countries is that many of these rights are not always backed by legislation, as such it is not protected or is its tradable collateral. Rights to land too can change relatively quickly affecting the security of those working it. Moreover, in some postcolonial areas, there still exists residual policies of pseudo-feudal systems with vast tracts of land owned by a privileged few whose rental or terms of lease has led to long-term tension and animosity (Maxwell and Wiebe, 1999). Added to this are the failed structural adjustment policies and a leave-it-be attitude to change in the face of mounting political pressures of land titling, particularly in Africa, that have collectively compounded to leave a legacy of land vulnerability in many parts of the world. Things are beginning to change, however, and in response to these challenges, the United Nations introduced its land policy model, which is predicated on five equitable constructs that aimed to set rigid guidelines in the hope of shaping future usage and management of land: these include land distribution, land utilization, land tenure security, land administration, and land adjudication (ECA/SDD, 2004). One of the major driving forces behind the promotion of land rights and a keen area of debate is the commercialization of agriculture. On the one hand, privatizing land in the commercialization of agriculture (especially for export) sees a general reduction in subsistence food at the household level and a concomitant increase in cash income. However, it is also argued that this reliance of export earning actually increases market vulnerability. On the other hand, it is said that this very model of integration into the exchange or globalized economy is a prerequisite for future sustained growth and development. These two opposing arguments have fueled much literature and are beyond the remit of this book.

14.4 Food and global governance

The idea of governance in food matters is a complex one that has shoots going back centuries. This resulted from the many multifarious trains of thought and their indelible criss-crossing paths that culminated in many international agreements and instruments of peace and social rights. As such governance can be thought of as responsibility, guidance, and/or oversight bound up in a mix of social accountability, political ideology, and economic development paradigms, as well as increasing moral and ethical influences (FAO, 2009). In the arena of food, the notion of governance at the multilateral level properly evolved with the advent of the First World War. By this time food was being seen in many quarters as a part of a "whole package" of rights. This was further solidified with the inception of the League of Nations and then after the Second World War with the creation of the United Nations. During this period, three UN agencies—the Food and Agricultural Organization (FAO), the World Health Organization (WHO), and the then-called United Nations International Children's Emergency Fund (UNICEF)—were established to tackle the growing problem of hunger and malnutrition. In terms of governance, it was initially difficult as sovereign nations seemed uncomfortable with the notion of any international organization beyond the control of national governments. However, in spite of these misgivings, the United Nations eventually received wide acceptance and has since then had more to do with the development of food in all its many guises than any other agency or government (FAO, 1946; Phillips, 1981; Williams, 2005; Shaw, 2007; FAO, 2010).

Thus, from the social and political detritus of two World War issues of food, in particular vis-à-vis hunger and malnutrition was firmly and finally fully politicized. Indeed,

> Democratic peace, cosmopolitanism, and global governance are among the most powerful conceptual frameworks in contemporary world politics. ***Aksu (2008), pg 368.***

Good stewardship facilitates progress and with the right balance of institutional governance extolling notions of human rights; health and international cooperation; development philosophies; greater public resources; improved physical infrastructure; and education, etc., more inroads can be made into many of the issues evolving around food.

As the decades passed, knowledge and research became invaluable tools in understanding the many facets of food and food-related issues, and in time new agencies and bodies such as the European Economic Community (later the EU), the UN's Research Institute for Social Development (UNRISD), the UK-based Institute of Development Studies (IDS), and the

Overseas Development Institute (ODI), as well as numerous other institutions, charities, and civil organizations joined the long list of governance-type organizations.

However, governance especially in the face of competing political ideologies is a fragile overlord leading many to suggest that present global governance on food issues is fragmented at best. Furthermore, it has also been suggested that with so many stakeholders it becomes practically impossible to develop common coherent policy objectives that many food issues are unlikely to be adequately addressed (Marzeda-Mlynarska and Curie-Sklodowska, 2009; RTFN, 2009). In particular, when it comes to issues of food security, even the FAO acknowledges the lack of unified global governance when it offered

> We recognize that there is a lack of coherence and efficiency in the current governance of world food security. The system is poorly organized and each institution operates to a large extent separately despite important progress in coordination. Responding to the global food insecurity crisis in an effective and sustainable way requires not only strong leadership and relevant policies, strategies and programs, but also coordinated implementation and monitoring capacities. **FAO (2009).**

In another example, the United States, being the world's premier superpower and the largest donator of humanitarian aid, is uniquely positioned to set the right course in food governance providing best practice in international governance and multilateral cooperation. Yet in this role, the United States is sadly lacking. The trouble, according to some, is that while US hegemony is felt around the world, the majority of its actions are motivated by self-interest. Many reasons have been cited for this state of affairs from partisan politics to conflicts of domestic interests; however, the UN feels the biggest single reason still remains the simple lack of political will (UN News Center, 2010).

While this is a poor indictment of the current state of food issue governance, it is a tacit and welcome admission of the difficulties facing the various stakeholders. And indeed, only with this type of open and forthright self-awareness can progress be made in such varied issues such as food wastage and security of food, etc.

In a similar vein, an important but not so widely discussed area of food supply centers on the type of governance within a country.

14.5 Regulation and trade

In the field of politics and a country's food supply, there is the question of whether to access the globalized market and if so to what degree—for all a country's food needs, some, most? Once this question has been satisfied there is then the question of whether this access is to be on a level playing field via the free market, or indeed whether a country intends to safeguard its domestic agricultural sector and associated industries through regulation and quota systems—effectively protectionism. Furthermore, there are also issues of food safety and regulation and whether or not to look favorably on such things as the import and export of genetically modified foods, etc. These questions are not solely politically motivated either—take the case for organic food, food sovereignty (see later sections), slow food, food miles, and carbon dioxide among a plethora of other socially led movements. While some of these are now incorporated to some degree or other within legislation, many of these started out as social movements; awareness campaigns that might have started small becoming politicized as social momentum achieved critical mass.

The following looks at the idea, in principal, of free trade and protectionism and explores whether or not the two can mutually coexist in the food supply chain ideology of a country.

14.5.1 Free trade versus protectionism

Globalization and the notion of free trade are both well established and are fairly well aligned—you often find the two working well together. To place food and free trade in perspective, we can see that globally traded food in 2010 stood at approximately 9.6% of total merchandise traded (WTO, 2010). However, there are some who feel that free trade is a misnomer or more judgmentally an oxymoron. The reason for such views lies in the fact that running parallel with the ideals of free trade, there commonly exist frequent barriers to the free movement of goods and services across international borders and trading blocks. Indeed, as a result of decades of such practices, some degree of national protectionism is seen as the norm throughout the world. To understand more of the debate, we need to understand something of comparative advantage as the fundamental ideologies of free trade and comparative advantage are two economic models often seen as complimentary. Comparative advantage is a simple notion, it exists in one country utilizing different skill sets and natural resources becomes more efficient and perhaps cheaper in certain industries than others.

Free trade then, pioneered by the likes of esteemed economists like Adam Smith, Robert Torrens, and David Ricardo, is predicated on the movement of goods and services without outside influence or impediments from governments or other bodies. Free trade also presupposes certain comparative advantages and given these conditions the idea supposes both sides mutually benefit from trade. In this freely functioning system, the supply and demand for food is theoretically in equilibrium and subsequently able to reflect the true needs of populations. On top of this there is also evidence to suggest that globally integrated trade further benefits countries in terms of increases in both economic terms and through better standards of living (IMF, 2010). Just as importantly though, global competition through a free trade also promotes competitive prices and equally the rational use of resources making such trade more environmentally friendly and sustainable.

The downside of free trade, however, lies firmly in its application. According to some, free trade does not or has it ever truly existed. The problem is one of protectionism through such instruments as the EUs Common Agriculture Policy (CAP)[1] or the US Farm Bill, for instance, continues to artificially distort a country's domestic agricultural advantage by providing generous subsidies, preferential agreements, or through import and export quotas. Such programs create unfair advantages which makes it difficult for developing countries specially to trade on a level playing field.

[1]Total subsidies paid to OECD countries in agricultural protectionism in 2007 was reported to be in the region of $365 billion per annum while an IMF report estimated the benefits of globally traded merchandise, were all barriers removed, to be in the region of $250 billion to $680 billion per year Harvey (2010). "An abbreviated UK/EU agricultural policy history." Retrieved 23rd Nov 2019, 2019, from http://www.staff.ncl.ac.uk/david.harvey/AEF372/History.html., IMF (2010). "International Montetary Fund Website." Retrieved 4th Jan 2019, 2019, from http://www.imf.org/external/.

These potential difficulties were recognized early on and resulted in the UN initiated General Agreement on Tariffs and Trade (GATT) in 1947. The treaty's purpose aimed to reduce global tariffs and other trade barriers as well as to eliminate trade preferences so everyone would benefit from equal opportunity. Over the years, the GATT held many rounds of talks which between them have tackled literally thousands of obstacles in the quest for true free trade. GATT was eventually taken over by the World Trade Organization (WTO) and in respect of food it was only by the 1992/95 Uruguay round that agriculture was finally fully admitted into negotiations. While these talks have indeed reduced many such barriers, there still commonly exist numerous preferential trading agreements, import tariffs, quotas, and subsidies by or between countries and regions. Indeed, the hypocrite here is that during these rounds and through certain international financial lending instrument's conditions, developing countries are encouraged to liberalize their markets in favor of free trade. This was perhaps one of the major failings of the Uruguay round as before Uruguay (which disallowed the practice of protectionism) it had been hoped that the success of the EU CAP model in raising living standards and production within its borders could be mimicked by other less developed countries worldwide. Adding to this hypocrite is the fact that at the time the EU refused to unravel its own trade distorting policies (Warnock, 1997; SED, 2004; Barnes, 2006; Harvey, 2010; Murphy, 2010). For the sake of completeness though, such practices were not seen in isolation within the developed world; rather numerous countries, particularly in Africa and the Middle East, for instance, also operated and continued to do so protectionist policies that continue to marginalize their global participation in free trade (IMF, 2010).

It is this dichotomy of "do as I say and not as I do" politics that many see as one of the overriding barriers to real progress free trade and indeed unless the status quo is challenged many are skeptical of the ultimate benefits of so-called liberalization. Furthermore, hypocrite and protectionism suggest critics; it favors rich countries and large corporations and producers while continuing to subjugate impoverished countries. Moreover, on the subject of the natural resources, it has been said that free trade powered by privatization and fueled by profit is charged with reinforcing short termism and promoting a laissez-faire attitude toward environmentalism running contrary to the good stewardship needed to maintain sustainable production methods (Harvey, 2010).

Thus, cynically it could be argued that it is not so much perhaps, nation states' global political and economic hegemony that is the current driving force behind societal change, but rather the collective social and ideological values of billions of global consumers. In this picture, globalization is an economic and political force all of its own, and one which is firmly in the hands of the people; it could also be considered the dominant vehicle of global governance when it comes to issues of food.

14.6 Safety nets and food reserves

Food, being the staff of life, is not always available—at times it is subject to economic drivers; price volatility; to spoilage; and to loss via natural or man-made emergencies; and even political whimsy among other things. Indeed, in the words of Sophia Murphy, lands cannot be moved, harvests are unpredictable, and consumption is neither elastic nor optional

(Murphy, 2009). Consequently, the idea of food safety nets is as old as civilization and since ancient times people have stockpiled grain and other foods for anticipated leaner times ahead. Recently, however, partly as a backlash at the food mountain horrors amid world starvation the practice has dwindled somewhat. Instead the trend is toward a more pragmatic market-led system. This has been shown during times of peace to work relatively well with any emergency shortfalls quickly offset by international aid. That was until the financial crisis of 2008 which prompted a revival of the idea of food reserves and food safety nets at the G8 meeting in 2009 (EC/FAO, 2008).

Such programs would be of particular benefit to vulnerable groups such as those unable to work for one reason or another or perhaps those affected by recession or natural disasters. Food safety nets include those direct feeding programs such as soup kitchens, free school meals, food fortification, provisioning for expectant and nursing mothers as well as for the under-fives, etc. Alternatively, in lesser developed countries, food safety nets also manifest in certain food-for-work programs, which also have the added bonus of helping to support local community projects such as irrigation, roads, or buildings like schools and health centers. Yet other safety nets take the form of income-transfer programs such as cash or in-kind payments like food stamps, subsidized rations, and other targeted measures. Lastly, food safety nets also include things like agricultural input subsidies or crop insurance; all in all, there are many things responsible governance can do to ensure adequate food provisions to the most vulnerable of its citizenry (SOFI, 2008).

14.7 Globalization: hegemony, multilateral cooperation, and people power

The march toward globalization has been relentless. For two, arguably three, centuries, advances in media, travel, and technology, all facilitated by conducive politics, have resulted in the relative ease of movement of goods and services across the globe (Gibson, 2016). However, within this context, food is a relative newcomer. In fact regard to food, despite the overt political will that has been bandied about since the 1940s or so, it was only by the 1990s that the General Agreements on Trade and Tariffs (GATT) talks that food was finally admitted into the global free-trading arena. In theory, this effectively allowed countries to buy the food they needed at the best possible price on the international market, supposedly unimpeded by economic or political barriers (caveats aside). That said, globalization has brought about economic and cultural liberalization as never before— a new orthodoxy (Vaidya, 2006). Much of this, according to Kennedy et al. (2004), is being spurred on by urbanization and, as far as food is concerned, by effectively transforming the entire food supply chain from production, processing, retailing, and marketing through to consumers all around the world, making it more accessible and generally more affordable to most (Kennedy et al., 2004). In fact, what was once the preserve of governments and multinationals alone is now more open to small to medium enterprises (SMEs). In this way it seems, more and more countries are taking back control over their own FSCs than ever before. There are many other benefits of this new global market too. One such advantage has to do with livestock and quality control standards. In dealing on the international market, for instance, bound up within increased global food trade is a trend for better quality control standards, especially within the meat and livestock sector. Inherent in this upward trend, there also exists the possibility of translating such coveted standards into better animal welfare values and by extension the potential of reducing disease and increasing food safety. There are also economic and social benefits of globalization too—as societies are becoming increasingly politically and culturally integrated, so the increased flow of commodities services, labor, and capital means more opportunities for many more individuals and countries. Consequently, greater international trade often raises domestic incomes and standards of living helping both the financially poor with improved food security (Parfitt et al., 2010).

Benefits aside, global trade in food and agriculture also brings with it, its own challenges. When it comes to animals for instance, the same strict quality controls that look to raise standards might also act as barriers. Also, when it comes to a finite food supply and with so many countries now looking to be fed from the same trough, the many global inequalities in terms of trade are potentially set to increase with certain established interests looking to benefit considerably more than others. Furthermore, on a political dimension, the FAO, Parfittet al (2010) also warn that with increased integration and a potential converging global agrarian policy, the rewards might come at a high price as reflected in the increased vulnerability of global production, the lack of development of individual internal markets, as well as subsequent price volatility (McMichael, 1994; FAO, 2003; Parfitt et al., 2010). Another cautionary note too, says Kennedy (2004), and one that counters the benefits of SMEs as suggested above, concerns the vast amounts of direct foreign investment by large multinational food companies as well as retailers. While no doubt this investment is resulting in cheaper food, greater availability, and more diversity. However, such "advances" are inducing fundamental changes in traditional production, procurement, and distribution systems often, suggests Kennedy, at the expense of smaller local agents and long-established food outlets. Moreover, as international trade tends to favor big business and centralized procurement systems, it further acts as an exclusionary hurdle to smallholders or small producers (FAO, 2005). Lastly, there also appears to be evidence that globalization is likewise bringing about a gradual shift—a convergence if you like—toward a more universal food culture (Kennedy et al., 2004; Dimitri et al., 2005; Scheuerman, 2008; Walker, 2008).

Another important driver of change within the food supply dynamic is increasing urbanization and a trend that is only going to continue.

14.8 Land grabbing

One growing trend faced by poorer economies is the trend of selling or leasing domestic agricultural land to foreign entities. While on the one hand such practices provide much needed income, on the other there are fears that it could lead to land conflict, water and other natural resource competition, evictions and increased land prices (RTFN, 2010). The trend is well-established; multinationals, governments, and investment funds have been buying up vast tracts of land to either ensure sufficient food supplies in their own countries or simply as tools of profit. This is perhaps not surprising when you consider some lands in the cheapest areas (especially in Africa) can be leased for as little as $1 a year. This trend is challenging in both moral and practical terms. Take Ethiopia for example, paradoxically the government, in the face of some of the worst food insecurity in the world, is offering up to 3 million hectares of its most fertile land to rich countries, and while some argue this is selling of the natural resource base, others suggest it brings much needed foreign investment in place of domestic measures which have ultimately failed (Vidal, 2010). This practice is not confined to Africa either; in Romania for instance, about 1 million hectares of arable land (around 12% of its national stock) has been sold to foreign interests (The Diplomat, 2011). In Saudi Arabia too, water stresses are encouraging the government to reduce domestic cereal production by at least 12% a year, and in efforts to conserve its valuable water resources is offering vast subsidized loan agreements to companies interested in buying and cultivating land overseas.

It's not only food pressures that are driving force behind this practice. China, for instance, cultivates 2.8 m ha of land in the Democratic Republic of Congo to produce palm oil for its biofuel industry. In the European Union, biofuel companies too have acquired large swathes—about 3.9 m ha—in Africa for commercial purposes.

Indeed the practice of land grabbing is so widespread involving numerous countries and organizations with mind boggling amounts of land that it is still something of an unknown quantity (von Braun and Meinzen-Dick, 2009). In fact, so quick has been the phenomenon emerged and so large the scale,[2] that the practice is receiving more than a measure of criticism amid fears of disruption to local food security for one thing. Of course, however, as with everything else it is not always clear cut and there are benefits and disadvantages; on the plus side such acquisitions might be seen in the light of much needed agricultural and rural investment in terms of finance and technology. This suggests proponents, creates both on and off-farm jobs for local communities while also providing much needed yield increases and promoting better land management practices. Also, of benefit are the many trickle-down benefits such as the rural development projects like new schools or health clinics that sometimes accompany such deals. However, on the downside there is concern in some quarters concerning the terms of these contracts and whether or not they benefit the investors more than the recipients; and in turn whether or not the recipient's terms benefit local communities or their own self-interests. Further tied up in these concerns is the tangible fear that such practices will also have an adverse, rather than beneficial effect on the local environment as well as people's livelihoods (von Braun and Meinzen-Dick, 2009).

14.9 Food sovereignty

One alternative view to the current food supply paradigms is that of food sovereignty. However, it must be noted at the outset that food sovereignty is not the same as self-sufficiency, rather it is an alternative view of the food supply chain that bucks prevailing trends of globalization. Introduced in 1993 in Belgium by a group of farmers' representatives calling themselves "La Via Campesina" the idea of food sovereignty was globally launched at the World Food Summit in 1996. In response to the fact that more and more of agriculture nowadays is influenced, governed or legislated at the international level, the movement

[2]The full extent of land grabbing is not fully known however, the FAO have suggested that over the last few years, perhaps as much as 20 million hectares in Africa alone might have been acquired by foreign interests RTFN (2009). Who controls the governance of the world food system? Right to Food and Nutrition Watch. Germany, Brot für die Welt (Bread for the world). **2009**. RTFN (2010). Land grabbing and nutrition: Challenges for global governance. Right to Food and Nutrition Watch. Germany, Brot für die Welt (Bread for the world). **2010:** 90.

acts as a voice for the millions of small and medium-size farmers, landless people, peasants, indigenous people, and others from around the world, in actively engaging in decision-making processes that affect all their lives. Food sovereignty is in part a backlash against modern farming practices, but more importantly it seeks to take back control of trade, agriculture, and land policies and ultimately food supply away from multinational corporations and place responsibility back in the hands of national interests. It is fundamentally opposed to transnational companies and corporate-driven agriculture favoring instead national self-determined food policies.

At the heart of the movement are the welfare, equity, and governance of the millions of livelihoods supported by small-scale farming, peasantry, and herding livelihoods that its proponents see as being challenged by the trend of globalization. This is evidenced in one example by the Uruguay round of trade negotiations in which liberalizing sovereign agricultural policies in favor of freer more open international trade became prescriptive at the national level. Another example can be seen in many developing countries external debt positions which, as part of international loan conditions, see countries having to initiate World Bank and the IMF structural adjustment programs containing, in many instances, mandatory conditions in such areas as the opening of international markets for agricultural products, etc. However, perhaps more emotive is the strong negotiating position of many foreign multi-nationals many of which wield significant power beyond their political weight and which is brought to bear in influencing terms of trade and sometimes by default, domestic policies.

In retaliation to these challenges the Food Sovereignty framework promotes the devolution of the centralized decision-making process away the global and back to the national. Furthermore, in keeping with their ideals, the Food Sovereignty framework prioritizes local food production and consumption while ensuring that the balance of power resides with the people who produce the food rather than the corporate behemoths (Windfuhr and Jonsén, 2005; Clements, 2009; La Via Campesina, 2011).

References

Aksu, E., 2008. Perpetual peace a project by europeans for europeans? Peace & Change: J. Peace Res. 33 (3), 368–387.

Barnes, D.G., 2006. A History of English Corn Laws: From 1660–1846. Routledge, London.

Burke, E., 1790. Reflections on the revolution in France, and on the proceedings in certain societies in London relative to that event. In: A Letter Intended to Have Been Sent to a Gentleman in Paris. Pall Mall, London (Printed for J Dodsley).

Cattell, C.C., 1874. Co-operative Production. G. H. Reddalls, Birmingham.

Clements, R.E., 2009. Scoping Study into the Impacts of Bioenergy Developmenton Food Security. Policy Innovation Systems for Clean Energy Security (PISCES), Nairobi, Kenya.

Dimitri, C., et al., 2005. Economic Information Bulletin: Number 3: The 20th CenturyTransformation of U.S. Agriculture and Farm Policy. United States Department of Agriculture, USDA.

EC/FAO, 2008. An Introduction to the Basic Concepts of Food Security. EC - FAO Food Security Programme, Rome, p. 3.

ECA/SDD, 2004. Land Tenure Systems and Their Impacts on Food Security and Sustainable Development in Africa. Addis Ababa. UN Economic Commission for Africa, Ethiopia.

Engel-Janosi, F., 1941. Reflections of Lord action on historical principles. Cathol. Hist. Rev. 27 (2), 166–185.

FAO, 1946. Report of the First Session of the Conference Held at the City of Quebec, Canada, 16 October to 1 November, 1945. FAO, Washington, p. 89.
FAO, 2003. Trade Reforms and Food Security: Conceptualizing the Linkages. Food and Agriculture of the United Nations, Rome.
FAO, 2005. The Globalizing Livestock Sector: Impact of Changing Markets. Committee on Agriculture Report. Food and Agriculture Organisation, Rome.
FAO, 2006. Policy Brief: Food Security. Food and Agriculture Organisation, Rome, p. 4.
FAO, 2009. Secretariat Contribution to Defining the Objectives and Possible Decisions of the World Summit on Food Security on 16, 17 and 18 November 2009. Food and Agriculture Organisation, Rome, p. 10.
FAO, 2010. The Food and Agriculture Organization of the United Nations. Retrieved 26th Jan 2019, 2019, from http://www.fao.org/about/en/.
Food Summit, 1996. Rome Declaration on World Food Security World Food Summit. Food and Agriculture Organisation, Rome.
Gibson, M., 2016. The Feeding of Nations: Re-defining Food Security for the 21st Century. CRC Press, Boca Raton, Florida.
Halhed, N.B., 1776. A Code of Gentoo Laws, or, Ordinations of the Pundits : From a Persian Translation, Made from the Original, Written in the Shanscrit Language (London).
Hartmannshenn, T., 2004. FOOD SECURITY: Guidelines for the Promotion and Execution of Food Security Projects by German Agro Action. German Agro Action, Bonn.
Harvey, D., 2010. An Abbreviated UK/EU Agricultural Policy History. Retrieved 23rd Nov 2019, 2019, from http://www.staff.ncl.ac.uk/david.harvey/AEF372/History.html.
Hasan, M.A., 2001. Role of Human Capital in Economic Development: Some Myths and Realities. Development Research and Policy Analysis Division, ESCAP, Bangkok, Thailand.
Hurlbut, E.P., 1847. Essays on Human Rights and Their Political Guaranties. Maclachlan, Stewart and Co, Edinburgh.
IMF, 2010. International Montetary Fund Website. Retrieved 4th Jan 2019, 2019, from http://www.imf.org/external/.
Ishay, M.R., 2008. The History of Human Rights: From Ancient Times to the Globalization Era. University of California Press, Berkeley and los Angeles.
Johns, C.H.W., 1904. Library of Ancient Inscriptions: Babylonian and Assyrian Laws, Contracts and Letters. Charles Scribner's Sons, New York.
Kennedy, G., et al., 2004. Globalization of Food Systems in Developing Countries: Impact on Food Security and Nutrition: A Synthesis of Country Case Studies. FAO, Rome.
La V. Campesina, 2011. La via Campesina: The International Peasant's Voice. News & Views. Retrieved 21st Feb 2019, 21st Feb 2019, from http://www.viacampesina.org/en/index.php?option=com_content&view=frontpage&Itemid=1.
Locke, J., 1714. The Works of John Locke Esq: In Three Volumes. Paternoster, London (printed for John Churchill at the Black Swan).
MacAuslan, I., 2009. Hunger, discourse and the policy process: how do conceptualizations of the problem of 'hunger' affect its measurement and solution? Eur. J. Dev. Res. 21, 397–418.
Marzeda-Mlynarska, K., Curie-Sklodowska, M., 2009. Multilevel governance of food security: theoretical model or new international practice. In: 4th Annual Conference of the GARNET Network. GARNET Network of Excellence, Rome.
Maxwell, D., Wiebe, K., 1999. Land tenure and food security: exploring dynamic linkages. Dev. Change 30 (4), 825–849.
McMichael, P., 1994. The Global Restructuring of Agro-Food Systems. Cornell University Press, Ithaca, NY.
Mettrick, E.F., 1929. Population, poverty, and ethical competence. Int. J. Ethics 39 (4), 445–455.
Murphy, S., 2009. Strategic Grain Reserves in an Era of Volatility. Institute for Agriculture and Trade Policy, Minneapolis, Minnesota.
Murphy, S., 2010. Trade and Food Reserves: What Role Does the WTO Play? Institute for Agriculture and Trade Policy, Minneapolis, Minnesota.
O'Niel, O., 2002. Second Reith Lecture. BBC Radio, London, p. 4, 10th April.

Orr, J.B., 1939. Diet and nutrition: nutrition problems, dietary requirements for health. Can. Med. Assoc. J. 41 (1), 78–80.

Parfitt, J., et al., 2010. Food waste within food supply chains: quantification and potential for change to 2050. Philos. Trans. Royal Soc. 365 (1554), 3065–3081.

Phillips, R.W., 1981. FAO: Its Origins, Formation and Evolution 1945-1981. Food and Agriculture Organisation, Rome.

RTFN, 2009. Who Controls the Governance of the World Food System? Right to Food and Nutrition Watch. Brot für die Welt (Bread for the world), Germany, 2009.

RTFN, 2010. Land Grabbing and Nutrition: Challenges for Global Governance. Right to Food and Nutrition Watch. Brot für die Welt (Bread for the world), Germany, p. 90, 2010.

Scheuerman, W., 2008. Globalization. In: Zalta, E.N. (Ed.), The Stanford Encyclopedia of Philosophy. Stanford University, Stanford, CA.

SED, 2004. European Trade Barriers and Developing Countries. Netherlands Ministry of Foreign Affairs Sustainable Economic Department.

Shaw, J., 2007. World Food Security: A History since 1945. Palgrave Macmillan, UK.

Sherlock, T., 1718. A Vindication of the Corporation and Test Acts: In Answer to the Bishop of Bangor's Reasons for the Repeal of Them. To Which Is Added: A Second Part, Concerning the Religion of Oaths. London Printed for J Pemberton at the Buck and Sun.

SOFI, 2008. The State of Food Insecurity in the World 2008. Food and Agriculture Organisation, Rome, p. 59.

SOFI, 2009. The State of Food Insecurity in the World 2009. Food and Agriculture Organisation, Rome, p. 59.

The Diplomat, 2011. Foreign farmland ownership rises over ten per cent. The Diplomat: Bucharest 7 (2).

UN, 1948. Universal Declaration of Human Rights. United Nations, U.N. A/RES/217 A (III).

UN, 2000. United Nations Millennium Declaration. United Nations, UN. A/RES/55/2.

UN, 2004. Voluntary Guidelines to Support the Progressive Realization of the Right to Adequate Food in the Context of National Food Security: Adopted by the 127th Session of the FAO Council November 2004. Food and Agriculture Organisation, Rome.

UN News Centre, 2010. Somalia and US Should Ratify UN Child Rights Treaty – Official. UN News Centre, UN News Service.

Vaidya, A.K. (Ed.), 2006. Globalization: Encyclopedia of Trade, Labor, and Politics. ABC-CLIO, Santa Barbara.

Vidal, J., 2010. The Guardian Profile: Amartya Sen, Food for Thought. The Observer. Guardian News and Media Limited, London.

von Braun, J., Meinzen-Dick, R., 2009. Land Grabbing" by Foreign Investors in Developing Countries: Risks and Opportunities. The International Food Policy Research Institute, Washington, D.C, 9pp.

Walker, P., 2008. The Origins, Development and Future of the International Humanitarian System: Containment, Compassion and Crusades. ISA's 49th Annual Convention, Bridging Multiple Divides. International Studies Association (ISA), San Fransisco, CA.

Warnock, K., 1997. Third world: feast or famine? Food security in the new millennium (commentary). J. Race Class 38 (3), 63–72.

Williams, A.J., 2005. Reconstruction' before the marshall plan. Rev. Int. Stud. 31 (3), 541–558.

Windfuhr, M., Jonsén, J., 2005. Food Sovereignty: Towards Democracy in Localized Food Systems. Warwickshire, ITDG/FIAN, p. 70.

WTO, 2010. International Trade Statistics 2010. World Trade Organisation, Geneva.

Further reading

Birchall, J., 2002. Evolutionary Economics. Retrieved 15th Jan 2019, 2019, from http://www.themeister.co.uk/economics/evolutionary_economics.htm.

D'Souza, F., 1994. Democracy as a cure for famine. J. Peace Res. 31, 369–373.

EIU, 2010. Democracy Index 2010: Democracy in Retreat. Economist Intelligence Unit, London, p. 46.

Freedom House, 2011. Website of Freedom House. Retrieved 21st Jan 2019, from http://www.freedomhouse.org/template.cfm?page=1.

INSCR, 2011. The Integrated Network for Societal Conflict Research (INSCR). Center for Systemic Peace.

Mukherjee, S., 2011. Blame the Rulers Not the Rain: Democracy and Food Security in Zambia and Zimbabwe. South African Institute of International Affairs (SAIIA), South Africa.

Roth, M., Myers, G., 2010. Issue Brief: Land Tenure, Property Rights, and Food Security: Emerging Implications for USG Policies and Programming. United States Agency for International Development, Washington D.C.

Scanlan, S.J., 2004. Feast or Famine?: Food Security, Democracy, and Information Technology in Less Industrialized Countries. In: Annual meeting of the American Sociological Association, San Francisco.

CHAPTER

15

Food and social trends

Lei Weng Si (Clara), Loi Kim Ieng (Connie)

Institute for Tourism Studies, Macao (IFTM): Colina de Mong-Há Macao, China

OUTLINE

Food and society have been interwoven in many ways and in both directions. Food is an indispensable part of the social culture, which in turn exerts salient influence on food evolution and trends throughout the years. This chapter is divided into four sections. First, the social function of food will be discussed. Second, the two main forms of dining styles will be illustrated, followed by how food is served and consumed in different social occasions. The chapter will further illustrate how and why different food items are prepared to commemorate various festivals in the old and modern times of China with reference to the unique solar terms. The chapter is concluded by recognizing the important relationship between food and society.

Food and Society
https://doi.org/10.1016/B978-0-12-811808-5.00015-5

15.1 Food as an important social function

The relationship of food and social activities goes back a long way in history. There is a famous quote from an eminent politician of China during the Spring and Autumn Period (770–476 BC)—"To a ruler, the people are foremost; to the people, food is foremost" 王以民為天, 民以食為天. The famous Chinese philosopher Confucius (551–479 BC) also postulated that food and gender were the two most important human desires 飲食男女，人之大欲存焉. Food is therefore an important element among people and society as long as human history goes. It is not only limited to the Chinese culture but also to all others.

According to the Intangible Cultural Heritage listing of UNESCO (United Nation Education Scientific and Cultural Organization), one can easily discover that there is quite a number of food and cuisine-related items being enlisted, for example, French gastronomy, traditional Mexican cuisine, the Mediterranean diet of Cyprus, South Korea's kimchi, cuisines from Croatia and Turkey, and the traditional Japanese cuisine "washoku." Among all these, South Korean kimchi expresses this important social function food as well as that of the French, just in a different way. French gastronomy serves as an important social custom with an emphasis on the enjoyment of the meal consumption, while the tradition of Kimchi making illustrates the importance of social function through the preparation process. The production process of kimchi is known as "kimjang" in Korean. It is a process starting from selecting, collecting, marinating, and preserving the vegetables. Every Korean family makes kimchi and the scale of production goes even to a village or a community. The entire process boosts cooperation among families, villages, and communities, contributing to social cohesion and harmony (UNESCO website). Friends and relatives share know-how of kimchi making. It is because of this process of making kimchi which enhances social bonding and interpersonal relationship. French gastronomy is described as "a social custom to celebrate the most important moments in the life of an individual or a group, such as weddings, birthdays, and so on." Occasions for celebration include child birth, birthdays, weddings, anniversaries, achievements, and reunions. The structure of a standard French gastronomic meal contains four courses. They are a starter, fish and/or meat with vegetables, cheese, and dessert. The meal always starts with an aperitif and ends with liqueurs. The essences of the meal are togetherness, drawing circles of friends and family closer together and strengthening social ties. It definitely is one of the best representations of such an important function of food—social.

15.2 Food and dining styles

In general, dining styles are categorized into two different ones: individual and communal (Gong; Mars and Mars, 1993). For communal dining style, diners normally sit together in a round or long table setting. Food is served in serving bowls and placed in the middle of the dining table. Every diner takes food from the serving bowls to their own plates or bowls for consumption. For individual dining style, diners are as well sitting together in a round or long table setting; however, each diner has his/her own plate of food to consume. Floral centerpiece is possible for individual dining, while centerpiece is displayed before meal starts and is taken away when meal starts for communal dinning. Those cultures which normally practice communal dining styles include Chinese, Korean, Japanese, Vietnamese, and Thai, just to name a few, while British, Canadian, American, German, French, Italian, and other western cultures typically practice individual dining styles more frequently than the others.

Communal dining style can be dated back to the Tang Dynasty (618–907 AD) (Gong). Communal dining has always been practiced whether at daily meals, family and friends' gatherings, village feasts, festival celebrations, or official banquets. Serving spoons and cutlery is always used for communal dinning. In addition, a rotating "Lazy Susan" was developed to facilitate communal Chinese dining, in which dishes are placed by the edge of the Lazy Susan for rotating to serve everyone sitting around the table. Lazy Susan is particularly efficient and practical when more than six diners are eating together. Condiments are also put on the Lazy Susan for diners' consumption. Furthermore, potluck is also a kind of communal dining style that each diner is expected to prepare and bring a dish to the party for sharing. By hosting pot luck party, the burden of food preparation and food cost is shared among diners. It is a common dining style among religious group, students, village, and community gatherings. Regardless of the dining styles, when people eat, people tend to stay together and share conversation and stories. Food thus has a very crucial role to draw people together to socialize.

15.3 Food and social occasions

15.3.1 Official occasions

During the Zhou Dynasty (1046–256 BC), Zhou Li (Rites of the Zhou), Yi Li (Protocol and Rites), and Li Ji (The Book of Rites) were developed. These three ancient classics offer many details of protocol and rites, such as food protocol to hosting guest, holding official banquets, having daily meals, and serving food to the goddesses. For example, status of a diner is reflected by his/her seating order. Normally, the eldest and the most honored guest is seated at the head of the table. If the guests sit facing each other along the north and south sides of the table, the west side is considered as the head of the table. If the guests are seated along the east and west sides of the table, the north side is considered to be the head. During a State Banquet in the ancient time, food service had clear divisions between ranks, from the emperor on down to the dukes, lords, and various officials. Ranks and status were reflected not only in food but also in the tableware used. For example, dings (tripod vessels) inlaid with solid gold were used only by the emperor, while dukes used bronze dings inlaid with silver, lords used bronze dings, and officials used iron dings. Similar practice was also applied in England during Tudor times (1485–1603); everything from the number of dishes eaten to the ways in which food was served, was dictated by status (Mason, 2014). For example, pottage was normally consumed by both laborers and the rich. However, the rich ate pottage containing almonds, ginger, and saffron, as well as wine, whereas the laborers consumed pottage with some barley or oats. The statuses reflected in England during Tudor times were among the different social classes, while the statuses reflected in the ancient China time were among emperor and different ranks of officials.

One of the most formal official occasions is probably the "State Banquet." According to Oxford Dictionary, State Banquet is a formal, ceremonial dinner hosted by a head of state, now especially one to which a visiting head of state is invited for the purpose of renewing or celebrating diplomatic ties. Today, State Banquet is one of the most solemn and significant occasions to host countries' leaders. State Banquet is another illustration to depict the social function of food in the highest order of a country and undoubtedly important in diplomacy of any regime. Countries usually present their best and most representative cuisine in a State Banquet, with the aim to showcase a country's unique culinary culture and folklore. In the following, State Banquets of two countries are discussed, one representing the west and the other representing the orient. They are the Great Britain and the People's Republic of China.

15.3.1.1 The People's Republic of China

The first State Banquet hosted in Yuhuatai Restaurant included eight cold dishes, six hot dishes, and four snacks. Cold dishes were cucumber with salt and sesame oil, jelly fish with sesame oil, duck's feet with mustard, among others. Hot dishes were abalone soup, Dongpo pork, steamed pork meat balls, steamed crab meat balls, sliced bean curd in chicken broth, sautéed shrimps with broccoli and hotchpotch. Four snacks were fried rice cake, Huangqiao sesame cake, steamed rice cake with sweet stuffing, and Huaiyang soup dumplings. "Lion's Head" was one of the favorite State Banquet dishes of Chinese first Premier—Zhou Enlai. It is made of the finest pork which is minced and shaped as a ball and served in a soup which is boiled hours from Chinese cabbage. The dish is named as Lion's Head because the meat ball is shaped as the head of the lion and vegetable is shaped as the lion's mane. Beijing roast duck is another popular State Banquet dish. It was among one of the favorite dishes of American former president Bush when he visited China. The duck is roasted with wood fire, in which fruit tree branches, such as pear, peach, or date branches are used to set the fire. Compressed air is injected into the duck during roasting to make the skin shiny and sleek. The crispy duck skin is peeled off skillfully by the chef typically on table side (gueridon) as a type of showmanship and is consumed with green onion, fresh cucumber, shallot, and soybean paste wrapping together in a thin Chinese pancake.

15.3.1.2 Great Britain

The first State Banquet hosted by Queen Elizabeth II was to welcome King Gustaf VI of Sweden during his visit to the United Kingdom of Britain on June 28, 1954 (Salter, 2015). The state visit of King Gustaf VI lasted for 4 days and marked the first of Queen Elizabeth II's reign. A State Banquet was held in the Ballroom of the Buckingham Palace. According to Anna Reynolds, the curator of Buckingham Palace summer exhibition, "The state visit is one of the most formal occasions of the Queen's year; it is the one most steeped in tradition." Nevertheless, a State Banquet was not a formal occasion during the period of Queen Victoria's reign. Queen Victoria wrote in her diary about entertaining Napoleon III of France in 1855 "in the usual dining room." The Ballroom of the Buckingham Palace was added during a massive renovation, and State Banquets have been held in the Ballroom of the Buckingham Palace since 1914.

The long history of British State Banquet makes its planning process much more complicated than the Chinese one. It takes normally 1 year to plan a State Banquet. Once visit confirmation of a head of a state is confirmed, the Master of the Household in the Buckingham Palace will start a seating plan. The Master of the Household has to go through a very detailed and careful thinking on seat assignment. Who is sitting next to whom, if the neighbor has met with each other before, and the languages each guest command are all under consideration.

The British State Banquet normally serves a four-course meal and speeches to commence the banquet. The first and second courses are usually fish followed by meat, pudding (dessert) prepared by the Royal Pastry Chef, and fruit courses are served on a porcelain. Each guest is served five different wines, starting with champagne, red and white wine, sweet wine, and port wine at last. The menu follows French gastronomic planning. Besides the menu, the table setting is also very comprehensive and specified. The Master of Household describes that advanced preparation is required in a few months' time to polish thousands of silver-gilt cutlery and glass items. The table setting for each guest's seat is checked with a measuring stick to ensure that there is at least 18 inches of space between guest's knife and fork. Queen Elizabeth II comes to check out the Ballroom before banquet starts. The banquet service is orchestrated—one footman serving nine guests. There are hidden lights in the ceiling of the banquet hall to signal the footmen for putting down dishes and cutlery. The serving process is described as a symphony. High level of attention to details, for example, the measurement of a chair to the dinner table is seamless for the hundred chairs lining up in the banquet hall. Cutlery and glasses are all in line with each other on the dining table, too. Each person gets 10 cutleries and 6 glasses to be used for the banquet. The butter is served individually to each guest and the butter is stamped with a royal crown.

The details and standard of a State Banquet, no matter that it is from the West or from the East, are present with rigorous specification and requirement. Food served in official occasions plays an important role to express a country's hospitality and respect to a visit of head of state.

15.3.1.3 Private occasions

Besides official circumstances, people gather for various private occasions too, from socialization, celebrating festive seasons, wedding, birthday party, and the like. Forms of gathering are plentiful, from dining in a restaurant to self-gathering activities. The following paragraph illustrates three kinds of popular ways of gatherings for food. They are hot pot from China, barbeque, and picnic from the western countries.

15.3.1.3.1 Hot pot

Hot pot 火鍋 is a very popular way of private gathering nowadays in different parts of China. Hot pot seems to have originated long time ago, dated back in the Qing Dynasty (1644–1912) in China. The traditional way of having hot pot is burning coal or wood fire to heat up a steam bowl. By burning coal or wood to simmer the bowl, different food ingredients are put to cook inside the bowl. Food is to be cooked one by one. Diners normally sit in a round table and the hot pot is placed in the middle for cooking .Diners have their own cutlery to cook and eat. A hot pot dinner can easily last for hours because of its slow cooking and slow way of food consumption. Diners normally share personal updates and stories during hot pot time. Typical hot pot food ingredients include thin-sliced meat such as beef, pork, and fish, shellfishes, leaf vegetables, dumplings, and noodles. Today, in many modern Chinese homes, coal or wood fire steam bowl was replaced by electric stove or gas cooker. Hot pot with two compartments for two kinds of broth is widely used as shown in photo below (Fig. 15.1).

FIGURE 15.1 An electric hot pot. *Source: https://asianinspirations.com.au/recipes/white-pork-bone-soup-for-yin-yang-hot-pot/.*

15.3.1.3.2 Barbeque

Cooking on open fires is probably the most ancient and earliest cooking method in human history. Barbeque (outdoor grill) is an open-air cooking method. It usually uses charcoal or wood to set up fire and slowly grill or spit-roasting poultry or seafood. According to food historian Davidson (1999), the word barbeque comes from the word of "barbacoa" in Spanish, denoting a structure on which meat could be dried or roasted. The word had first entered in the English language during 17th century, meaning a wooden framework could be used for storage or sleeping on, without a culinary context. The word barbeque has taken its present meanings only by the 18th century. By 1733, the word barbeque had taken on the implications of a social gathering (Mariani, 2014). It became commonplace during late 20th century (Davidson, 1999).

Barbeque nowadays is undoubtedly a form of social gathering. People gather outdoor and spend time together to enjoy this slow cooking method for not only poultry and seafood but also vegetables and even marshmallow and bread. Many countries and cities provide barbeque stoves or facilities in parks and public leisure area for people usage. In Australia, barbeque is even considered as a national social activity, in which Australians do barbeque frequently to celebrate various social occasions, such as national day, Christmas, and family reunion. Aussie barbie, the colloquial way to call Australian barbeque, is commonly referred. Aussie barbie has been a long tradition practiced by the indigenous people. New expressions emerged as part of the barbeque culture, such as "throw a shrimp on the barbie" (Wells, 2015). Barbeque stove styles vary from a simple wooden one, a gas type, to outdoor oven styles, all of which are popular items in most of the Australian households. Barbeque has become an Australian lifestyle that Australians meet, greet, and party with friends and family for food cooked alfresco and on a slow fire.

15.3.1.3.3 Picnic

Picnic also has a long tradition. According to food historians, picnic has been practiced since the mid of 16th century (Olver, 2000; Butler, 2013). The English word "picnic" comes from the French word "pique-nique." The French word was used to describe gourmands who brought their own food and wine when dining out. Picnic was a glamorous outdoor affair during the Victorian era in Britain. Picnic was depicted in many English classic books and even illustrated among paintings of great painters. It has been a social and fashionable entertainment in the old times until today. According to the Oxford English Dictionary, the word "picnic" denotes a social event at which each guest contributes a share of the food. Picnic usually takes place in countryside, outdoor, or at leisure locations where people can enjoy the environment and scenery.

Picnic is a popular activity in America too, as a celebration of human spirit, culinary diversity, and adventure. Picnics are very much personal. People choose their own foods, dining partners, and venues. Picnics can be planned or impromptu; they are very different from public outdoor dining events or festivals, for example, community feasts (New England clambakes, Texas barbeques, New Orleans shrimp boils) or alfresco dining (trendy waterfront bistros, central city cafes). In America, picnic can be in any form of the following (Olver, 2000):

- "traditional American foods" prepared at home and served on a blanket in a local park
- ethnic cuisine celebrated by an extended family in an urban riverfront location
- an artfully presented basket of gourmet delights served on fine linen and china
- box lunch obtained from a convenience store consumed at the beach
- bread, cheese, and grapes shared by best friends in a canoe
- a family passing peanut butter crackers and bottled water at a highway rest stop
- a child serving imaginary cakes to stuffed animals beneath the protective branches of the family's backyard tree.

Most of time, food comes second in a picnic but its spirit of social gathering and being together with families and friends is paramount.

15.4 Food, solar terms, and festivals in China

The role of food is most prominent when it comes to celebration of festivals all over the world. Among the many different cultures, Chinese can be considered as one that places particularly high emphasis on the use of various food items; their symbolic meanings in festivals and celebrations; and their relationship with the seasons. Chinese philosophy stresses heavily on eating "seasonal food." Among the many rules of Confucius about his food as stated in the Confucian Analects **論語**, one of which was "he did not eat what was not in season" **不時不食** and this has become an important saying in the Chinese food culture. In addition to the four universal seasons, the Chinese also look at 24 "solar terms" **廿四節氣**, which is a gross name of the system that comprises 12 major solar terms and 12 minor solar terms interlaced with each other in the calendar. There is one major and one minor solar term in each calendar month. Developed through observation of changes of seasons and the movement of the sun, the solar terms are used by farmers and the community to guide agricultural production and daily routines of people. Moreover, some rituals and festivities in China are closely associated with the solar terms. The categorization of solar terms is so unique that the United

Nations Educational, Scientific, and Cultural Organization (UNESCO) inscribed it on the Representative List of the Intangible Cultural Heritage of Humanity on November 30, 2016.

Each of these solar terms carries special meaning(s), characteristics, and seasonal food items. The combination of festivals, symbolic meanings of food, and seasonality results in a whole calendar punctuated with opportunities to eat and celebrate. Table 15.1 provides a summary of these 24 solar terms:

In each season and corresponding to the 24 solar terms of Chinese, there are different culture festivals. These festivals are important occasions for families and friends getting together to reunion and socialize in China. The types of foods used and cultural meanings behind those foods are discussed below under each iconic festival. They are Chinese New Year, Dragon Boat Festival, Mid-Autumn Festival, Chongyang Festival, Qingming Festival, and Winter Solstice Festival. In fact, there are many more festivals in Chinese culture. The following are selected for discussion because they are the most representative ones in a particular season or 24 solar terms.

15.4.1 Chinese New Year

Chinese New Year is no doubt one of the most important festivals in the Chinese culture. It indicates the beginning of one prosperous year ahead and having a good start is essential for all Chinese. This is usually a 16-day festival beginning with the Chinese New Year Eve 除夕, which is believed to bring good luck for the entire upcoming year and ending on the Lantern Festival 元宵節, which is the 15th day of the Chinese New Year month (first full moon of the New Year). Chinese families around the world return to their homes during these 2 weeks for family reunion and bonding. In the parts of China where migrant working culture is prevalent, this festival is often the only time of a year when people get to see their friends and relatives in their hometowns. As with many other cultures, food is the cornerstone of the celebrations and special food/dishes will be served during the entire festive period.

Families will gather together over a reunion dinner 團年飯 on the Chinese New Year Eve with a full table of sumptuous and auspicious dishes, usually of the even numbers "6," "8," or "10" to indicate continuity (6), fortune (8), and completeness (10). The selection of dishes is not haphazard either as Chinese believe a lot in auspicious symbolic meanings of food/dishes which are often based on pronunciations/homonyms (words that share the same pronunciation but have different meanings) or appearance. In fact, certain food items or dishes that carry lucky symbolic meanings will be served during the entire 2-week festival either at the meal tables or altars when people make offerings and prayers to their ancestors and gods. Below is an extracted list of food with auspicious symbolic meanings, which you often see during Chinese New Year period.

Fish: In Chinese "fish" 魚 has a similar pronunciation as "surplus" 餘 and is a symbol of prosperity. Chinese often will not finish all food items on their tables, especially during Chinese New Year period, as they always like to have a surplus at the end of the year which is an indication of wealth and prosperity.

Chinese dumplings: The shape of the Chinese dumplings resembles that of a gold or silver "ingot" 元寶 (boat-shaped, oval, and turned up at the two ends Fig. 15.2) in the Chinese culture and thus Chinese dumplings carry the symbol of wealth.

TABLE 15.1 Solar terms.

Date[a]	Solar terms	Seasons (key Festivals)	Representative in-season food (Vegetable/Fruit/Seafood)
February 4	Spring commences 立春	Spring (Chinese New Year)	Spring onion/strawberry/mackerel
February 19	Spring showers 雨水		Burdock/orange/pomfret
March 5	Insects waken 驚蟄		Chive/loquat/oyster
March 21	Vernal equinox 春分		Asparagus/star fruit/skipjack
April 5	Bright and clear (Qingming) 清明		Pea/plum/sakura shrimp
April 20	Corn rain 穀雨		Crown daisy/banana/flying fish
May 5	Summer commences 立夏	Summer (Dragon Boat Festival)	Okra/lychee/beltfish
May 21	Corn forms 小滿		Water bamboo/peach/stonefish
June 6	Corn on ear 芒種		Soya bean/pineapple/dolphinfish
June 21	Summer solstice 夏至		Eggplant/watermelon/squid
July 21	Moderate heat 小暑		Corn/avocado/mussel
July 23	Great heat 大暑		Peanut/grape/squid
August 7	Autumn commences 立秋	Autumn (Mid-Autumn Festival)	Lotus root/longan/swordfish
August 23	End of heat 處暑		Celery/yellow pear/cuttlefish
September 8	White dew 白露		Taro/pomelo/shrimp
September 23	Autumnal equinox 秋分		Green bamboo shoot/chestnut/anchovy
October 8	Cold dew 寒露		Cabbage/lotus seed/clam
October 23	Frost 霜降		Bell pepper/persimmon/snapper
November 7	Winter commences 立冬	Winter (Winter Solstice)	Broccoli/custard apple/amberjack
November 22	Light snow 小雪		Mushroom/tomato/red bigeye fish
December 7	Heavy snow 大雪		Carrot/wax apple/mullet
December 22	Winter solstice 冬至		Leaf mustard/honeydew melon/red seabream
January 5	Moderate cold 小寒		Roselle flower/date/pomfret
January 20	Severe cold 大寒		Onion/kumquat/swordfish

[a]*The exact dates of the 24 solar terms vary within a narrow range from year to year depending on the solar movements. The dates provided here are based on the 2018 calendar announced by the Hong Kong Observatory official website (Hong Kong Observatory, 2017).*

Glutinous rice cake (Nian gao): The Chinese pronunciation of "nian" 年 means "year" and "gao" 高 means "higher/increasing." It is considered good luck to eat nian gao because it has the symbolism of increasing prosperity every year. The New Year greeting "Nian Nian Gao Sheng" (年年高升) is to wish people "advance toward higher positions and prosperity step by

FIGURE 15.2 Picture of a golden ingot. *Source: https://lovepik.com/sousuo-413413-0-2-0-0-0.html#.*

step" It is mainly sweet and steam-wrapped in leaves. The main ingredient is glutinous (sticky) rice, which is a symbol of family togetherness too. Because "gao" (cake) (Fig. 15.3) carries such a strong symbolic meaning, one can always come across various types of "gao" such as the gao in the picture as well as turnip cake and taro cake.

Dried black moss (fat choy): "Kung Hei Fat Choy" is the standard greeting phrase during Chinese New Year and it is for the same reason why dried black moss is often a must-see on the dining table as it indicates wealth and prosperity. Fat choi gets its name from its resemblance to hair, literally translating in Chinese as "hair vegetable." "Fat" means both hair and fortune, while "choy" (vegetable) also means "fortune," so Kung Hei Fat Choy means wishing your good fortune and wealth. In fact, all types of vegetables are welcome during this festival because they all carry the meaning of "fortune" (choy) (Fig. 15.4).

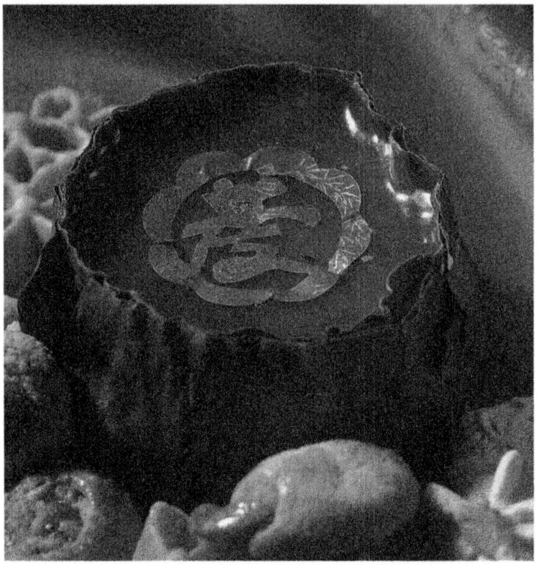

FIGURE 15.3 Nian gao. *Source: http://www.echoskitchen.com/2016/02/traditional-chinese-new-year-cakenian.html.*

FIGURE 15.4 Fat choy. *Source: https://www.elementseafood.com/recipe-hoe-see-fat-choy-dried-oysters-with-hair-seaweed/.*

Dried oyster (Ho Si): In the southern part of China where Cantonese is the prevailing language, Fat Choy is often served with dried oysters (ho si, which sounds like "good matters" in Cantonese) and the dish is called Fat Choy Ho Si (fortune and good matters). What better wish you can make to anyone during the Chinese New Year than this?

Sweet rice dumpling (tangyuan): Tangyuan is the main food for China's Lantern Festival; however, in southern part of China, people eat them throughout the Chinese New Year Festival. The pronunciation and round shape of tangyuan are associated with family reunion and togetherness, which are two main features of this important festival.

Noodles: Long noodles represent longevity and it is important that they are served uncut if they were to carry this auspicious meaning. Chinese people usually eat rice more often than noodle as their main daily source of carbohydrate, but long noodles are often served during Chinese New Year and birthday celebration because of its meaning of "longevity."

Chicken and tangerine: Chicken 雞 sounds similar to tangerine 橘 in the Chinese language, which in turn is similar to luck 吉. Whole chicken (including head, limbs, and organs) and big whole tangerines (usually in pairs) will be served to symbolize good luck for the whole year.

Certainly, the list is longer than this but this serves as a useful and (hopefully) interesting summary of the food tradition in this uniquely important Chinese festival.

15.4.2 Dragon boat festival

The Dragon Boat Festival, also called Duanwu Festival 端午節, is a traditional holiday observed annually over 2000 years in China to commemorate the famous patriotic Chinese poet Qu Yuan 屈原 (340–278 BC) who gave his life for his country by drowning himself in the Miluo River located in the Hunan province. Because of this origin, Dragon Boat Festival enjoys higher popularity in southern areas, such as Jiangsu, Zhejiang, Guangdong, and Fujian provinces than the north. On October 30, 2009, it was added to the UNESCO Representative List of the Intangible Cultural Heritage of Humanity. Besides the famous tradition of dragon boat race, the festival is also celebrated with food.

FIGURE 15.5 Zongzi. *Source: https://archive.shine.cn/city-specials/hangzhou/Sampling-Zhejiangs-local-zongzi-varieties/shdaily.shtml.*

Leaf-wrapped rice dumpling (zongzi): Zongzi is a kind of sticky rice dumpling wrapped in bamboo leaves. It is made from sticky rice with various fillings that can be savory or sweet. The tradition of eating zongzi is said to begin with people throwing rice/food wrapped with bamboo leaves into the Miluo River to protect Qu Yuan's body from being eaten by fish. Gradually, it has become a tradition to eat zongzi during the Dragon Boat Festival in honor of his patriotism. Nowadays, zongzi (Fig. 15.5) already becomes a common food, which can be easily found in supermarkets or restaurants.

Rice field eel: Eel in the rice field is most tender and nutritive at the time of Dragon Boat Festival and is treated as important a festive food as zongzi for people in southern China. Eel is usually stewed with tofu (bean curd) and mushrooms. Braising with garlic is also another popular way to cook eel.

Deep-fried rice dumpling (Jian dui): Jian dui is round deep-fried ball made with sticky rice flour and covered with sesame. The legend goes that the rainy season before the Dragon Boat Festival was actually caused by a hole in the sky. Therefore, people eat glutinous rice balls to mend the hole and stop the rain. In Fujian Province in southeast China, the celebration of eating jian dui is widely followed.

15.4.3 Mid-Autumn festival (Moon festival)

The Mid-Autumn Festival (Moon Festival) falls on the 15th day of the eighth lunar month. It takes its name the date falls in the middle of the autumn season. Although the moon is full on the 15th day of every lunar month, it is generally believed that the moon is the roundest and brightest during this time of the year. Chinese people celebrate the Mid-Autumn Festival with many meaningful activities, such as having reunion dinner with family, hanging

FIGURE 15.6 Moon cakes: Fat choy. *Source: https://www.timeout.com/hong-kong/restaurants/11-best-mooncakes-to-try-this-mid-autumn-festival.*

lanterns, guessing lantern riddles, and worshipping the moon. This festival is celebrated with significance because the ancient Chinese observed that, on top of solar movement, the movement of the moon also had a close relationship with changes of the seasons and agricultural production. Hence, to express their gratitude to the moon and celebrate the harvest, they offered a food sacrifice to the moon in this festival.

Moon cake: The tradition of eating moon cake is so representative that sometimes this festival is even called the "Mooncake Festival." Moon cake is usually round to take after the shape of the full moon and to symbolize reunion, with fillings inside (egg yolk, lotus seed paste, red bean paste, ham, nuts, etc.). Moon cakes are made differently from region to region. Besides self-consumption, people also send moon cake as gift when they visit friends and relatives around that time of the year (Fig. 15.6).

Harvest fruit and vegetables: Together with moon cakes usually come various types of fruit and vegetables which are in harvest during that time of year, for example, pomelo, pumpkin, yellow pear, and grape (fruit); lotus root and taro (vegetables), which are in their freshest and most nutritious condition.

Outdoor picnic/barbeque: One essential feature of the festival is the full moon which symbolizes family union. Therefore, many people opt for an outdoor picnic dinner or family barbeque under the moon so that they can appreciate the full moon while enjoying family bond. Autumn is also said to be the best season for outdoor picnic or barbeque.

15.4.4 Chongyang festival/Qingming festival (tomb sweeping festivals)

These two are the two festivals in the Chinese calendar, which are dedicated to the remembrance of ancestors spread across two seasons (autumn and spring, respectively).

Chongyang is also known as "Double Ninth Festival" because it falls on the ninth day of the ninth lunar month. In Chinese, "nine" is regarded as a "Yang" number (which means masculine as opposed to Yin, which is feminine). The ninth day of the ninth month is the day that has two Yang numbers, and "chong" in Chinese means double, which is how the name Chongyang was created. People usually climb mountain (or a modern reinterpretation is to climb anything of a certain elevation such as a tall building), pay homage to chrysanthemums, and sweep tombs. This time of the year is when the chrysanthemums are in full bloom so people also drink wine made with chrysanthemums. People will also eat Chongyang cake, which is a kind of steamed layered cake with nuts and jujube (date) paste as filling in between. As previously mentioned, cake in Chinese is pronounced as "gao," which means "high/increase"; people consider climbing a high mountain to be the same as eating cake, symbolizing that one wishes to make improvements by advancing to a higher level.

15.4.4.1 Qingming

Qingming (bright and clear), being a traditional Chinese festival, is also one of the 24 solar terms described in the section above. Qingming Festival is a time of many different activities, among which the main ones are tomb sweeping, taking a spring outing, and flying kites. People eat differently during Qingming in different regions, but the main foods are Qingming cake, peach blossom porridge, and snails. Qingming cake (sazi) is crispy fried food made of wheat flour or glutinous rice flour, eggs, sesame, onion, salt, and other ingredients. Spring is the blooming/harvest season for peach blossom and snails, so they are also popular Qingming food consumed by the people.

A rooted culture for respecting and remembering their family ancestors is in the heart of the Chinese filial piety. Therefore, it is an important ritual for Chinese to pay respect to their ancestor by tomb sweeping during Chongyang and Qingming festivals. As part of the ceremony they will offer food sacrifice, which mainly include suckling pig, tea/rice wine, fruit, poultry, and any other food items that their ancestors liked.

15.4.5 Winter Solstice festival (Winter festival)

Winter Solstice is one of the 24 solar terms and falls on the day in the 11th lunar month when the day is shortest and night is longest. At the same time, it is one of the most important Chinese festivals, second only to the Chinese New Year. In fact, for many Chinese families, the Winter Solstice Festival is even "more important than the Chinese New Year" as represented by the Chinese saying 冬大過年. This is because the Winter Solstice Festival was traditionally the time by which farmers and fishermen had to prepare for the coming colder months. This is also the time for another family reunion just before the winter as Chinese typically regard the winter months as a harsh and difficult period and thus a family reunion before it is important to wish for smooth sailing through this hardship. It also symbolizes that the warm months will come soon after winter. Since after this day, the daytime will become longer and night will become shorter, representing gradual improvement of weather.

Traditions and customs for the Winter Solstice Festival vary in different areas of China. In the northern parts of China, dumplings, wonton, and mutton are standard items used for celebration which are considered to be able to keep one warm in this extremely cold weather. Moreover, the dumpling is the main food for people in northern China. It is believed that after the Winter Solstice, the weather will become colder and colder; and by eating dumplings which take the shape of people's ear (on top of the shape of ingot as previously mentioned), people believe that their ears can be protected from frostbite. In the southern parts, sweet rice dumplings (tangyuan) and noodles are the popular festival food, carrying the same auspicious meanings as mentioned in the Chinese New Year section.

15.5 Conclusion

The chapter has illustrated food as a social tool in different kinds of occasions. Through understanding of its social functions, dining styles, and food roles to play during official and private social occasions, one can see that food is a powerful tool for not only socialization but also communication and even diplomatic relationship development. Furthermore, the extent to which food playing a key role in cultural representation is well depicted in the celebration of festivals in China. The most important Chinese cultural festivals were selected to display how one culture utilizes food in the society and how people socialize and connect with each other over food gatherings and sharing. Recently, public health and policy experts across the world also see the importance of food being a social tool, particularly for young children bonding with their families. Family dinner is considered as "a vital tool to address social, emotional, physical, and mental health for kids and their parents," stated by Dr. Grace Freedman, the founder of eatdinner.org. To this end, we would like to suggest families to treat mealtime a sacred ritual to develop bonding; meanwhile we all should have food gathering with families and friends more often.

References

Butler, S., 2013. History in a Basket: It's Picnic Time! Retrieved September 13, 2016, from: http://www.history.com/news/hungry-history/history-in-a-basket-its-picnic-time.

Davidson, A., 1999. Oxford Companion to Food.

Gong, W. Food in Chinese Culture. Beijing, Ministry of Culture of the People's Republic of China.

Hong Kong Observatory, H.K.O., 2017. Date and Time Information of the 24 Solar Terms in 2018. Retrieved November 8, 2017, from: http://www.hko.gov.hk/gts/astron2018/Solar_Term_2018_e.htm.

Mariani, J.F., 2014. Encyclopedia of American Food and Drink. Bloomsbury Publishing USA.

Mars, G., Mars, V., 1993. Two contrasting dining styles: suburban conformity and urban individualism. Food Culture Hist. 1, 49–60.

Mason, E., 2014. Tudor Dining: A Guide to Food and Status in the 16th Century. Retrieved 23 August, 2016, from: http://www.historyextra.com/feature/tudors/tudor-dining-guide-food-and-status-16th-century.

Olver, L., 2000. Food Timeline. Retrieved 10 August, 2016, from: http://www.foodtimeline.org/.

Salter, J., 2015. Dinner Fit for a Queen: The Secrets of Buckingham Palace's Royal Receptions. Retrieved 31 August, 2016, from: http://www.telegraph.co.uk/news/uknews/queen-elizabeth-II/11757987/The-secrets-of-Buckingham-Palaces-Royal-receptions.html.

Wells, K., April 7, 2015. Australian Food and Drink. Retrieved August 30, 2016, from: http://www.australia.gov.au/about-australia/australian-story/austn-food-and-drink.

CHAPTER

16

Trending foods and beverages

Lei Weng Si (Clara)

Institute for Tourism Studies, Macao (IFTM): Colina de Mong-Há Macao, China

OUTLINE

In this chapter, the recent food and beverage trends will be discussed. In particular, a brief tour on the different types of food, such as whole food, raw food, superfood, and medicinal food, which consumers coming across all the time in restaurants, cafes, and supermarkets, is also discussed. Readers will be able to grasp a fundamental idea of what these foods are and why they have become trendy in modern times. Furthermore, the beverage trend of handcrafted drinks and fermented drinks will be introduced. The chapter concludes with the knowledge and information of different food labeling, which are indispensable for foodies to recognize and distinguish.

Food and Society
https://doi.org/10.1016/B978-0-12-811808-5.00016-7

16.1 Whole foods and raw foods

The terms whole food, raw food, superfood, and medicinal food are on the public's minds everywhere. Today, consumers encounter those food names in all kinds of food packaging, menus, advertising campaigns, magazines, online food forums, and many other mediums. What they are and are they all different or the same are the questions that hopefully the readers of this chapter will apprehend afterward.

16.1.1 Whole foods

Amazon's acquisition of Whole Foods Market, which has more than 400 stores in North American and the United Kingdom, in June 2017, was headline news (Valinsky, 2018). The acquisition has drawn a lot of consumers' attention on whole foods. Together with the slogan—"Whatever Makes You Whole"—of the advertising campaign by Amazon, the trend of whole foods has emerged into Europe and Asia. One could even get an option of light whole food meal when taking a flight from Europe to Asia (Lufthansa, 2018).

According to the dictionary of Merriam Webster, the definition of whole food is a natural food and especially an unprocessed one (such as a vegetable or fruit), while Cambridge dictionary defines whole food as food that has not had any of its natural features taken away or any artificial substances added. Despite the dictionary definitions of whole foods, the Whole Foods Market has developed a set of quality standards to define whole foods (WholeFoodMarket, 2018). They include (1) unacceptable food ingredients, such as hydrogenated fats, bleached flour, foie gras, and so on; (2) responsibly grown food, e.g., addressing pesticides, farmers' welfare; (3) animal welfare and meat quality, e.g., grass-fed beef and free-range chickens; (4) seafood sustainable, e.g., seafood traceability and sustainable wild-caught seafood; (5) safe body care, e.g., organic skin care and body care free from harmful substances, e.g., microbeads and BHA; (6) nontoxic cleaning products, e g., free from unacceptable ingredients, such as methanol, perfume, phosphates, and so on; (7) cage-free eggs, e.g., no antibiotics given to hens, hens have room to roam and scratch about, to supply their 365 everyday value eggs. All these standards developed by Whole Foods Market indeed educate consumers on the concept of whole food and food safety. Meanwhile, they do not only serve as good marketing tactics to the company but also spread the concept of whole food and enhance the awareness of certain level of food sustainability issues in the modern days.

16.1.2 Raw foods

On the surface, consumers may have doubts about the differences between raw food and whole food. In fact, they differ very much from each other. Raw food is not just about eating fruit and salad. Raw food refers to eating uncooked foods. Raw food is anything that has not been refined, chemically processed, or even canned and has not been heated above 48°C. The raw food movement has been rising up a lot in recent years. Advocates claim that heating food up more than 48°C will destroy or even remove many of the natural enzymes which compose food nutrition (e.g., Casupei and Kaupert, 2011, p. 3). To the opposite, antagonists show the benefit of cooked food. As reported by Steven Salzberghe Bloomberg Distinguished

FIGURE 16.1 Raw cacao nibs. *Source: Lei, Weng Si.*

Professor of Biomedical Engineering, Computer Science, and Biostatistics at Johns Hopkins University (Salzberg, 2016), a 2011 study by Chris Organ and colleagues at Harvard University pointed out that the ancestors of modern humans who invented food processing (including cooking) gained critical advantages in survival and fitness through increased caloric intake. In addition, an *Escherichia coli* outbreak was reported in 2018, tied with romaine lettuce which consumers normally use as salad (New York Times, 2018). Although raw food is trendy lately, consumers should be well aware of its potential risk, such as salmonella infections and the danger of using unpasteurized milk or cheese. Moderation of choosing some foods to be consumed raw, for example, fruits of all kind, is very much healthy; nevertheless, pesticides used in agriculture is always an issue. To this end, the benefit of going raw food is still questionable until further scientific evidence is available (Fig. 16.1).

16.2 Medicinal and superfoods

16.2.1 Superfoods

Superfood has been a trendy word appearing in many food packagings nowadays but what exactly are superfoods. According to Cambridge Dictionary, superfood is a food that is considered to be very good for your health. Merriam Webster defines superfood as a food (such as salmon, broccoli, or blueberries) that is rich in compounds (such as antioxidants, fiber, or fatty acids) considered beneficial to a person's health. ORAC (oxygen radical absorbance capacity) unit, a unit of measurement for antioxidant content developed by the National Institute on Aging (NIA) at the National Institutes of Health (NIH), is a common gage to measure how good a food is for one's health (David and Haytowitz, 2010). Normally, the higher the ORAC scale, the better the food in terms of having good effects on human body. In other words, many scientists theorize that foods higher on the ORAC scale may be more effective at neutralizing free radicals. Slowing down oxidative processes may contribute to postponing aging process, although further medical evidence is needed. In addition, Carlsen et al. (2010) have compiled a list of more than 3100 foods' antioxidant content. However, the list is very much Scandinavian-oriented. It serves still as a good reference on antioxidant foods. Below are some foods that contain high level of antioxidant content (Table 16.1) (Carlsen et al., 2010; David and Haytowitz, 2010):

When reading the above tables, particularly from the column of fruits and vegetables, the darker the color of the fruits and vegetables, the higher the antioxidant content it seems containing, for example, acai berry and red cabbage. It is also well displayed in restaurants and cafes nowadays with acai smoothies, matcha shakes, charcoal bun, beetroot soup, etc. Celebrity chef Jamie Oliver has joined the superfood movement and has written a cook book on superfoods as well (Oliver, 2016). In addition, superfoods are also available in different forms, such as dried and powder as the illustration below shows (Fig. 16.2).

According to Pratt and Matthews (2004, p.20), superfoods are based on a very simple concept: some foods are better than others for your health. Thus, consumers are facing tough choices every day when choosing the best or the most superfood to consume to fully capitalize the health benefits on food consumption. There are food producers who label their products as superfood to attract consumers' attention and interest. As the photo shows, superfood has been labeled on the left top corner of the packaging. In fact, this labeling assist consumers to select food items and enhance their awareness on food information and knowledge. Later part of this chapter includes a section on food labeling.

TABLE 16.1 Examples of superfoods.

Herbs and Spices	Fruits	Vegetables	Nuts and Seeds
Fresh sage, marjoram, dried basil, oregano, rosemary, ground cinnamon	Acai, acai powder Wild blueberries goji berries Dried rowanberries Pomegranate	Artichokes Raw beets Broccoli Red cabbage Eggplant Radish seeds Spinach Sweet potato with skin	Almonds Hazelnuts Walnuts
Beverages	Legumes	Baked products	Sweets
Cabernet Sauvignon wine, green tea	Black bean Kidney bean Lentils Peanut butter Soybeans	Butternut wholegrain bread	Unsweetened baking chocolate Unsweetened cocoa dry powder Sorghum Sumac

Source: Compiled by Lei, Weng Si Clara.

FIGURE 16.2 Acai powder; So-called superfoods. *Source: Lei, Weng Si.*

16.3 Medicinal or health foods

Medicinal food is not just a popular term appearing in food packaging—it is a field of science. There is a medical journal named the "*Journal of Medicinal Foods*." The foods that we consume regularly, for example, raspberries, rice cookies, Korean kimchi, grape seed, etc., are being studied scientifically. Different effects on human beings' bodies are examined and identified. Although there is not any official definition of medicinal food from Cambridge Dictionary or Merriam Webster, foods providing different therapeutic or preventive functions are often called medicinal foods. For example, a study conducted by Onuegbu et al. (2018) revealed that consumption of soymilk reduces lipid peroxidation.

Today's dietary patterns, particularly in the United States, have differed a lot from the old days. The shift is believed to originate from many lifestyle choices and diseases. This includes such things as lifestyle diseases including heart diseases, cancer, and chronic lung disease (Greger and Stone, 2016, p. 5). In the book of "*How Not to Die*" by Greger and Stone (2016), many different types of food were discussed and their therapeutic functions were proved by many medical science studies. For example, flaxseed is described as "miraculous" in fighting against breast and prostate cancers (Greger and Stone, 2016, p. 155). Table 16.2 illustrates examples of medicinal food. The evidence is supported by different medical studies as stated in the book of Gerger and Stone. Other types of food, such as coffee, goji berries (Greger and Stone, 2016, pp. 175, 279), garlic, and many others, are all having different medicinal functions. To fight with the lifestyles disease as mentioned earlier, people could learn about medicinal foods and their functions, to make more informed and better food choice every day.

TABLE 16.2 Examples of medicinal foods.

Flaxseed	Turmeric	Dandelion	Goji berries

Beetroot	Basil	Coriander	Hibiscus tea

Furthermore, another category of food—fermented food—has gained considerable attention not only in the medical science field but also in many health foods stores. Examples include kefir, kombucha, kimchi, cheese, etc. According to the book by Spector (2015), gut health is being recognized as an important key to overall health. Microbiomes, living in humans' gut, and the composition of microbiomes within a human body contribute a lot to one's health. According to the dictionary of Merriam Webster, microbiomes are defined as a community of microorganisms (such as bacteria, fungi, and viruses) that inhabit a particular environment and especially the collection of microorganisms living in or on the human body. The Cambridge dictionary defines microbiomes as the microorganisms, living things too small to be seen, that exist in a particular environment or in the human body. The recent scientific discovery of the importance and significance of microbiomes opens up the causes of many human conditions, for example, obesity, inflammation. The new wave of medical science draws a lot of attention on holistic benefits of fermented food on peoples' gut and overall health. Fermented food, for example, kefir,

has been examined and studied by many medical science researchers, for example, the study by Bourrie et al. (2016) on kefir's microbiota and health promoting characteristics. According to the dictionary of Merriam Webster, kefir is defined as a beverage of fermented cow's milk. Cambridge Dictionary gives a more detailed definition—a sour drink made from fermented milk, originally from Russia and Central Asia. There are many different sayings regarding the origin of kefir. A "Gift from the Gods" is a very common saying and kefir grains are believed to be occurring in the nature. It is still impossible to be replicated completely in an industrial laboratory for mass production. Nevertheless, kefir is available in many supermarkets in Europe nowadays, as shown in a photo below (Fig. 16.3). Its composition, microbial cultures, and biological activities are often examined and researched, for example, the study by Prado et al. (2015).

Fermented food products are currently widely available in many healthy food stores and supermarkets. The latest book by Derek Dellinger, *The Fermented Man: A Year on the Front Lines of a Food Revolution*, is a good read for readers who are interested in fermented foods; recipes are included, too. To conclude this section, one could easily spot that there are many duplications between the discussed food categories. For example, goji berry is under the category of superfood and medicinal food at the same time. This is not surprising because many food items were first discovered and considered as superfood. When its therapeutic function is proved, then the food item will become a medicinal food. Normally, superfoods and medicinal foods are mostly whole foods and they are recommended to be consumed raw or with minimal cooking heat to preserve its nutrient or enzyme contained.

16.3.1 Veganism

Being a vegan is one of the latest trends among foodies from the viewpoint of the Western diet—although perhaps a rather simple generalization. Vegan Society was founded back in 1944, in England. The Society defines veganism as a philosophy and way of living which seeks to exclude—as far as is possible and practicable—all forms of exploitation of, and cruelty to, animals for food, clothing, or any other purpose. This, by extension, promotes

FIGURE 16.3 Medicinal food: Kefir. *Source: Lei, Weng Si.*

the development and use of animal-free alternatives for the benefit of humans, animals, and the environment. In dietary terms, it denotes the practice of dispensing with all products derived wholly or partly from animals (VeganSociety, 2018). Meanwhile, a vegan is a person who follows the diet rule of veganism. According to the Cambridge dictionary, veganism is the practice of not eating or using any animal products, such as meat, fish, eggs, cheese, honey, or leather. In practice, a vegan does not wear any clothing products made by wool and silk, too. Merriam Webster Dictionary defines a vegan as a strict vegetarian who consumes no food (such as meat, eggs, or dairy products) that comes from animals, as well as one who abstains from using animal products (such as leather). Among type of vegetarian, there are vegan as mentioned: lacto-vegetarian, ovo-vegetarian, and lacto-ovo vegetarian. Lacto-vegetarians eat plant-based foods and dairy products. Ovo-vegetarians eat plant-based foods and eggs. Lacto-ovo vegetarians eat as well no animal food but dairy products and eggs. Yet one thing all vegetarians or vegans have in common is a plant-based diet avoiding all animal foods such as meat (including fish, shellfish, and insects).

Being a vegetarian or vegan is never awkward or wired nowadays. There are many ways to embrace vegan or vegetarian living. School refectories, companies' canteens, and hospital restaurants organize different green days of a week, when the establishment offers only vegetarian food. The new wave of vegetarian or vegan option calls even upon the attention of Michelin star restaurants. These restaurants offer vegetarian options too, for example, Alyn Williams at The Westbury and Vegetarian Tasting Menu at Michelin-Starred Galvin La Chapelle, Bishopsgate, London.

Vegan or vegetarian diets are also getting into professional athletes' arena. There are ultramarathon runners, for example, the legendary Scott Jurek is vegan or vegetarian as well. Vegan triathletes and Olympic sprinter, for example, Ariel Rosenfeld, claims that being a vegan helps him to tackle extreme distance training reaching 200 km. Fiona Oakes, a woman champion of North Pole Marathon, believes that being a vegan carries no burden of guilt, hinting not eating meat as an important benefit of being a vegan. It is a common perception that you must eat poultry to feed your body with sufficient protein to sustain your body for endurance or resistance training. In fact, protein could be absorbed from numerous plant-based foods, such as beans, tofu, seeds, legumes, and nuts. The Daily Dozen designed and advocated by Dr. Michael Greger (Greger and Stone, 2016, pp. 309–311) is highly recommended for vegans and different types of vegetarians. It composes of three servings of beans, fruits, and whole grains; two servings of green and vegetables; one serving of berries, cruciferous, flaxseeds, nuts, and spices; and five servings of beverage—lastly, one serving of exercise referring to 90 minutes of moderate-intensity activity, for example, brisk walking. Dr. Greger also suggests vegans to take vitamin B12 supplement which is an essential nutrient that vegan diet may not provide sufficiently.

16.4 Beverage trends

16.4.1 Handcrafted beverages

Albeit perhaps in low numbers, there appears to be a general malaise or a sort of backlash from many beverages that entail mass production processes. This is especially so as microbreweries continue to flourish and bloom in many regions. Microbreweries can even be found in tourist areas in Vietnam. In addition, craft spirits are rising and catching up with the antimass production companies too. Craft beer and spirits are considered more authentic and unique. They are mostly produced in small quantities. Small groups of consumers that gather to consume craft beers or spirits are considering it a quality social experience, rather than just alcohol consumption.

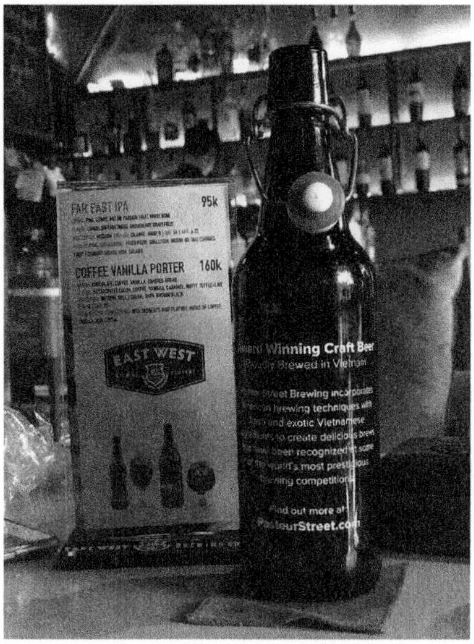

Handcrafted beverages are not only limited to beers and spirits but also include favorites such as coffee and tea. Handcrafted coffees or teas in general refer to a drink made fresh at the point of sale. They are not bottled or canned beverages. The coffee that one grabs on the go is a typical handcrafted beverage. Handcrafted coffee has been trendy in the United States ever since Starbucks started its business in 1971. In fact, handcrafted coffee has been served in cafés all over Italy since the first caffee or café opened in Venice around 1683 (Delonghi, 2019). Having an expresso or cup of cappuccino is a daily must-have to most of the Italians. Caffee or cafés appear in all neighborhoods and they are the social gathering places for areas most of the time. People gather and share daily lives, happenings of a community, and have conversation. Similar trends happen in the United States after Starbucks introducing similar Italian coffee house experience first in Seattle dated back in 1971. Nowadays, many more different coffee houses operate around the globe, and handcrafted coffee has become a daily ritual of people from many different places. Nevertheless, handcrafted tea is the next trend spreading from Asia to the rest of the World. Handcrafted tea originated from Taiwan since the beginning of the early 80s. Oriental tea is freshly brewed at stores or take-away counters. On top of the freshly brewed teas are often contained pearls or bubbles, which are chewy balls or beads made with tapioca and which are added in the tea for textual contrast. The sweetness and amount of ice cubes for iced tea or the amount of pearls are all personalized. There are sometimes five different levels of sweetness that customers can order from as well as a measurable amount of ice cubes and pearls/bubbles. As pearls and bubbles are getting very popular in Asia, so other options besides bubbles/pearls, such as puddings, fruit jelly, agar jelly, and grass jelly were introduced to add dimension to the tea, among which the most popular one has always been the black pearl milk tea or green pearl milk tea. With the acquisition of Teavana by Starbucks (Fox, 2012), Starbucks has added freshly brewed teas in their shop menu. Starbucks also has introduced tea with jelly to capture the Asian trend of freshly brewed tea with chewy balls or jellies.

16.4.2 Nonalcoholic beer

Nonalcoholic beverages including beer and naturally fermented fizzy drinks are rising stars in the industry too. With consumers becoming more aware of the effects of alcohol on health, nonalcoholic beers are a growing industry in the market. Indeed, the trend is increasing so much that the oldest and biggest German brewery, for example, Weihenstephaner, is producing alkoholfrei—alcohol-free products—unheard of many years ago. The beer tastes very similar to those containing alcohol and is welcomed more and more these days. Another German brewery, Paulaner, has produced an alkoholfrei product, which the company claims is "isotonic" and which contains electrolytes, minerals, and other nutrients. It is probably a sign that this nonalcoholic beer is trying to compete with mineral water and sport drinks markets in the near future.

16.4.3 Mocktails

Mocktails are also getting into the restaurant and bar scene. Nowadays, it is very common to find mocktails' menu existing alongside cocktails' menus. According to the Cambridge dictionary, a cocktail is a drink, usually an alcoholic one, made by mixing two or more drinks together. Cocktails usually mix one type of alcohol or spirit with sugar, tonic water, juices, soft drinks, tea sometimes, cream, herbs, or other adventurous ingredients. Mocktails too are mixed with similar combinations but generally exclude any alcohol or spirit. In addition, mocktails often share the same name as their cocktail counterparts. Examples include the Singapore Sling, Margarida, where the word "Virgin" is nearly always added in front to signify that the drink is nonalcoholic. A typical mocktail recipe Spiced Cider Punch contains, two quarts' apple juice, one cup sugar, one teaspoon cinnamon, one teaspoon ground allspice, one can (12 oz) orange juice concentrate, and 1-quart chilled ginger ale. There is no alcohol involved at all.

16.4.4 Fermented drinks

Fermented drinks, such as kombucha, kvass, or water kefir, are emerging and taking prominence in certain locations. Their rising popularity can also be determined from their increasing availability from such places as health food stores, or their emergence in news articles, books, and online discussion forums as well.

Kombucha is made by fermenting black or green tea by using tea mushrooms or SCOBY (an acronym for Symbiotic Culture of Bacteria and Yeast). The formal name of tea mushrooms is Medusomyces gisevii. SCOBY contains *Saccharomyces cerivisiae* and other species of bacteria. Kombucha's flavor varies depending on the type of tea used and the fermentation duration. It can be augmented by adding fruits and herbs, such as hibiscus kombucha or blueberry kombucha. Kombucha tastes a bit acidic and tangy and comes with small bubbles of gas. There are various claims of health benefits from taking kombucha, such as cure asthma, diabetes, and help lessen herpes and insomnia; however, scientific evidence is not consensual. Often times, kombucha is home-brewed; however, due to its popularity and the potential health benefits, there are companies now brewing it for wider consumption in the marketplace too.

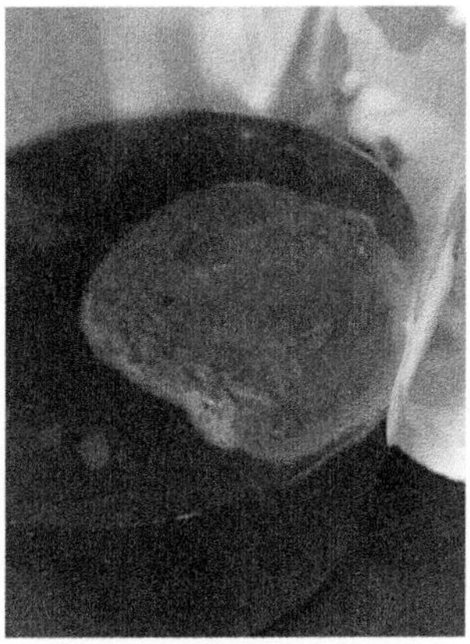

Water kefir is another type of fermented drink, which is simply made with kefir grain (the grain itself contains different kinds of bacteria) as well as yeast, sugar, water, fruit juice, or coconut water and is then fermented for 24 h. As mentioned earlier, kefir (the diary version) offers many different health benefits, many of which are backed up by scientific evidence. Fortunately, water kefir too offers similar benefits as its dairy-based counterpart. People enjoying water kefir often finds better digestion and experiences less stomach problems. As the grain of water kefir appears nicer than tea mushroom or SCOBY, people who are interested in trying out fermented drinks may find them easier to start with.

16.5 Brunching is the trend

The question of how many meals a day is appropriate or healthy for human beings always forms great discussion from different perspectives, ranging from gender differences, diet preferences, sport preferences, weight loss versus weight gain preferences, and various health concerns. It is not surprising to see all kinds of suggestions and recommendations on the number of meals we should take in a day. Of course, there is no conclusive answer to the question; nonetheless, more meals per day are taken in comparison to older times. Brunch, tea, and supper menus are available in restaurants. For example, brunch is a relatively new and the word is a portmanteau of breakfast and lunch, and it likely originated in England in the late 19th century and became popular in the United States in the 1930s. According to the Cambridge Dictionary, brunch is defined as a meal eaten in the late morning that is a combination of breakfast and lunch. Normally, it is served between 10 a.m. and 3 p.m. In addition, it is served differently and comprises different elements; for instance, in one example it is served in a buffet style with alcoholic drinks, champagne, and cocktails. Brunch nowadays is not only popular among church goers, who would gather for a meal after attending church services—"Church Brunch"—but also very popular among young people, who are willing to spend a premium on a good Sunday brunch. Chic and expensive brunch menus are available in many classy hotels and trendy cafés in many cities, for example, New York and London. A brunch costs range from US$20–70. A 2017 survey conducted by Mintel Research revealed that 40% of the 1670 consumers surveyed considered brunch as a time to communicate and socialize with family and friends (Mintel, 2017). Brunching seems to be trend carrying upward and serves as an important social occasion.

16.6 Food labels

Food labels and packaging are the two most direct and obvious ways in providing food ingredients to the consumer, for instance, where a product was made, nutritional values, allergy warnings, organic, fair trade, vegan, halal or kosher, and packaging material used. However, trustworthy consumers aware of all this information are now being told that some of this information's veracity is now being called into question. A report by Mintel (2018) report illustrated that many geo-varied consumers around the world are lacking when it comes to trust in manufacturers and certain food regulatory systems. In particular, many are mistrustful of certain quoted ingredients used, production processes, and supply chains themselves. According to Mintel's research in 2018, 65% of Thai, 64% of Australian, and 61% of Indonesian adults in major metropolitan areas feel cheated when a company is not particularly clear about the high sugar content of some of its products (Mintel, 2018). Such research results are alarming and offer great challenges to those food manufacturers that are not necessarily forthcoming about food information and disclosure. Consumers are smart nowadays and reading food labels is becoming common practice. Food information transparency is soon to become the norm, and consumers demand the real picture of the food purchased. After all, consumers require information to make important decisions about health and fitness, sustainability, origin of main product, additives, etc. This does not even begin to cover the various awarding bodies such as USDA Organics, Fair trade, The Rainforest Alliance Certified, Vegan Society, Vegan Action, Halal, Kosher, Marine Stewardship Council—some of these are looked at in the following.

16.7 USDA organic

USDA (the United States Department of Agriculture) is an organization that certifies organic foods and beverage products in which at least 95% of the product's ingredients have to be free from synthetic additives, for example, pesticides, chemical fertilizers, dyes, and the products are produced under approved methods, for example, no industrial solvents are allowed. As stated in the official website of the USDA (McEvoy, 2012) (Organic 101: What the USDA Organic Label Means), produce can be called organic if it's certified to have grown on soil that had no prohibited substances applied for 3 years before harvest. As for organic meat, regulations require that animals are raised in living conditions accommodating their natural behaviors (like the ability to graze on pasture), fed 100% organic feed and forage, and not administered antibiotics or hormones. According to the USDA, there are a total of 41,198 certified organic operations or businesses that exist by the end of 2017. Of these, more than 50% of these operations (26,401) are found in the United States of America, while Mexico comes in second with 2331.

16.8 FAIRTRADE

The FAIRTRADE Mark is a registered certification label used for products that represent a fair deal for producers in developing countries. It is about fair prices and working conditions, sustainability of production (through pricing among other things), and aims for nondiscrimination, enabling them essentially to have more control over their lives. Being ethical is at the core of fair trade. The FAIRTRADE initiative started first in the Netherlands in the 80s and mainly concentrated on coffee bean products (Fairtrade, 2018). Bean growers were offered a price which was more than the market price to ensure growers' fundamental living standards. FAIRTRADE certification thus provides a level of assurance to coffee bean products, which enabled growers to make a better standard of living out of growing beans. Nowadays, FAIRTRADE products reach a wider audience and include cocoa, fruit, tea, cotton, sugar, flowers, and others.

16.9 The rainforest alliance certified

The Rainforest Alliance Certified seal is awarded to farms, forests, and businesses that meet rigorous environmental and social standards. The logo assures consumers that purchases are produced by companies which follow environmental and socially responsible practices. For example, traceability of the products' entire supply chain is an important criterion to earn the certification. Cocoa, tea, banana, and coffee are the products that you will find mostly labeling with such certification.

16.10 Vegan logos

One of the main functions of various vegan logos is to assure vegan consumers that the products themselves are suitable for consumption according to vegan requirements. Certified logos generally signify products that do not contain any animal or animal-related products. It also indicates that the product has not been used for any animal testing. There are currently different types of vegan certifications. Confusingly there are many such bodies that potentially not all will share the same criteria for certification. Of the few out there, the Vegan Society is registered in Europe, USA, Canada, Australia, and India. It is based in the United Kingdom and is a nonprofit making organization founded in 1944. For any company that wishes to register its product(s) with the organization and use the Vegan Society's logo on the company's packaging, marketing materials, and other medium, the product has to pass certain. This includes animal or animal ingredients, no animal testing, and genetically modified organisms during product development and manufacturing and must have a certain level of kitchen and hygiene standards, in which separate set of utensils for preparing vegan food should be used and should not mix with nonvegan utensils and dishes. The logo is being used and displayed on over 22,000 products according to the organization's website (VeganSociety, 2018). Vegan action, also known as the Vegan Awareness Foundation, is a nonprofit organization with the aim of educating the public about veganism and assisting vegan-friendly businesses. It's headquartered in Virginia (USA) and it follows similar principles as the Vegan Society. Certified product beverages must not contain any animal-related ingredients, including animal by-products, such as eggs, milk, honey, and sugar filtered from bone char (charred animal bones). Beverages must also undergo no filtering or clarification process involving animal products. According to the organization's website, the certified logo is currently displayed on thousands of types of products produced by more than 800 companies (Vegan.org, 2018).

16.11 Halal

Halal foods are generally associated with Muslims and others who follow a Muslim diet. Halal certification is a must-have requirement to import food products into Muslim countries. The meaning of the word Halal comes in many interpretations: "permitted, allowed, authorized, approved, sanctioned, lawful, legal, legitimate, or licit." By displaying the logo of Halal on meat or meat products, it refers to meat which has been slaughtered according to the rules of the Quran and Sunnah and is thus permitted for consumption by Muslims and others. Meat or other ingredients must be processed and produced by using machinery, equipment, and utensils that have been cleansed according to Islamic Law as well. Halal Certification Authority (HCA) was established in 1993 to certify food this meeting the Halal standard (HalalCertificationAuthority, 2018).

16.12 Kosher

Kosher food refers to all food and food ingredients including the preparation processes of the said food products. In kosher foods, products must be orthodox kosher-compliant. This refers to a set of intricate biblical laws that detail the types of food that a Jewish person may eat and its preparation process. Among different kosher certification bodies in different regions, the K kosher certification is relatively widely adopted. The organization was founded in 1935 and has certified approximately 3000 production facilities, including some giant American food producers, such as Snapple, Tropicana, Maxwell House, Kellogg's, and many others (KosherCertification, 2018).

16.13 Marine Stewardship council

The Marine Stewardship Council (MSC) fisheries standard has three overarching principles that every fishery must meet. They are as follows: Principle 1: Sustainable fish stocks—the fishing activity must be at a level which is sustainable for the fish population. Any certified fishery must operate so that fishing can continue indefinitely and is not overexploiting the resources. Principle 2: Minimizing environmental impact—fishing operations should be managed to maintain the structure, productivity, function, and diversity of the ecosystem on which the fishery depends. Principle 3: Effective management—the fishery must meet all local, national, and international laws and must have a management system in place to respond to changing circumstances and maintain sustainability (MarineStewardshipCouncil, 2018). In short, the label of MSC indicates that the fish are traceable to a sustainable source of fishing activity (Table 16.3).

TABLE 16.3 Food labeling.

Organization	Logo
USDA Organics	
Fair trade	
The Rainforest Alliance Certified	

(Continued)

TABLE 16.3 Food labeling.—cont'd

Organization	Logo
Vegan Society	Vegan
Vegan Action	CERTIFIED VEGAN VEGAN.ORG
Halal	حلال HALAL
Kosher	K
Marine Stewardship Council	CERTIFIED SUSTAINABLE SEAFOOD MSC www.msc.org

References

Bourrie, B.C.T., Willing, B.P., Cotter, P.D., 2016. The microbiota and health promoting characteristics of the fermented beverage kefir. Front. Microbiol. 7, 647. https://doi.org/10.3389/fmicb.2016.00647.

Carlsen, M.H., Halvorsen, B.L., Holte, K., Bøhn, S.K., Dragland, S., Sampson, L., et al., 2010. The total antioxidant content of more than 3100 foods, beverages, spices, herbs and supplements used worldwide. Nutr. J. 9 (1), 3.

Casupei, J., Kaupert, V., 2011. The Art of Raw Food: Delicious, Simple Dishes for Healthy Living. North Atlantic Books, Berkeley, California.

David, B., Haytowitz, S.B., May 2010. USDA Database for the Oxygen Radical Absorbance Capacity (ORAC) of Selected Foods, Release 2. Retrieved from: http://www.orac-info-portal.de/download/ORAC_R2.pdf.

Delonghi, 2019. History of Coffee. Retrieved from: http://www.thegreatitaliancafe.com/history4.html.

Fairtrade, 2018. Retrieved from: https://www.fairtrade.org.uk/.

Fox, E.J., November 14, 2012. Teavana up 50% on Starbucks Acquisition. Retrieved from: https://money.cnn.com/2012/11/14/news/starbucks-teavana/index.html.

Greger, M., Stone, G., 2016. How not to Die: Discover the Foods Scientifically Proven to Prevent and Reverse Disease. Pan Macmillan.

HalalCertificationAuthority, 2018. Retrieved from: https://halalauthority.org/.

KosherCertification, 2018.

Lufthansa, 2018. Special Meals. Retrieved from: https://www.lufthansa.com/us/en/Special-meals.

MarineStewardshipCouncil, 2018. Retrieved from: https://wsc.org/.

McEvoy, M., March 22, 2012. Organic 101: What the USDA Organic Label Means. Retrieved from: https://www.usda.gov/media/blog/2012/03/22/organic-101-what-usda-organic-label-means.

Mintel, 2017. Restaruants Breakfast and Brunch Trends 2017. Retrieved from London.

Mintel, 2018. Global Food and Drink Trend 2018. Retrieved from: http://www.mintel.com/global-food-and-drink-trends/.

NewYorkTimes, April 19, 2018. E-coli Roamine Lettuce. Retrieved from: https://www.nytimes.com/2018/04/19/us/e-coli-romaine-lettuce.html.

Oliver, J., 2016. Everyday Super Food. Penguin UK.

Onuegbu, A.J., Olisekodiaka, J.M., Irogue, S.E., Amah, U.E., Okwara, J.E., Ayelagbe, O.G., Onah, C.E., 2018. Consumption of soymilk reduces lipid peroxidation but may lower micronutrient status in apparently healthy individuals. J. Med. Food 21 (5), 506–510. https://doi.org/10.1089/jmf.2017.0094.

Prado, M.R., Blandón, L.M., Vandenberghe, L.P.S., Rodrigues, C., Castro, G.R., Thomaz-Soccol, V., Soccol, C.R., 2015. Milk kefir: composition, microbial cultures, biological activities, and related products. Front. Microbiol. 6, 1177. https://doi.org/10.3389/fmicb.2015.01177.

Pratt, S.G., Matthews, K., 2004. SuperFoods Rx. William Morrow, York.

Salzberg, S., 2016. Whats Really in Raw Food. Retrieved from: https://www.forbes.com/sites/stevensalzberg/2016/02/22/whats-really-in-raw-food/#3b5154be18d7.

Spector, T., 2015. The Diet Myth: The Real Science behind what We Eat. Hachette UK.

Valinsky, J., February 9, 2018. 5 Ways Amazon Has Already Changed Whole Foods. Retrieved from: http://money.cnn.com/2018/02/09/news/companies/amazon-whole-foods-changes/index.html.

Veganorg, 2018. Vegan. Retrieved from: https://vegan.org/.

VeganSociety, 2018. The Vegan Society. Retrieved from: https://www.vegansociety.com/about-us/history.

WholeFoodMarket, 2018. Whole Foods Standard. Retrieved from: https://www.wholefoodsmarket.com/mission-values/core-values.

CHAPTER

17

Food, culture, and food tourism

Lei Weng Si (Clara), Ubaldino Couto

Institute for Tourism Studies, Macao (IFTM): Colina de Mong-Há Macao, China

OUTLINE

Food, culture, and tourism appear to be strange bedfellows at its first glance, but on a closer look, their relationship is inseparable from each other. We argue that they are even indispensable of each other in today's tourism when increasingly more and more modern travelers seek not only the sun, sea, and sand but also an enriching, renovating, and educational cultural experience through their travels.

Food is a very solid form of cultural identity; it is a reflection of the use of local ingredients, prevailing cooking methods, spices, and herbs, as well as wisdom culminated over decades, if not centuries, of people cooking and preparing food. Modern tourists who thirst for enriching and inspiring cultural experience find local food a welcoming, accessible, and almost irresistible first encounter to a new culture. This chapter builds on this simple relationship between food, culture, and tourism; world examples are extensively used to illustrate the concepts discussed and a featured case study of Macanese cuisine concludes the chapter by drawing

Food and Society
https://doi.org/10.1016/B978-0-12-811808-5.00017-9

all the main points covered in this chapter, highlighting the interesting and indispensable nature that connects food, culture, and tourism.

17.1 Food and culture

Cultural values and beliefs impose strong influences on people lives and ingrain deeply in food choice and consumption. Rice is core food for most of the Asians, while wheat and potatoes are for Europeans and North Americans. Cassava is such an important staple for the Amazonians; meanwhile, quinoa is essential for the highlanders inhabited in South America. The culture that one lives in exerts considerable guidance on one's food choice. For instance, cassava or yucca is the ultimate food to start a day of the Amazonians. Women rise up before dawn to prepare cassava breakfast and guayusa tea for their husbands to consume before they start their hunting activities of the day. Guayusa tea is naturally caffeinated. In the west, people typically start their day with a cup of coffee, and guayusa tea starts the day of Amazonians (Fig. 17.1).

According to the Foods Model proposed by Mintz and Schlettwein-Gsell (2001), core food always serves with complementary food items to enhance scrumptiousness. In the foods model, there are four types of food: (1) core foods; (2) complementary foods; (3) secondary foods; and (4) peripheral foods.

The category food of legumes and vegetables sometimes work together as complementary or secondary foods, the consumption of which differs greatly from various cultures in terms of types and cooking styles. For instance, Chinese rice with stir-fried tofu, Japanese rice with brown miso soup, Peruvian quinoa pancake with corn kernels, Italian capellini with tomato and mozzarella, and Middle Eastern couscous with roasted winter squash are all good examples. Peripheral foods are not core category and can be easily replaced as food consumption behavior changes due to availability and food trends, for instance, herbs and spices used, snacks, and cooking styles.

FIGURE 17.1 Photo of guayusa tea cooking in Amazon village. *Source: Lei Weng Si.*

17.2 Food functions and beliefs

People from different cultures possess varies beliefs on food. Beliefs include food functions, medicinal values, and impacts on human bodies. In Chinese culture, foods are generally categorized into five types, namely "hot," "warm," "neutral," "cool," and "cold" food. It is believed that the type of food a person consumes should match with the person's type of body and condition at a specific period of time. For instance, if a person is identified as having a "cold" body type (normally the body type is identified by Chinese medicinal doctor), the person should avoid food which are categorized as "cold" and should consume "neutral" or "warm" food to balance the body. Examples of "cold" food are banana, turnip, and watermelon. Examples of "hot" food are lychee and chili (Table 17.1).

According to traditional Chinese medicine, the month after a woman giving birth to a baby is of critical importance to the woman's health in her life. It is believed that the month offers a golden opportunity to the women to transform her health and body condition by having proper diet and medicinal treatment. Tea made with lightly tossed rice is suggested to be taken regularly to boost up the Qi of the body. Qi is a vital life energy that is believed to flow through the body. In addition to the rice tea, different diet plans are prescribed based on the actual laboring condition of the women by a Chinese medicinal doctor. Plans are very different for women who have natural labor versus caesarean. For example, ginger is highly recommended for women having natural labor, while it is completely forbidden for women having caesarean. The reason is because ginger is considered as a type of warm to hot food which stimulates blood flow and it is dangerous for women after a caesarean procedure. Furthermore, another food belief from the southern part of China surly surprises the modern western diet, in which animal organs are considered unhealthy to consume. Southern Chinese people believed that organs of animal will supplement or supply nutrition to the same organ of human being that is consumed. For example, using pig lung, dried

TABLE 17.1 Food classifications.

Types	Grains	Meat	Vegetables
Neutral	Corn, soya bean, black bean	Pork, beef, chicken, goose, carp, shellfish, sea cucumber, octopus	Potatoes, sweet potatoes, yam, sesame, olive, carrot, apple, star fruit, sugarcane, grapes, lotus seed, dates, plum
Cool	Millet, barley	Nil	Cucumber, spinach, eggplant, green bean, mushroom, tofu, lotus root, lemon, tomato, pear, melon, asparagus, tea leaves
Cold		Duck, crab, clams	Bamboo shot, seaweed, winter melon, Chinese cabbage, grapefruit, pomelo, coconut, watermelon, turnip
Warm	Glutinous rice, sorghum, peanut oil	Lamb, shrimp, eel	Ginger, garlic, chives, papaya, longan, guava, hawthorn, lychee, peach, almond, chestnut, coffee, vinegar, durian, chocolate
Hot	Chilli, mustard, raw garlic, pepper, dried longan, sesame oil, hard liquor, deep fried food, deep fried peanut		

Source: Kaohsiung Medical University Chung-Ho Memorial Hospital.

vegetables, and almonds to make soup is very common in the southern part of China. This soup is perceived as supplying nutrition to the lung and moisture human's skin during autumn and winter.

In the Southern part of Chile, Argentina, Paraguay, Uruguay, southern and Central-Western Brazil, the tea of yerba mate is very popular and one will find more brands of mate than coffee in a local supermarket. Many South Americans consume yerba mate daily during breakfast and many consume a few cups the whole day. Yerba mate is believed as an energy drink to start a day. In fact, yerba mate contains caffeine and no doubt it is widely consumed early in the morning. Drinking yerba mate is a leisure and domestic activity in South America. The traditional way to consume yerba mate is to drink from a container made with a dried calabash by using a metal straw, called bombilla in Spanish. In the mountain of Patagonia, al gauchos, the traditional cowboys inhabited in Argentina, southern part of Chile, Uruguay, and Brazil are often enjoying and sharing yerba mate in a calabash container. Yerba mate is also used to welcoming new friends. When a new friend is introduced among a group of indigenous South American, yerba mate will be prepared and served in the calabash container with a metal bombilla. The whole group will use only one calabash and one straw to drink and share the yerba mate. It is a very common social practice that a person will have a zip of yerba mate from the bombilla (Fig. 17.2), then refill it with hot water, wipe the tip of the bombilla with a fragrant leave of a local herb, and pass to the next person to enjoy. If a person does not share or pass on the yerba mate, it hints that friendship is not built or discontinued.

From the above two examples of how pregnant women practicing traditional Chinese medicine and how yerba mate is served in South American countries, food functions and

FIGURE 17.2 Yerba mate tea cup and straw. *Source: Photo by Lei Weng Si.*

its beliefs, including impacts and value on people, are illustrated. These are only examples among numerous practices of different cultures. More importantly, food choice and consumption practices have been passed on from generations to generations and have been an essential element forming a particular culture. In the following section, how cultures contribute to the different food consumption habits and etiquettes will be discussed to further elucidate the entwined relationship between food and culture.

17.3 Food habits

The type of food people consume for breakfast, lunch, dinner, and the meals schedule tell a lot about a city or a country's culture and lifestyle. Meal patterns and meal cycles of different countries vary a lot. In the United States, for instance, a proper meal should consist of a main course and side dishes, for example, a meat, vegetable, and starch. In China and many parts of Asia, rice is a must to be served for a proper meal, while bread is essential in Europe. Bread and cereal are common breakfast in the west. In Asia, rice, noodles, dumplings, and steamed buns are widely consumed for breakfast. Furthermore, the sequence of food to be consumed also varies a lot. In the United States, appetizers are served before soup and salad, followed by the entrée and then by dessert. Everyone gets his/her own plate to eat. In China and some Asian countries, all dishes are served at the same time and to be shared among all diners. A proper Chinese banquet has the following sequence in serving dishes. Cold dishes are served before any hot dishes. Among hot dishes, style of stir-fried dishes will be served first and followed by braised dishes. Light flavored dishes are always served before strong flavored dishes, for example, stir fried sea limpet with celery and cashew nut is served before braised abalone and seasonal vegetables. Rice and noodles will be served after hot dishes and desserts and fruits are to be served at last. In addition, certain Asian cultures like Chinese and Japanese typically serve food according to the season. It is believed that the body is best to consume seasonal food and ingredients are at its finest when they are in season.

The portions of each meal to be consumed differ very much among different cultures. When understanding the proverbs of different cultures, one can understand how food habits are very much influenced by the culture ingrained in their day lives. According to the western food proverbs stated in Pollan (2009), there are two proverbs below:

> Eat breakfast like a king, lunch like a prince and dinner like a beggar.

> Eat breakfast alone, share lunch with a friend, and give your dinner to your enemy.

They illustrate the portion of each meal should be consumed by people in the West. Breakfast is considered as the most important meal of a day and should have the biggest portion among daily meals. In contrast, dinner is advised to be smaller comparatively to other meals or to be skipped.

In contrast, the Middle East has a very different view on dinner as illustrated by this proverb:

"أصل خراب البدن ترك العشاء (aslu kharab albudn tarak aleasha)"—The origin of the destruction of the body is the removal of dinner.

A lot of emphasis is put on dinner in the Middle East and normally Middle Eastern cultures leave their biggest meal for dinner time. In China and Japan, the proverbs suggest people never fill the stomach completely full, as these advise the following:

"七分飽 (Qi fen bao)"—Eat until you're only 70% full.
"腹八分目/はらはちぶんめ (Hara hachi bu)"—Eat until you're only 80% full.

Besides the types of food and portions of meals, eating etiquettes of different cultures deserve much attention for tourists and people learning different cultures nowadays as international travels become a phenomenon. In Asia, chopsticks are widely used to eat. Manipulating a pair of chopsticks is a real challenge for first timers. In addition, there are a few etiquettes to practice when using chopsticks:

- Never stick chopsticks in a rice bowl when resting the chopsticks or put the chopsticks across on a rice bowl. This is similar to rituals when offering to ghosts and spirits. People may rest the chopsticks on a plate or the whole pair by one side of a rice bowl or on a plate.
- Never pass food from a pair of chopsticks to another pair of chopsticks. This is simply considered unlucky.
- Lift the rice bowl up when you have to get rice with your chopsticks and put into your mouth. This is considered a practice by the less fortunate.

In addition, when consuming steamed fish, the fish is not to be flipped even when one side is finished. This is particularly prevalent in the southern part of China with many related to the fishing industry as flipping the fish is similar to a fisherman's boat capsizing. In Korea, you should not start eating until the eldest or the most senior among the group has started. Respecting the elders is a very important culture in Korea. If an older person offers you a drink or hand you something, you must lift your glass to receive it with both hands.

In South America, do not be surprised a dinner is served only after 10 in the evening if you are invited to your friends' home, and restaurants open only at 9 o'clock in the evening for business. A dinner starting at 10 o'clock in the evening is very normal and dinner ends only after 1 or 2 o'clock in the midnight. In Chile, you are expected to eat everything with your cutleries, even fries. Do not pick up any food with your fingers. In Mexico, you are expected to eat tacos or burritos with your hands instead of a fork and a knife. Using cutleries to eat taco is considered inappropriate and silly in Mexican culture.

In Europe, do not order a cappuccino after noon time or after dinner. Italians believe that a late-day cappuccino upsets stomach. Similarly, orange juice is taken only before afternoon in Portugal. Portuguese consider that orange juice is gold and is beneficial for the body only before noon time. In France, do not eat your bread as an appetizer before a meal. Bread is consumed mostly as a supplement to your food or to the cheese platter at the end of a meal.

In the Middle East, India, and some part of Africa, you should not eat with your left hand or even passing documents or things to people. It is because left hand is used to do the cleaning ritual after using bathroom. Meanwhile, shaking your coffee cup to signal your host that you do not want more coffee from Bedouins is an important etiquette; otherwise your host will continue to refill coffee for you. However, in many cultures, it is mandatory to eat with both hands, typically holding cutlery. It is considered impolite to eat with only one hand while the other is resting. Therefore, it is important for today's global travelers to appreciate different food cultures and not to show any disrespect to local traditions and customs, however unintentional. As the saying goes, "When in Rome, do as the Romans do."

17.4 Food and tourism

Research studies on local food, food habits, and culinary have been branches of sociology and anthropology in the 90s (e.g., Bell and Valentine, 1997; Maines, 1997; Warde and Martens, 2000; Beardsworth and Keil, 2002). Food as tourism products (e.g., Hjalager and Richards, 2003; Meler and Cerovic, 2003), tourists' food consumption pattern (e.g., Tse and Crotts, 2005; Chang et al., 2010), tourists' dining experience (e.g., Chang et al., 2011), and tourists' interest and motivation to attend food and beverages related events in a destination (e.g., Galloway et al., 2008) have been studied since the millennium. There has been considerable growth of food tourism-related studies and publications. Furthermore, the World Tourism Organization (UNWTO) launched the first Global Report on Food Tourism in 2012, which was a cooperation of UNWTO Affiliate Members Program. As the world travel phenomenon arises and tourists are constantly seeking new destination experiences, food/gastronomy tourism has emerged, and no doubt it has become the latest trend in the tourism sector. In this section, food/gastronomy tourism and cultural capital will be discussed. The highlights of the Second Global Report of Gastronomy Tourism, trends, and its importance, as well as the creation of UNESCO Creative Cities Network, in the field of Gastronomy are depicted.

Food tourism, domestically and internationally, has been further fueled by the "foodie" culture of particularly young people who travel to different places (cities within the same country, but more typically international travel), try out different eateries, and share photos and videos on social media such as Facebook, Twitter, and most commonly for such purpose, Instagram, affectionally called "Foodstagrams." Research has shown that foodie culture gives additional meanings to food consumption (Watson et al., 2008), which in turn brings positive economic benefits such as electronic word of mouth on social media. Travelers also increasingly refer to online reviews such as Trip Advisor and food blogs instead of guidebooks, which makes food and tourism inseparable from each other. Appreciating food no longer means conventionally and literally "to love" and "to enjoy" food, but rather to embrace the whole consumption experience with food as the center, complemented by the ambience and service, as well as a photograph to be shared on social media.

17.5 Food identity, culture, and tourism

Although the value that connects food identity and tourism is indispensable, food identity to a place does not need to be a culmination of culture and heritage even if it is historically rooted to a destination. One example is the tourism industry in Amakusa, a small town in the Kumamoto Prefecture, Japan. The city of Nagasaki and various districts in the Kumamoto Prefecture, including Amakusa, were the first spots the Portuguese landed in China and was central for its missionary work in Japan. However, the practice of Christian faith was severely oppressed and devotees moved inner land and smaller islands to continue their faith generations to generations. Amakusa contains important historical artifacts that show this period of oppression of the Japanese Christians. In an ongoing bid to become a UNESCO World Heritage site, the local tourism industry reacted by reintroducing Portuguese cultural elements into Amakusa to synergize the positive effects brought forth by tourism to the region to produce Japanese's very own Portuguese food souvenirs. These included products made from figs and traditional sweets and even wines, as well as textiles.

In 2015, one of the authors was invited as a part of a delegation to provide advice to the local tourism businesses on how to proceed with this endeavor, with different experts giving advice to the local businesses on developing Portuguese-style souvenirs (particularly food). In this example, identity was rooted in this destination, but neither the Portuguese culture nor heritage was practiced in the region. It is through the potential benefits that tourism may bring that local businesses reacted proactively.

17.6 Gastronomy tourism and cultural capital

According to the Global Report on Gastronomy Tourism by the World Tourism Organization in 2017, an upward trend of tourists spending on food and beverage has been observed in different countries and cities, for example, in Ireland. Although food and beverage consumption is a must-do or even necessity of any tourists traveling, the broad spectrum of food tourism often refers not only patronization at high-end restaurants, large-scale food festivals, but anything from street food sellers to produce markets (Yeoman and McMahon-Beatte, 2016). The World Food Travel Association defines food tourisms as

> The pursuit and enjoyment of unique and memorable food and drink experiences, both far and near.

Nowadays, traveling to France and dine in a Michelin three-star restaurant; traveling to places where food ingredients originated or beer locally brewed; traveling across town to a new coffee shop for a home-roasted Arabica coffee; or traveling deep in the Amazon forest and experience a local lunch served by indigenous people have become more popular and much more accessible than in the old days. Food and traveling are surely inseparable, and the increase of world travel serves as catalyzer to blossom the development of Gastronomic Tourism. In addition, as stated by UNWTO Secretary General Taleb Rifai in the foreword of the Second Global Report on Gastronomy Tourism:

> Linking gastronomy and tourism also provides a platform for the promotion of cultures through their cuisine. This not only assists in destination branding, but also helps to promote sustainable tourism through preserving valuable cultural heritage, empowering and nurturing pride amongst communities, and enhancing intercultural understanding …

As seen from his statement, one can visualize that food, tourism, and culture are tightly linked together and they are supplementing each other. With proper planning, a destination will enjoy the benefits from Gastronomy Tourism economically and socially. The study of Mak et al. (2012) suggested that food consumption in tourism fits well in the embodied state of cultural capital proposed by Bourdieu (1984). Cultural capital refers to a stock of knowledge and experience people acquire through the course of their lives that enables them to succeed more than someone with less cultural capital. There are three states in cultural capital, namely the embodied state (the knowledge and skills an individual possesses), the objectified state (expressed in a form of cultural goods), and the institutionalized state (represented by actual documents and other proof of cultural status). As food consumption or participating in gastronomy tourism embeds in the embodied state of cultural capital, for example, joining a cooking class of authentic northern Thai cuisine when traveling to Chiang Mai, Thailand, tourists will learn the authentic way in cooking Thai green curry. And the tourists will later cook a Thai curry upon returning home to showcase the cooking skills and Thai food knowledge earned from the cooking class. As Mak et al. (2012) stated, this kind of cultural capital can be used on return to home or be used immediately within the destination visiting to detect the authenticity of food consumed.

17.7 Trends and its importance

The first Global Report on Food Tourism was launched by the UNWTO Affiliate Members Program under the World Tourism Organization (UNWTO) in 2012. The report reveals the new trend in food tourism and movements from different types of entities, namely destination management organizations, business organizations, and education organizations. The second Global Report on Gastronomy Tourism was published 5 years after the first report in 2017. The report is a fruit of the UNWTO Affiliate Members Program with case studies from 60 contributors including UNWTO Member States, Affiliate Members, and partner organizations. The second report is much richer than the inaugural one in 2012. Conferences and activities of gastronomy tourism taking place of the last 5 years were also reported. The report addresses the importance of gastronomy in tourism destination development as well as the global trends on gastronomy tourism. Case studies on gastronomy tourism are incorporated. Experiences and best practices in gastronomy tourism from around the globe are included. Furthermore, findings of a survey aiming at seeking an in-depth understanding of gastronomy tourism development in different destinations are reported too. The pool of respondents consisted of UNWTO Members from 29 countries between the periods of June to September 2016. A total of 77 responses were received in which 67.5% of them were from Europe, 22% from the Americas, 7.8% from Asia and the Pacific, and 2.6% from Africa. The majority of respondents represented governmental departments (41.5%) and not-for-profit organizations (28.5%), as well as educational institutions (15.5%) and private sector businesses (14.5%). The key points are as followed:

- The majority (87%) of responding organizations believe that gastronomy is a distinctive and strategic element in defining the image and brand of their destination;
- Gastronomy ranked the third main reason why tourists decide to visit a destination;
- Approximately 70% of respondents have already targeted gastronomy tourists as a market segment;
- The potential of gastronomy to enhance the livelihood of the local community has been valued at 8.53 points out of 10;
- 80.5% of destinations rating gastronomy tourism potential with 8, 9, and 10 points.

The findings project the important trend of gastronomy tourism, particularly on its potential contribution to local community and destination image. Additionally, research conducted by the World Food Travel Association - 2016 Food Travel Monitor depicts that 93% of travelers can now be considered food travelers. The association defined "food travelers" as travelers who had participated in a food or beverage experience other than dining out, at some time in the past 12 months, including activities as followed:

- Visiting a cooking school;
- Participating in a food tour;
- Shopping in a local grocery or gourmet store;
- Joining tours at food or beverage factories;
- Participating in wine/beer/spirits tasting;
- Eating out in unique or memorable foodservice establishments.

The association also describes food travelers as explores who are passionate to get off the beaten path and discover new, unique, or undiscovered experiences. An increase from 8.1% to 18% food travelers expressed an interest in Gourmet experiences, as their primary interest between 2010 and 2016 was seen in the study of Psycho Culinary research, conducted by the same association. It illustrates a clear bright future for gastronomy tourism development.

17.8 UNESCO Creative Cities Network—Gastronomy

Branding of gastronomy has reached a new global level of attention when UNESCO (United Nations Educational, Scientific and Cultural Organization) launched its Global Alliances Creative Cities Network in 2005. The Network recognizes cities worldwide with reputations as centers of excellence in seven areas, namely Crafts and Folk Art, Design, Film, Gastronomy, Literature, Music and Media Arts. The UNESCO Creative Cities Network was created to promote cooperation within and among cities that have recognized creativity as a strategic tool for sustainable development in the area of economic, social, cultural, and environment.

To be qualified in the Network for gastronomy, cities have to demonstrate that they have met the following criteria: evidence of traditional food markets, a traditional food industry and a tradition of hosting gastronomic festivals. To date, there are a total of 180 cities in 72 countries designated in the Network. Among the new addition of creative cities in 2017, there are eight cities designated under the field of gastronomy. They are Alba (Italy), Buenaventura (Colombia), Cochabamba (Bolivia), Hatay Metropolitan Municipality (Turkey), Macao Special Administrative Region (China), Panama City (Panama), Paraty (Brazil), and San Antonio (United States of America). Together with the new designation of Macao, there are three cities in China: Chendu, joined in 2010; Shunde, joined in 2014; and Macao are in the field of gastronomy of the UNESCO Creative Cities Network.

17.9 The recent case of Macao Special Administrative Region, China

Destinations are in fierce competition to attract tourists worldwide nowadays. Gastronomy tourism offers new path to destinations in showcasing not only the authentic dishes and beverages but also roots, stories, and histories go along with meals serving on a dining table. As suggested by the World Food Travel Association, the benefits that a city or community could reap from gastronomy tourism are ranged from social to economic and even environmental. Below is a short list of benefit in activating gastronomy tourism, provided by the association.

- Increase visitor arrivals
- Increase sales (rooms/airplane seats/restaurant meals/wine/beer/car rentals)
- Increase media coverage
- Provide a new competitive advantage or unique selling proposition
- Increase government authorities tax revenue
- Enhance community awareness about tourism in general
- Enhance community pride about, and awareness of, the area's food and drink resources.

Furthermore, as acknowledged by Yolanda Perdomo, Director of the Affiliate Members Program, in the Second Global Report of Gastronomy Tourism:

> Gastronomy tourism has been continuously increasing and its importance has been recognized by many destinations; in addition to its economic benefits, gastronomy has demonstrated to be a major source of identity and pride for local communities, drawing on an intangible heritage. It is also able to attract tourist flows to less visited regions, consequently generating spending in a way that is more equitable throughout the entire region. This has a widespread impact throughout the value chain and contributes to economic development and opens doors to new opportunities.

The Historic Center of Macao consists of 25 historic buildings, which was enshrined on the World Heritage List in 2005. A year later, in 2006, Macao adopts the UNESCO Convention of Safeguarding Intangible Cultural Heritage (UNESCO, 2003), a commitment by the government to ensure the continuity of particularly Portuguese-related heritage in Macao. The Intangible Cultural Heritage is classified into five domains: oral traditions and expressions; performing arts; social practices, rituals and festive events; knowledge and practices concerning nature and the universe; and traditional craftsmanship. Although "food" does not fall into a single domain, it encompasses in all five domains in different manifestations.

In a recent successful endeavor, Macao initiated the application to be qualified and listed in the UNESCO Creative Cities Network, in field of Gastronomy at the end of 2015. Macao becomes the third Creative City of Gastronomy in China, after Shunde and Chengdu, in 2017. The designation of Macau has displayed the convergence of food, culture, and identity of a city through a potential realization of gastronomy tourism development.

The designation of Macao as a Creative City in the field of Gastronomy reinstates Macao's unusual history and rich heritage of the first and last European outpost in Asia. The recognition of built, intangible, and food heritage by world experts through UNESCO encourages and rejuvenates the idea of strategically position Macao as a food destination using its unique food identity—the Macanese food.

17.10 Featured case study: Macanese cuisine and identity

The term "Macanese" is heavily disputed among the Macanese community, hereby referred to a community of Eurasians born in Macao to both parents of both Portuguese and Chinese ancestries. Macanese does not refer to the people of Macao—ethnically

Chinese—who make up the majority of Macao's population. Because of the rich and long influence of the Portuguese in Macao, the Macanese has its own food, language, and customs which reflect the Macanese identity and very different from the other cultural groups in Macao. The reason the Macanese gains prominence in Macao was because of their socioeconomic and political background, but further discussion on this is outside the scope of this chapter (for further reading into the Macanese identity, see http://www.macstudies.net/engaging-the-macanese-a-case-study/). Regardless, the Macanese has a strong identity with food as the center of all celebrations—tea parties, festivals, religious events, birthdays, and weddings. With the designation as a Creative City of Gastronomy, it is possible to further promote Macanese cuisine and leverage its benefits through tourism.

The Macanese food is central to all celebrations and gatherings because it reinstates and reinforces the diversity and variety of cultures that make up today's Macanese—probably one of the world's oldest forms of fusion cuisine. As simply noted by Cíntia Serro in her famous cookbook (Serro, 2013) "food is the focal point of any event that brings Macanese people together, anywhere in the world." Interestingly, although food is the focus of the Macanese gastronomy culture, the identity is also manifested in other forms related to food. One example is the typical Macanese wedding banquet centerpiece which is decorated by a homemade bridal doll, and wedding favors made up of traditional special sweets (particularly of those made of coconut), as well as the Macanese cake (rum cake). The coconut is a reminder that Macanese people have Chinese ancestral roots, as the pronunciation of the word "coconut" (ye-zi) is similar to "grandfather–son," thus meaning births to come, particularly of male offspring, an auspicious symbol in a wedding. The bridal doll is traditional given to the youngest girl sharing the same table (Fig. 17.3), the one who is still single or the next to be married (there is no tradition; as it is illustrated in the following, the traditional Macanese culture bears no institutionalized form).

The Macanese cuisine is a culmination of food wisdom—ingredients, methods, customs, and practices—of the Portuguese maritime route reaching the Far East. For instance, the

FIGURE 17.3 Homemade bridal doll and Macanese cake are typical table centerpieces and wedding favors in a modern Macanese wedding banquet. *Source: Photo by Ubaldino Couto.*

FIGURE 17.4 Galinha à Portuguesa; A typical Macanese tea party. *Source: Photo by Florita Alves.*

Macanese cuisine incorporates spices typically used in Africa and India such as turmeric and paprika, incorporates Malaysian characteristics such as coconut and stews reminiscent of those from its Portuguese origins. One example is the *Galinha à Portuguesa* (Portuguese-style Chicken), which was never originated from Portugal but very much a Macanese dish. It is essentially a chicken and potato stew in tomato sauce and turmeric (and/or curry), with a coconut milk finish to thicken the sauce, typically garnished with coconut flakes and Portuguese chouriço (sausage) (Fig. 17.4). Many family recipes would oven bake or grill for a few minutes before serving to give a slightly burnt finish.

There is no standard in Macanese cuisine. Each family has its own secret recipe and rarely Macanese families share with outsiders; recipes were typically passed on from generation to generation, and rarely in full—most recipes even passed down from elders in the family, require significant adjustments and improvisation. This is reflective of the so-called "origins" of Macanese cuisine according to many Macanese old wives' tales. The story has it that when the Portuguese men arrive in the Far East on naval ships (and along the route from Portugal, circumnavigating Africa, India, and the Malaysian peninsula to reach Macao), many became too weak to travel back to Portugal or simply they were mesmerized by the beauty of both the Far East lands and women. Consequently, many remained in Macao and married local women, who obviously did not cook anything remotely similar to Portuguese food or Macanese dishes. The husbands would describe fondly the food from home and what they had along the different ports before reaching Macao, and the housewives would share recipes and advices. As a result, families typically have different recipes even for the same dish as the Macanese cuisine is essentially a fusion food, made up of memories, shared wisdom, and improvisation of local and seasonal ingredients, as well as the ethnic roots of the Macanese family (some men and women who arrived in Macao later were second generations born in Africa, India, and Malaysia).

One example of variety across families is *minchi* (sautéed minced meat; the word is neither Portuguese nor Chinese but patuá, a creole spoken by the Macanese people), a somewhat signature and favorite of the Macanese people. In its purest form, *minchi* is typically sautéed pork and beef mince in onions, with Chinese dark soy sauce to taste and fried potato cubes to garnish, typically served with a fried egg and plain boiled rice. Some families would insist the "original" has Portuguese chouriço julienne to finish while others would argue theirs with diced Chinese dried mushrooms. Conversely, some families proudly share their "original" recipes with a spicy kick (by adding curry powder when stir-frying the meat) and others would have diced bitter melon to replace the fried potato cubes but in a shrimp paste sauce. The variety of *minchi* is endless and the differences are not subtle as evidenced by the few examples illustrated (Table 17.2). Indeed, Macanese families respect the diversity and variety of Macanese cuisines among families and often strive to learn from each other to improve their own family recipes.

Unlike many other traditional cuisines which are institutionalized and typically come with a standard recipe, Macanese cuisine—although bearing a few hundred years of history—is rather secretive and informal. Fortunately, there have already been efforts in institutionalizing Macanese cuisine before the UNESCO Creative City of Gastronomy designation. One example is the establishment of the *Confraria da Gastronomia Macaense* (CGM, Macanese

TABLE 17.2 Recipe of minchi.

Ingredients: Corn starch 1 tsp Dark soy sauce 1 tsp Garlic 4 cloves Green onion 1 sprig Cooking oil 1 tbs Light soy sauce 3 tsp Minced pork 500 g Pepper $^1/_2$ tsp Potato 4 Salt Shallot 4 bulbs Sugar 1 tsp Vinegar 2 tsp	In a pan, sauté minced pork (can modify recipe to 50–50 pork and beef) for 5 min; drain and set aside in a bowl, discard liquid. In the same pan, heat cooking oil and sauté garlic, shallot, and green onion. Add minced pork and both light and dark soy sauces. Add salt and pepper, sugar, and vinegar to taste. Cover and cook in medium heat for 15 min, stirring occasionally to keep the bottom from sticking. Add corn starch when the meat is cooked (this can be omitted but drain excess liquid; the mince must be dry). Before serving, garnish with fried potato cubes. To prepare fried potato cubes, peel and dice potatoes (1/2-inch cubes) and let them stand in cold salted water for an hour. Dry well. Deep fry and leave to drain excels oil on paper towel.

Source: Photo and recipe by Florita Alves. Recipe adapted from Serro.

Gastronomy Association, web http://www.confrariamacaense.com). According to its statues, CGM aims to research and promote the Macanese cuisine heritage; in addition, the statues also ensure the authenticity of Macanese cuisine and protect the natural evolution of recipes, thus addressing the problem of variety of the same dish in Macanese cuisine.

There are obviously many more dishes in Macanese cuisine and the Macanese people are most proud of its *Chá Gordo* (fat tea, tea parties typically for children birthday but can also be modified into wedding receptions, baptism celebrations, baby showering, and so on). These events are typically filled with many Macanese snacks and dishes, sometimes in the form of a "potluck" where different families bring different dishes to show off their culinary skills and best dish (Fig. 17.5).

As we pointed out at the beginning of the chapter, food tells a lot about a culture—there are also many more stories about the Macanese culture and identity through food. It is therefore imperative for the government and tourism authority of Macao to promote the Macanese food culture, among others, so as to diversify the touristic offerings in Macao. Today's travelers are different from the past—they seek novelty, truth, and authentic cultural experiences (Table 17.3). It is possible for tourism authorities to unleash food tourism for the greater benefits of tourism to a destination.

Food, after all, connects us all, regardless of which culture we are in. In the selected recipes presented herein these chapters, the authors have heard comments from both extremes, ranging from "hey, that's just cheese on toast" and "it's same as my grandma's chocolate brownie" to "that tastes so different from the food we have" and "don't you serve these with mashed potatoes?" It is exactly these comments when we as travelers feel and say as we sample local and traditional foods. While food brings out the unique identity of different cultures, it also reinstates the very basic connection among us all.

FIGURE 17.5 Chá Gordo: A typical Macanese tea party. *Source: Photo by Florita Alves.*

TABLE 17.3 Recipe of bolo de chocolate.

Ingredients: Butter 340 g Sugar 454 g Eggs 6 (50 g each) Baking powder 3 tsp Cocoa 55 g Flour 395 g Salt $^{1}/_{4}$ tsp Fresh milk 250 mL Pure vanilla extract 2 tsp	In a bowl, sift and combine baking powder, cocoa, flour, and salt. In another bowl, combine milk and pure vanilla extract. Mix well sugar and butter and add eggs one at a time. Add contents from the two bowls little by little and mix well. Pour batter into a cake pan greased with butter and dusted with flour. Preheat oven to 175°C. Bake for 50 min approximately. Remove to a rack to cool. Cracks appearing at the top are normal and a symbol that the cake is ready. To check, poke in a long toothpick; if it is clear after removing, the cake is ready.

Source:Photo and Recipe by Florita Alves.

References

Beardsworth, A., Keil, T., 2002. Sociology on the Menu: An Invitation to the Study of Food and Society, Routledge.

Bell, D., Valentine, G., 1997. Consuming Geographies: We Are where We Eat. Psychology Press.

Bourdieu, P., 1984. Distinction: A Social Critique of the Judgement of Taste. Harvard university press.

Chang, R.C., et al., 2010. Food preferences of Chinese tourists. Ann. Tourism Res. 37 (4), 989–1011.

Chang, R.C., et al., 2011. Attributes that influence the evaluation of travel dining experience: when East meets West. Tour. Manag. 32 (2), 307–316.

Galloway, G., et al., 2008. Sensation seeking and the prediction of attitudes and behaviours of wine tourists. Tour. Manag. 29 (5), 950–966.

Hjalager, A.-M., Richards, G., 2003. Tourism and Gastronomy, Routledge.

Maines, D.R., 1997. Kitchens: the culture of restaurant work. Soc. Forces 75 (4), 1479.

Mak, A.H., et al., 2012. Globalisation and food consumption in tourism. Ann. Tourism Res. 39 (1), 171–196.

Meler, M., Cerovic, Z., 2003. Food marketing in the function of tourist product development. Br. Food J. 105 (3), 175–192.

Mintz, S.W., Schlettwein-Gsell, D., 2001. Food patterns in agrarian societies: the"Core-fringe-legume hypothesis"A dialogue. Gastronomica 1 (3), 40–52.

Pollan, M., 2009. Food Rules: An Eater's Manual, Penguin.

Serro, C.C., 2013. Traditional Macanese Recipes from My Auntie Albertina. Macao, International Institute of Macau.

Tse, P., Crotts, J.C., 2005. Antecedents of novelty seeking: international visitors' propensity to experiment across Hong Kong's culinary traditions. Tour. Manag. 26 (6), 965–968.

UNESCO, 2003. Convention for the Safeguarding of the Intangible Cultural Heritage. Retrieved from: http://unesdoc.unesco.org/images//0013/001325/132540e.pdf.

Warde, A., Martens, L., 2000. Eating Out: Social Differentiation, Consumption and Pleasure. Cambridge University Press.

Watson, P., et al., 2008. Online communities and the sharing of extraordinary restaurant experiences. J. Foodserv. 19 (6), 289–302.

Yeoman, I., McMahon-Beatte, U., 2016. The future of food tourism. J. Tour. Futur. 2 (1), 95–98.

CHAPTER

18

Feeding the future: challenges and limitations

OUTLINE

Despite the apparent abundance of the food supply, we cannot remain complacent, as an ever-increasing global population provides us with a formidable challenge in the coming decades. Currently as we have already noted, global population is estimated to be in the region of 7 billion, and with growth rates of about 80 million people per year this figure is set to rise to about 7.94–8.33 billion by 2030 before trailing off at 2050/60 to between 9.2 and 9.5 billion (ESA, 2007; GEO, 2007; Zhang, 2008). In this regard the true test of our ingenuity will be the extent to which we can find innovative sustainable approaches that will enable us to provide the 70% increase required in agricultural production demand alone to feed these numbers. This is, indeed, a weighty challenge by itself, and to add to this the effect of current trends away from traditional grain-based diets in favor of more meat and animal products and the difficulties are compounded. A change some have suggested could result in a global average per capita calorific increase of 40% from 440 to 620 kcal per day (Smeets et al., 2006). Of course, this is subject to regional variation where through to 2030 Europe and North America's per capita meat consumption is projected to rise to 14% (by weight), while in East Asia consumption these increases will potentially rise by 55% and sub-Saharan Africa by 42% (IMechE, 2013).

While historically we have risen to the occasion of these challenges, we have done so through collaborative advancements in science and technology accompanied by a conducive political will, which gave rise to the green revolution. However, many have criticized this not

https://doi.org/10.1016/B978-0-12-811808-5.00018-0

so silent revolution as being of the time yet unsustainable in the long term. Instead, many now are looking to a new, more "evergreen" revolution that takes a more holistic consideration of environmental and sustainability issues.

A great example of this new philosophy in action is through the New York Columbia University Millennium Villages project, which has seen a tripling of wheat yields in sub-Saharan Africa. This has taken wheat yields from historic averages in the region of one tons per hectare to over three simply by addressing issues of soil nutrient depletion (Earth Institute, 2009; Sanchez, 2010). Through closer, more sustainably sympathetic controls of agricultural inputs, and the introduction of high-yielding cultivars and updated technologies, some of these areas are now, for the first time, enjoying annual calorie production in excess of needs. This is just one example and there are many others too benefitting from the application of sustainable ecologically sound "evergreen" principles.

Yet standing in the way of future progress are many obstacles, chiefly among them are the strains on natural resources especially of land and water. This is particularly relevant in regard to the potential reduction in available land due to such factors as environmental degradation; climate change stresses; restrictions aimed at preserving ecosystems; and the inevitable competition with other land-use demands of biofuels, urbanization, industrial, and leisure needs, etc. Likely further hindering future progress too will increase energy costs resulting from greater demand as well as problems of reduced agricultural labor workforce as nations come to rely more and more on industrialization of the sector while the younger generations are simultaneously finding it less and less attractive (IMechE, 2013).

Importantly, however, all of this of course is predicated on increasing overall food production volumes, and, as has been suggested, increases in food availability can also effectively be achieved through reductions in the 30% or so of global food wasted annually.

18.1 Food confused?

Imagine the following: one is having dinner with friends and is about to tuck into a keenly anticipated meal of free range roast rump of New Zealand lamb accompanied by a decadent French foie-gras "pâté en croute," complemented with UK-grown organic baby carrots, green beans from Kenya, and sauté Ackee from the Caribbean all topped with a subtle truffle and smoke-roasted jasmine tea "jus" from Italy and Japan, respectively. All this global food is paired with a very reasonably priced full-bodied Bordeaux Pauillac red wine. But wait! It doesn't stop there; for later, the mouth salivates at the thought of fair trade chocolate mousse from South America, Korall shrimp cheese from Russia and a slightly salted African Roumy cheese; then to follow is the hosts' favorite imported heartwarming, 10-year-old vintage (1992) single-cask Glenmorangie whiskey from Scotland followed by medium-roast Arabica coffee from Vietnam. So, among all of this, one might reasonably ask: where am I? Well, in truth one could be anywhere in the world because the food chain is now a globally functioning network—a largely borderless enterprise where such dinners described above are not uncommon. Indeed, a plateful of sustenance like the one described above can cost anywhere up to tens of thousands of those environmental currency units—"food miles" (Rae Chi et al., 2009).

Still, as one is about to get stuck into a thoughtfully choreographed, culinary ballet, when a voice from across the table changes the focus of the evenings' soiree with the oft-quoted and—yes in part—smugly annoying "... do you know how far this food has traveled to be here?"

The room quietens and the mood changes. At once mouthwatering anticipation now ceases and saliva dries up while guilt constricts the stomach, reducing one's appetite. Firstly, one is annoyed as, by simply raising the question itself, the guest has assumed the moral high ground. Does he/she think they are the only one who cares about the environment? Indeed, somewhere in the back of one's mind, gremlins are studiously calculating the environmental, social, and ethical impact of this little soiree; however, that's where it was—at the back of the mind—and that's where it was supposed to remain—unencumbered by the reality of this little night's festivities! The dilemma though—should one sit politely and say nothing—holds one's own council while the voice externalizes its own moral fortitude, allowing the speaker to ramble on while patting themselves on the back as they do so? Or should one engage ... hmmm decisions! But it's too late as others have now taken the bait. The dinner dynamics has now changed, and what promised to be a gathering of friends for a light, albeit, slightly extravagant bite and drinks was fast sliding into an ad-hoc save-the-world rally ... the dinner had now been officially hijacked! Yet, the trouble with the topic was it was a free and, dare one say, an easy platform to occupy – everyone knows the validity of the argument ... indeed, there is no argument ... food miles! Indeed, didn't everyone know and care about the food miles' concept. You, me, and the next person, ad infinitum, are all aware and certainly care, but one doesn't need it rammed down one's throat every time he/she sits down to an evening's innocent/guilty pleasure.

"Ah but that's all very well ...," another voice pops up "... if not us, then who would buy this produce from Kenya, from Columbia, who would support these farmers?" The detractor continued:

> ... can you not see, my friends, food choices are no longer the sole responsibility of the flag-flying environmental stalwarts. And what of fair-trade? The poorer developing countries? And what of the processers, the packers, the importers, the exporters? Would you deny these people a fair and legitimate livelihood? No, my friends, food choices have long-since not been as straightforward as they once were. Over the last few years, they have become social, political and economic bones of contention.

The room was quiet and the message seemed all the more potent for it. And yet despite oneself, one's initial annoyance morphs to one of interest and sincerity as our curiosity is piqued. Others too seemed to take note; their interest also piqued. The rebuff might be onto something here, as it now appears that food miles might not be the only game in town after all. This was the opening gambit and not surprisingly with such an emotive topic, conversation was lively as the party parried and parlayed into the night.

Today's food choices, it would seem, are increasingly seen as convoluted and as complex as the food chain itself. Such choices—once largely based on economic criteria (cheap and value for money)—seem to be ever increasingly wrapped up in wider societal aspirations. What, where, and how we buy our food and how we use it is increasingly indicative of what is important to us in terms of social and ethical values as much as the nutritive value of the food itself.

Take the often-debated food miles as mentioned above for example; on the one hand, consider loud noises of staunch environmentalists whose mantra is to avoid, or at least drastically reduce food miles and its associated carbon footprint, at all costs. On the other hand, we judiciously balance this position with equally self-righteous indignation at the conspicuous lack of "true" global free trade and developmental growth as well as fair and equitable practices as we aggressively encourage the free movement of food trade around the world. So, there are delicate and fine line choices to make before chowing down on the above international gourmet. Indeed, one routinely considers the carbon footprint of the several thousand-mile journeys the carcass has traveled as well as factoring in the various carbon footprint trade-offs between intensive and organically farmed animals; but that's a separate issue, isn't it? Well, "no" is the short answer. This is because even this apparently simple trade-off is not that straightforward; many are now realizing that the distance a food item travels—the "food miles"—is only one way to quantify the environmental impact of food produce. And it may be flawed in some very important areas. In fact, of the UK's food chain—including production, processing, and distribution, for example—only a smallish percentage of total greenhouse gas emissions (including its carbon footprint) are attributed to transportation (Rae Chi et al., 2009). Moreover, the "food miles" mantra has led to a global proliferation of "local is best" whereby many farmer markets and the like are popping up at ever increasing rates. Yet, can one be sure the food miles approach is a suitable means of measuring the products environmental friendliness? Take for instance food, which has been grown or reared in foreign (high food miles) free-range or organic pastures without pesticides, fertilizer use, soil degradation, intensive green housing, and associated labor costs, and which also provides a sustainable living to millions of farmers as well as to those associated people/industries along the entire food chain. Compare this with the same produce grown locally, especially "out-of-season" or "nonnative" produce, which is grown utilizing the latest technology and/or intensive or non–greenhouse gas emissions (GHG) friendly methods and one can reasonably argue that the former scenario, i.e., the all-inclusive approach, is in some (or even many) cases more societally and environmentally friendly than just the food miles approach alone would suggest (Rae Chi et al., 2009). Indeed, while the global food sector accounts for around one-third of total globally produced GHG emissions, transportation of food products in actual fact accounts for only a small proportion of such emissions. So, although without doubt very useful within an arsenal of overarching metrics, the preoccupation with food miles is often misleading as it tends to focus largely on carbon dioxide emissions associated with travel and other lesser voluminous, but no less potent GHG emissions.

In reality, the ecological footprint is another metric that comprises so much more than just a foods CO_2 position; instead it takes into account emissions of nitrous oxide, methane, and industrial gases such as hydrofluorocarbons as well as environmentally friendly concepts (ecological accounting), sustainability, fair trade, organic, local versus global many other concepts. As far as agricultural GHG is concerned, Fig. 18.1 gives a general overview of agricultural GHG emissions by type. Ultimately when it comes to environmental metrics, many researchers and policymakers are questioning whether the "food miles" approach is suitably robust to judge whether what's on one's plate is in fact socially and environmentally friendly or not. Ecological accounting might very well be the better option.

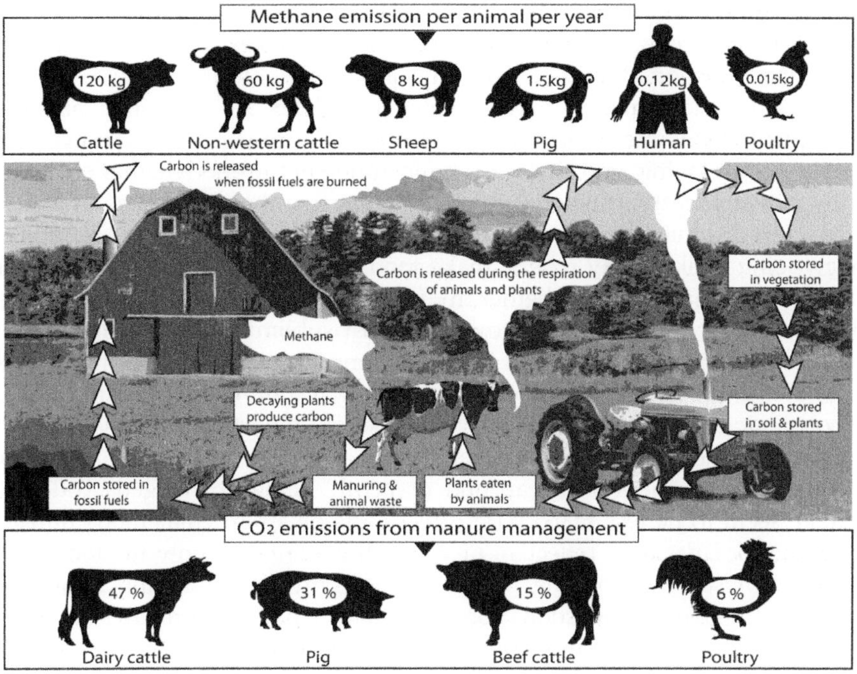

FIGURE 18.1 Agricultural greenhouse gases emissions by source. *Source: Courtesy of Pat Newsham*

In light of such challenges, OXFAM along with the International Institute for Environment and Development (IIED) is modifying the food miles concept in favor of a "fair miles" approach. This relatively new concept takes these challenges and recognizes that there is an environmental–economic trade-off, which on occasion suggests that local is not always or necessarily better (DeWeerdt, 2009; Williamson and Young, 2012). This leads us to the question of whether or not we, as consumers, are willing participants, exercising free choice, or simply sheep towing the line in an ever-changing social landscape. The following section elaborates.

18.2 Food choices: evolution or moral and ethical blackmail?

Although, as a counterbalance to the previous view, one needs to be aware that when it comes to food and society, as of the last two decades or so, people are becoming more knowledgeable, more discerning, and certainly more interested in the values, cultural diversity, and health and sustainability of food and food products among other things. In short there has been something akin to a social awakening regarding such considerations. Indeed, this "awakening" incorporates a much broader scope in which food's environmental impact is taken into account vis-à-vis other such laudable aims such as: food and economics; sociocultural food; food in relation to population size and growth; food and globalization; urbanization; wastage; environmental resources concerned with food; and human capital development among others. All in all, food is having something of a mini social revolution. Indeed, so much has changed when it comes to food choices, that people are nowadays routinely making selections based not only on price or value for money but also on the many ethical and environmental aspects some of which are described above (Gibson, 2016). Consequently, this means that people are now more engaged than ever in the role of food and society. Having said that, one could rationally argue whether such choices are made wholly independently or whether such apparent free choice has been morally hijacked by an increasingly "opinionated" population at large. Take the following example from the author's viewpoint: I, like many others, make food choices daily—but in this regard I find myself pondering the question of whether my food-based decisions are evolving or whether they are being morally hijacked by a growing almost subliminal moral/ethical movement? Indeed, one only needs to look around at the current state of affairs in the environmental impact of food, and it becomes fairly obvious to see that food choice decisions, once largely driven by economic considerations, are now increasingly reflecting wider society's collective social aspirations. And yes of course, even now I—"Homo Ignoramus"—find my ever-mindful food shopping mantra "cheap and cheerful" being challenged. Today, the food choices I make seem to be less to do with direct costs or a particular foods nutritive value per se, but rather, increasingly reflective of where it came from; how the food was grown; and what my purchases say about me. Indeed, vis-à-vis my food shopping, I now find myself asking some very challenging questions about whom or what my purchases support in terms of financial, political, ethical, cultural, and social aspects and of the environment as a whole. Yes, I fear to add that my food shopping has fast become a not-so-silent exercise in political activism. Yet within these considerations, whatever my choice trends, I still expect safe, wholesome, and healthy food (cravings notwithstanding). However, with all of this, nowadays my choices, once unencumbered by moral and ethical dilemmas, are not necessarily what they seem and sometimes the trade-offs leave me feeling a touch confused.

When it comes to *nutrition*, for example, I might want food that is sometimes low in salt and fat and sometimes with minimum or at least reduced additives, depending on how I feel I might even question the inclusion of genetically modified (GM) ingredients such as pesticide and herbicide residues—all this on top of the nutritive values of the food in question. Referring back to food additives, it is understood that this particular trade-off is not always that straightforward as some of these additives are natural commodities and can be very beneficial; while even some GM products although perhaps are socially distasteful among the wider population, in some instances they might also very well be ethically up there with fair trade. Continuing on the *ethical* front too, I might consider animal welfare or organic and free-range standards when it comes to my food choices as they are safe bets: Aren't they? Well yes and no; on the plus side, good animal welfare practices ensure a higher standard of quality of life throughout the animals rearing, while, on the downside, farm animals, particularly the bovine species, are major contributors of GHG (methane gas). Also, on the ethical front, I occasionally opt for the safe, fair trade choice which is usually ethically correct but one which might also be rather heavy on the food miles! This brings me to the "local versus global" debate? As mentioned previously, while sourcing local produce in support of reduced food miles and adding to the intrinsic benefits of local and national businesses, sometimes such choices come, confusingly, with a higher environmental price tag. On the flip side, globally sourced produce can and does (under the right circumstances) support poorer farmers and others along the food chain in developing countries. There are also considerations of *cultural* aspects too, which might be influenced by the economic maturity of a region or country, etc. And how do I contend with the perception of quality and freshness too as manifest in the simple appearance of food? In this little trade-off, it is worth noting that the more demanding our standards are in terms of uniformity of say fruit and vegetables, the greater the potential for more waste at the end of production of the chain. What is the paradox of slow traditional foods versus modern fast foods? Do we have the luxury of time in this particular trade-off or, in spite of the semantics, is "time" not the issue here? And how tempting is it to take up the offer of "buy one get one free," only to see the second item rot in the fridge, only to be thrown away at a later date? This also touches on the *economics* of food choices, which might be concerned with the cost of food comparative to our income, etc. In this scenario, the question here is can I comfortably buy the cheaper version when I know for a couple of cents/pence more, can I help the local butcher remain in the high street or the organic farmer stay in business or the African grower earn a living. When it comes to *environmental* factors these also weigh quite heavy on the social conscience. Such issues cover a plethora of related topics, including things like the ecological footprint, food miles, land degradation, climate change, water resources, and agricultural inputs such as fertilizer use, herbicides, pesticides, and other associated inputs within the sector. Moreover, the profligacy of our packaging habits or perhaps other general, natural resource squandering habits that seem pervasive might also be called into question in the choices of the food we buy.

Yet even with a smidgen of social awareness or more simply a fair understanding of what good choices are, where is my guiding light in all of this? Once I have deliberated, cogitated, and agonized over the many, often contradictory or mutually exclusive options and finally made my food choices, in whom can I trust to provide me with honest, unbiased environmentally friendly advice? Do I rely on government labeling initiatives, and if so which government or agency do I take notice of? Local? The exporting government? Are standards comparable across the varied regions and even countries? Or do I pay attention to benchmarks of the various stakeholder industries and NGO's like the World Health Organization (WHO); the Food and Agricultural Organization of the United Nations (FAO); the American Food and Drug Administration (FDA); the Food Standards Agency (FSA). And if so, are any of these standards national or global in coverage? Do I trust a fair trade labeled product to live up to its promise and ensure more income is shared out among the primary producers? or the locally grown stamp? the organic standard? the free-range mantra? Farm Assured Labels? The Red Tractor promises? The Leaf Mark? The Plan A? Freedom Food? and so on. Furthermore, do I trust in hearsay or anecdotal evidence or are the reams of so-called "factual and honest" advertisements and websites to be trusted? Indeed, while I am not wholeheartedly a cynic in these matters, I am also not naive enough to realize that the some of these self-aggrandizing messages are simply mere marketing tools or marketing spin—spin that ultimately shapes, affects, and dare I say hijacks my food choices with an emotional barrage of "do-the-right-thing" messages.

Yet, even in spite of all of the above, what of my guilty craving for chocolate?—Especially when, knowing such desires are contrary to those high standards we, I or society, self-righteously set for ourselves. When knowing this I still chow down on that scrumptious morsel of 77% pure cacao chocolate. In such cases, am I deliberately contributing toward a descending spiral of societal responsibility? Or am I (or we) left feeling guilty; and if so, guilty of what? Increasing GHG emissions? a burgeoning waste line? or can I judiciously throw caution to the wind and enjoy my chocolate—guilt free? Is this why such cravings are called guilty pleasures?

In the end, I, like many others, desire food that not only tastes good but also makes me happy, without excessively troubling my health or, more importantly these days, my conscience. In all of this, the question begs: Are the days of guilt-free, cheap food, fast coming to an end? Because, if I were to take all these choices into consideration in everything I bought—food-wise, I might be forgiven for failing under the weight of indecision or, in the words of Lucas Hollweg:

> Everything is fraught with ethical complications. If I tried to follow all of them, I'd end up an oxygenarian [eating] ... nothing but air. ***Hollweg (2007).***

"In the end, the dilemma boils down to several key questions; firstly, can I contentedly accept the fact that on certain levels I can be hypocrite? Like enjoying a bowl of shark's fin soup while simultaneously espousing the virtues of animal welfare? Or can I seriously chow down on a treat of foie gras pâté giving little thought to the cruel way in which the animal is reared? Furthermore, while taking this standpoint, on many of the other issues at hand, can I convincingly adopt the highest of moral standards? Or are these ideological polar opposites genuinely mutually exclusive?

So, where does it go from here? While I am loath to generalize, on this occasion, I must make a rare exception. We humans are not devoid or ignorant of the big picture, nor are we inconsiderate sentient beings. Yet even in spite of such generous human traits, we are not blameless either. As such we are confronted daily with sustenance versus moral and ethical trade-offs, many of which have repercussions and consequences beyond our immediate satiation. In this respect I cannot advocate a one-size-all policy. In the end, it is your food and hence your choice!

Yet, having outlined many of the seemingly contradictory issues at hand, when it comes down to it, I am still left with one nagging doubt—an uncomfortable feeling that will not go away. That is, even for those of us who can afford to spend an increasingly smaller proportion of our salary on food than in the developing countries, and in spite of the fact that we often know or can guess what the right choices are, I am still left wondering just how many of us still make many of our food selections based largely on cost and economic considerations rather than the more moral and ethical taxing ones? For me—for my money—ashamedly I still do. Does that make me, you, and us hypocrites? Sadly, I think it does? But that's just my thoughts—what about you? Well, in answer, I reiterate an earlier point—it's your food, it's your choice."

Source: WorldChefs.org; (Gibson, 2012).

References

DeWeerdt, S., 2009. Is Local Food Better?, by ww-Admin. World Watch Magazine. Online. World Watch Institute, p. 22.

Earth Institute, 2009. Development in Rural Africa the Millennium Villages after Three Years. Columbia University, New York, p. 111.

ESA, 2007. World Population Prospects: The 2006 Revision. UN Department of Economic and Social Affairs:Population Division, New York.
GEO, 2007. Global Environment Outlook 4. Environment for Development, Malta (United Nations Environment Programme).
Gibson, M., 2012. Food choices evolution or blackmail? WorldChefs E-Mag. 5 (12).
Gibson, M., 2016. The Feeding of Nations: Re-defining Food Security for the 21st Century. CRC Press, Boca Raton, Florida.
Hollweg, L., 2007. Help! I'm Food Confused. The Times, UK (Times Newspapers).
IMechE, 2013. Global Food Waste Not, Want Not: Feeding the 9 Billion: the Tragedy of Waste. Retrieved from: http://www.imeche.org/knowledge/themes/environment/global-food.
Rae Chi, K., et al., 2009. Big Ideas in Development: Fair Miles Recharting the Food Miles Map UK. International Institute for Environment and Development; OXFAM, p. 48.
Sanchez, P.A., 2010. Tripling crop yields in tropical Africa. Nat. Geosci. 3 (5), 299–300.
Smeets, E.M.W., et al., 2006. A bottom-up assessment and review of global bioenergy potentials to 2050. Prog. Energy Combust. Sci. 33 (1), 56–106.
Williamson, D., Young, L., 2012. Food: WWF-UK viewpoint. In: WWF: For a Living Planet. World Wide Fund For Nature, UK, p. 21.
Zhang, W., 2008. A forecast analysis on world population and urbanization process. Environ. Dev. Sustain. 10 (6), 717–730.

PART FOUR

Food, Service, and other Deliverables in the Hospitality Sector

One cannot think well, love well, sleep well,
if one has not dined well.
Virginia Woolf

CHAPTER

19

The sociology of the chef: a new theoretical proposition from the open professional kitchen

David Graham

Department of Service Sector Management; Sheffield Hallam University

OUTLINE

https://doi.org/10.1016/B978-0-12-811808-5.00019-2

19.1 Introduction

This chapter discusses the origins of the traditional western restaurant kitchen and its social construction as a closed place of work typical of the manufacturing economy. This environment is a manifestation of a highly aggressive and masculine world, which contrasts starkly with the open kitchen, where employees are expected to focus on and engage with customers, resulting in a shifting sociology of the chef. The chapter identifies the changing and changed nature of the restaurant kitchen and its emergence into the world of the service economy transforming into the experience economy as it is reorientated from closed to open and the requirements of the chef who work in it.

19.2 The restaurant in the industrial era

Lashley and Morrison (2000) synthesize a number of restaurant definitions to describe a restaurant as a public space for the consumption of food and drink, where, following the service delivery, an exchange of money takes place. As Graham (2017) identifies, a restaurant service space traditionally hides the chef as a production craftsman from the customer, it is a stark contrast to the contemporary open restaurant where the kitchen is designed to be openly viewed from the customer service area, thus adding excitement and value to the customer experience (Frable, 1998). This is reiterated by Strong (2002), who describes the dining room and restaurant as being spatially designed with elaborate and ornate sideboards and separate doors for the servant (waiter) and guest. Professional cooking was seen as a craftsman trade for the wealthy to benefit from, with the practicalities of the production process not being something the customer wanted to view, explore, or understand. The finished item was the narcissism of style, with food displayed on elaborate buffets, often with pillars and carvings to emanate the Baroque style of the times. The age of railway travel brought the grand railway termini, and with this came opulence in hotel building, with the central aim of publicizing the grandeur of the Victorian railway companies (Bradley, 2007). Such hotel and restaurant designs incorporated the hiding away of the kitchen, reinforcing the segregation of working class production from upper class consumption, and reflecting the period of the grand country estate (Hembry, 1997; May 1998). Grand dining palaces required professional chefs with the skills to lead large kitchen brigades and the creativity to produce new and innovative dishes. The newly emerged hotel restaurant industrial cooking emanated the French style, adopting the technological developments of the time, and cast the chef in the grand hotel restaurant as a craftsman. To fill the gap in the "*hard skills*" needed for the new cooking techniques, the recruiting of French chefs became the norm. Alternatively, chefs who had the experience of having worked in France on their curriculum vitae were employed (Mennell, 1996). Demand in Great Britain for this French haute-cuisine was epitomized at the Reform Club, London, a kitchen which was led by Alexis Soyer (Brandon, 2009; Cowen, 2010), a French celebrity chef of the time, who was later eclipsed by Escoffier (1846–1935), who was perhaps the best known early celebrity chef, Escoffier reduced the visible Baroque format of the chef's work by casting aside the traditional grand style of cooking and the ornamented displays introduced by Careme. Who believed in the use of architectural pieces of food in the restaurant, creating "*waves of the sea, waterfalls and rivers with a photographic concern for instantaneity"* (Weiss, 1998, p. 63). It was during this period that the rich country house estate, the cultural capital owner of dining etiquette, where the cook was below stairs and the butler acted as the dispenser of the food on offer (Powell, 2011), began to lose its power as employment in the grand estate houses shrank. The growing middle classes, the brokers, merchants, civil servants, industrialist, and bankers, with their increased wealth, demanded superior hospitality and fine restaurants that emanated the style of the grand estate, with the food being served by waiting staff and the chef/cook being unseen (Short, 2007).

It was during this period that Escoffier applied his past knowledge and experience to radically challenge and change the manner in which the closed kitchens were operating, recognizing that sections within the kitchen were often replicating work processes. He instilled greater professionalism through the banning of alcohol while at work and introduced a barley drink to be made available to all chefs in the kitchen to quench their thirst in the heat, claiming that it *"was to be healthy and restorative"* (Rossant, 2004, p. 79). Escoffier contended that drunkenness led to bad language, and he did not allow vulgar street language in the kitchen, nor did he permit the kitchen staff to treat the younger members of the kitchen brigade brutally. Escoffier's legacy of traditional French cuisine remained unchanged well into the 1970s. His writings articulated the strict discipline and severity of his French cuisine training and its ongoing usage. The professional kitchen was a microcosm of industrialization, a small world, devoid of the customer, with work rules and norms devised and reiterated by the chefs working within them. This did not wholly reflect the *"industrial values of mechanization, standardisation, and time thrift for food processing"* (Fantasia, 2010, p. 34) as practiced in the factories of Britain but rather the organization of *"male artisans distinct from the female purveyors of domestic cuisine"* (p. 34), thus perpetuating the idea that the professional skilled craft job of cooking was a masculine occupation. Escoffier wrote in his diaries that the public had little regard for, or understanding of, the work of the chef. High society still saw themselves and the chefs as being in the master and servant role, with the lavish surroundings and the maître d'hôtel (head waiter) being the center stage of the restaurant and the chef being hidden behind the scenes. However, some did challenge Escoffier's orthodox thinking based on his complex style of cookery, with its use of heavy sauces and lavish presentation on service flats, culminating in food that looked appealing but often lacked flavor. This cookery style was challenged because of the development of nouvelle cuisine, an approach to cooking in the 1970's that emphasizes natural ingredients and the delicate flavors of the food, which is presented on the plate with symmetry and flair. This food movement was led by the progenies of Fernand Point (1897–1955). At this time, *"Careme`s la cuisine modern had become la cuisine classic in light of nouvelle cuisine and the disciples of Paul Bocuse, Pierre Troisgos and Michel Guerard"* (Trubek, 2000, p. 13). As food production systems changed to accommodate nouvelle cuisine, the chef was thrust into the limelight, and with the emergence of the chef as a restaurant celebrity, the role of the waiter diminished.

19.3 The changing orientation of the restaurant

The traditional restaurant had theatrical décor, incorporating marble, mosaics, grand staircases, mirrors, chandeliers, and silk wall coverings, creating opulent and decadent surroundings as an escape from the reality of the home. At this time, it was felt that observing the practice of cooking would ruin the experience for the customer. Following the traditions of industrialization, the kitchen, (the work place), was a dirty world of masculine production, and the chef, (the worker), was to remain unseen a world recognized by Fine (1996) in his ethnographical study of chefs in the United States of America. The total dining experience depended on the quality and standards of the fixtures and fittings and the staff service and interaction (Wood, 2000; Gillespie, 2001). If food presentation and cooking was to be viewed, it was via the layout of the food on the plate brought from the kitchen to the guest by the service staff, either on grand service flats to be spoon and forked (silver service) or on the elaborate buffet table. The culinary kitchen was still dominated by French cuisine, and those who worked in the kitchen used techniques and cooking terminology that harked back to the era of haute cuisine because many still regarded France as the epicenter of professional cookery. As Grimod (1802 cited in Schehr and Weiss (2001), p. 62) found:

> "Although French cuisine is without contest the best in the world, we think it could be enriched with a great deal of foreign dishes and appropriate them while perfecting them ... Similarly, if France has become the supreme arbiter in the art of taste, it is greatly due to the care it has taken to reject foreign discovery."

The chefs who worked in a traditional closed kitchen were so alienated from the customers, never seeing their reactions to, or an appreciation of the dishes created in the kitchen created a certain apathy. This apathy was reinforced by the old traditions of haute cuisine, which involved the chef being shut away and producing dishes in the way they chose, regardless of the needs and desires of the customer. However, the public was now demanding a different kind of cuisine that mirrored new cooking techniques and practices (Kelly, 2003). This lighter cuisine attracted the title of new cooking or "*nouvelle cuisine*," and with it came a restaurant movement centered on smaller and lighter portions arranged on the plate by the chef (Ladenis, 1988) and delivered by waiting staff to the customer.

19.4 The restaurant kitchen in the service economy

The chef orchestrated the presentation of the food served to the customer and due to being central to the process began to influence the orientation of the food service style in the restaurant (Lane, 2014). Ross (1992, p. 100) refers to:

> "...a rich seam of genuinely food-orientated innovation, springing partly from nouvelle cuisine, which returned structure and composition to restaurant food after the shapeless Mediterranean stews of the 1960s, and partly by the new world possibilities that ethnic cuisines brought with them."

A repercussion of nouvelle cuisine, which placed the chef central to the service through the plating of the food, was the deskilling of the waiter as their job now merely involved carrying the completed dish out to the customer. The philosophy of "*a la russe*" silver service, which had been lost, was that the chef was regarded as a production element in relation to food service. The service from the silver flat, cutting and carving at the table, offered a level of theater and gave the customers the opportunity to interact and discuss the food (Cousins et al., 2014). Food service showmanship was a key element of the restaurant theater, and this was forgotten in the era of nouvelle cuisine in the 1970s. As Sloan (2004, p. 72) states, "*...a 100 and 50 years after Brilliant-Savarin the restaurant is still theater,...*" and as such it required a new vehicle of delivery. The food and the decor became the key attributes that the restaurateur now focused upon. The showmanship of the waiter and the entertainment offered by the head waiter had been a key aspect of the customer's dining experience. The theatrical aspect of dining was resigned to history as the traditions of food service were slowly removed, the essence of the restaurant being forgotten. The dining experience was reduced to a sterile focus on the gastronomic features of the food, with nouvelle cuisine being linked to the plated food arrangements in Japanese cookery, both serving healthy food in small quantities that was exquisitely arranged (Fuller, 1992). So began a battle between the waiters and chefs, with the chefs arguing that the traditional silver service and gueridon service were slow in serving the food to the customer and that the food thus often arrived cold (Graham, 2001). Furthermore, chefs claimed that the manner in which the waiters presented and served the food to the customer did not match the skilled manner in which the chefs had prepared the dish. The nouvelle cuisine food movement was the catalyst for the erosion of the waiting profession and customer service interaction, and with this the chef was beginning to be launched into the restaurant (Graham, 2006; Graham and Dunning, 2011).

19.5 The open kitchen and the questioning of closed kitchen work

Plated food has become the norm for food service delivery, deskilling the waiter's role and thus reducing pay and career prospects (Fuller, 1992). The structural change to the server's job has led to a far greater usage of part time staff, offering the restaurant operation flexibility, reduced server hours, and consequentially greater productivity (Wood, 2000). Although service staff voiced complaint, they were largely ignored as the chef patron and the celebrity chef embraced the concept of food being delivered from the kitchen as a composition on the plate, extracting the server wherever possible.

However, high-profile headwaiters (maître d'hôtel), such as Marjan Lensnik from Claridges, London, did voice concerns in the Caterer and Hotelkeeper (July 1, 1988) cited in (Fuller, 1992, p. 8):

> "He welcomed a further new style, `cuisine moderne` with adaptations that do not detract from the waiters' skill. Commenting on lovely porcelain oval dishes with everything arranged on it beautifully served on the plate by the waiter .... No it won't look such a pretty picture but this is a restaurant not an art gallery."

Marjan Lensnik recognized that the waiters' contribution to the dining experience was the communication between the waiter and customer and the visual showmanship, which together created an experience that was memorable because of its uniqueness. Not until the maturity of nouvelle cuisine plated food would the restaurateur begin to recognize and appreciate that both the food delivery service method and the service staff interaction played a key role in the dining experience and contributed to customer satisfaction (Hansen et al., 2005). The pleasure associated with the dining experience was not solely dependent on the food being produced, but rather on the amalgamation of the various tangible and intangible factors (Wood, 2000) that created the "*servicescape*" (Wakefield and Blodgett, 1996; Lin, 2004), which chef practitioners had overlooked. The servers created a level of excitement in the restaurant (Mars and Nicod, 1984) as they engaged in a performance that was as central to the restaurant visit as the food (Graham, 2001). A whole generation of seminal restaurant research (Whyte, 1947; Mars and Nicod, 1984) on the importance of the server was forgotten as the chef took center stage and the kitchen was omitted from the performance.

However, some chef patrons and restaurateurs failed to fill the interaction void that plated food service brought and continued to produce food in the traditional closed kitchen in the manner Escoffier had advocated. The social construction of the restaurant remained one of hidden food production, out of sight from the customer. The impression of high-quality food and the chef preparing food in a high-pressure environment could be maintained and fueled by the masculinity of craft employment. Closed kitchens facilitated the concealment of the chef and prevented chef interaction with the customer, with the chefs constantly venting their anger or frustration on the servers while they queued at the hotplate (Mars and Nicod, 1984; Fine, 1996) and the use of "*dirty work*" practices. The continued use of the closed kitchen enabled the waiter and chef aggressive relationships to be confined to the back-of-house environment. The decoupling of the chef from the customer was about to be challenged as the restaurant service evolved into one homogenized service involving artistically plated food emerging from the kitchen. In practice, as a result of nouvelle cuisine food being plated in the kitchen, chefs had inadvertently reorientated themselves closer to the customer. The waiter was, in effect, no longer acting as the sole intermediary food production service worker; the chef had become engaged in food service work through creating the meal on the plate and being put on show when placed in the open kitchen.

19.5.1 The open kitchen or theater kitchen

The open kitchen concept is not a wholly new idea; Italian pizzerias have used this format for many years, and it is a key design feature of the Japanese Teppanyaki kitchen (Fang et al., 2013; Norii, 2015). It is in the evolution of the traditional British and French mid- and upscale restaurant that the greatest growth has been seen in the open kitchen operational style. In the traditional restaurant, it was historically traditional for the chef to enter into the dining area to carve the meat and serve it at the buffet table. As Frable (1998) (p. 5) explains:

> "Exposing food preparation to diners has remained popular for more than fifteen years because it creates culinary and visual excitement for the guest and reduces wasted back-of-the-house space because the pickup area is shared with the dining room circulation."

The development of the open kitchen offers the chef an opportunity to show off his production skill, the freshness of the ingredients, and the cleanliness of the kitchen (Graham, 2006a; Snaith and Pitham, 2006). The open kitchen provides entertainment for the diners through suspense and action, for example, from flashes of fire, the sounds of cooking food, and the chef's chopping skills. "*Now the open kitchen has evolved into entertainment, a frenzy of excitement just a few feet from the table*" (Petrowski, 1999, p. 171). However, the open kitchen is not to everyone's taste. Petrowski (1999) goes on to say that the open kitchen has as much atmosphere as a hospital emergency room. They are loud, bright, and full of odors that can be good or bad, although some of this can be overcome with sliding glass doors to keep the kitchen noise to a minimum. Sheridan (2001, p. 85) contends that:

> "Open kitchens are a big source of noise. Unless the dining room space is large, keep the kitchen behind closed door, or counsel the cooks to work and speak as quietly as possible."

Open working environments have created a fundamental change in working practice for the chef. For some chefs, working in front of the customer is a way of promoting their talent and skills, and they can get a "*buzz*" from doing it. For others, it can be their worst nightmare as some chefs are customer averse and do not even want to be seen (Graham, 2006). Furthermore, chefs are often renowned for their boisterous and aggressive behavior in the kitchen, in keeping with the masculine identity of the role (Meloury and Signal, 2014) promoted by the celebrity chefs, for whom working with food is a creative art which can only occur when high emotions are involved and displayed (Dorenburg and Page, 1996). Chefs work in an aggressive manner and have a reputation for shouting and swearing, with this persona reinforced through the high-profile celebrity chef. "*Cooks fight with servers. The tales of conflict are myriad, but the causes are common – lack of communication and an ability to empathize*" (Lorenzini and Johnson, 1995, p. 148).

19.5.2 The experience economy and the development of the contemporary open kitchen

A key reason theater cooking was introduced was to meet the growing demand from customers for appetizing presentation of meals alongside restaurant entertainment (Graham, 2001). Mintel (2006), in the study Eating Out: 10 Year Trends, found that the media, with its cooking and food provenance programs and celebrity chefs, has increased the consumers' interest in food. In turn, with a greater number of restaurants designed with open kitchens and the reintroduction of theater into the dining experience, Graham (2006a) argues we have returned to a "*new*" gueridon service (Cousins et al., 2014). Throughout the 1980 and 1990s, with the growth of the celebrity chef, personality, and entertainment, the location of the kitchen has become increasingly important in restaurant design (Pratten and O'Leary, 2007). Rohatsch et al. (2007, p. 135) found that since the early 1990s there has been a growing trend to shift kitchen functions closer to the customers and celebrate food preparation in front of them:

> "Within the restaurant, the design influence is generally based on the market, the demographic of the target consumers, and their expectations along with a belief…

> … that dramatizing the service performance is the best way to gain sustainable competitive advantage."
> ***Morgan et al. (2008), p. 111.***

Bruni (2005) notes that one chef believes that the open kitchen has been developed without considering how overimposing it is, arguing that the concept has been overdone, with the driving force being the aim of merely entertaining the customer. Research undertaken by

Chow et al. (2009, p. 101) identified customers who believed that the mystique of cookery was lost with an open kitchen, saying *"you don't want to know how the magician does his trick, you just want to be entertained."* Other respondents raised fears about bad language and communication issues between front and back of house. Even with such negative findings, the overall conclusion is that the benefits of the open kitchen far outweigh the disadvantages cited in the negative responses. However, Baraban and Durocher (2010, p. 12) do offer a word of caution, one that emphasizes the true reason for the success of a restaurant. They say that *"whilst the food may not initially be what brings the customer to a restaurant,"* it is the food that will ensure repeat business. Franck (2002, p. 83) confirms this, saying *"no restaurant could lure customers with good dishes alone these days, but it is the food that keeps them coming back."*

The open kitchen has created an experience as part of the customer journey, one that provides lasting memories of something special, with the aim of generating new and retaining old business (Walter et al., 2012). In effect, the open or theater kitchen is being used to enhance the dining experience by bringing the cooking to the front of house and introducing theatrical elements into the cooking process. The chef becomes part of the atmosphere as one of the new service staff and a central part of the entertainment on offer in the restaurant environment adding to the hospitality experience (Lugosi, 2008; 2014). The idea of theater cooking is not new as food service has always *"contained elements of theatrical performance"* (Morgan et al., 2008, p. 112) and *"food paraded"* through the restaurant has been acceptable practice (Alston, 2015, p. 50). For example, gueridon and silver service styles, where food can be, and still is, carved and flambéed in front of the guest, using finishing processes of food preparation and cooking (Graham, 2006a). In food service dining, the quality and the prices of food have been the decisive factors in determining which restaurants have prospered and which have not (Kotler, 1973; Wood, 2000).

The visual and verbal engagement of the chef with the customer adds to the cacophony of stimulating sounds, sights, and smells in such an environment, with the open kitchen chef working in view of the customer taking great care with the visual and audio aspects of cooking (Franck, 2002), thus communicating the restaurant image and values to the guest Bruner (1990). As Katsigris and Thomas (2009, p. 223) note, *"open kitchen, noise levels have become part of the atmosphere to the extent that the open kitchen and associated noise is part of the design and concept."* Anderson and Mossberg (2004), Hansen et al. (2005), Heide and Gronhaug (2006) Morgan et al. (2008), and Katsigris and Thomas (2009) suggest that the tangibles of theater cooking can add to the theatrical nature of the experience, entertaining customers and easing their fears about hygiene. Jennings (2011) believes that as open kitchens have become so common, *"they are no longer about the `wow factor` but rather used as a vehicle to communicate business core values"* (*p. 32*), placing the chef as an actor on the stage to communicate with the customer.

19.6 Emotional labor

Frontline workers are required to display certain types of emotions, such as friendliness, warmth, politeness, confidence, enthusiasm, or cheerfulness, while interacting with the customer (Soares, 2003) and change or control their emotions when interacting with the guest in exchange for a monetary reward (Grandey et al., 2013). This is a new representation which the chef in the open kitchen will now need to display. An emotional laborer can now be described as someone who interacts with a customer and has to use their emotional skills to provide a positive interaction between themselves and the customer in accordance with the management's job role performance requirements (Cole et al., 1994).

To be classified as an emotional laborer, Haynes and Kleiner (2001) contend that workers must be in an occupation that possess three characteristic: (i) it requires the employee to make face to face or voice to voice contact with the public, (ii) it requires the employee to produce an emotional state in the customer, and (iii) it allows the employer the opportunity to exercise a degree of control over the emotional activities of employees, either through training or supervision. Under these circumstances, as Grandey and Gabriel (2015, p. 324) emphasize, the employee's emotional labor *"is sold for a wage and therefore has exchange value."* The open kitchen chef meets these criteria as an employee who now has to perform for the customer's entertainment in line with the organization's expectations of the chef while on show in the open kitchen. Barsade (2002, p. 646) describes emotions as *"intense but relatively short term reactions to specific stimulus,"* while Vincent (2011, p. 1369) claims that emotions at work can fall into three categories, which are *"feelings we cannot control; feelings that result from our emotional ability to evoke, manipulate and suppress our feelings; and feelings that are affected by morals, values, attitudes and dispositions."* Emotional laborers engage in communication that results from either the expression of felt emotions or a decision to disguise or manage them (Fiebig and Kramer, 1998) so that the customer has a positive engagement with the server. It is deemed by many employers to increase customer satisfaction (the overall feeling of contentment with the interaction) and thus improve revenue and sales, resulting in increased repeat business and financial success (Rafaeli and Sutton, 1987, 1991; Lee and Ok, 2014).

These factors underline how and why employees' emotional displays are twofold: (i) the employees who participate in customer contact are the interface between the guest and the organization and therefore represent the face of the business. Negative interactions from the employee will leave a poor impression of the company; (ii) due to the *"unique"* attributes concerning the nature of the restaurant service engagement, it is necessary that the industry establishes policies, display rules, and procedures to govern the standardization of their product and service (Wong and Mei, 1999; Diefendorff et al., 2005). The behavior of the service deliverer, how they manage the interaction (Gulati, 2007), and how the relationship develops (Parvatiyar and Sheth, 2001; Koopmann-Holm, 2011) strongly determines the customer's perception of the product and service quality (Johnson and Grayson, 2005). Bowen and Schneider (1988) and Brown et al. (1991) suggest that the concept of emotional labor has a particular relevance to service encounters because frontline service personnel are naturally situated at the organization–customer interface and therefore represent the face of the organization to the customer. This performance put on for the customer requires employees *"to produce an emotional state"* (Wharton, 2009, p. 157), suggesting that there are two ways that employees may engage in emotional labor with customers, which are *"surface acting"* and *"deep acting,"* through which *"acting occurs when we actually deceive ourselves as much as we deceive others"* (Taylor and Tyler, 2000, p. 77). Hochschild (1983) believes that while both forms are internally false, the motives behind them differ. Employers attempting to control workers interaction (Belanger and Edwards, 2013) can impact on the employees' sense of self, thus creating threats to their identity (Wharton, 2009). When employees smile and convey friendliness, their apparent emotions can impact on the emotions of the customer, who may associate this with good service. However, this works both ways as a customer's negative emotions can affect the employee's emotions.

19.6.1 Different emotional acting in closed and open environments

According to Ashforth and Humphrey (1993, p. 98), *"emotions are an integral and inseparable part of everyday organisational life."* Putnam and Mumby (1993, p. 39) point out that in organizations, emotions are *"consistently devalued and marginalized while rationality is privileged as an ideal for effective organisational life."* Only a limited range of emotional expressions tend to be socially acceptable in the workplace. An emotional outburst out of sight of the customer might be socially acceptable, but it would be frowned upon in public. Within the open world, displays of negative emotion, such as fear, anxiety, and anger, tend to be unacceptable, as do expressions of intense emotion (Ashforth and Humphrey, 1993). These norms of acceptability have directly influenced the craft worker in open environments, who while on view must suppress emotions that would normally show when under pressure. Working practices have had to become more professional in terms of language and communication, production skills, personal hygiene, and personal appearance (*esthetic labor*) (Warhurst and Nickson, 2001).

19.6.2 The current emotional labor literature focus

Customer satisfaction can be described as a customer's overall evaluation of their purchase (Cronin and Taylor, 1992). According to Korczynski (2003, p. 57), customers *"are increasingly seeking service quality."* Delivering service effectively to satisfy customers requires a humanistic intervention and a display of positive emotions by staff in many service occupations (Tsai, 2001). Research undertaken by Bolton and Boyd (2003) found that service organizations require their employees *"to do more than simple surface acting, they need to invest in the performance"* (p. 300). Hochschild (2003) agrees that *"surface acting"* is not sufficient in the contemporary service interaction and authenticity needs to be provided through *"deep acting"* orientations.

Not all customers are positive and engaging toward the server in service organizations, and there is little doubt that service providers have to deal with rude and demanding customers, for whom the scripted engagement process may be inadequate (Bolton and Houlihan, 2005). A customer's mood can affect how they respond to a particular experience, and people's moods can be ameliorated by both social and environmental conditions. It is for the service worker, and in this case the new craft worker, to identify the various customer moods and customer types to engage with the guest accordingly. Barsade (2002) reinforces this when he contends that *customer service jobs may be very stressful, not only because of overt conflict but because of the continuous low-grade effect of catching customers' negative moods.* Bolton and Houlihan (2005) note that customers are like customer-service workers in that they are many-sided, complex, and sophisticated actors who may not always behave socially as they do when they interact with the service provider, believing that the current generation of consumers have much higher expectations than previous generations. While it can be argued that Hochschild (1983) is correct in her argument that the exchange is unequal,

Bolton and Houlihan (2005) findings suggest that customer sovereignty may be mythical as neither producers nor consumers believe it to be true. While consumers can be demanding, this is not due to a sense of divine right or to demean the service worker. The craft service work over the last two decades has had far greater recognition through the media. The greater interest in food and the growth in leisure or hobby cooking has created a mystique around the skill, which when coupled with the interest in the work and craft that the chef deploys has led to a respect for the chef and customers viewing them as being at least their equals (Graham, 2006).

Brook (2009a; 2009b) states that Hochschild has been criticized as customer service interactions are in fact double edged and have the potential to both satisfy and distress the worker. As a school of thought, this partly rejects the notion that having one's emotions commoditized is alienating and uncomfortable for the worker. Contradicting Hochschild (1983) in her original research, which found that cabin crew who put on a service act unconditionally altered themselves for the role, Bolton and Boyd (2003), Sheehan (2012) and Tungtakanpoung and Wyatt (2013) in their studies of cabin crew found that rather than employing *"deep acting"* the workers gave *"empty performances"* to satisfy the targets set by the company without ever *"buying-in"* and that the employees did not need to love or believe in the product to sell it effectively.

Lopez (2010) further extended the work of Jenkins et al. (2010) by putting forward the concept of the triangle of power, which encompasses the worker, manager and customer. Lopez argues that even if companies control their employee's emotions when interacting with customers, some workers still choose to go that extra mile during the service encounter. He asks whether employees go the additional mile for purposes of self-satisfaction or to impress management and whether they make the extra effort because of their personality, upbringing or even education. Jenkins et al. (2010) believe that working with customers is not only about feelings, but also about the physical interaction between humans as social beings.

Further to this, Bolton (2009) argues that although company guidelines are used in the delivery of the service product, within this delivery process the individual server will also be inclined to add their own individual delivery perspective, insisting that skill can be objectively measured across two dimensions, *"task and discretionary content"* (Bolton, 2004, *p. 26),* and that if a job scores highly on both of these measures, it can be classed as skilled emotional work. This concept of work was adapted from Litter (1982, p. 8) and further developed by Bolton as The Dimensions of Emotional Work, from which a framework was developed, as illustrated in Figs. 19.1.

According to Bolton, box A (Standardised Services) includes the *"emotional proletariat"*, who are described as the workers undertaking *"mundane, routine, low skilled work and most importantly are tightly controlled* via *scripts" (2004, p. 26).* Within the hospitality industry, these can be identified as the fast food and casual dining restaurant staff, who can be grouped together and described as adhering to the *"have a nice day culture, where niceness is routinely delivered"*. As such, the level of skill that they have in customer interaction is one that is not valued or classed as a skill. The work is so highly scripted that the service encounter cannot always deal effectively with customer uncertainty as it gives the service worker little flexibility. Box C (Specialist Services) comprises, *"call center or the retail and catering style market*

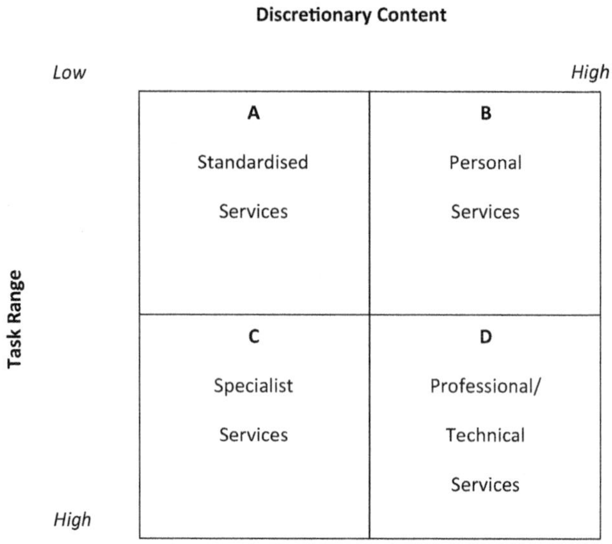

FIGURE 19.1 The dimensions of emotional work. *Source: Bolton, S.C., 2004. Conceptual confusions: emotional work as skilled work. In: Warhurst, C., Grugulis, I., Keep, E. (Eds.), The Skills That Matter. Palgrave MacMillan, Basingstoke, Hampshire, UK, pp. 19–38.*

who may be placed higher in the hierarchy" (p. 26). The work that they undertake involves high levels of technical or specialist knowledge along with a greater component of discretionary content than box A work. Bolton states that such workers have *"limited autonomy as direction is only granted to workers with the right attitude"* and goes on to contend that they can be relied on to express real feelings in the interests of creating the right emotional climate to improve customer service (Bolton, 2004). It is apparent that the skilled craft work of the open kitchen chef does not fit into this box as their work comprises a high task range and a relatively high discretionary content, and as such they generally fit better into box D (Professional/Technical Services). Bolton argues that box B (Personal Services) and D represent those workers with high levels of discretion in the customer interaction. Box B represents personal carers, nursing auxiliaries, child carers, security and distribution services. Those in box D are allowed self-determined interactions, which are determined by the *"professional ethos"* (Bolton, 2004, p. 28) of the job. This is associated with their widely recognised qualifications, which indicate the specialist knowledge in the area in which they engage with the customer. Bolton states that this group of workers is comprised of the medical profession, legal services, education and social services. Arguably, the skilled craft worker as the chef gravitates toward box D, which sets the chef in particular apart from the restaurant server in box C, but it does not wholly identify the reality that a chef is a skilled craftsman with high levels of autonomy and discretionary content.

Building on Bolton's work, Rose and Wright (2005), Grugulis (2007) and Payne (2009) have all noted that referring to emotional Labor as a skill would be difficult to achieve. Payne (2009) argues that there is no denying that many jobs entail emotional Labor and the concept is a fundamental feature of all jobs naturally learnt through one's cultural assimilation. Payne further suggests that skill remains in the eye of the beholder and that by labeling emotional work as skill and paying workers based on their levels of such a skill, there would be no shortage of its applicability to the *"skilled"* emotion work in low end service jobs. It is a challenge to orientate the chef as a skilled craft worker into the categorizations that Bolton (2004, p. 26) puts forward. The chef can arguably be suspended between the personal service or expert service worker and the professional/technical worker. The identification has not been discussed in the literature, and to date, the craftsman has not been identified in the new service role paradigm of customer contact and never defined within these parameters. As a group of employers they have, through operational work design, undertaken a transformation from closed to open work and have only now through this chapter discussion and the authors published work (Graham, 2015, 2017) been identified as emotional Laborers and researched in-line with the integrated three component model (Grandey and Gabriel, 2015).

The existing emotional labor employee types previously identified and researched have not had to undergo the fundamental change in their work form that the chef has, from being hidden in employment in the old world of work to now being exposed to customer contact. The customer is now able to observe these employees as *"back office"* workers and see through a *"window"* into a hidden world, whereas the employee's perception is of being a front office worker. The employee is actually positioned as an *"intermediary"* service worker, on the *"back stage"* for the customer and the *"front stage"* for the employee with the expected requirement to engage in customer interaction. The nature of the craft element and the production interface with the customer actually sets this group of workers apart as a new category of emotional laborer. The debate within the arena of this new transformational worker group type, who can be identified as those whose role is significantly low routine with high levels of craft engagement to create a unique service tangible product (Graham, 2006). In part because this group of employees have traditionally been seen as back-of-house employees and not as front office interactive service workers, they have been overlooked in the current emotional Labor literature and not Identified as a research group who have worked in a private space having transferred into a public viewed space. The employee holds a level of discretionary content and task range but the Laborer is skilled in a craft which is separate from the technical or professional services. This group of workers have facilitated Graham's (2015; 2017) research into a new worker group of emotional Labor in order to understand the transition of the chef and the going *"round these roadblocks"* that Grandey and Gabriel identify (2015, p. 342) and so *"take the road less traveled and drive emotional labor forward" (2015, p. 342)* which is at the "crossroads." Furthermore, if one accepts Litter's (1982) claim that the focus of emotional work is simultaneous production and consumption, it becomes clear that the open kitchen chef is a new exemplification of emotional employment, one that the hospitality and generic business literature has to-date omitted.

19.6.3 Esthetic Labor

Esthetic labor is the concept that every frontline service job requires the individual member of staff to *"look good and sound right"* (Warhurst and Nickson, 2001, p. 1; Karlsson, 2011, p. 51) and fit with the organizational values and image. In essence, the worker who is employed in a customer facing role is required to embody the customer's preconception of the worker type and be able to engage with the customer in a manner that they would expect him or her to (Warhurst et al., 2000; Pettinger, 2004). The employee is required to be well groomed, wear a uniform and communicate in a manner that the customer is able to relate to and hence enter into a dialogue with at an identifiable level (Warhurst and Nickson, 2005). The esthetic laboring literature argues that the service worker (p. 4) is the *"mobilization, development and commodification of embodied dispositions"* in that the worker is *"selling"* ones *"class"* or *"taste"* for the corporate good. The service worker is employed by the organization for the way that they sound and the manner in which they effectively communicate (Butler, 2014) along with their physical attributes (Harvey et al., 2014). Examples of this are given by Karlsson (2011), who discusses the manner in which staff employed in an up market retail shop use words such as *"exquisite and luxurious"* and rather prosaic terms such as *"nice and lovely" (p. 54), thus* identifying with the organization (McIntyre, 2014).

Telephone call center workers require language which is complementary to the customer class level that they serve. Clarke (2014) found that for some this creates a barrier to employment, for example in the offshore call centers (Taylor, 2005; Derry et al., 2013). Warhurst et al. (2000) discuss the existence of a *"style labor"* market, which is comprised of the designer retailers, boutique hotels, style bars, cafes and restaurants, where staff are employed to fit the brand. Karlson (2011) found that in some style retail organizations this is taken further; staff are not supplied with a uniform but instead the clothes that they wear for work are expected to fit within the image of the outlet or they select discounted clothes from the current range, which reflects the image of the corporate brand. Witz et al. (2003) underpin this by reporting that even staff who wish to cut their hair or dye it in a drastic manner are expected to discuss their fashion image with their manager first. The use of cosmetics while at work and the manner in which they reflect the values and image of the organization is a concern. Makeup and tattoos are related to the social class and gender of the staff, and (Trimming, 2014) organizations are keen to ensure that these are in line with the social expectations of the customer type they serve (Williams and Connell, 2010). It is argued that labor is no longer performing in the *"experience economy"* (Pine and Gilmore, 1999) but rather in the *"esthetic economy"* (Postrel, 2003) in which the *"the employee's look can be as much a part of the atmosphere as the grain of the furniture or the beat of the background music" (p. 127).*

To enable service staff to be effective in the service encounter, they need to be skilled in approved social attributes. Sheane (2011, p. 147) argues that *"emotional Labor and aesthetic Labor are concepts relying on social, presentational rules that are cultural, situational and learned."* Such employment is linked to a service interaction the laborer feels comfortable in and can relate to (Schaubroeck and Jones, 2000). This is enhanced by the worker customizing the interaction to match the level of the individual guest within an organization that allows *"the laborer to shape the service interaction"* (Warhurst and Nickson, 2007, p. 791). Sheane (2011) puts forward that the emotional and esthetic literature has for too long been focused on the employer–worker relationship and the emphasis should now be on the worker–customer interaction. As such, *"communication, comes to the forefront and this makes room for an autonomous subject who makes contextual aesthetic and emotional choices based on temporal and situation conditions" (p. 153).* This reiterates Goffman's (1959) claim in his study of hairdressers regarding the significance of self-presentation and the importance of emotions, esthetics, and body techniques, as well as the employer's appreciation of the value of the staff member who is able to interact in a service encounter that fits with the values of the organization. The chef as a traditional craft worker has normally been viewed as a *"back office"* employee, who is employed for skill levels in the production area and not necessarily within a customer service engagement role.

Closed hidden craft employment is changing in the service economy, and the level of emotional and esthetic labor required has increased according to the design of the experience or the esthetic craft environment, and this has been overlooked by the literature. The traditional orientation of the chef employed in a masculine world was one which required a craft skill level for the task, with no regard for customer interaction and with-it limited levels of esthetic or emotional Laboring, which is no longer appropriate. Goffman (1959) asserts that the apprentice or working class employee is left to their own devises in the acquisition of the social and style capital and unless they are able to access the style capital they are unable to access the high-end and better paid jobs. He contends that lower class craft employees have to learn on the job to access the mobility of employment and move up in status through the work outlets (fast food; casual dining; fine dining). Warhurst and Nickson (2007) argue that unless this can be achieved they are unable to swell the ranks of a new labor aristocracy as they are missing the *"soft skills"* that connect them to the customer from higher social class and never take advantage of the democratization between the worker and the customer.

For the chef, production skill is the key to employment in this job role while social skills are an additional enhancement useful in customer engagement. Less training is required in the *"soft skills,"* as these skills are already socially instilled. This has led to a greater interest in service staff being recruited from a middle-class background, challenging the employment position of the working-class craftsman. Until the development of the open production area, this was never an employment issue for the craftsman, a worker who was traditionally drawn from the working and lower middle classes, where masculinization of the job identity prevailed, and who was traditionally hidden from view. It can be suggested that the transformation to the new open production area will begin to impact on this group of workers as emotional and esthetic labor skills become a work requirement. In the craft service encounter, this offers the worker in higher status style operations who exhibit a high level of *"soft skill,"* the opportunity to assert their knowledge capital and close the gap between the worker and the customer and become the new *"labor aristocracy"* (Warhurst and Nickson, 2007, p. 739).

The craftsman as the chef has not traditionally been subject to this gentrification, in the main due to the closed nature of craft work and societal expectations that closed craft work is still the main stay of the working classes (McIvor, 2013). The new craft service aristocracy is a product of the mechanism of socialization and interaction at work in the open production environment. The craft worker is employed for his skill and handicraft, with customer interaction and social skills not being salient selection criteria. This is reiterated through Goffman's (1959) work, which claimed that staff develop the social skills required for the job in a learning process through interaction with and having to face middle class customers. Via craft workers in the new open production world experiencing exposure to the customer and engaging with them, they are learning the social skills required for their new role. Warhurst and Nickson (2007, p. 794) conclude that there is today a linkage between occupational change and class, just as there was in the past during the manufacturing era (Goldthorpe et al., 1969). However, a new relationship now exists, with the production craft worker having to engage with the customer in the new service or experience economy as an interactive service worker, one who has not yet been explored through research or discussed in the literature.

19.6.4 The changing masculine identity of the chef through public display

The closed kitchen pertains to the world of the industrial era when production was largely deemed to be a male occupation (Green and Owen, 1998) and a masculine activity (Cook, 1996). The chefs demonstrated this "*macho*" culture in the closed kitchen through laddish activities such as throwing food items and playing practical jokes, enhancing the masculine identity of the trade. Being on the receiving end of the jokes as such appears to have formed part of a rite of passage ritual for the new recruit. Alexander et al. (2012) refer to this in the title of the article "*He just didn't seem to understand the banter; bullying or simply establishing social cohesion?*" and as Bloisi and Hoel (2008) discuss in their review of the literature that it is the socialization process that creates the "*hardness*" (p. 649) required to be able to operate effectively and be successful in a commercial kitchen. It seems this macho behavior was being used to bolster the prevailing belief that household cooking was women's work and that the closed restaurant kitchen was no place for females. Laddish behavior and male-orientated work antics and games appear to have been central in creating and reinforcing the rules (Connell, 1995, 2000) of the masculine culture that prevailed in the closed kitchen (Blauner, 1964) and as such an occupational community (Hill, 1976), a male dominance of space which Robinson and Beesly (2010; Robinson & Barron, Developing a framework for understanding the impact of deskilling and standardisation on the turnover and attrition of chefs, 2007) identify as man's work. These masculine games and rules excluded females, and those women that did enter into the closed world of the kitchen were expected to adopt similar male values and join in (Segal, 1997). The levels of masculinity varied between kitchens, a feature which concurred with the research by Collinson and Hearn (1996) and Watson (2000) on how males reinforce their male identity through differing and increasingly offensive masculine behaviors, leading to the intimidation of women (Sims, 2012), a position which has been an expected part of kitchen culture (Bloisi and Hoel, 2008).

Approaches to work and the entrenchment of these attitudes toward women in other masculine-centered service tasks is discussed by Simonton (1998), who investigated furniture retailing, which is an area where males seem content to take on service roles. Simonton argues that such service roles are attractive to males as they give the employee a relatively high level of power over the customer because of their perceived skills and knowledge. Furthermore, Simonton (1998) contends that males are drawn to such service work as a result of the socially constructed message that furniture is an extension of carpentry, which is perceived as skilled man's work. When aligned with the kitchen environment, this goes some way toward supporting the prevailing idea that an environment that involves "*hard skills*" and dirty work, such as a restaurant "hot" kitchen, must be dominated by males, while one where *"soft skills"* are needed, such as a restaurant dining room, it is more suited to females (Nixon, 2009). This seems to lead to the self-constructing belief that the chef in the kitchen is of far greater importance than the waiting staff in the dining room and that females are somehow not capable of being chefs (Robinson and Barron, 2007; Bloisi and Hoel, 2008; Robinson, 2008; Robinson and Beesley, 2010). Such a view enabled the male respondents to legitimize their feelings of power and superiority over the female chefs and demonstrate their masculinity through aggressive and controlling behavior, whenever possible banishing the women to the cold kitchen or the pastry section, as Bourdain (2000) states and other celebrity chefs' testify in their biographies (Ladenis, 1988; White and Steen, 2006).

The direct engagement of the customer with the chef has removed the chef from the *"back office"* of service into the *"intermediary"* service space (Grayson, 1998) and with it a reorientation of the work place has occurred. This encroachment of the experience economy into the kitchen has transformed the hitherto closed world of the kitchen from a manufacturing or production arena into a space that is now available for consumption in the same manner as other traditional service experiences. Such a changed orientation has altered perceptions of the chef's role in the wider service industry from simply being a production function to now being directly involved in customer interaction (Bolton, 2004). This shift in the position of the chef to that of a directly accountable service worker has thrust the kitchen into direct customer engagement with the customer and welded together the production and service process (Frable, 1998). This reorientation of the kitchen into an element of direct service delivery is because the customer engagement involved created a *"soft skill"* requirement along with the existing *"hard skills"* already associated with restaurant service production. The *"hard skill"* element of the job role remains but with a new requirement to acquire the *"soft skills"* of hiding emotions (Burns, 1997) and masking aggressive thoughts (Korczynski, 2005, 2013), features which Bolton argues are *"the skills that matter"* in the service economy (Bolton, 2004).

The presence of females in the closed kitchen did not lead to a reduction in the level of masculine behavior and indeed that the male chefs merely further asserted their masculinity (Donkin, 2001; Bunting, 2004). This was a challenging situation for most females as male chefs continued to assert their power in the environment (Nixon, 2009). As Sims (2012) discusses, leading female celebrity chefs have had to overcome this macho male challenge to be accepted and succeed in a society which, since the era of the manufacturing economy, has created a service world in which closed and traditional environments are still socially perceived as being dirty, male occupations (Roper, 1994; Roberts, 2012; McIvor, 2013). Kitchen laboring in such traditional harsh environments of *"working in crappy conditions, in spaces with poor kitchen design for long hours, under significant pressures"* (Robinson, 2008, p. 408), a space which Robinson and Barron (2007, p. 915) discuss as being *"both physical and psychologically straining environments."*

The opening up of the kitchen to public scrutiny has been a key factor in the reducing of macho performances as the chef now has to enter a different world of social acceptability and service work (Fillby, 1992) and is expected to interact with customers using a *"softer skill"* delivery (Bolton, 2004). The open kitchen has begun to erode the traditional masculine kitchen behavior; however, they said that their masculinity and dominance is still retained whenever possible through performing antics which were previously so obvious in the closed kitchen in a more subdued way. The masculinity of the traditional kitchen together with the normative male dominance behavior continues to be practiced in a far more subtle and potentially subversive manner, in an attempt to continue to exercise domination in the working environment.

This level of masculinity demonstrated in the kitchen has been reduced because of the direct impact of the external environment and the customer observation of male chefs together with increasing numbers of women being attracted to open kitchen work (People 1st, 2014). This suggests a potential change to the traditional male dominance in the kitchen, with hot steamy kitchens no longer being an exclusively male domain and the heavy lifting of pans no longer being seen as man's work. This notion is now clearly being challenged, perhaps as a consequence of the increased level of direct service contact and the need for traditional male work to employ feminized "*soft skills*" (Drucker, 1994; Ashforth et al., 2007). The juxtaposition of the chef and customer in the new open kitchen development in a direct manner was almost never required in the manufacturing economy era.

19.7 Theoretical considerations debated

This discussion has brought to the fore that the concepts of emotional and esthetic labor are actually linked and intertwined with one another and that both of these conceptual structures are mutually inclusive as discussed by Sheane (2011), Grandey and Gabriel (2015), and Warhurst (2015). In particular "*surface acting*" being attributed to stress and burnout leading to job dissatisfaction and the increased levels of labor turnover. The negative consequences of emotional laboring are mitigated by the antecedents and moderators, which reduce the impact that emotional laboring has on the individual employee. The literature reaffirms the contention that emotional laboring draws heavily on the internal feelings of the worker having to mask how they feel to fit in with the expectation of the interaction, a performance expected by the organization in exchange for a wage. The level of emotional labor that the chef in the open kitchen is required to perform is related to their exposure and interaction with the guests, where for the first time in their employment, they have to now be perceptive toward the customers' needs. The amount of acting that was required and the pressure that this placed on the employee was identified as a key reason why some chefs left the open kitchen and why those that remained had to rapidly develop coping strategies, such as set piece conversations, and draw upon their reservoir of "*hard skills*" and technical knowledge to engage in sustainable customer contact. The chefs' internal feelings had to be masked to present a façade of "*looking good and sounding right*" to meet the requirements of the customer.

19.8 Concluding remarks

This chapter has discussed the changing nature of kitchen work with the opening up of the closed kitchen thrusting the chef into public view. As a world of work that has had a fundamental impact on the chef and with this, the employee becomes a true emotional and esthetic labor type. The impact of such working practice has seen the reduction and removing of the traditional masculinity of the kitchen, a social constructed norm which was a manifestation of the manufacturing world and with it the "feminizing" of the work space. A kitchen space, which now requires the employee to have a greater level of interactive customer service skills, complemented with the traditions of hard cooking skill. For some chefs migrating from the closed to the open kitchen has been a challenge, work practice which has fundamentally shifted, and one that does not necessarily resonate with them. The challenge for employees and educators is to recognize the changing nature of open kitchen work and identify how the individual chef can be trained, developed, and inducted into this new world of work. By effectively doing so, it enables them to remain longer and be more contented in such employment and not to be diverted back into the traditional employment of the closed kitchen and all that it represents. The professional kitchen is witnessing the dawning of a new era in work practice and with it the welcoming of a new generation of female chefs as the old world of masculinity in kitchen work is slowly being dismantled.

References

Alexander, M., MacLaren, A., O'Gorman, K., Tahari, B., 2012. "He just didn't seem to understand the banter"; bullying or simply establishing social cohesion? Tour. Manag. 33 (5), 1245–1255.

Alston, A., 2015. Performing labour in look left, look right's. Above and beyond. J. Appl. Theatre Perform. 20 (1), 50–61.

Anderson, T.D., Mossberg, L., 2004. The dining experience; do restaurants satisfy customer needs? Food Service Technology 9 (3), 171–177.

Ashforth, B.E., Humphrey, R.H., 1993. Emotional labour in the service roles; the influence of identity. Acad. Manag. Rev. 18, 88–115.

Ashforth, B.E., Kreiner, G.E., Clark, M.A., Fugate, M., 2007. Normalizing dirty work: managerial tactics for countering occupational taint. Acad. Manag. 1 (149–174), 149–174.

Baraban, R.S., Durocher, J.F., 2010. Successful Restaurant Design, third ed. John Wiley and Sons, New Jersey.

Barsade, S.G., 2002. The ripple effect; emotional contagion and its influence on group behaviour. Adm. Sci. Q. 47, 644–675.

Belanger, J., Edwards, P., 2013. The nature of front-line service work: distinctive features and continuity in the employment relationship. Work Employ. Soc. 27 (3), 433–450.

Blauner, R., 1964. Alienation and Freedom; the Factory Worker and His Industry. University of Chicago Press, Chicago.

Bloisi, W., Hoel, H., 2008. Abusive work practices and bullying among chefs: a review of the literature. Int. J. Hosp. Manag. 27 (1), 649–656.

Bolton, S.C., 2004. Conceptual confusions: emotional work as skilled work. In: Warhurst, C., Grugulis, I., Keep, E. (Eds.), The Skills that Matter. Palgrave MacMillan, Basingstoke, Hampshire, UK, pp. 19–38.

Bolton, S.C., 2009. Getting to the heart of the emotional labour process: a reply to Brook. Work Employ. Soc. 23 (3), 549–560.

Bolton, S.C., Boyd, C., 2003. Trolley dolly or skilled emotion manager? Moving on from Hochschild's managed heart. Work Employ. Soc. 17 (2), 289–308.

Bolton, S.C., Houlihan, M., 2005. The (mis)representation of customer service. Work Employ. Soc. 19 (4), 685–703.

Bourdain, A., 2000. Kitchen Confidential: Adventures in the Culinary Underbelly. Bloomsbury Publishing, New York.

Bowen, D.E., Schneider, B., 1988. Services marketing and management: implications for organisational Behaviour. Res. Organ. Behav. 10 (1), 43–80.

Bradley, S., 2007. St Pancras Station. Profile Books Limited, London.

Brandon, R., 2009. The People's Chef: The Culinary Revolution of Alexis Soyer. Bloomsbury Publishing, London.

Brook, P., 2009a. In critical defence of 'emotional labour': refuting Bolton's critique of Hochschild's concept. Work Employ. Soc. 23 (3), 531–548.

Brook, P., 2009b. The alienated heart: hochschild's 'emotional labour' thesis and the anti-capitalist politics of alienation. Cap. Cl. 33 (7), 7–31.

Brown, S.W., Gummerson, E., Edvardsson, B., Gustavsson, B., 1991. Service Quality: Multidisciplinary and Multinational Perspectives. Lexington Books, Lanham, Maryland, USA.

Bruner, P.M., 1990. Music, mood and marketing. J. Mark. 54 (4), 94–104.

Bruni, F., 2005. Yes, the Kitchen's Open Space. Retrieved March 22nd, 2012, from Bruni, F. (2005, July 27). The New York Times. Retrieved Mar 22, 2019. http://www.nytimes.com/2005/07/27/dining/27open.html?pagewanted=all.

Bunting, M., 2004. Willing Slaves: How the Overwork Culture Is Ruining Our Lives. Harper Collins, London.

Burns, P.M., 1997. Hard-skills, soft-skills: undervaluing hospitality's 'service with a smile'. Prog. Tour. Hosp. Res. 3 (3), 239–248.

Butler, C., 2014. Wanted, straight talkers: stammering and aesthetic labour. Work Employ. Soc. 28 (5), 718–734.

Chow, A.J., Alonso, A.D., Douglas, A.C., Douglas, A.C., O`Neil, M.A., 2009. Exploring open kitchens' impact on restaurant cleanliness perceptions. J. Retail Leis. Prop. 39 (2), 233–254.

Clarke, J., 2014. Pre-employment training for the unemployed: a case study of a call centre foundation programme. Local Economy 29 (1–2), 113–128.

Cole, P.M., Michel, M.K., Teti, L.O., 1994. The development of emotion regulation and deregulation: a clinical perspective. Monogr. Soc. Res. Child Dev. 59 (3), 73–100.

Collinson, D., Hearn, J., 1996. Men at work: multiple masculinities/multiple work places. In: Understanding Masculinities, pp. 61–76 (Buckingham).

Connell, R.W., 1995. Masculinities. Polity Press, Cambridge.

Connell, R.W., 2000. The Men and the Boys. Polity Press, London.

Cook, P., 1996. The Industrial Craft Worker. Mansell, London.

Cousins, J., Lillicrap, D., Weekes, S., 2014. Food and Beverage Service, 9 ed. Hodder Education, Abingdon, Oxford.

Cowen, R., 2010. Relish: The Extraordinary Life of Alexis Soyer, Victorian Celebrity Chef. Hachette, London.

Cronin, J., Taylor, S.A., 1992. Measuring service quality: a re-examination and extension. J. Mark. 56, 55–68.

Derry, S., Nath, V., Walsh, J., 2013. Why do off-shore Indian call centre workers want to leave their jobs? New Technol. Work. Employ. 28 (3), 212–226.

Diefendorff, J.M., Croyle, M.H., Gosserand, R.H., 2005. The dimensionality and antecedents of emotional labour strategies. J. Vocat. Behav. 66 (2), 339–357.

Donkin, R., 2001. The History of Work. Palgrave Macmillan, Basingstoke.

Dorenburg, A., Page, K., 1996. Culinary Artistry. Wiley, New York.

Drucker, P.F., 1994. The age of social transformation. Atl. Mon. 274 (5), 53–80.

Fang, C., Peng, P., Weita, P., 2013. Does using an à la carte or combo set menu affect the performance of a Teppanyaki-style restaurant. Int. J. Contemp. Hosp. Manag. 25 (4), 491–509.

Fantasia, R., 2010. Cooking the books' of the French gastronomic field. In: Silva, E., Warde, A. (Eds.), Cultural Analysis and Bourdieu's Legacy. Routledge, New York, pp. 28–44.

Fiebig, G.V., Kramer, M., 1998. A framework for the study of emotions in organizational contexts. Manag. Commun. Q. 11, 336–372.

Fillby, M.P., 1992. The figures, the personality and the bums; service work and sexuality. Work Employ. Soc. 1 (6), 23–42.

Fine, G.A., 1996. Kitchens; the Culture of Restaurant Work. University of California, Berkley and Los Angeles.

Frable, J., 1998. Planning for display and open kitchens requires special attention to detail. Rest. Nat. News 34 (35).

Franck, K.A., 2002. Design on the plate. Architect. Des 72 (6), 58–67.

Fuller, J., 1992. Advanced Food Service. Stanley Thorne Publishers, Cheltenham, England.

Gillespie, C., 2001. In: Cousins, J. (Ed.), European Gastronomy into the 21st Century. Butterworth-Heinemann, Oxford.

Goffman, E., 1959. The Presentation of Self in Everyday Life. Doubleday, Garden City, NY.

Goldthorpe, J.H., Lockwood, D., Bechhofer, F., Platt, J., 1969. The Affluent Worker in the Class Structure. Cambridge University Press, Cambridge.

Graham, D., 2001. The changing face of the traditional theme restaurant. In: CHME (Ed.), 10th Annual Research Conference. Council for Hospitality Management Education.

Graham, D., 2006. The re-orientation of the professional chef; an exploratory study. Int. J. Manag. Cases 8 (2), 23–34.

Graham, D., 2006a. The return of gueridon service. Hospital. Rev. 8 (3), 5–11.

Graham, D., 2015. The Chef as a Emotional and Aesthetic Labourer - an Employee in Transition. Sheffield Hallam University, Sheffield.

Graham, D., 2017. Autumn). Views from the open kitchen. Hosp. Q. 1 (1), 44–46.

Graham, D., Dunning, J., May 11-12, 2011. The professional chef from behind the scenes to front of house; an amalgam of roles. In: CHME (Ed.), 20th Annual Research Conference (Pp. 11-12th May). Leeds Metropolitan University: Council for Hospitality Management Education.

Grandey, A.A., Gabriel, A., 2015. Emotional labour at a crossroads: where do we go from here? Ann. Rev. Organisat. Pschol. Organ. Behav. 2, 323–349. http://www.researchgate.net/publication/265547029_Emotional_labor_at_a_crossroads_Where_do_we_go_from_here.

Grandey, A., Diefendorff, J., Rupp, D., 2013. Emotional Labour in the 21st Century; Diverse Perspectives on the Psychology of Emotional Regulation at Work. Routledge, New York.

Grayson, K., 1998. Customer responses to emotional labour in discrete and relational service exchange. Int. J. Serv. Ind. Manag. 9 (2), 126–154.

Green, A.E., Owen, D., 1998. Where Are the Jobless? Changing Unemployment and Non Employment in Cities and Regions. Joseph Rowntree Foundation, York.

Grugulis, I., 2007. Skills, Training and Human Resource Development: A Critical Text. Palgrave Macmillan, London.

Gulati, R., 2007. Silo busting, how to execute on the promise of customer focus. Harv. Bus. Rev. 98–108.

Hansen, K.V., Jensen, O., Gustafsson, I.B., 2005. The meal experiences of a`la carte restaurant customers. Scandinavian J. Hospital. Manag. 5 (2), 135–151.

Harvey, G., Vachhani, S.J., Williams, K., 2014. Working out: aesthetic labour, affect and the fitness industry personal trainer. Leis. Stud. 33 (5), 454–470.

Hayes, S., Kleiner, B.H., 2001. The managed heart: the commercialisation of human feeling and its dangers. Manag. Res. 24 (3/4), 81–85.

Heide, K.V., Gronhaug, K., 2006. Atmosphere; conceptual issues and implications for hospitality management. Scandinavian J. Hospital. Manag. 5 (2), 135–151.

Hembry, P.M., 1997. British Spas from the 1815 to the Present: A Social History. Associated University Press, Cranbury.

Hill, S., 1976. The Dockers: Class and Tradition in London. Heinemann, London.

Hochschild, A.R., 1983. The Managed Heart. University of California Press, Berkley CA.

Hochschild, A.R., 2003. Afterword to the twentieth anniversary edition. In: Hoschild, A.R. (Ed.), The Managed Heart. University of California, London, pp. 199–207.

Jenkins, S., Delbridge, A., Roberts, A., 2010. Emotional management in a mass customised call centre: examining skill and knowledge-ability in interactive service work. Work Employ. Soc. 24 (3), 546–564.

Jennings, L., September 26, 2011. In the Kitchen Show off Your Chef, p. 32.

Johnson, D., Grayson, K., 2005. Cognitive and affective trust in service relationships. J. Bus. Res. 58 (4), 500–507.

Karlsson, J.C., 2011. Looking good and sounding right. Econ. Ind. Democr. 33 (1), 51–64.

Katsigris, C., Thomas, C., 2009. Design and Equipment for Restaurants and Food Service. John Wiley and Sons, New Jersey.

Kelly, I., 2003. Cooking for Kings; the Life of Antonin Careme the First Celebrity Chef. Heinemann, London.

Koopmann-Holm, B., 2011. Values and display rules for specific emotions. J. Cross Cult. Psychol. 42 (3), 355–371.

Korczynski, M., 2003. Communities of coping: collective emotional labour in service work. J. organ. 10 (1), 55–79.

Korczynski, M., 2005. Skills in service work; an overview. Hum. Resour. Manag. J. 15 (2), 3–14.

Korczynski, M., 2013. The customer in the sociology of work: different ways of going beyond the management-worker dyad. J. Employ. Soc. 1–7. E-special issue.

Kotler, P., 1973. Atmospherics as a marketing tool. J. Retail. 49 (4), 48–64.

Ladenis, N., 1988. My Gastronomy. Headline Book Publishing, London.

Lane, C., 2014. The Cultivation of Taste: Chefs and the Organization of Fine Dining. Oxford University Press, Oxford.

Lashley, C., Morrison, A., 2000. In Search of Hospitality, Theoretical Perspectives and Debates. Butterworth Heinemann, Oxford.

Lee, J.H., Ok, C.M., 2014. Understanding hotel employees' service sabotage: emotional labor perspective based on conservation of resources theory. Int. J. Hosp. Manag. 36, 176–187.

Lin, I.Y., 2004. Evaluating a servicescape; the effects of cognition and emotion. Hospita. Manag. 23, 163–178.

Litter, C., 1982. The Development of the Labour Process in Capitalistic Societies. Heinemann, London.

Lopez, S.H., 2010. Workers, managers and customers; triangles of power in work communities. Work Occup. 10 (37), 251–271.

Lorenzini, B., Johnson, B., May 1, 1995. Restaurant wars. Rest. Instiut. 148–153.

Lugosi, P., 2008. Hospitality spaces, hospitable moments: consumer encounters and affective experiences in commercial settings. J. Foodserv. 19 (2), 139–149.
Lugosi, P., 2014. Mobilising identity and culture in experience co-creation and venue operation. Tour. Manag. 40 (1), 165–179.
Mars, G., Nicod, M., 1984. The World of Waiters. Allen and Unwin Publishers, London.
May, T., 1998. The Victorian Domestic Servant, vol. 388. Osprey Publishing, London and New York.
McIntyre, M.P., 2014. Commodifying passion the fashion of aesthetic labour. Journal of cultural economy 7 (1), 79–94.
McIvor, A., 2013. Working Lives; Working in Britain since 1945. Palgrave Macmillan, Basingstoke.
Meloury, J., Signal, T., 2014. When the plate is full: aggression among chefs. Int. J. Hosp. Manag. 41, 97–103.
Mennell, S., 1996. All Manners of Food: Eating and Taste in England and France from the Middle Ages to the Present. University of Illinois Press, Illinois.
Mintel, 2006. Eating Out 10 Year Trends. Mintel.
Morgan, M., Watson, P., Hemmington, N., 2008. Drama in the dining room; theatrical perspectives in the food service encounter. J. Foodserv. 19, 111–118.
Nixon, D., 2009. 'I can't put a smiley face on': working-class masculinity, emotional labour and service work in the 'new economy'. Gend. Work. Organ. 16 (3).
Norii, T., 2015. The sushi chef. J. Acad. Emerg. Med. 22 (2), 248–249.
Parvatiyar, A., Sheth, J.N., 2001. Customer relationship management: emerging practice, process and discipline. J. Econ. Soc. Res. 3 (2), 1–34.
Payne, J., 2009. Emotional labour and skill; a reappraisal. Gend. Work. Organ. 16 (3), 348–367.
People 1st, 2014. State of the Nation Report 2013; an Analysis of Labour Market Trends, Skills, Education and Training within the UK Hospitality and Tourism Industries. People 1st, London. Retrieved from: https://www.bighospitality.co.uk/Article/2013/03/21/People-1st-State-of-the-Nation-report-2013.
Petrowski, E.M., September 1999. The open kitchen as theatre. Architect. Rec. 187 (9), 171.
Pettinger, L., 2004. Brand culture and branded workers: service work and aesthetic labour in fashion retail. Consum. Mark. Cult. 7 (2), 165–184.
Pine, J.B., Gilmore, J.H., 1999. The Experience Economy: Work Is Theatre and Every Business a Stage. Harvard Business School Press, Harvard.
Postrel, V., 2003. The Substance of Style. Harper Collins, New York.
Powell, M., 2011. Below Stairs; Memoirs of a 1920's Kitchen Maid. Pan Macmillan, London.
Pratten, J., O'Leary, B., 2007. Addressing the causes of chef shortages in the UK. J. Eur. Ind. Train. 31 (1), 68–78.
Putnam, L., Mumby, D.K., 1993. Organisations, Emotion and the Myth of Rationality. Sage, New York.
Rafaeli, A., Sutton, R.I., 1987. Expresssion of management as part of the work role. Acad. Manag. Rev. 12 (1), 23–37.
Rafaeli, A., Sutton, R.I., 1991. Emotional contrast strategies as means of social influence. Acad. Manag. J. 34 (4), 749–775.
Roberts, S., November 12, 2012. Boys will be boys, won't they? Change and continuities in contemporary young working class masculinities. Sociology 47 (4), 671–686.
Robinson, R.N., 2008. Revisiting hospitality's marginal worker thesis: a mono-occupational perspective. Int. J. Hosp. Manag. 27 (1), 403–413.
Robinson, R.N., Barron, P.E., 2007. Developing a framework for understanding the impact of deskilling and standardisation on the turnover and attrition of chefs. Hosp. Manag. 26 (1), 913–926.
Robinson, R.N., Beesley, L.G., 2010. Linkages between creativity and intention to quit: an occupational study of chefs. Tour. Manag. 31 (1), 765–776.
Rohatsch, M., Lemme, F., Neumann, P., Wagner, F., 2007. Professional Kitchens, fourth ed. Huss Medien, Berlin.
Roper, M., 1994. Masculinity and the British Organisation Man since 1945. University Press, Oxford.
Rose, E., Wright, G., 2005. Satisfaction and dimensions of control among call centre customer service representatives. Int. J. Hum. Resour. Manag. 16 (1), 136–160.
Ross, C., 1992. London Eats Out; 500 Years of Capital Dining. Museum of London: Philip Watson Publishers.
Rossant, J., 2004. Super Chef; the Making of the Great Modern Restaurant Empires. Free Press Simon and Schuster Inc, New York.
Schaubroeck, J., Jones, J.R., 2000. Antecedents of workplace emotional labor dimensions and moderators of their effect on physical symptoms. J. Organ. Behav. 21 (2), 163–183.

Schehr, L., Weiss, A., 2001. French Food; on the Table, on the Page and in French Culture. Routledge, New York.
Segal, L., 1997. Slow Motion: Changing Masculinities, Changing Men. Virago, London.
Sheane, S.D., 2011. Putting on a good face; an examination of the emotional and aesthetic roots of presentational labour. Econ. Ind. Democr. 33 (1), 145–158.
Sheehan, C., 2012. Hard labour at 35,000 Feet; a reconsideration of emotional demands in airline service work. Hosp. Soc. 2 (1), 99–116.
Sheridan, M., November 2001. Good vibrations, 111. Restaurant and Institutions, pp. 85–86.
Short, E., 2007. Dinning out. In: Freedman, P. (Ed.), The History of Taste. California Press, California, pp. 301–332.
Simonton, D., 1998. A History of European Woman's Work, 1700 to the Present. Routledge, New York.
Sims, F., 2012. Here Come the Girls - Sphere Magazine. Winter Issue 5.4. Retrieved 10 21, 2019, from: http://www.spherelife.com/here-come-the-girls/.
Sloan, D., 2004. Culinary Taste, Consumer Behaviour in the International Restaurant Sector. Elsevier, Butterworth Heinemann, Oxford.
Snaith, J., Pitham, P., August 17, 2006. Inside Kitchens: On with the Show. Caterer And Hotel Keeper.
Soares, A., 2003. Tears at Work; Gender, Interaction, and Emotional Labour, vol. 2. Department of organisation and human resources, pp. 36–44 (Spring).
Taylor, p., 2005. `India calling to the far away towns' the call centre labour process and globalization. Work Employment and Society 19 (2), 261–282.
Taylor, S., Tyler, M., 2000. Emotional labour and sexual difference in the airline industry. Work Employ. Soc. 14 (1), 77–95.
Trimming, A.R., 2014. Visible Tattoos in the Service Sector: A New Challenge to Recruitment and Selection. Retrieved Dec 21, 2019, from: http://wes.sagepub.com/content/early/2014/07/30/0950017014528402.abstract.
Trubek, A.B., 2000. Haute Cuisine: How the French Invented the Culinary Professional. Pennsylvania Press, USA.
Tsai, W.C., 2001. Determinants and consequences of employee displayed positive motions. J. Manag. 27 (4), 497–512.
Tungtakanpoung, M., Wyatt, M., 2013. Spirituality and cultural values in the reported cognitions of female cabin attendants on Thai airways. J. Air Transp. Manag. 27 (1), 15–19.
Vincent, S., 2011. The emotional labour process; an essay on the economy of feelings. Hum. Relat. 64 (10), 1369–1392.
Wakefield, K.L., Blodgett, J.G., 1996. The effect of the sevicescape on customers' behavioural intentions in leisure service settings. J. Serv. Mark. 10 (6), 45–61.
Walter, U., Edvardardsson, B., Ostrom, A., 2012. Drivers of customers' service experiences; a study in the restaurant industry. Manag. Serv. Qual. 64 (10), 1369–1392.
Warhurst, C., 2015. Service economy. In: The Wiley Blackwell Encyclopedia of Consumption and Consumer Studies, ed. Wiley Blackwell, London, pp. 1–4. Published online: 15 Jan 2015.
Warhurst, C., Nickson, D., 2001. Looking Good, Sounding Right; Style Counselling in the New Economy. The Industrial Society, London.
Warhurst, C., Nickson, D., 2005. Becoming a class act? Reflections of aesthetic labour and routine interaction service. In: 23rd Annual Labour Process Conference. Strathclyde University, Glasgow. Retrieved 2018. https://www.strath.ac.uk/.
Warhurst, C., Nickson, D., 2007. A new labour aristocracy? Aesthetic labour and routine interactive service. Work Employ. Soc. 21 (4), 785–797.
Warhurst, C., Nickson, D., Witz, A., Cullen, A.M., 2000. Aesthetic labour in interactive service work: some case study evidence from the `new' Glasgow. Serv. Ind. J. 20 (3), 1–18.
Watson, J., 2000. Male Bodies: Health, Culture and Identity. Open University Press, Buckingham, UK.
Weiss, A.S., 1998. Edible architecture, cannibal architecture. In: Scapp, R., Seitz, B. (Eds.), Eating Culture. New York Press, New York, pp. 161–180.
Wharton, A.S., 2009. The sociology of emotional labour. Annu. Rev. Sociol. 35 (1), 147–165.
White, M.P., Steen, J., 2006. White Slave - Marco Pierre White the Godfather of Modern Cooking. Orion Books, London.
Whyte, W., 1947. The social structure of the restaurant. Am. J. Sociol. 54 (4), 302–310.
Williams, C.L., Connell, C., 2010. "Looking good and sounding right": aesthetic labor and social inequality in the retail industry. Work Occup. 37 (3), 349–377.

Witz, A., Warhurst, C., Nickson, D., 2003. The labour of aesthetics and the aesthetic of organization. Organization 10 (1), 33–54.
Wong, A., Mei, O., 1999. Analysing service quality in the hospitality industry. Manag. Serv. Qual.: Int. J. 9 (2), 136–143.
Wood, R., 2000. Strategic Questions in Food and Beverage Management. Butterworth Heinemann, Oxford.

CHAPTER

20

Food ingredients

David Wiley, Christy Ng Yen Nee
Macao Institute for Tourism Studies (IFTM), Macao, China

OUTLINE

Food and Society
https://doi.org/10.1016/B978-0-12-811808-5.00020-9

20.1 Food additives

Food additives have been added to food since the beginning of time, as humans had to find ways to preserve and store food and the addition of flavors. Throughout time, the need to add food additives has changed from the 1800s when producers saw a decline in personal accountability and a lack of government regulation. There was an increase in adulteration of food, where chemicals or molecules were added to foods to increase the quantity of the food item, which could then result in bad quality. Today there are well over 3000 food additives that have been approved by governing bodies in both the European Union (EU) and the United States of America which are the two largest regulators. Food additives are defined as

> ... any substance not normally consumed as a food by itself and not normally used as a typical ingredient of the food, whether or not it has nutritive value, the intentional addition of which to food for a technological (including organoleptic) purpose in manufacture, processing, preparation, treatment, packing, packaging, transport or holding of such food results, or may be reasonably expected to result (directly or indirectly), in it or its by-products becoming a component of or otherwise affecting the characteristics of such foods. ***Codex Alimentarius* (2018)**

The Food and Agriculture Organization (FAO) of the United Nations and the World Health Organization (WHO) are responsible for presenting the definition of food additives and had been revised with the most recent revision taking place in 2012. The Codex Alimentarius is also known as "Food Code," and it collects the data surrounding standards, codes of practice, and guidelines about food and its processing. The Codex Alimentarius is governed by an intergovernmental body consisting of over 180 members with the framework of the Joint Standards Program established by the FAO and WHO aiming to ensure consumers are receiving food that is protected and that manufacturers are using fair practices in the food trade.

Food additives are required to be tested extensively to determine how much of each additive may be added to food over an amount of a specific time. Food additives are allowed to be added to food based on a measurement of any additive in milligrams per kilogram per body weight per day. The acceptable daily intake (ADI) is the "amount of an additive in food that could be ingested orally on a daily basis over a lifetime without appreciable health risk" (Carocho et al., 2014). The parameters are set based on testing observed in susceptible animal species and are known as the no-observed-adverse-effect level (NOAEL). The additives are given "quantum satis" or no maximum numerical level that can be added and expressed as milligrams of additive per kilogram body weight (mg/kg bw) or are given a specific limit per kilogram of body weight.

20.2 Food additives today

People depend on food additives more today than at any other time in history. It is said that the industrialized world would not have been possible without them. The need for food additives stems from the need to process and transport food long distances from manufacture or harvest to the consumer. With the demand for rising food needs, there is also demand for more food additives to assist with the special requirements needed for packaging and conditions to reduce spoilage and prevent contamination.

If the use of food additives that reduce spoilage or reduce contamination were eliminated, there would be an increase in the already significant numbers of reported foodborne illnesses. The Centers for Disease Control and Prevention in the United States reported having an estimated 76 million people get sick from food-related illnesses with 300,000 of them being hospitalized and 5000 deaths as a result of contaminated foods or beverages. While not all occurrences of foodborne illness are related to food additives from processing and distribution, additives do help to reduce the outbreak of any of the 250 different foodborne illnesses.

Food additives are divided into 26 functional classes depending on the functions in the EU and the United States; they are more narrowly classed into six distributed groups—preservatives, nutritional additives, coloring agents, flavoring agents, texturizing agents, and miscellaneous agents—and can be mentioned in more than one class. For this text and to simplify the information gathered we have adopted the Food and Drug Administration (FDA) classifications but will also include the European adaptations. The food additives are distinguishable between where they are produced by the edition of "E" before a number indicating the additive is derived in Europe.

20.3 Preservatives

Preservatives are divided into three subcategories: antimicrobial, antioxidant, and anti-browning additives. They are added to reduce natural food spoilage, oxidation, and extend the shelf life of processed foods.

20.3.1 Antimicrobials

Antimicrobials have two main purposes in food preservation: the first is to control natural food spoilage and the other is to avoid/control contamination by microorganisms. Potassium acetate, calcium acetate, lactic acid, and carbon dioxide are of quantum satis status with benzoic acid and benzoates, sorbic acid, and sorbate having restricted use ADI.

20.3.2 Antioxidants

Antioxidants are an essential preservative for extending the shelf life of foodstuffs used to prevent oxidation and do not add any additional flavors or odors to the food they are preserving. Antioxidant is commonly used in confectionary items and breakfast cereals for improving the shelf life of a product. Ascorbic acid, sodium ascorbate, and calcium ascorbate are examples of antioxidants that are quantum satis level while butylated hydroxyanisole a common antioxidant additive has an ADI of 0.5 mg/kg bw.

20.3.3 Antibrowning

There are two types of browning that occurs in food processing that can happen at separate times during manufacturing. The first type is enzymatic, which occurs through the conversion of polyphenols to quinones that create the darkening of foods. The second is nonenzymatic, which is the browning of caramelized sugar and the Maillard reaction. Sulfites are the most used antibrowning agents, but natural compounds can be used as well such as ascorbic acid–based formations.

20.4 Nutritional additives

Nutritional additives should be considered as food enhancers as they can contribute vitamins, amino acids, fiber, fatty acids, and polyphenols. The sources of the enhancements can be from plants, fungi, animals, or synthetic.

20.5 Coloring agents

Food dyes are used to make food more pleasant looking for consumers. There is only one food dye that is listed as quantum satis, calcium carbonate, which adds white color to food.

20.6 Flavoring agents

Flavoring agents are generally added to improve the taste or aroma of a food item. They are the largest number of additives added to our food with hundreds of varieties of flavorings available. As with all food additives, there are international standards that are monitored by the WHO and FAO. Flavoring agents have been divided into three subcategories: sweeteners, natural and synthetic flavors, and flavor enhancers.

20.6.1 Sweeteners

Sweeteners as the name imply add sweetness to food and add nutritive value or can be nonnutritive. Nutritional sweeteners include sucrose, fructose (high-fructose corn syrup), and glucose. Nonnutritive sweeteners include saccharin, aspartame, acesulfame K, and sucralose (Carocho et al., 2014). There are no sweeteners that have quantum satis level.

20.6.2 Natural and synthetic flavors

Natural and synthetic flavors are used to substitute the flavors of foods by using a mixture of chemicals that mimic the flavor of natural flavors, such as lemon or vanilla extracts. More than 1700 compounds are available and can be multifunctional acting as gelling agents or acting as an antimicrobial in food. A large number of natural and synthetic flavors are at the level of quantum satis and are organic acids such as acetic, lactic, citric, and malic acids.

20.6.3 Flavor enhancers

Flavor enhancers are added to foods to magnify, supplement, or enhance flavors in food but do not contribute any added flavor of their own. Flavor enhancers do not have a specified ADI and include glutamic acid and monosodium glutamate as the most popularly used flavor enhancers.

20.7 Texturizing agents

Texturing agents modify the mouthfeel and textures of foods through the use of chemicals. There are main two subcategories used in texturizing agents: emulsifiers and stabilizers.

20.7.1 Emulsifiers

Emulsifiers are used to combine two immiscible liquids and to maintain good dispersion such as oil and vinegar in salad dressing. Lecithin, calcium alginates, and alginic acid are examples of emulsifiers that are quantum satis; other more known additives are locust bean gum, agar, gum arabic, and pectin.

20.7.2 Stabilizers

Stabilizers and emulsifiers work the same way, and most are used to prevent evaporation and deterioration of flavor oils and maintain desired textures in food. Pectin, which is a natural gelling agent, is an example of both an emulsifier and stabilizer. Other examples are alginic acid, carrageenan, locust bean paste, xanthan gum, and invertase, all of which are quantum satis.

20.8 Genetically modified organisms

Plant breeding has been applied to farming for millennia giving way to plants that are suited for food, fiber, feed, and energy production (Barrows et al., 2014). Farmers use plant grafting as a means of creating a stronger or more beautiful crop by growing two of the same plants: one known as the rootstock and the other known as the scion, which features the size, flavor, and color of the particular crop or plant. The rootstock is grown for the viability and strength of the crop; the two plants are meticulously cut and joined together as seedlings through their vascular systems to create the modified plant.

Farmers that use conventional methods of breeding plants have yielded crops that have unforeseen mutations and is proven to be a slow and random process that is constrained by the availability of desirable traits in other plant species that may be closely related (Barrows et al., 2014). The use of newer biotechnologies allows scientists to use attributes from plant species from distant relatives to use the desirable genetic markers in other plant forms to create an improved plant.

The first genetically engineered (GE) food was the Flavr Savr tomato developed in 1982; it was not allowed to be released for field testing, however, until 1985 by the United States Department of Agriculture Animal and Plant Health Inspection Service (APHIS), which is responsible for approving new plant seed species to be distributed for commercial use. The Flavr Savr tomato is categorized in the second-generation trait of three generational traits for GE crops. The first-generational traits are those crops that have enhanced traits that may include "herbicide tolerance, better insect resistance, and better tolerance to environmental stress" (Schneider et al., 2017). The second generation of GE crops comes with value-added traits where scientists were able to add or "improve nutrient enhancement of animal feed" (Schneider et al., 2017). The third generation of GE crops is mostly used for pharmaceuticals or other products that are processed for biofuels and crops beyond food and fiber.

Manufacturers and farmers are needed to find ways to increase production, lower costs, and reduce pollution to the air and soil surrounding farms while trying to protect the environment. Genetically modified organisms (GMO) or genetically modified (GM) foods are a result of gene splicing in plant species that helps to alleviate the need for or a reduction in toxin-producing pesticides. The process of genetically modifying foods results from the use of recombinant DNA biotechnological procedures that changes the genetic makeup of a food or organism (Schneider et al., 2017, p.1). The process takes desired DNA traits from one plant or organism and attaches it to another desired attribute of an organism to a host organism which produces a new plant or organism in a controlled environment differing from the farm-raised grafted plant. The resulting plants produce crops such as delayed-ripening tomatoes, pest-resistant crops, and herbicide-tolerant crops.

20.8.1 Common genetically modified foods

While there have been many advances to GE foods, not all of them have been commercially viable as food fit for human consumption. Most of the GE crops are regulated and are forced to go under rigorous testing before being released into food production for humans. Companies that produce GE seeds can petition the government for deregulation of their seed after testing and finding that the seed and resulting crops are successful and commercially viable. The following chart depicts the most commonly deregulated and farmed GM crops.

20.9 Internationally deregulated and farmed genetically modified crops

See Table 20.1.

TABLE 20.1 International Service for the Acquisition of Agri-biotech Applications (2016).

Genetically modified crops	Number of varieties
Corn	231[a]
Apples	3
Soybeans	25
Papayas	4
Tomatoes[b]	11
Alfalfa	5
Potatoes	48
Sugar beet	3
Total	**380**

[a]*Only 57 varieties available for commercial production.*
[b]*Tomatoes are no longer being grown and produced.*

20.9.1 Apples

GM apples have been field grown and tested in Canada before being approved by the FDA for production and farming in America. As of 2016, the first harvest for commercial use was planned for an early 2017 release in the American market. Ojanagan Specialty Fruits Inc. is the company responsible for the apple which has been given a trait that prevents apples from browning after being cut. The company has planted more than 70,000 trees across North America with hopes of increasing planting to 300,000 and 500,000 of Golden and Granny Smith verities, respectively, with Fuji in the pipeline for future growth.

20.9.2 Papayas

Researchers at Cornell University and the University of Hawaii have tacked the papaya ringspot virus (PRSV), which has affected a significant amount of Hawaii's growth of papayas in the early 1990s. They were able to develop a PRSV-resistant papaya in 1997, and it became commercially viable and available in 1998. GM papayas are now being grown in Canada and Japan where it was added to the Japanese GM fresh food quality labeling system in 2011.

20.9.3 Alfalfa

Alfalfa is the fourth largest crop grown in the United States covering more than 8.4 million hectares of land. The alfalfa grown in the United States is mostly used for animal feed with only 7% of production being used as sprouts for human consumption. Herbicide-tolerant and low lignin alfalfa are covering 1.2 million and 20,00 ha, respectively. The crops were planted for hay, haylage, and green chop, all being used as animal feed.

20.9.4 Sugar beets

Sugar beets are only grown in the United States and Canada and are also the only crop to be adopted by 100% of growers, with 472,000 ha grown in the United States and 8000 ha grown in Canada. The modified beet offers a weed control trait and is more cost effective and easier to cultivate. The sugar beet has been commercially available since 2012.

20.9.5 Tomatoes

The Flavr Savr tomato was developed by Calgene now owned by Monsanto, one of many agricultural science companies producing GM crops. It was produced from 1994 until 1997 with traits to give the tomato vine-ripened taste that could withstand the shipping process. According to Schneider et al. (2017), the Flavr Savr tomato was discontinued because of public concern and specialized transportation equipment. There are no known GMO tomatoes on the market today in North America or Europe.

20.9.6 Corn

Corn is the most widely grown GE crop. It was first introduced in 1996 as BT (*Bacillus thuringiensis*) corn and can produce an insecticide to stave off insects such as the European corn borer, the fall armyworm, and corn earworm. The varieties and their insect and herbicide resistances and tolerances have grown exponentially to include resistances to butterflies and moths, corn rootworms, and crops drought tolerant. Farmers are currently growing GE corn in 16 countries with Vietnam being the most recent to join the group with America, Brazil, and Spain leading the group (ISAAA, 2016). GE corn helps to produce food ingredients used in food processing such as high-fructose corn syrup and cornstarch.

20.9.7 Soybeans

Soybeans are responsible for the most used edible oil in the United States, and they are the second largest crop grown in the United States covering 83 million acres of farmland. GE soybeans are comprised of 93% of sustainably grown soybeans in production now. There are 11 countries currently growing soybeans including America, Brazil, Argentina, and Paraguay. The soybean varieties can withstand herbicides, moths, and butterflies and has been bred to improve the fatty acid profile of the bean. Soybeans are used in many other products that include lecithin, which is found in chocolate and some other confectionary items.

20.9.8 Potatoes

GM potatoes were deregulated in 1995 with a coleopteran-resistant, an insect species. The potatoes have since gone through various stages of deregulation and commercial sale of the seeds that include late blight resistant traits, low-acrylamide potential traits, a chemical that forms when starchy foods are heated to high temperatures, and Colorado potato beetle resistant. The potatoes are also resistant to potato virus Y and potato leafroll virus and black spot and bruise resistant versions.

Although the potatoes were deregulated in the mid-90s, they did not hit the commercial market for consumers until 2015 by the J.R Simplot company, a producer of frozen potato products. The company has caused some controversy over their use of GM potatoes in food for human consumption. The Ranger Russet, Russet Burbank, and Atlantic varieties are being grown and distributed by the J.R Simplot company. It has been noted that one of America's largest fast food chains is the most prominent customer of the J.R Simplot company and have rejected the use of any GM food from the company (ISAAA, 2016).

20.9.9 Canola

GM canola is currently being grown in Canada, Australia, and the United States. Chile produces GM canola for the seeds that are then grown elsewhere. Annually, 36 million hectares are being grown globally, and of that only an estimated 24% are GM and of that 1% is used in the production of biofuels. Canola is also known as rapeseed; it is an oilseed crop which is cultivated for its high-quality oil. The variations were created to have an impact on weed growth and quantity or higher yield for farmers.

20.10 Food labeling

20.10.1 Legislation

In the United States at least, the packaging and labeling of food is subject to regulation in most regions/jurisdictions, both to prevent false advertising and to promote food safety. The US FDA announced the new nutrition facts label for packaged foods to reflect new scientific information, including the link between diet and chronic diseases such as obesity and heart disease on May 20, 2016. The new label will make it easier for consumers to make better informed food choices. FDA published the final rules in the Federal Register on May 27, 2016, which are listed below.

20.11 Highlights of the final nutrition facts label

a. Features a refreshed design
- **i.** The "iconic" look of the label remains, but FDA is making important updates to ensure consumers have access to the information they need to make informed decisions about the foods they eat. These changes include increasing the type size for "calories," "servings per container," and the "serving size" declaration and bolding the number of calories and the "serving size" declaration to highlight this information.
- **ii.** Manufacturers must declare the actual amount, in addition to percent Daily Value of vitamin D, calcium, iron and potassium. They can voluntarily declare the gram amount for other vitamins and minerals.
- **iii.** The footnote is changing to better explain what percent Daily Value means. It will read: "*The % Daily Value tells you how much a nutrient in a serving of food contributes to a daily diet. 2000 calories a day is used for general nutrition advice."

b. Reflects updated information about nutrition science

i. "Added sugars," in grams and as percent Daily Value, will be included on the label. Scientific data shows that it is difficult to meet nutrient needs while staying within calorie limits if you consume more than 10% of your total daily calories from added sugar, and this is consistent with the 2015–20 Dietary Guidelines for Americans.

ii. The list of nutrients that are required or permitted to be declared is being updated. Vitamin D and potassium will be required on the label. Calcium and iron will continue to be required. Vitamins A and C will no longer be required but can be included on a voluntary basis.

iii. While continuing to require "total fat," "saturated fat," and "trans fat" on the label, "calories from fat" is being removed because research shows the type of fat is more important than the amount.

iv. Daily values for nutrients such as sodium, dietary fiber, and vitamin D are being updated based on newer scientific evidence from the Institute of Medicine and other reports such as the 2015 Dietary Guidelines Advisory Committee Report, which was used in developing the 2015–20 Dietary Guidelines for Americans. Daily values are reference amounts of nutrients to consume or not to exceed and are used to calculate the percent Daily Value (% DV) that manufacturers include on the label. The % DV helps consumers understand the nutrition information in the context of a total daily diet.

c. Updates serving sizes and labeling requirements for certain package sizes

i. By law, serving sizes must be based on amounts of foods and beverages that people are actually eating, not what they should be eating. How much people eat and drink has changed since the previous serving size requirements were published in 1993. For example, the reference amount used to set a serving of ice cream was previously 1/2 cup but is changing to 2/3 cup. The reference amount used to set a serving of soda is changing from 8 ounces to 12 ounces.

ii. Package size affects what people eat. So, for packages that are between one and two servings, such as a 20-ounce soda or a 15-ounce can of soup, the calories and other nutrients will be required to be labeled as one serving because people typically consume it in one sitting.

iii. For certain products that are larger than a single serving but that could be consumed in one sitting or multiple sittings, manufacturers will have to provide "dual column" labels to indicate the amount of calories and nutrients on both a "per serving" and "per package" or "per unit" basis. Examples would be a 24-ounce bottle of soda or a pint of ice cream. With dual-column labels available, people will be able to easily understand how many calories and nutrients they are getting if they eat or drink the entire package/unit at one time.

20.12 Compliance date

d. The FDA extended the compliance dates for the Nutrition Facts and Supplement Facts label final rule and the Serving Size final rule, from July 26, 2018 to January 1, 2020, for manufacturers with $10 million or more in annual food sales. Manufacturers with less than $10 million in annual food sales would receive an extra year to comply—until January 1, 2021.

e. In May 2016, the US FDA finalized the Nutrition Facts and Supplement Facts Label and Serving Size final rules and set the compliance date for July 26, 2018, with an additional year to comply for manufacturers with annual food sales of less than $10 million. After those rules were finalized, industry and consumer groups provided the FDA with feedback regarding the compliance dates. After careful consideration, the FDA determined that additional time would provide manufacturers covered by the rule with necessary guidance from FDA and would help them to be able to complete and print updated nutrition facts panels for their products before they are expected to be in compliance.

20.13 Information shown on the food label in the United Kingdom

1. The name of the food
2. A "best before" or "use-by" date (or instructions on where to find it)
3. Any necessary warnings
4. Net quantity information
   - **a.** Net quantity in grams, kilograms, milliliters, or liters must be included on the label of
     - **i.** packaged food over 5g or 5 mL
     - **ii.** packaged herbs and spices
   - **b.** Solid foods packed in a liquid (or an ice glaze) must show the drained net weight.
   - **c.** The net quantity must be close enough to the name of the food that one can see all this information at the same time. This also applies to the alcoholic strength for alcoholic drinks.
   - **d.** Do not show the weight or volume on foods sold by number, for example, two bread rolls, provided that you can clearly see the number of items inside the packaging.
5. A list of ingredients (if there is more than 1)
   - **a.** For food or drink product that has two or more ingredients (including any additives), one must list them all. Ingredients must be listed in order of weight, with the main ingredient first. Percentage of an ingredient need to be showed if it is
     - **i.** highlighted by the labeling or a picture on a package, for example, "extra cheese"
     - **ii.** mentioned in the name of the product, for example, "cheese and onion pasty"
     - **iii.** normally connected with the name by the consumer, for example, fruit in a summer pudding
6. The name and address of the manufacturer, packer, or seller

TABLE 20.2 Special product considerations: Part 1.

• Bottle water	• Bread and flour
• Cocoa and chocolate products	• Fats and oils
• Fish	• Fruit juices and nectars
• Honey	• Jams and preserves
• Meat and meat products	• Milk and milk products
• Soluble coffee	• Sugar

Source: Compiled by David Wiley & Christy Ng.

7. The country of origin for the following items. One must also show the country of origin if customers might be misled without this information, for example, if the label for a pizza shows the leaning tower of Pisa but the pizza is made in the United Kingdom.
   a. Beef, veal, lamb, mutton, pork, goat, and poultry
   b. Fish and shellfish
   c. Honey
   d. Olive oil
   e. Wine
   f. Fruit and vegetables imported from outside the EU
8. The lot number or use-by date
9. Any special storage conditions
10. Instructions for use or cooking, if necessary

One must also show this information if they apply to the product:

- A warning for drinks with an alcohol content above 1.2%
- A warning if the product contains GM ingredients, unless their presence is accidental and 0.9% or less
- A warning if the product has been irradiated
- The words "packaged in a protective atmosphere" if the food is packaged using a packaging gas.

Also, there are special rules about what one has to show on the label if they supply any of the following (see Table 20.2).

20.14 Allergens

According to the United Kingdom's food labeling and packaging legislation, one must highlight allergens on the label using a different font, style, or background color. They must also list them in the ingredients. The allergens needed to be highlighted and listed are shown in the table below (Table 20.3):

TABLE 20.3 Special product considerations: Part 2.

• Celery	• Cereals containing gluten, including wheat, rye, barley, and oats
• Crustaceans, including prawns, crab, and lobster	• Eggs
• Fish	• Lupin
• Milk	• Mollusks, including squid, mussels, cockles, whelks, and snails
• Mustard	• Nuts
• Peanuts	• Sesame seeds
• Soybeans	• Sulfur dioxide or sulfites at levels above 10 mg per kilogram or per liter

Source: Compiled by David Wiley & Christy Ng.

20.15 Food and drink warnings

When it comes to labeling food and drinks which contain specific ingredients that may or may not trigger specific allergens, then the ingredient must be seen on the label (see Table 20.4):

TABLE 20.4 An appropriate warning must be shown on the label if the food contains certain ingredients.

Ingredient	Wording must be used
Allura red (E129)	"May have an adverse effect on activity and attention in children"
Aspartame	"Contains a source of phenylalanine"
Caffeine over 150 mg/L	"Not suitable for children, pregnant women and persons sensitive to caffeine"
Carmoisine (E122)	"May have an adverse effect on activity and attention in children"
Liquorice	"Contains liquorice" (you may need extra wording for confectionery or alcohol containing liquorice)
Polyols	"Excessive consumption may cause a laxative effect"
Ponceau 4R (E124)	"May have an adverse effect on activity and attention in children"
Quinoline yellow (E104)	"May have an adverse effect on activity and attention in children"
Raw milk	"This milk has not been heat-treated and may therefore contain organisms harmful to health"
Skimmed milk with nonmilk fat	There's no fixed wording, but you must show a warning that the product is unfit or not to be used for babies.
Sulfur dioxide over 10 mg/l	"Contains sulfur dioxide (or sulfites/sulfites)"
Sunset yellow (E110)	"May have an adverse effect on activity and attention in children"
Sweeteners	"With sweetener(s)"
Sweeteners and sugar	"With sugar and sweetener(s)"
Tartrazine (E102)	"May have an adverse effect on activity and attention in children"

Source: Compiled by David Wiley & Christy Ng.

20.16 Hong Kong–Macao labeling/marking requirements

Hong Kong's nontariff barriers related to labeling requirements, standards, and other requirements are generally minimal but are expected to become more restrictive for supplements. Pharmaceuticals and over-the-counter drugs must have indications of dosage and frequency on the label in both Chinese and English.

Hong Kong's food labeling regulations are developed and administered by the Food and Hygiene Environmental Department's Center for Food Safety. According to the Food and Drugs (Composition and Labeling) Regulations (Cap. 132 W), the following information should be provided on prepackaged food labels:

Name of the food

- List of ingredients (including allergens)
- Indication of "use-by" or "best before" date
- Statement of special conditions for storage or instructions for use
- Name and address of manufacturer or packer
- Count, weight, or volume of food
- Nutrition information stick-on product labels are allowed to remedy any noncompliance issues on preexisting labels and can be applied by Hong Kong importers with permission of the manufacturer. The labels may be provided in English, Chinese, or both languages. If both languages are used on prepackaged food, the name of the food, ingredient list, and nutritional information must be provided in both languages.

20.16.1 Labeling for biotech food

The Hong Kong government (HKG) does not require labeling of biotech food products. In 2006, the government issued guidelines for voluntary labeling of biotech foods containing 5% or more of an individual biotech food ingredient. Negative labeling is not recommended.

20.16.2 Organic certification

Hong Kong recognizes USDA organic certification, so products are sold in Hong Kong with the USDA organic logo.

20.16.3 Others

Hong Kong regulates the use of sweeteners, preservatives, coloring matters, metallic contaminants, and pesticides. In the absence of a particular regulation, the HKG refers to Codex standards and/or conducts a risk assessment to determine applicable import requirements. For details on Hong Kong's general import regulations for food products, please consult Public Health and Municipal Services Ordinance and the Food Safety Ordinance.

20.17 Expiration and use-by dates

Expiration date can be defined as the last date recommended for a product to be at peak quality. The "best before" date is appropriate for the vast majority of foods and indicates the period for which a food can reasonably be expected to retain its optimal condition (e.g., it will not be stale) and relates to the quality of the food.

The "use-by" date is the required form of date mark for those foods which are highly perishable from a microbiological point of view and which are likely after a relatively short period to present a risk of food poisoning and relates to the safety of the food.

References

Barrows, G., Sexton, S., Zilberman, D., 2014. Agricultural biotechnology: the promise and prospects of genetically modified crops. J. Econ. Perspect. 28 (1), 99–120.

Carocho, M., Barreiro, M.F., Morales, P., Ferreira, I.C., 2014. Adding molecules to food, pros and cons: a review on synthetic and natural food additives. Compr. Rev. Food Sci. Food Saf. 377–399.

Codex Alimentarius. (2018). Retrieved from Food and Agriculture Organization of the United Nations: http://www.fao.org/gsfaonline/additives/index.html?showSynonyms=1.

International Service for the Acquisition of Agri-biotech Applications, 2016. Annual Report. http://www.isaaa.org.

Rules and Regulation, 2016. Federal Register, vol. 81 (103). Retrieved from: https://www.regulations.gov/document?D=FDA-2012-N-1210-0875.

Schneider, K.R., Schneider, R.G., Richardson, S., February 2017. EDIS. Retrieved from University of Florida IFAS Extension: http://edis.ifas.ufl.edu/.

Further reading

Food Additives. January 31, 2018. Retrieved from World Health Organization: http://www.who.int/news-room/fact-sheets/detail/food-additives.

Food information to consumers – legislation – Food Safety – European Commission. 2018. Retrieved from: https://ec.europa.eu/food/safety/labelling_nutrition/labelling_legislation_en.

Food Legislation/Guidelines. January 25, 2018. Retrieved from: http://www.cfs.gov.hk/english/food_leg/food_leg.html.

Food from Genetically Engineered Plants. January 04, 2018. Retrieved from: U.S Food and Drug Administration.

GM Approval Database. 2018. Retrieved from: International Service for the Acquisition of Agri-Biotech Applications.

Government Digital Service, December 04, 2014. Food labelling and packaging. Retrieved from: http://www.gov.uk/food-labelling-and-packaging/food-labelling-what-you-must-show.

Hong Kong-Macau – Labelling/Marking Requirements Hong Kong-Macau – Labelling. February 6, 2019. Retrieved from: http://www.export.gov/article?id=Hong-Kong-Macau-labeling-marking-requirements.

Jain, A., Mathur, P., 2015. Estimation of food additive intake: overview of the methodology. Food Rev. Int. 31, 355–384.

CHAPTER

21

The role of information technology in the food industry

Chammy Lai Peng Tai[1], *Ricky On Peng Sou*[2], *Cindia Ching Chi Lam*[1]

[1]Institute for Tourism Studies, Macao (IFTM): Colina de Mong-Há Macao, China; [2]General Manager (IT and Operation) of TEDMEV Macau EDI Van, S.A. Macao, China

OUTLINE

21.1 Introduction

In the past decades, the food industry depended highly on static marketing through the push theory, mostly using advertising channels such as televisions, magazines, and leaflets. The information is channeled to the mass public in a spoon-fed manner. The food providers will decide on the type of information they want to provide to the customers, when fair

Food and Society
https://doi.org/10.1016/B978-0-12-811808-5.00021-0

benchmarking information is seldom on the strategic list. For customers, they highly depend on the marketing information for purchase decisions, while information from word of mouth is normally restricted to friends and relatives. Upon reaching a new stage, information technology (IT) has overturned this status quo between the food industry and customers.

In the contemporary era, customers are not satisfied with static marketing information provided by the food companies and advertisements are no longer the main source of information that customers will refer to. They prefer to check photos and comments from: social media or blogs; they want to know not only the ingredients of the dish but also the source of the food; the transportation and preparation methods; and even the cooking philosophies of chefs. Diners will share their dining experiences online, with known and unknown people. They will share what they see and what they think, turning them into compliments and complaints, which will impose positive or negative impacts on the food companies.

In the past, customers make food consumption decisions mostly based on the brand name, experience, word of mouth, price, and taste. However, rise in food allergies, ingredient-specific diets, and general awareness on food safety and health have alerted the concern of customers about the food they consume. Customers are now seeking for information on the ingredients and the nutrition, the source, existences of harmful or chemical elements, the cooking process, and even how the ingredients are delivered and stored. Food transparency is no longer just nice to have but has become vital.

Take McDonald's in Canada as an example—as a response to customer requests, they produced a video to show how the chicken nuggets were made and have developed a website "Our food. Your questions" (https://yourquestions.mcdonalds.ca/) to answer customer queries about their food. In another example, McDonald's in the United Kingdom has launched a virtual reality experience for customers on their behind-the-scenes supply chain processes. It portrayed three key stages of the supply chain, farming, production, and cooking, to settle the customer concerns on the ethics and safety of their supply chain management.

The increased accessibility to people and information has enabled the food providers to reach their target customer segments instead of the mass public; in turn, customers can obtain information from a variety of sources, tapping into a wide information base tapping into an almost infinite. The customer relationship management (CRM) has changed from static to interactive. The food providers can now provide information to customers based on their need and preferences, while customers can benchmark the available providers and receive both marketing and factual word of mouth on the aspects of services, food quality, esthetics, and alike. These changes are brought on through the rocket transformation of IT. This chapter will discuss the major means of technological platforms that cultivate the interactive relationship and the available IT tools that the food industry can employ to utilize the available resources, improve effectiveness, reduce cost, and enhance customer experiences. The chapter will conclude with the discussion on how Internet of Things (IoT), big data, and the Cloud can be used as differentiation opportunities for the food industry.

21.2 Means of interactivity

Effective customer communication is one of the critical success factors in the digital era. From mere basic marketing of food and services, contemporary marketing will differentiate the food companies, position them within their segments, and furnish customers with unique offerings, weekly specials, competitive information, and even behind-the-sense stories. Among all IT measures available this means, utilizing responsive websites, social media, blogs, as well as mobile apps and loyalty programs to take the message to the people.

21.2.1 Responsive websites

With improved connectivity and the advances in smartphone technology, most people now search online for restaurants and information on their food, the menu, location, and photos. A website is a kind of portfolio or online business card that officially presents the business. Responsive website applies smart artificial intelligence (AI) that will display the contents differently for individual gadgets such as desktop computers, tablets, and mobile phones. This type of website is normally informative in nature and covers the company background, its products and services, special events, photos, and videos for differentiation, market positioning, and promotion. These websites ranged from simple self-developed types to sophisticate professional designer versions and can be managed by user-friendly website interfaces called content management systems (CMS). The CMS is used as a tool for website content management and for pushing news to different social media on a timely basis.

To enhance customer relationships, a section for customer suggestions and comments is normally available. This is a type of important word-of-mouth management, when compliments should be treated as employee incentives, while complaints should be met with timely responses to avoid complaints going viral. The responsive website is also an important source of information for CRM, through add-on functions like integrating with online reservation systems, loyalty programs, and digital communication systems. The data collected can be analyzed for strategic planning and benchmarking.

21.2.2 Social media and blogs

Posting news on social media has become the trend for restaurant advertising (Toast Blog, 2017). Restaurants are now turning to websites and social media to stay in touch with customers and attract their attention. They will share restaurant news and menu's, events, and other activities, while chefs will post cooking videos and share tips in blogs to maintain closer customer relations. Some marketing campaigns are designed to be available only on social media; the most common ones include offering discounts or free dishes when customers share an event organized by the restaurant, share their dining experience, tag the restaurant, or like the post of the restaurants. This will enable a wider reach of potential customers through connecting to the friends of the existing customers.

According to BrightLocal (2018), around 85% of customers trust online reviews and consider the information as trustworthy and comparable to personal recommendations. The management of online customer reviews has grown in importance, partly because the extent of customer reach can cover a far more substantial customer base than traditional communication platforms like television or radios ever could. Moreover, while word of mouth can be compliments or complaints, can act not only as promotion tools but also as attacks. According to Gregoire et al. (2015), there are six major types of social media crisis and each can become either an opportunity or threat to the company, when the outcome of this duality depends on the strategic actions and responses of management.

21.2.3 Mobile apps

Once the idea or need to purchase arises, customers are changing the way they shop. Indeed the wide usage of mobiles and mobile devices is altering customer purchasing practices, from sourcing information through third parties to self-sourcing and from until-available to instant. The demand for one-stop source information seeking or purchasing on mobile devices has grown exponentially and this has given rise to the existence of mobile apps. Through these mobile apps, customers can check the menu, place the order, input personal preferences, settle payment, join loyalty programs and enjoy CRM benefits, track purchase history, develop personalized favorite lists, and even share their purchase and dining experiences or comment on the food and service quality. Restaurants can also subscribe to mobile apps from third-party service providers or develop their own apps. Starbucks launched their mobile apps since 2016 (Mobile Order & Pay). Together with their loyalty program, order, and payment system, Starbucks has achieved a significant increase in their online order sales, reduced queues, and motivated customer loyalty.

As mentioned, mobile apps from third-party service providers are taking off rapidly. Restaurant apps such as Yelp in the United States, Open Rice in Hong Kong, and Aomi (澳覓) in Macau are widely used by customers. Aside from restaurant location and information on cuisines, these apps also allow customers to make reservation, receive recommendations, provide online coupons, hold promotional events, and incorporate personalized functions through filters such as location, type of cuisine, price, or customer ranking. There are also add-ins like the remote queue function that provides added convenience to customers. For some of these third-party apps, customers can even create their own log books, like on Food Guru and Feedme Guru. Among the various mobile apps in the food business, delivery apps are the most demanded service by customers.

A delivery app can provide takeaway service like the Starbucks app that allows customer to make purchases online and complete advanced payment, before picking up the food from the company, or third-party mobile apps like UberEATS and Ele.me (e.餓了麼) that provide not only mobile order and payment but also delivers the food directly to the customers. Both of these serve similar purposes at increasing customer convenience and this is where opportunities lie. Nevertheless, mobile apps can also be a challenge. In the case of Starbucks, the company faced operational challenges when crowding swapped from long queues at shops to congestion at the handoff point due to the high online traffic volume (Forbes, 2017). A strategic plan has to be in place to capture the opportunities while avoiding the risks. For this case, a tailor-made menu can be designed for delivery service, to avoid providing food and beverages that require complicated preparation and cooking; design packing that can maintain the quality, taste, and freshness; re-engineer the workflow process; and set up a packaging delivery team or adding a kitchen to separate the delivery services from serving sit-in customers.

21.2.4 Loyalty programs

Loyalty programs are part of the CRM and have long been used in business and marketing. Through the use of technology, loyalty programs improved the interrelationship between the food industry and their customers and allow them to better understand the needs of their customers. In the past, customers were normally asked to present loyalty cards to enjoy discounts or record purchase for loyalty points. When customers forget to bring along the cards and present it for benefits, they will feel disappointed and may have negative impressions on the service provider, bearing thoughts that they are setting up barriers to prevent them from enjoying the benefits. Under these situations, the loyalty program will fail its original objective on customer relationship building but turned into a costly tool that disappoints the customers.

With new technology, physical loyalty cards are no longer a requirement, when they are replaced by electronic ones. Customers can enjoy the loyalty membership benefits and earn the purchase credits using the electronic card on the mobile apps or on the web. They can enjoy discount coupons, check the bonus points, and update their membership status. They can also personalize their membership by marking special dates such as birthday or anniversary and using personalized background. The food industry is able to operate more effective CRM strategies with the digitized customer information (Fig. 21.1). They can closely connect with customers by sending messages on special promotions, coupons, or recommendation plans. Moreover, as purchase transactions are linked to point of sales (POS) systems, this allows better understanding of customer purchase behaviors and preferences, to tailor personalized correspondence and marketing strategies. The use of electronic cards to replace physical ones can also contribute to environmental friendliness.

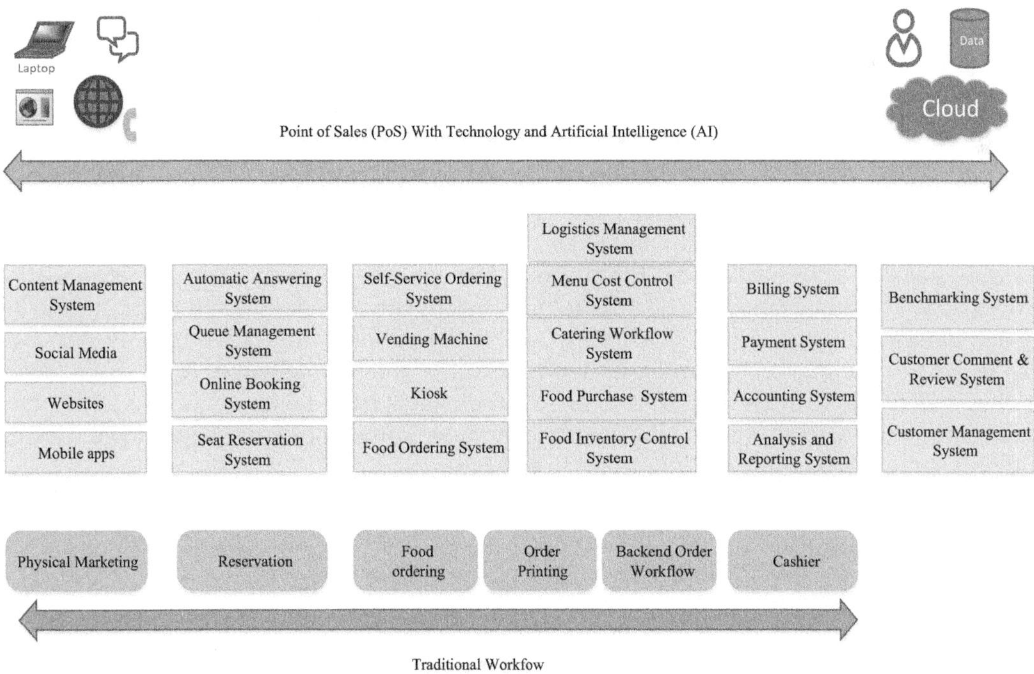

FIGURE 21.1 Technology and artificial intelligence in the food industry. *Source: Chammy, Ricky and Cindia*

21.3 IT tools for food industry

The food industry is a labor-intensive industry and service quality is the major value added that can differentiate food companies from their competitors. Moreover, laborers are human beings who have emotions and will be affected by their emotional and health status. The memory system of human beings is also complicated to the point that it affects the understanding of skills learned and the type and quality of service being delivered. To improve the effectiveness of the labor force and to minimize human error, IT has been widely applied in the food industry. The application of IT evolved from the simple POS system—an auto-vending machine that became a valuable aid to the various meal order and reservation systems. In the past, IT systems mostly focused on frontend operations, such as food ordering and bill payments at cashier. Nowadays, through the passage of time, IT is now involved in nearly every step of both the frontend and backend processes of many food-related businesses, even before and after customer purchases, from prepurchase marketing to postpurchase customer review. A competent catering system will collect a wide range of data from restaurant operations, which include sales details, purchases, inventory status, etc. These data are useful in the analysis of the overall operations within the restaurant including the sales, guest visit frequency, attendance rates at different time spans, waiting time, and duration of meals, etc. It also provides information on the sales of new dishes, on how far business has boosted by promotions or discounts, on the usage of check out rates and payment methods, etc., when all these can better serve to control costs and enhance the overall cost efficiency of the business.

21.3.1 Point of sales

The POS system is the most widely used tool for the sales of goods and services. In the food industry, the POS system can show the seat floor picture of the restaurant to customers and allow interactive pick-a-seat function. It also provides real-time information on seat availability, so frontend servers can easily make seat arrangement for customers. For customer orders, the receipts are printed with a barcode, when the cashier can check the bills simply by pressing the POS screen and scan the barcode; this speeds up the billing process and improves accuracy. In addition, a competent POS system can also be used to order food with mobile devices or smart tools. Smart tools are used not only to improve effectiveness but also as part of the marketing strategies with respect to the brand image. Aside from smart phones and iPads, with the application of AI, smart tools have taken a new era with robot attendants and robot chefs. The AI smart tools have commenced the experimental stage and will surely become the leading trend in the future.

21.3.2 Reservations

Crowding is one of the most common challenges for the food industry, when there are peak and slack time periods. For better reservation scheduling, managing available space resources, and allowing customers to have the correct expectations, restaurants can use the online booking system and queue management system (QMS). The online booking system enables customers to make self-reservations online and receive instant confirmation on availability, while the reservation will be instantly recorded in the reservation database. For the QMS, it is mostly used for the controlling of queues. It will provide a queue number (ticket) to the customer and when table is available, the customer will be arranged to a table based on the number order. For advanced QMS, it can even provide the estimated waiting time as an added value.

The use of these tools can reduce manual manpower and simplify the processes, provide service providers with more information to predict the volume of customers with higher accuracy, and also assist with the food procurement and labor arrangement plans. These tools will also allow the customers to have factual information on the availability and waiting time required before service is available. These will improve the effectiveness of the service and minimize potential service quality gaps between customer expectation and perception, before they receive the service and on the actual service received (Fitzsimmons et al., 2014).

21.3.3 Placing orders

Traditionally, frontend servers will take down orders manually and send the orders to backend kitchen for processing. Meal (food) ordering system makes use of electronic devices; when the frontend servers or customers can place order directly on the device, the orders will then be printed out instantly and also directly in the kitchen for processing. At the same time, the orders are transmitted to cashiers for bill checks and records. This can improve the efficiency of the kitchen processes and reduce errors and omissions between order placements and checkouts caused by illegible handwriting. The orders can be revised and canceled, when instant updates will be recorded in the respective systems. Moreover, the time and date of order will be recorded together with the number of guests, membership, and the type and quantity of food and beverages ordered. The management can then analyze the data to better understand the preference of customers, perform segmentation, provide personalized service, and design high-return loyalty programs.

Some catering management systems support self-service meal ordering, through self-service vending machine (kiosk). Ordering is performed on the self-service meal ordering machine or electronic devices. This can help to reduce the workload of frontend servers and serve similar purposes as the meal ordering system. Logistics management system is also crucial as it handles the complicated logistics workflow of the food industry. It includes mainly the menu cost control system and the food and beverage inventory control system.

21.3.4 Payment

Mobile payment is always an embedded function of mobile apps. While WeChat pay and Alipay are growing rapidly in Mainland China, there are various digital payment platforms such as Apple Pay, Google Wallet, and others joining the market. The convenience of cashless payment has remarkably increase sales. A study by USA Technologies and Michigan State University indicated that customers in general spend 37% more when they pay by credit cards instead of cash (Maras, 2019). In Mainland China, mobile payment transactions have outreached US$13 trillion in early 2018 (SCMP, 2018). People in China are using mobile payment from street food to fine dining through Alipay and WeChat pay by scanning a QR code.

This practice is affecting other countries, when it is found that 65% outbound Chinese tourists paid by mobile payment, according to the outbound Chinese Tourism and Consumption Trends report by Nielsen Company (2018). Nowadays, mobile payment has also incorporated different add-ins, such as bill splitting, to better replicate customer actions at physical settings.

21.4 The new opportunities

From improving customer satisfaction and operation efficiency, IT will lead the food industry into a new epoch with foci on food safety, transparency, food wastage control, logistics management, and data usage for CRM. Among all, the IoT, the Cloud, and big data will certainly play an important role.

21.4.1 Internet of Things

IoT is a network of devices to gather and convey data via the Internet, which can facilitate the food industry to trace, monitor, and control the food supply chain activities. Commencing from the source, sensors are equipped to monitor the ambient conditions, transportation time, and the physical production conditions. These can safeguard the supply chain to comply with local and global regulations, while the sensors will collect data at each stage of the process for analyses, so problems can be identified for solution planning. The Hazard Analysis and Critical Control Points (HACCP) can be automated within the IoT to identify safety hazards and critical control points and for comparative analyses between stores, location, and even countries. During the production stage, the sensors will control the quality, track the products, monitor the activities, and undergo real-time analysis, so immediate actions can be taken at failure points. These sensors can also determine the standard of colors, specks, and moisture content of ingredients. For the distribution chain, the GPS systems and transmitters will monitor the delivery and storage processes from production house and at the sales points. This can automate the processes and provide data transparency to both management and customers. Meaningful data will be collected to identify the incompetency of the system and preferences of customers and minimize surplus. The IoT network can effectively reduce costs, wastage, and risks of the food supply chain.

21.4.2 The Cloud

Cloud computing provides a comprehensive coverage to almost all IT equipment and allows the access of information without limits through the use of different smart tools. Cloud has its basis on the concept of resource sharing pool (resource pooling). It provides services and assigns or reassigns companies and virtual resources according to the customer requirements by means of multitenancy mode. The Cloud allows users to easily access IT resources and services in accordance with personalized needs; it can flexibly adapt to demand and quickly adjust to the size of resources with ample grounds for expansion. The capacity and speed can also be increased in response to business demands, so resources will not be wasted. The Cloud services can help companies, especially small and medium enterprises (SMEs), to reduce investment costs and expenses and save the cost of installing, maintaining, and upgrading of hardware and software. They can use Cloud services at low cost to improve operational efficiency and customer service. Traditional and mature industries can also benefit from the use of Cloud services to improve the inadequate communication problems between the supply chain and the demand chain. These inevitably improve the competitive edges.

Despite the powerful functions, there are, however, security threats in using Cloud computing services:

- Malicious internal employees

  The internal employees of Cloud service providers are most likely to have access to the Cloud data, but how they are regulated and being recruited is a mystery. The information on the Cloud platform is voluminous and it covered an extensive number of companies; therefore, the risk of internal staffs stealing data is relatively higher than that of general corporate organizations.
- Shared environment

  The key of using Cloud computing to reduce costs is by optimizing the use of resources. From the facet of users, Cloud appears to have independent computing environments and data storage space, but in reality, users are just sharing a main computer with others by means of virtualization technology. Cloud generates multiple virtual spaces in the actual environment by means of virtualization technology to provide customers multiple virtual spaces. In this instance, there are security risks associated with data storage and protection, isolation, and cross access of user resources.
- Data loss or leakage

  For the food industry, the impact of data loss or leakage is absolutely not only monetary but also invisible losses such as goodwill or customer confidence. The Cloud environment is affected by the platform architecture, and in sense of operation it is subject to high risk, while the threat of data leakage is also mounting. To reduce the risk of losing or divulging data in the Cloud, it is necessary to evaluate the Cloud service provider's verification, authorization and ensure that the auditing control measures are complete, the encryption technologies applied are legal, the data deletion methods are secure, and the disaster recovery are capable.
- Insecure interfaces and application interfaces

  Cloud service providers provide Internet-enabled interfaces for users to operate Cloud services; therefore, the security of these interfaces is closely related to the security and applicability of Cloud services. If the Cloud platform provides value-added services to third-party suppliers, the use of these interface services may increase the complexity and risks to the original application interface.
- The account or service being stolen

  Hackers stealing user accounts is one of the most common internet threats. Attacks are nothing more than fishing, fraud, and the use of software loopholes; however, in the Cloud environment, these problems have greater impact as users do not have control over the hardware and cannot implement remedial measures to reduce the loss from stolen account or service. When being hacked, they may even need to show evidence as a legitimate user of an account or service, otherwise, the hacker is likely to completely replace the original user.

- Audit and forensic
  Under the traditional IT infrastructure, it is not easy for companies to perform internal audits and collect computer evidence because of the complicated procedures. This situation becomes more complex with the information infrastructure located in the Cloud, as on how to obtain useful and effective information, in particular, when it involves a legal dispute. The Cloud service provider should have a corresponding mechanism for the company to prove the validity and legitimacy of the data.
- Unregulated evaluation
  The most common discussions about the Cloud are commonly on its applications, especially when emphasis is placed on its features, benefits, and functions. The deeper tiered information such as the internal security processes, platform infrastructure, patch formulas, audit systems, log records, data storage processes, and access rights seldom attract public attention. Ironically, this deeper tiered information is widely used to evaluate the security of the Cloud platform, when in fact there is no public supervisory mechanism on these areas. Without such information, it cannot effectively evaluate the risks of Cloud security.
- Legal compliance
  Cloud service providers are likely cross-border companies bound by dissimilar laws and regulations, including the security of personal data. If the country where the Cloud server is registered does not attach importance to personal data protection, it will inevitably erode the protection originally enjoyed by the users.

21.5 Big data

For the service-oriented food industry, the excavation and application of big data has become a priority to achieve continuous operation growth and effective management. Big data, officially interpreted as an information asset, refers to a collection of data that cannot be captured, managed, and processed with software tools within an affordable time frame. Traditionally speaking, it is a large amount of diverse data that is difficult to handle in traditional databases. It has a "3V" characteristic: large volume of data (Volume), high data growth rate, high processing speed (Velocity), and a variety of data types (Variety).

Big data is the collective name of all data, including structural, semistructured, and unstructured data. All kinds of data and consumption records in the food industry can be classified as big data. When these scattered data accumulate to a certain number, analytical methods for big data can be applied to identify underlying information that general data analysis methods will neglect.

21.5.1 Data collection

It is not difficult for the food industry to collect big data, especially with the application of commercial Wi-Fi. When customers use the Wi-Fi networks of the company, the hardware will require them to sign in to the social network account before they will be authorized. From there, company can control the various big data, such as the customer age through the customer social network account, habits, places of visit, and even the preferences of customers on products and services to the preferred brands.

Another important source of data is the POS system that records all the details of the customer order. The data collected will be analyzed to allow better understanding on customer preferences and market trends for business strategic planning. The customer management system (CRM) can also collect data from customers and summarize the preferences of different types of customers, such as the customer consumption frequency, consumption area, preferences, and consumption amount.

21.5.2 Opportunities

- Being the lead in consumption

  The power of big data is that it can convert customer spending habits, user preferences, and alike into a transparent data display that allows restaurants to more accurately grasp the market trend. In the era of big data, when customers refer to big data to select restaurants, restaurants are also using big data as a guide to perform their business models. For example, when customers order takeaway from a restaurant at an ordering platform and then revisit the platform after a period of time, the platform will "choose" from the platform menu the types of food similar to their previous orders and bring to the referral list. This kind of referral guide is the result of "match with preference" after using big data to analyze customers' preferences.
- Accurately plan marketing strategy

  From big data collection to analysis, company can understand the consumption pattern of customers in terms of time, amount, type, and all other available criteria; this enables the establishment of an informative customer database. The database will allow the segmentation of customers based on demographics or customer management criteria. The company will have better understanding of the customer consumption behavior, interest preferences, and the status of the market reputation.

  In this sense, company can forecast more dependable sales budget for the design of accurate and precise marketing strategies to meet the preference of customers. They can also make timely responses to the demand through price leveraging and differentiated pricing to increase market returns and revenue. The better match of marketing targets to products can also utilize resource allocation efficiency and effectiveness.
- Improve operational efficiency and market competitiveness

  For both short-term operation and long-term strategic planning, big data can provide comprehensive data support from a holistic view, such as market structure, products, customers, and marketing strategies, which enables the management to make scientific decisions. The basics of business operations is a contest of benefit effectiveness, cost reduction, increase in work efficiency, lock in customer demands, create profitable space, etc. The capability of big data to assist the management at achieving these is one of its greatest values.

21.5.3 Challenges

- Reliability of data source

  The amount of big data can be very huge. How can we extract valuable information from the sea of data? It is a challenge! For the characteristics of big data, besides having huge quantities, the speed of its production is also very fast, and the variety of data is substantial. The reliability of the data is therefore in question, and it is important to determine the quality of data and strategically identify the methods to refine its value.

- Lack of big data analysts

  Big data is developing rapidly but specialized big data analysts are extremely rare. The threshold for becoming big data analysts is very high, as apart from being familiar with statistics, they need to have programming skills, big data–type library, and machine learning knowledge. Big data analysts should be competent enough to have thorough understanding in the different industries and companies, possess ability to create detailed analysis with the quantitative information, make logical analysis, and provide decision-making recommendations. These constraint the future development of big data.

21.6 Conclusion

The need to apply IT in the food industry is a must. The potential benefits not only allow short-term improvement in efficiency and customer satisfaction but also in the long-term, it will become a competitive edge of the company. The food industry is undergoing dramatic changes with the use of technology. From tradition marketing through advertisement on TV and magazine, newsletters, fliers, and trade shows, it is now necessary to understand the heave between push and pull marketing when self-service kiosks, tablets, social media, blogs, and mobile apps are picking up the leads. To maintain the pioneer role, food companies should consider new opportunities through high technologies such as 3D food printing, virtual reality, AI, and cloud services.

References

BrightLocal, 2018. Local Customer Review Survey. https://www.brightlocal.com/research/local-customer-review-survey/.

Frobes, 2017. Starbucks' Success with Mobile Order and Pay Is Too Much of A Good Thing. https://www.forbes.com/sites/greatspeculations/2017/02/02/starbucks-success-with-mobile-order-and-pay-is-too-much-of-a-good-thing/#1bfc66cc43da.

Fitzsimmons, J.A., Fitzsimmons, M.J., Bordoloi, S.K., 2014. Service Management: Operations, Strategy, Information Technology, eighth ed. McGraw-Hill International Edition, Singapore.

Gregoire, Y., Salle, A., Tripp, T.M., 2015. Managing social media crises with your customers: the good, the bad, and the ugly. Bus. Horiz. 58, 173–182.

Maras, E., 2019. Cashless Payment Capability Boosts Both Cash and Card Sales in Vending. https://www.mobilepaymentstoday.com/articles/cashless-payment-capability-boosts-both-cash-and-card-sales-in-vending/.

Nielsen Company, 2019. 2018 Trends for Mobile Payment in Chinese Outbound Tourism. https://www.nielsen.com/content/dam/nielsenglobal/cn/docs/2018-trends-for-mobile-payment.pdf.

South China Morning Post (SCMP), 2018. China Pulls Further Ahead of US in Mobile Payments with Record US$12.8 Trillion in Transactions. https://www.scmp.com/tech/apps-gaming/article/2134011/china-pulls-further-ahead-us-mobile-payments-record-us128-trillion.

Toast Blog, 2017. The Top 5 Restaurant Digital Marketing Trends in 2017. https://www.restaurantsupply.com/blog/post/the-top-5-restaurant-digital-marketing-trends-in-2017/.

CHAPTER

22

Food service in hospitality management

Grant O'Bree

Institute for Tourism Studies Macao (IFTM): Colina de Mong-Há Macao, China

OUTLINE

Modern food service operators compete in a dynamic environment, where competition is intense and consumers have a multitude of choices in both cuisine and establishments to choose from. To compete, food service operators have to find new ways to differentiate from their competitors, showcasing culinary arts with interactive open kitchens, celebrity chefs and their endorsements, highly rated social media reviews, coveted Michelin star ratings, and a host of modern and traditional preparation and presentation styles (O'Fallon and Rutherford, 2011). However, operational costs continue to grow and the modern food service manager must be adept at controlling their operations in an efficient and effective manner, particularly in the hotel environment where there is immense pressure for the food service operation to achieve and maintain its profit center status (Iredell, 1963; Rodgers, 2007).

Food and Society
https://doi.org/10.1016/B978-0-12-811808-5.00022-2

Food service operators need to combine several intricate aspects of their operations to be successful—after all, it's not just about great food. They need to be highly effective in combining food, service, ambiance, and, if relevant, entertainment to provide guests with a complete dining experience (Barrows et al., 2012). Behind the scenes, food service managers need to be competent in the management of expenses, supplier relationships, and minimizing waste and labor costs (Brien and Smallman, 2011; Dopson and Hayes, 2011).

While many hotels struggle to make their food service operations profitable, there are many instances where the food service operations of a hotel provide a significant revenue stream through a profitable F&B service offering. There are significant opportunities for hotel food service operators to generate revenue that goes beyond the typical in-house guest. Local residents combined with effective and well-timed promotions can improve revenue streams from repeat business (Bowden, 2009). Accessing this revenue stream is often overlooked by hoteliers where their primary focus is on in-house guests and attracting new business. Admittedly, there is no one-size-fits-all approach to hotel management. However, a holistic approach to revenue generation across all departments is essential. Here are two very different scenarios:

22.1 Scenario 1

Hotel A is a five-star hotel in the CBD; they have an all-day-dining restaurant with a seating capacity of 350 people. They are typically full on Friday evening, Saturday lunch and dinner, and Sunday brunch. However, their Monday to Friday lunch and Monday to Thursday dinner business averages 14%–20% of capacity. They have limited promotions that are cyclic, with varying minimal variations and are limited to in-house guests and the occasional walk-in guest.

22.2 Scenario 2

Hotel B is identical to Hotel A. However, they have made a significant effort in regard to internal and external promotions and knowing *who* their customers are through the creation and management of an active guest database. They plan a host of seasonal menu's, celebrity chef events, wine pairing events, and other promotions a year in advance. Every month, there is something new. Like hotel A, they are full every Friday evening, Saturday lunch and dinner, and Sunday brunch. However, their average Monday to Friday lunch and Monday to Thursday dinner business is 36% higher than Hotel A. The additional business comes primarily from local residents within a 50 km radius of the city, which provides a significant revenue stream to the F&B department and additional incremental revenue to the hotel through room nights, spa packages, upgrades, and, most importantly, long-term, repeat customers. Additionally, partnering with financial institutions to target credit card holders (for example, VISA, AMEX, MasterCard, Alipay) can also boost both covers and revenue. If the hotel restaurant staff are able to impress with high levels of service

TABLE 22.1 Hotel A versus hotel B's proactivity in revenue generation through promotion.

				Hotel A			Hotel B		
Day	Period	Capacity	Price Per person	Percentage of capacity	Pax	Revenue	Percentage of capacity	Pax	Revenue
Monday	Lunch	350	$68.00	20%	70	$4760.00	56%	196	$13,328.00
	Dinner	350	$88.00	20%	70	$6160.00	56%	196	$17,248.00
Tuesday	Lunch	350	$68.00	20%	70	$4760.00	56%	196	$13,328.00
	Dinner	350	$88.00	20%	70	$6160.00	56%	196	$17,248.00
Wednesday	Lunch	350	$68.00	20%	70	$4760.00	56%	196	$13,328.00
	Dinner	350	$88.00	20%	70	$6160.00	56%	196	$17,248.00
Thursday	Lunch	350	$68.00	20%	70	$4760.00	56%	196	$13,328.00
	Dinner	350	$88.00	20%	70	$6160.00	56%	196	$17,248.00
Friday	Lunch	350	$68.00	20%	70	$4760.00	56%	196	$13,328.00
	Dinner	350	$98.00	100%	350	$34,300.00	100%	350	$34,300.00
Saturday	Lunch	350	$88.00	100%	350	$30,800.00	100%	350	$30,800.00
	Dinner	350	$98.00	100%	350	$34,300.00	100%	350	$34,300.00
Sunday	Lunch	350	$88.00	100%	350	$30,800.00	100%	350	$30,800.00
	Dinner	350	$68.00	20%	70	$4760.00	56%	196	$13,328.00
		Revenue per week				$183,400			$279,160
	Revenue per month (28-day cycle for 13 months)					$733,600			$1,116,640
		Annual				$9,536,800			$14,516,320

Source: Grant O'Bree.

and the kitchen can produce tantalizing culinary delights, then these customers often become a key part of the hotels local database, enabling the hotel to run targeted marketing campaigns to encourage repeat visits on birthdays, anniversaries, and special events, a term that is often referred to as customer loyalty or relationship marketing (Bowen and Shoemaker, 2003; Narteh et al., 2013; Shoemaker and Lewis, 1999).

Proposing that these hotels were in the same city had the same product with the same price, the difference in revenue from one restaurant alone could be >$4 million dollars per year (see Table 22.1 below). In many cases, upper scale and luxury hotels will feature multiple F&B options and, if the scenarios mentioned above are applied in all instances, then revenue will receive a significant improvement for this key profit center.

22.3 CEOs, Director of Operations, and General Managers (operational level)

22.3.1 Senior management team

There is an abundance of research on management and leadership with the aim of identifying the key traits and behaviors that make a manager or leader successful (Robbins and Coulter, 2014). While theories abound, senior managers of a hotel require a broad range of skills to manage and lead their teams effectively. When most people consider the skills required to operate a hotel, they think of two departments: (1) rooms and (2) food and beverage. However, this is just the tip of the iceberg. Senior managers need to have exceptional soft skills; these are the skills that enable them to communicate and inspire their team members to achieve organizational goals and objectives. Additionally, they need to be great listeners and, where necessary, be able to show empathy (Barrows et al., 2012). Hotel managers must also be proficient across other departments and areas such as sales and marketing, finance, HR, engineering and maintenance, strategy, internal and external relationships, and stakeholders.

The core functions of management, in their simplest form, fall under four distinct categories: planning, organizing, leading, and controlling (see Fig. 22.1). These four functions are a cyclic, reiterative process, requiring managers to juggle the multitude of activities, systems, and processes on a day-to-day, month-to-month, and year-to-year basis. Ultimately, when effectively and efficiently combining each of the four core functions of management, higher organizational performance can be achieved. However, combining the four functions of management does not necessarily breed success, there are other key elements that managers must be proficient in and include areas such as employee motivation and efficient and effective decision-making. For the latter, it can be said that managers are in a constant stream of decisions, not all will be good ones and there will be many that we look back and say "I should have done this ... or could have done that..." In reflecting on past decisions we rely on another management tool commonly called hindsight, which facilitates learning.

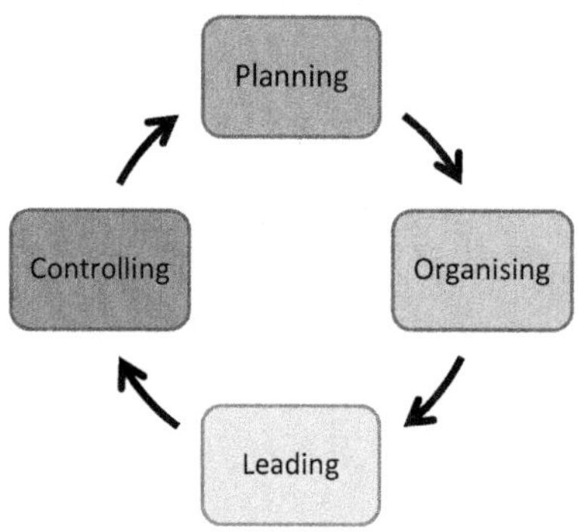

FIGURE 22.1 Reflecting on past decisions. *Source: Patrick Newsham*

22.3.2 General Manager

Traditionally, General Managers in hotels have spent many years developing their careers from the meager beginnings as a bellhop, guest services agent, or waiting staff in restaurant and have worked their way up through one of the two pathways: rooms division or food and beverage. Having experienced so many roles at all levels in the hotel provides a General Manager with a wealth of experience to provide effective leadership for the successful operation of the hotel. The role of the General Manager goes well beyond just leadership alone to include an array of management functions in areas of planning, organizing, controlling, decision-making, and motivation (Robbins and Coulter, 2014). The role of a General Manager is to orchestrate and facilitate the performance of the head of departments who, in turn, are responsible for driving the daily operations to achieve planned goals and objectives (Bardi, 2011). However, effective General Managers are not typically isolated from operations and will often attend a daily executive committee meeting to learn of key developments and or challenges in the hotel, providing advice, guidance, encouraging cooperation between departments, leadership, motivation, and decision-making as and when necessary.

Depending on how the hotel is organized, the General Manager may have a Hotel Manager or Resident Manager who handles the key day-to-day operations of the hotel (Barrows et al., 2012).

22.3.2.1 *Hotel Manager/Resident Manager*

Depending on the size and organizational structure of the hotel, there may be a Hotel Manager or Resident Manager who reports directly to the General Manager. The role of the Hotel Manager is to take full responsibility for the day-to-day operations of the hotel and work very closely with the senior management team to ensure that their respective departments are operating efficiently and effectively.

22.3.3 Director level

Director roles will generally be appointed for each of the key departments in a hotel. Again, this depends on the size of the hotel and the specific organizational design. However, the members of the senior executive team will report directly to the Hotel Manager and are responsible for the strategic management of their department (Figs. 22.2 and 22.3).

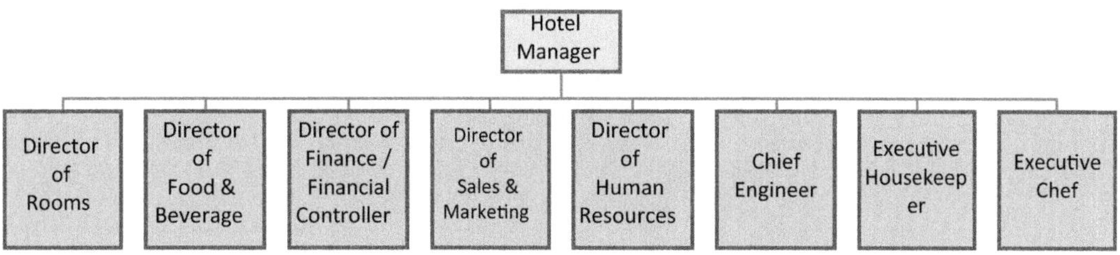

FIGURE 22.2 An executive team will normally report to the Hotel Manager. *Source: Patrick Newsham*

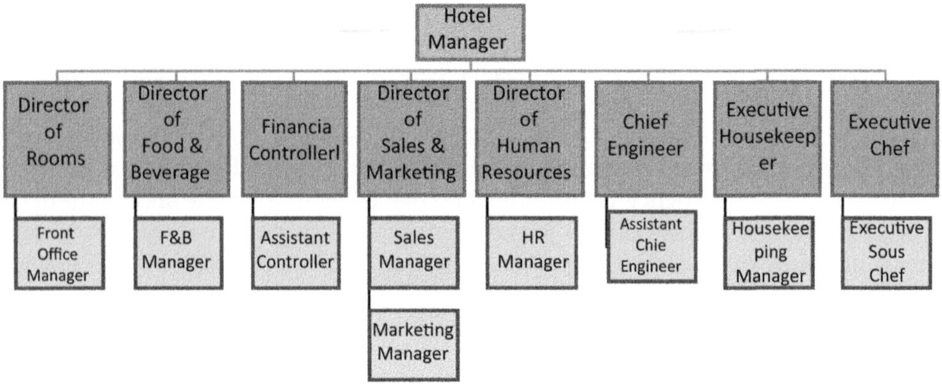

FIGURE 22.3 A typical departmental structure. *Source: Patrick Newsham*

Typically, each of the department directors or senior management team will be on the Executive Committee, which will generally meet every morning to discuss the performance of previous days, any significant challenges or successes and inform colleagues of any key operational items that day or for future operational requirements.

22.3.4 Managerial level

Managerial level employees report directly to their respective department directors.

22.3.4.1 Diversity

Workplace diversity has evolved from compliance to a strategic tool, particularly in the hospitality industry where the workforce is typically an eclectic mix of diverse cultures working together to achieve organizational goals (Kapoor and Madera, 2011; Robbins and Coulter, 2014; Sourouklis and Tsagdis, 2013). Evidence suggests that holistically embracing workplace diversity as a strategic tool results in greater organizational performance when compared with organizations that do not (Baum et al., 2007; Singal, 2014). Workforce diversity goes beyond just age or gender to include ethnic background, sexual orientation, and/or gender identity (Baum et al., 2007; Robbins and Coulter, 2014; Singal, 2014; Sourouklis and Tsagdis, 2013). Depending on local laws and immigration policies, workplace diversity in hospitality is significantly higher than other industries, which is a direct result of the broad customer base of international travelers that hotels accommodate and the ethnic background of employees, particularly expatriate employees (Singal, 2014). It is not uncommon to find a diverse mix of cultures in a hotel; for example, a typical ethnic background of a hotel team could look like (Fig. 22.4)

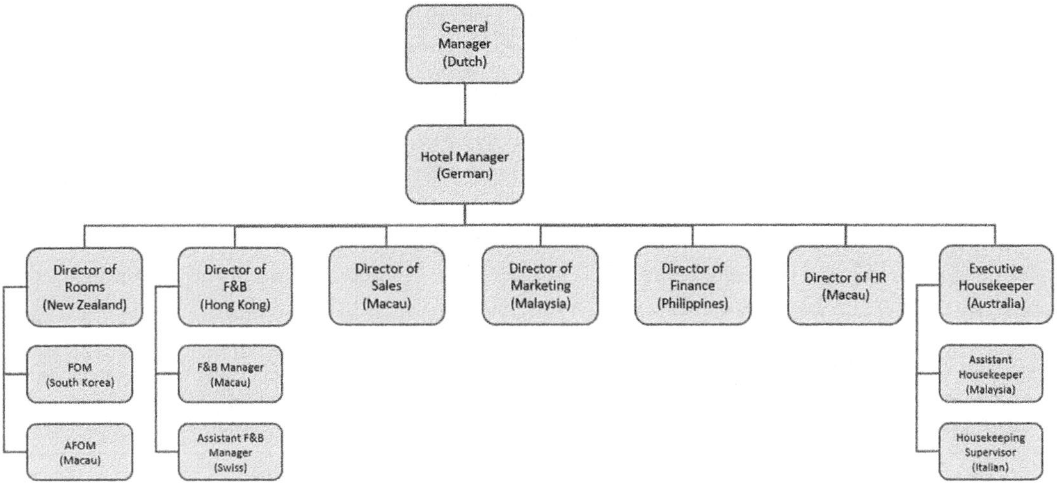

FIGURE 22.4 A typical hotel team. *Source: Patrick Newsham*

The example of ethnic diversity illustrated above extends right through to frontline, customer-facing staff and back-of-house staff. Managers need to be very skilled at managing these dynamic groups of individuals, which is not a one-size-fits-all approach to management and the success of the organization (Reynolds et al., 2014). Ultimately, a hospitality operator's key point of differentiation is through its people, attracting, training, retaining, and developing their talent, despite the diversity of their background to exceed guest expectations (Barney and Wright, 1998; Reynolds et al., 2014; Schneider and Bowen, 1993).

22.3.4.2 I guess you changed your mind

As employees move up the corporate ladder, they tend to be deskbound—the outcome is that they are seen by frontline employees. This lack of visibility results in a "disconnect" between management and frontline staff. Highly efficient and effective management requires managers to practice the art of management by walking around (MBWA). Breaking down this process, MBWA works from two perspectives; MBWA provides managers with a wealth of opportunity to learn about their people and their operations through direct observation. Additionally, when managers practice MBWA, they increase their visibility among employees which builds on perceived organizational support (Eisenberger et al., 1986; Johnson, 2012). This management practice is an effective form of control, allowing managers to monitor, identify, and, where necessary, make corrections to workplace activities (Robbins and Coulter, 2014). Let us look at the following example of a General Manager and the principle of using *MBWA*; *Miguel* is the General Manager of a hotel. His normal working hours are 9 a.m.–6 p.m., Monday to Friday, and he will often drop in to work on a Saturday when the need arises. However, Miguel is an avid believer in the principle of MBWA. Waking up at 5.30 a.m., Miguel heads to the hotel gym for his morning workout, engaging in conversation with gym staff and the occasional regular guest/gym member. After the workout he showers and changes into his suit, has his breakfast in the hotel restaurant, again, engaging with restaurant staff. Once Miguel finishes breakfast he spends the next 60–90 min walking around the hotel, engaging in conversations with staff who are finishing the night shift. Miguel then heads to his office to check emails in preparation for the day ahead. Miguel attends the EXCO meeting at around 9.30 a.m., where he gets a chance to learn key operational success and failures from the day before and any important areas of business for the coming day/week. Miguel will spend time in the administrative (back-of-house) areas of the hotel before or after lunch and will do his "rounds" for 60–90 min again at the end of the day. Miguel encourages senior and middle managers to practice MBWA in their daily work habits as a means to drive organizational performance.

22.4 Customer-oriented (operational level)

The capacity of frontline or customer-facing employees to deliver high levels of service is imperative to the ongoing success of service-based organizations, providing a significant opportunity to gain market share through differentiation (Berry, 1995; Bitner et al., 1990; Hennig-Thurau, 2004; Parasuraman et al., 1985; Schneider et al., 2009; Susskind et al., 2017). However, delivering quality service consistently every day to a broad range of customer wants and needs is no easy task. Frontline customer-facing employees must have the appropriate skills and attributes to fulfill the broad, constantly changing expectations of customers and the strategic goals of the organization (Hennig-Thurau, 2004; Schneider and Bowen, 1985; Zeithaml et al., 1988). Typically, the majority of frontline employees are newcomers to the industry, who receive low-pay, work long or odd hours, lack relevant experience, and receive either inadequate or infrequent training to achieve the high service standards as set by the organization (Berry et al., 1990; Bitner et al., 1990). To improve service quality, it is important to first consider the effectiveness of the human resource systems and practices that drive service delivery outcomes (Bowen and Ostroff, 2004; Kale and Spence, 2009). Essentially, it is the management's responsibility to ensure customer-facing employees are paid fairly, treated well, and have had the appropriate quantity and quality of training and ongoing development to effectively and efficiently engage with customers (Bowen and Ostroff, 2004; Li and Huang, 2017). Heskett et al., (1994) proposed the Service Profit Chain model that includes many of the aforementioned elements through a linkage of internal service quality, employee satisfaction, and customer satisfaction. Evidence suggests that an antecedent to achieving high levels of customer satisfaction is improving the internal service climate, which is "employee perceptions of the practices, procedures, and behaviors that get rewarded, supported, and expected with regard to customer service and customer service quality" (Schneider et al., 1998, p. 151). For managers, understanding the importance of the internal service orientation of an organization can provide significant insight into actual customer satisfaction. For example, Kralj and Solnet, (2010) researched service climate in a gaming resort and identified differences in the attitudes of casino, hotel, and food and beverage employees, whereas Huang et al. (2016) identified that in Chinese hospitality organizations service climate improved significantly in-line with the intensity of competition. Research by Eisenberger et al. (1986) suggests that organizational goals can only be met if the organization takes care of its employees first. Evidence suggests that employees will form their own perception of how well "the organization values their contributions," which in turn determines the employee's efforts and commitment to the organization (Aselage and Eisenberger, 2003; Eisenberger et al., 1986, p. 500). This give–take relationship is called reciprocity, whereby two parties give and receive treatment that is equally fair and mutually favorable (Gouldner, 1960). This mutually beneficial relationship helps the organization in two key ways: greater performance and less employee turnover (Rhoades and Eisenberger, 2002). From the employee's perspective, they develop a level of trust in the organization, which evolves through the organization's capacity to effectively communicate and deliver support to the employee (Trembath and Herbert-Hansen, 2017). This becomes a delicate balancing act as service employee's experience significant workplace stressors due to the ever-increasing demands of both customers and management (Schneider, 1980). Building this "relationship" takes time, yet emphasizes the importance the management function has in forming the perception of employees to develop a unique service climate, resulting in higher customer satisfaction and organizational performance (Berkley and Gupta, 1995; Schneider et al., 1994). However, Schneider et al. (2000) suggest that an organization's service climate is not limited to customer-facing staff, it extends to back-of-house employees across all departments, suppliers, and contractors. Nevertheless, maintaining consistently high customer satisfaction levels is extremely difficult to achieve (Zeithaml et al., 1988).

The pressure on hospitality operators to maintain and improve financial performance has a tendency to overshadow the antecedents that drive performance outcomes (Presbury et al., 2005). The human resource policies and practices are a key antecedent to performance outcomes and will enable the organizational team members to attract, select, retain, and develop employees who are both willing and able to work in a high pressure, high volume environment that demands high levels of service (Chan and Kuok, 2011; Kale, 2007; Zeithaml et al., 1988). Provided that the human resource policies and practices are highly effective, they will facilitate employee capabilities in delivering high levels of consistent service and improve performance expectations and outcomes (Armstrong, 2010; Schneider and Bowen, 1999).

References

Armstrong, M., 2010. Armstrong's Essential Human Resource Management Practice: A Guide to People Management. Kogan Page, London.

Aselage, J., Eisenberger, R., 2003. Perceived organizational support and psychological contracts: a theoretical integration. J. Organ. Behav. 24 (5), 491–509. https://doi.org/10.1002/job.211.

Bardi, J.A., 2011. Hotel Front Office Management, fifth ed. John Wiley & Sons, Hoboken.

Barney, J.B., Wright, P.M., 1998. On becoming a strategic partner: the role of human resources in gaining competitive advantage. Hum. Resour. Manag. 37 (1), 31–46.

Barrows, C.W., Powers, T., Reynolds, D., 2012. Introduction to Management in the Hospitality Industry, tenth ed. John Wiley & Sons, Inc, Hoboken, New Jersey.

Baum, T., Dutton, E., Karimi, S., Kokkranikal, J., Devine, F., Hearns, N., 2007. Cultural diversity in hospitality work. Cross Cult. Manag. Int. J. 14 (3), 229–239. https://doi.org/10.1108/13527600710775775.

Berkley, B.J., Gupta, A., 1995. Identifying the information requirements to deliver quality service. Int. J. Serv. Ind. Manag. 6 (5), 16–35.
Berry, L.L., 1995. On Great Service - A Framework for Action. The Free Press, New York.
Berry, L.L., Zeithaml, V.A., Parasuraman, A., 1990. Five imperatives for improving service quality. Sloan Manag. Rev. 31 (4), 29–38.
Bitner, M., Booms, B., Tetreault, M., 1990. The service encounter: diagnosing favorable and unfavorable incidents. J. Mark. 54, 71–84.
Bowden, J., 2009. Customer engagement: a framework for assessing customer-brand relationships: the case of the restaurant industry. J. Hosp. Mark. Manag. 18 (6), 574–596. https://doi.org/10.1080/19368620903024983.
Bowen, D.E., Ostroff, C., 2004. Understanding HRM–firm performance linkages: the role of the "strength" of the HRM system. Acad. Manag. Rev. 29 (2), 203–221.
Bowen, J.T., Shoemaker, S., 2003. Loyalty: a strategic commitment. Cornell Hotel Restaur. Adm. Q. 44 (5/6), 31–46.
Brien, A., Smallman, C., 2011. The respected manager… the organisational social capital developer. Int. J. Hosp. Manag. 30 (3), 639–647. https://doi.org/10.1016/j.ijhm.2010.11.009.
Chan, S.H., Kuok, O.M., 2011. A study of human resources recruitment, selection, and retention issues in the hospitality and tourism industry in Macau. J. Hum. Resour. Hosp. Tour. 10 (4), 421–441. https://doi.org/10.1080/15332845.2011.588579.
Dopson, L.R., Hayes, D.K., 2011. Food and Beverage Cost Control, fifth ed. John Wiley & Sons, Hoboken, New Jersey.
Eisenberger, R., Huntington, R., Hutchison, S., Sowa, D., 1986. Perceived organizational support. J. Appl. Psychol. 71 (3), 500–507.
Gouldner, A., 1960. The norm of reciprocity: a preliminary statement. Am. Sociol. Rev. 25 (2), 161–178.
Hennig-Thurau, T., 2004. Customer orientation of service employees. Int. J. Serv. Ind. Manag. 15 (5), 460–478. https://doi.org/10.1108/09564230410564939.
Heskett, J.L., Jones, T.O., Loveman, G.W., Sasser, W.E., Schlesinger, L.A., 1994. Putting the service-profit chain to work. Harvard Business Rev. 72 (2), 164–174.
Huang, J., Li, W., Qiu, C., Yim, F.H.-k., Wan, J., 2016. The impact of CEO servant leadership on firm performance in the hospitality industry. Int. J. Contemp. Hosp. Manag. 28 (5), 945–968. https://doi.org/10.1108/ijchm-08-2014-0388.
Iredell, A.E., 1963. Control in food service operations. Cornell Hotel Restaur. Adm. Q. 4 (3), 39–45. https://doi.org/10.1177/001088046300400308.
Johnson, B., 2012. Graduate to leadership with an MBWA. Appl. Retail. 18 (4), 11-11.
Kale, S.H., 2007. Internal marketing: an antidote for Macau's labor shortage. UNLV Gaming Res. Rev. J. 11 (1), 1–11.
Kale, S.H., Spence, M.T., 2009. Casino customers in Asian versus western gaming jurisdictions: implications for western casino operators. Worldw. Hospit. Tourism Themes 1 (4), 320–331. https://doi.org/10.1108/17554210911002165.
Kapoor, C., Madera, J.M., 2011. Defining diversity: the evolution of diversity. Worldw. Hospit. Tourism Themes 3 (4), 284–293. https://doi.org/10.1108/17554211111162408.
Kralj, A., Solnet, D., 2010. Service climate and customer satisfaction in a casino hotel: an exploratory case study. Int. J. Hosp. Manag. 29 (4), 711–719. https://doi.org/10.1016/j.ijhm.2010.01.005.
Li, Y., Huang, S., 2017. Hospitality service climate, employee service orientation, career aspiration and performance: a moderated mediation model. Int. J. Hosp. Manag. 67, 24–32. https://doi.org/10.1016/j.ijhm.2017.07.012.
Narteh, B., Agbemabiese, G.C., Kodua, P., Braimah, M., 2013. Relationship marketing and customer loyalty: evidence from the Ghanaian luxury hotel industry. J. Hosp. Mark. Manag. 22 (4), 407–436. https://doi.org/10.1080/19368623.2012.660564.
O'Fallon, M.J., Rutherford, D.G., 2011. Hotel Management and Operations, fifth ed. John Wiley & Sons, Hoboken.
Parasuraman, A., Zeithaml, V.A., Berry, L.L., 1985. A conceptual model of service quality and its implications for future research. J. Mark. 49 (4), 41–50.
Presbury, R., Fitzgerald, A., Chapman, R., 2005. Impediments to improvements in service quality in luxury hotels. Manag. Serv. Qual. Int. J. 15 (4), 357–373. https://doi.org/10.1108/09604520510606835.
Reynolds, D., Rahman, I., Bradetich, S., 2014. Hotel managers' perceptions of the value of diversity training: an empirical investigation. Int. J. Contemp. Hosp. Manag. 26 (3), 426–446. https://doi.org/10.1108/ijchm-02-2013-0079.

Rhoades, L., Eisenberger, R., 2002. Perceived organizational support: a review of the literature. J. Appl. Psychol. 87 (4), 698–714. https://doi.org/10.1037/0021-9010.87.4.698.

Robbins, S.P., Coulter, M., 2014. Management, twelfth ed. Pearson Education, Essex.

Rodgers, S., 2007. Innovation in food service technology and its strategic role. Int. J. Hosp. Manag. 26 (4), 899–912. https://doi.org/10.1016/j.ijhm.2006.10.001.

Schneider, B., 1980. The service organization: climate is crucial. Organ. Dyn. 9 (2), 52–65. https://doi.org/10.1016/0090-2616(80)90040-6.

Schneider, B., Bowen, D., 1985. Employee and customer perceptions of service in banks: replication and extension. J. Appl. Psychol. 70 (3), 423–433.

Schneider, B., Bowen, D.E., 1993. The service organization: human resources management is crucial. Organ. Dyn. 21 (4), 39–52. https://doi.org/10.1016/0090-2616(93)90032-V.

Schneider, B., Bowen, D.E., 1999. Understanding customer delight and outrage. Sloan Manag. Rev. 41 (1), 35–45.

Schneider, B., Bowen, D.E., Erhart, M.E., Holcombe, K.M., 2000. The climate for service: evolution of a construct. In: Ashkanasy, N.M., Wilderom, C.P.M., Peterson, M.F. (Eds.), Handbook of Organizational Culture and Climate. Sage, Thousand Oaks, pp. 21–36.

Schneider, B., Gunnarson, S.K., Niles-Jolly, K., 1994. Creating the climate and culture of success. Organ. Dyn. 23 (1), 17–29. https://doi.org/10.1016/0090-2616(94)90085-X.

Schneider, B., Macey, W.H., Lee, W.C., Young, S.A., 2009. Organizational service climate drivers of the American customer satisfaction index (ACSI) and financial and market performance. J. Serv. Res. 12 (1), 3–14. https://doi.org/10.1177/1094670509336743.

Schneider, B., White, S.S., Paul, M.,C., 1998. Linking service climate and customer perceptions of service quality: test of a causal model. J. Appl. Psychol. 83 (2), 150–163.

Shoemaker, S., Lewis, R.C., 1999. Customer loyalty: the future of hospitality marketing. Int. J. Hosp. Manag. 18 (4), 345–370. https://doi.org/10.1016/S0278-4319(99)00042-0.

Singal, M., 2014. The business case for diversity management in the hospitality industry. Int. J. Hosp. Manag. 40, 10–19. https://doi.org/10.1016/j.ijhm.2014.02.009.

Sourouklis, C., Tsagdis, D., 2013. Workforce diversity and hotel performance: a systematic review and synthesis of the international empirical evidence. Int. J. Hosp. Manag. 34, 394–403. https://doi.org/10.1016/j.ijhm.2012.11.013.

Susskind, A.M., Kacmar, K.M., Borchgrevink, C.P., 2017. Guest–server exchange model and performance: the connection between service climate and unit-level sales in multiunit restaurants. J. Hosp. Tour. Res. 20 (10), 1–20. https://doi.org/10.1177/1096348016683512.

Trembath, J.-L., Herbert-Hansen, Z.N.L., 2017. Expatriate academics and perceptions of organisational support. Eur. J. Int. Manag.

Zeithaml, V.A., Berry, L.L., Parasuraman, A., 1988. Communication and control processes in the delivery of service quality. J. Mark. 52 (2).

CHAPTER

23

The power of service quality: front-of-house service skills

Florence Iok San Lam, Alan Wan Lok Cheng, Cindia Ching Chi Lam

Institute for Tourism Studies, Macao (IFTM): Colina de Mong-Há Macao, China

OUTLINE

23.1 Introduction

It will be naïve to think that anybody who makes a decision to walk into a restaurant holds only the intention to satisfy the basic needs—the need for food and drink. The consumers in today's world are becoming even more sophisticated and educated in the way they choose a product or service; they have broader experiences with food service operators and have developed a more varied palate and wider tastes in food and beverages. Together with the intrinsic desire to seek out new culinary experiences, the competitive pressure from different food service providers competing on different basis, and the emerging substitutes for restaurants (for example, home cooking and dining in as a modern quality lifestyle concept), the

Food and Society
https://doi.org/10.1016/B978-0-12-811808-5.00023-4

bottom line is that customer expectations are growing and they are getting increasingly difficult to impress.

Indeed, these customers are looking for a dining "experience," and it is people, the servers, who will bring this experience to life. One could imagine the dining experience as a theater performance, where the diners are the audience and the service personnel is the cast. The restaurateur will come up with the concept for the dining establishment, the interior design will give the space its intended vibe and atmosphere, and the chef will create a scrumptious menu. Once these are done, the imaginary theater is built and a good show will rely on the people on stage, the servers who will deliver the service, to set the right balance of "rhythm, timing, and flow" of the performance. To deliver pleasant and memorable dining experiences in the contemporary epoch, service quality plays a significant role with equivalent importance as delectable meals. This chapter will explicate the meaning and spirit of service quality and provide practical guidelines on good service practices for frontend operators.

23.2 Service and service quality

The meaning of service has been widely studied and interpreted. It brings out that the activities of a food service business are not limited to the production and provision of food and drinks but a large part of it consists of a hospitality or service component.

> "Services are those separately identifiable and essentially intangible activities that provide satisfaction of wants, and are not necessarily linked to the sale of a product or another service". ***Stanton (1986)***

> "An action or an activity which can be offered by a party to another party, which is basically intangible and cannot affect any ownership. Service may be related to a tangible product or an intangible product". ***Kotler (1997)***

It is not an understatement that the survival of a restaurant operation of any scale relies heavily on returning or repeat customers, and its service quality dictates the business' lifeline. While good service adds significant value to a business that translates into profitability through higher margins, poor service comes with a huge cost. Ron Kaufman, a service guru and author of "Uplifting Service" (2012) who had worked with the workforce in Singapore in cultivating their customer-focused service mindset in the years 1985–2010, when the country was transforming itself from a manufacturing economy to a service economy, said poor service can "mess things up" even when the product is great. Indeed, when put into the context of a food service business, excellent service can make up for bad food (at times) while the vice versa may not be so true. In most cases, it takes only one negative service experience to lose a customer—forever—even when food is exquisite.

Pine and Gilmore (1999) argued "there is a need to provide the customer with a memory of an experience that they can take away with them rather than just a good product or service." Quality service has that magic power to transform an average meal to a delightful and remarkable experience or turn around customer anger to customer loyalty.

> The Restaurateur who provides great service and value has a competitive advantage over those operators who do not. ***Stevens et al. (1995)***

On the contrary, the biggest cost linked to poor service is paid when negative reviews and word of mouth spread through social networks faster than any good words and exciting news about the business resulting in unpredictable lost in revenues and damage to the brand and reputation. In this perspective, the restaurateur is placing high stakes on its frontend service staffs who directly contact with the customers, while their service quality will impose influential effects.

Broadly speaking, the more upscale is the dining establishment, the more expectations a customer have in receiving quality service. But how should we measure quality, when quality is a complex, subjective concept, more difficult to precisely define and measure than in physical goods. We may at least attempt by saying that, in a broad sense, quality service in a sit-in restaurant does not mean providing food at the fastest possible speed, but rather involves delivering food at the right time desired by the customer, at the desired temperature, with the desired layout of tableware.

Parasuraman et al. (1985) had developed a conceptual model called "SERVQUAL" to measure service quality in a restaurant using five dimensions: reliability, responsiveness, empathy, assurance, and tangibles. These are referred as the determinants that put quality in service. Customers form their expectations by gathering information through word of mouth, past experience, marketing communications, personal needs, and desires for the service or product they will experience (Fig. 23.1).

In determining quality, customers compare their expectations about a service with the overall quality they perceive to have received during the actual service. Following the model, if the restaurant's service performance meets or exceeds these expectations, then they will have a happy, satisfied customer who will feel they have received "quality." If perception of the service received does not match the expectations, then there is a gap between customer expectations and satisfaction level. When this happens, that is when the customers will turn to competitive products or service.

Following this definition, quality can occur at any level of service, as long as expectations of that level of service are met. Food service operators, from fast food restaurant to fine dining, who understand quality service are those that endeavor to close the "gap" to achieve customer satisfaction by matching expected quality to perceived quality (actual service).

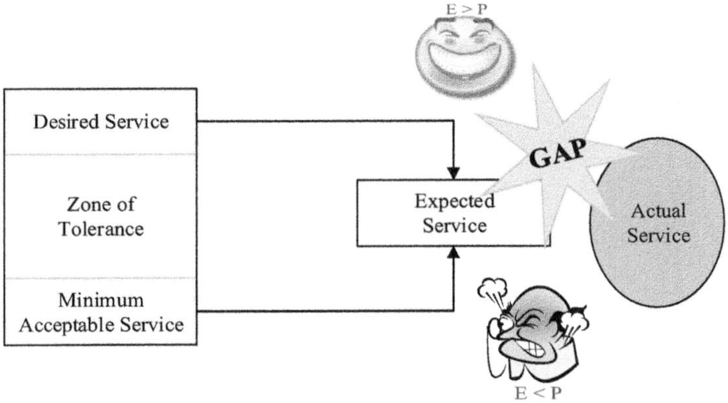

FIGURE 23.1 Service and service quality.

However, the fact is that even if expectations are met, customers may still go to new operators—there is nothing to lose. To pursue excellent quality, restaurateurs should look to exceed expectations, by giving what would surprise and what would be desired and not just what is expected.

23.3 Managing service quality

The advantages of quality service are evident—competitive advantage, long-term customer loyalty, reputation, and increased profitability. When the benefits are so great, why few food service operations become champions in delivering service standard consistently? And why problems as a result of negligence in customer service are notable in so many food service operations?

It is easy to discuss hospitality, but to practice it consistently is never an easy task. The characteristics of a service operation have restricted the food service industry from adopting a "pure" product-manufacturing operation process. This provides insights into the many challenges for food service operations at achieving quality consistency in their level of service. Lamb et al., (2011) suggested that service has unique characteristics, which differentiate it from a product: intangibility, involvement, inseparability, perishability, and heterogeneity. Let us understand each one of these characteristics and their specific implications for food service operations below.

Intangibility: We know diners seek for an experience. "Meal experience" as a term was first introduced by Campbell-Smith in 1967 but fast became the buzzword in the food service industry. Unlike some other industries, the food service operation has a production unit consisting of a product (i.e., food and drinks) and a service performance. Fig. 23.2 shows the

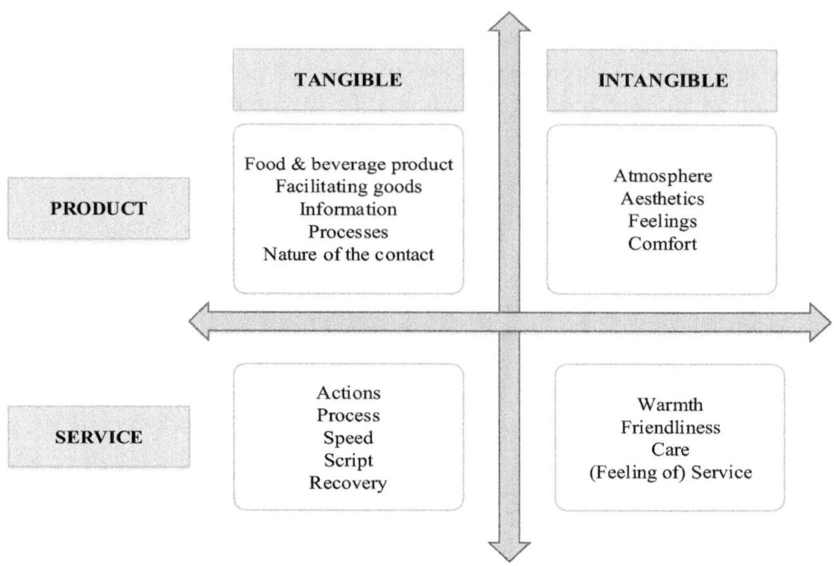

FIGURE 23.2 The product/service matrix. *Source: Adapted from Davis, B., Lockwood, A., Alcott, P., Pantelidis, I., Food and Beverage Management, fifth Edition, P.334.*

product/service matrix as in a dining experience characterized by a customer's encounter of the restaurant's products and service, both comprising of tangible and intangible factors.

The service aspect of a food service operation is largely intangible. Service is not something that can be seen, listened to, or touched. It is therefore difficult to quantify, measure, and evaluate service. An operator can measure the speed at which food is delivered to the table, listen and evaluate the effectiveness of a customer service script, or set out service actions and delivery procedures and see and assess how closely they are followed. However, neither the operator nor its customers will be able to tell service crew the exact level of warmth, empathy, and friendliness needed and desired in every single service encounter when the evaluations of these qualities can be very subjective.

Even the very tangible elements of the food service operation possess intangible factors, such as the atmosphere created and felt, esthetics of the food, which affect the interaction between the operator and the customers, and the perception of service quality delivered. It is easy to imagine that a service operator can control the tangibility factors. However, the intangibility part is often regarded by the customers as just as, if not more, important.

Involvement and inseparability: In a food service operation, the customer participates and is involved in the provision of the service (e.g., when the customer selects and orders a food item on the menu) that is consumed at the same time as it is produced. The customer may get personalized services according to his/her specific needs, but it also means that satisfactory service delivery depends on positive customer's participation that is voluntary and this is difficult to control.

Service cannot be taken away for future consumption or enjoyment. Both the service consumer and the service producer have to be present for service provision and consumption that are inseparable and happens almost if not simultaneously. Imagine that in the food service industry, the ingredients for meal can be purchased, prepared, cooked, served, and consumed within a day or even a few hours. The very little or absence of time interval between service production and consumption adds difficulties in managing consistency in service standards.

Perishability: Service is highly perishable, cannot be stored, and has virtually zero shelf life. In the context of a restaurant business, a vacant table during its operating hours means potential revenue lost forever. Unlike manufacturing physical goods, service cannot be produced ahead to the time of demand and stored, making it more challenging to implement control procedures and quality management.

Heterogeneity: Service has a large measure of human factors. As service in the food service industry is dominantly an activity performed by human beings and not machines, the standard of performance of a service staff may vary from day to day and from customer to customer. This characteristic of service allows service staff to modify the service according to the needs of a customer as every customer is treated as an individual with different needs and desires. On the downside, it is difficult to ensure consistent quality from a service member and even more so to avoid irregularities between service members in the service delivery process. This, however, directly affects what the customers receive. The food service industry is a demanding industry generally characterized by long, irregular working hours including working weekends and holiday seasons, comparatively lower wages, high pressure, and heavy workload. These unfriendly industry conditions do not make managing human resources and maintaining service quality in a food service operation easier.

23.3.1 Frontend service skills

Remarkable service requires a lot of tact and attention to details. Too much service will make customers feel demeaned, while inadequate service will make customers feel being neglected. The typical common vision of basically all catering organizations is to provide high-quality "catering products and services" to customers. As a restaurant operator, your "catering products and services" should provide the kind of experience to your customer that can satisfy both their physical and psychological needs, i.e., hunger and experiences.

The frontend operators, who can be waiters, managers, public relations officers, and even chefs, are usually described as "the first point of contact" for arriving customers. They are therefore regarded as the staff responsible for creating the first good professional impression on customers—remember, as a frontend operator, you represent your restaurant!

Customers may come to dine at your restaurant for pleasure, experience, or simply because of hunger. A true understanding of this may help you, as a professional frontend operator, provide the highest-level personalized customer services. You should be people-oriented and customer-focused. You should always feel happy and ready to assist or even solve problems for your customers. To perform these duties effectively and efficiently, you are expected to have good listening and communication skills and here lies the important role of language ability in communication. In fact, it is always highly desirable that you are able speak more than one language or dialect. You do not have to be perfect in each one of them. Sometimes, being able to greet customers in their native language is already good enough to "shine" your professionalism and great attitude toward work.

23.3.1.1 Communication skill

What does effective communication mean? It implies being able to get a message across from one person to another successfully. It is always "easier than done" because communication looks like a very common thing to most of us. This might be the reason why on rare occasions would a person pay close attention to how they could make continuous improvements on this. However, having a true understanding of effective communication skills will provide you with a personal motivation to proactively identifying your customers' needs so as to build a good rapport with them. A frontend service operator often meets with people face to face or over the phone. Most often you will be asked questions by your customers and you might have to keep on responding to the same questions. As a professional, every question should be treated as if you encountered it for the very first time. With this mentality in mind, it is very likely that you will be able to derive constant satisfaction from your work, which will maintain your growing enthusiasm about your profession.

In terms of communication, one of the main duties is to answer phone enquires. There are high expectations that they should handle telephone calls efficiently. For example, when you are taking down the full name of your prospective customer at the time of taking a table reservation, it is advisable that the phonetic alphabet be used to avoid possible mistakes in the spelling of their name, which is generally regarded as a kind of "must-avoid" error in front desk operations.

When communicating with customers over the phone, you must note the following:

1. The caller cannot see you because this is not a face-to-face interaction. They all rely on your voice to decide whether you are a professional.

2. You must always clearly identify your restaurant and yourself when answering at telephone call.
3. The phone connection might not be good enough and therefore it might be difficult to hear or understand the caller.
4. Try to understand or have empathy that the caller has a need and reason for calling.
5. You actually represent your restaurant each time when you answer a telephone call.
6. Note that if the call is not a toll-free one, the caller is paying for the call.

If you can truly understand and bear the above important points in mind, you will always find yourself in position to keep improving your communication techniques.

23.3.2 Guidelines on good telephone skills

1. Always answer your call with a greeting: Good Morning/Good Afternoon/Good Evening. Remember to greet the caller in a polite manner. Answer the call within three rings, or for consistency on the third ring.
2. Always identify your restaurant. Otherwise, the caller will not be sure if they have dialed the correct number or place.
3. Always identify yourself professionally: "This is John speaking. How may I help you? What can I do for you?"
4. When speaking on the telephone, make sure you speak politely. Also, pay special attention to the speaking tone you are using because your tone will tell the caller whether you are in a positive position to help.
5. Because this is a telephone conversation, make sure you speak clearly and slowly, should the connection or transmission appear not as good as desired or expected.
6. Because the customer(s) over the phone might not be familiar with the catering industry, you should avoid using jargon or technical terms should your guests find it hard to understand.
7. Make sure you listen attentively, which is the first and foremost thing to achieve in what we always call "effective communication."
8. Ask the caller for his/her last name politely and remember it. Try to use it at appropriate intervals to show your respect to the caller.
9. Take down important notes where necessary.
10. Make sure you understand the caller's request and then offer appropriate assistance.
11. Always end a telephone conversation in a professional manner by using some polite words and expressions such as "Thank you for your call" or "Have a good day," etc.
12. Remember to always hang up after the caller has done so.
13. A concluding remark: "Smile" through the telephone conversation!

23.3.2.1 Knowing your "catering product"

Remember, when we refer to "catering product," it means much more than the "food" or "beverage" itself. It extends to include other unique "features" of your restaurant. For example, the location of your restaurant is considered as part of your catering product. Your restaurant can be the vicinity of local tourist attractions and therefore it may be this particular "characteristic" of your restaurant that has attracted a particular group of customers. So, by familiarizing yourself with the restaurant you are working at, you will have a good knowledge of the "product" you are selling. This will certainly help project your professional image. So, what and how much do you need to know about your restaurant to be able to demonstrate your professionalism?

1. Make sure that you can accurately describe the location of your restaurant. You can do this by describing the distance to the airport, ferry terminals and major sites and buildings, neighboring landmarks, average travel by taxi, etc. Also, ensure you know all modes of transport to your restaurant.
2. Ensure that you can readily provide the address, telephone and fax numbers, e-mail address, and website of your restaurant. Be sure you know the area code of your country or city.
3. Because you might need to always communicate with your prospective customers about the types of food your restaurant offered, you should make sure you know the different dishes and their unique flavors. More importantly, you have to make sure you know and able to describe the chef recommendations, set meals, special promotion, operation hours, and restaurant rules and policies (e.g., whether smoking is allowed, if minimum charges apply to private rooms, charges on food and beverages brought in by customers).
4. Double check whether you have a good knowledge about the beverages offered: wine, beer, soft drinks. You should also ask yourself whether you can tell the name of chefs and if sommeliers are available.
5. Make sure that you know the availability of tables, private rooms, and availability for functions and events.
6. Make sure that you know what types of payment method your restaurant accepts, for example, local and foreign currency, credit cards (Visa, MasterCard, American Express, UnionPay, JCB, Diners, travellers' checks). You should also be able to give accurate information about your reservation policy, such as waiting time, cancellation policy, and no-show policy under a guaranteed reservation. For example, waiting time is normally 15 min from time of reservation.
7. Ensure you know the different services and facilities offered by your restaurant, for example, whether your hotel provides such services as valet services or pick-up service from the airport or ferry terminal; whether your restaurant has handicapped accessible facilities or wheelchair access, etc.

23.3.2.2 *Do's and Don'ts*

1. Always acknowledge your customers with polite greetings and good eye contact.
2. Always maintain good eye contact with your customers when serving them.
3. Always mention your customer's last name at least three times in one encounter.
4. When you hand an item to or receive it from your customer, always use both hands to demonstrate professionalism.
5. Always show a natural smile on your face. Good emotional quotient plays an important role in the service sector, especially when you come across a situation where you may need to deal with difficult customers.

6. Always appear well groomed. Remember, you are always supposed to project a professional image on behalf of your restaurant.
7. Always learn how to strike an appropriate balance between providing quality services to customers and complying with your restaurant policies. For example, customers must not smoke in a smoke-free restaurant. In a situation where you see a guest smoking in front of you, think about how you should deal with this in a professional manner.
8. Ensure confidentiality, privacy, and security.
9. Never have any "casual talk" with your colleagues when you are on duty, especially when you are serving a customer. You will look very "unprofessional" if you do this kind of thing when there is a customer in front of you or being served by you.
10. Always use the employee entrance when entering your restaurant.
11. Always remind the customer of the cancellation policy of your restaurant when a customer has chosen to make a guaranteed reservation.
12. Always learn how to "control the encounter" when serving your customers. One effective way of achieving this is to "do the asking" so as to proactively identify customers' needs, thereby providing them with a "Vow" staying experience with your property. As a frontend service person, you are not supposed to take a passive role in a service encounter. Taking a passive role in a service encounter may project an undesirable image that you are not enthusiastic about your work. More importantly, this may go against our common principle and practice of "courtesy" in the service industry.

Customers are always demanding but not difficult. Staff needs to be highly technically skilled, when one key to success is—empathy—to look at things from the perspective of the customers—focusing on the person, on settling the person's emotions first, and on understanding the customer's feelings and not rushing to resolve the situation. Servers should be trained to identify behaviors that indicate a service need or to predict a service need. Many scholars have studied the psychology of service to satisfy customers by satisfying their self-esteem and in pleasing the persons being served.

23.4 The way forward

A restaurant looking to meet or even exceed customers' expectations will achieve a "win-win" situation and in doing so will establish the reputation of an establishment. Constantly meeting these expectations reinforces that reputation but, before being able to do so, it is necessary to identify what the customers want in terms of service—e.g., what are the levels of service expected, the behaviors desired, the target market demands, the service features that are seen important by different segments such as young couples, families with children, retirees.

In hospitality, the term "moment of truth" refers to an instance of contact between a customer and the service provider that adds up to form a continuing impression on the customer. If any of the moments of truth fail an expectation—service gap develops, meaning there is a service shortfall and a negative impression is formed. In a dining experience, examples of moments of truth include table reservation, greetings and customer seating, menu presentation and dinner ordering, table service, after-dinner upselling, dinner quality, quantity and presentation, bill presentation, and more. Understanding the moments of truth that are important to an operation's target customers is the first step toward offering good customer service. Service provider leaders also believe that excellent service begins with people with a right attitude. Service-oriented attitude drives hospitable behavior. Managers can reinforce positive, desired service behaviors and correct negative ones through well-designed training and appropriate reward schemes. Yet, attitudes are something difficult, if not possible, to change.

The restaurant business is a people business. Those working in the front of house must be genuinely passionate about serving people. The service team, as an "immediate interface" for the food service operator and its customers, is at the heart of the success of any food service operation. Yet, service personnel are often not regarded as high-status employees. They are described as filling "boundary spanning roles" (Bowen and Schneider, 1988) and "subordinate service roles" (Shamir, 1980).

In recognition of the difficult role of a service operator, Bowen and Schneider (1988) highlighted the importance of careful selection of "people with a predisposition and a motivation to play the role in an appropriate way" as one of the key approaches toward service operation success. They also suggested that staff training, development, and socialization is one of the top priority duties of the management in a quality-focused establishment. Managers must train, supervise, and support service staff in the development of their technical skills and interpersonal skills, among many other skills and knowledge, to provide service in a hospitable way that will win and keep customers. Gardner and Wood (1991) also mentioned on the importance of getting the right people to do the right job. They say that new staff in a restaurant that has a vision of hospitality should be selected for their personality traits and not their basic skills, be it culinary or service-related skills. On the other hand, some restaurateurs will say that their method of selection is simply somebody who can smile at himself/herself.

Managers also have the duty to set out a clear order of service outlining the standard service sequence servers should follow in performing certain duties within their individual operation. The order of service will provide a framework for service standards consistency; the service procedures will help enhance the esthetics of service while promoting productivity and safety. They are a code of service that will guide all service members, although they should not be followed as rigidly and blindly as in a way that will impose inconveniences and discomfort on the customers.

In reality, it is impossible to directly manage the thousands of individual actions and interactions that a front end service operator engages in with a customer in a day-to-day operation. From the perspective of management, service personnel should be empowered appropriately to allow them to respond to situations instantaneously using his or her own judgment and ways of doing things in satisfying the customer's needs in a timely manner. Empowerment leads to motivation for service performance especially for those employees who are driven by intrinsic satisfaction. In promoting a culture of hospitality, a friendly manager will set a positive role model for the service team. In many cases, many food service managers are increasingly seen to be moving away from the traditional authoritarian approach toward embracing a management structure characterized by fewer levels of hierarchy. Conducting regular customer surveys and actively seeking opinions of front-line personnel will help uncover hidden service areas that need improvement.

The use of technology such as computerized order placing system is also getting popularity in modern restaurants. The system reduces unnecessary runs to the kitchen to check orders so service staff can be stationed at the front of house to provide uninterrupted attention to customers. It also helps reduce stress due to interdepartmental conflicts by reducing direct contacts between the chefs at the back of house and the service members, who have "different roles, different values, and, in many cases, different genders." All these to some extent go toward ensuring quality service for the customers at the end.

References

Bowen, D.E., Schneider, B., 1988. Boundary-spanning-role employees and the service encounter: some guidelines for management and research. In: Czepiel, J.A., Solomon, M.R., Surprenant, C.F. (Eds.), The Service Encounter: Managing Employee/customer Interaction in Service Business, fourth ed. Lexington Books, Lexington, pp. 125–147. (1988).

Campbell-Smith, G., 1967. The Marketing of the Meal Experience. Surrey University Press, Guildford.

Gardner and Wood, 1991. Theatricality in food service work. International Journal Hospitality Management 10 (3), 267–278.

Kaufman, R., 2012. Uplifting Service: The Proven Path to Delighting Your Customers, Colleagues, and Everyone Else You Meet, second ed. Evolve Publishing.

Kotler, P., 1997. Marketing Management: Analysis, Planning, Implementation, and Control. Prentice Hall.

Lamb, C.W., Hair, J.F., McDaniel, C., 2011. Essentials of Marketing. Cengage Learning.

Parasuraman, A.P., Zeithaml, V., Berry, L.L., 1985. A conceptual model of service quality and its implication for future research (SERVQUAL). J. Mark. 49, 41–50.

Pine, J., Gilmore, J., 1999. The Experience Economy: Work Is Theatre and Every Business a Stage. Harvard Business School Press.

Shamir, B., 1980. Between service and servility: role conflict in subordinate service roles. Hum. Relat. 33 (11), 741–756.

Stevens, P., Knutson, B., Patton, M., 1995. Dineserv: a tool for measuring service quality in restaurants. Cornell Hotel Restaur. Adm. Q. 36 (2).

Stanton, J.W., 1986. Fundamentals of Marketing. McGraw-Hill Companies.

CHAPTER

24

The financial environment: Culinary math

Miguel Gassmann de Oliveira

Institute for Tourism Studies, Macao (IFTM): Colina de Mong-Há Macao, China

OUTLINE

As in any business, it is vital that you understand your figures. You need to know how you are doing as well as where you went wrong. Although the financial reports cannot tell you everything that is wrong, especially as we are in an industry with many personal touches, with experience, you should be able to at least see where the problem is just by looking at the figures.

Food and Society
https://doi.org/10.1016/B978-0-12-811808-5.00024-6

24.1 Percentages (%)

Before you begin to analyze any figures in a food and beverage operation, you need to understand what percentage (%) means. Very often when asked what your results were, or how much your revenue increased from one year to the next, talking plain numbers does not do it (the following equation).

Equation 24.1: Percentage (%) means part of a whole 1

$$\frac{\text{Part}}{\text{Whole}} = \text{Percentage}(\%)$$

Compiled by Miguel Gassman from multiple sources.

For example, you have 10 fruits in total and 4 of those fruits are apples. So, apples represent 40% of the total amount of fruit you have (the following equation):

Equation 24.2: Percentage (%) means part of a whole 2

$$\frac{\text{Apples}}{\text{Total fruits}} = \frac{4}{10} = 0.4 \times 100 = 40\%$$

Compiled by Miguel Gassman from multiple sources.

It is difficult to visualize an increase in revenue in monetary terms, take for example Gibby's French Burgers which has two outlets: One in the Town Center and one on the Beach Front. They each had an increase in revenue of $500,000.00 in the year 2018 in comparison to last year. It would seem that they each had the same amount of increase in revenue as we only expressed the amount of revenue each increased (Table 24.1):

However, when you talk about the increase in revenue in percentage terms, we can see that the growth revenue for the Beach Front outlet has by far exceeded that of the Town Center. The Beach Front's revenue for this year has increased by 12% over that of last year. This is double of what the Town Center's increase in revenue, which was 6.25%.

To calculate the percentage of increase from one year to the next, we have to calculate the variance %, which brings us to the next step.

TABLE 24.1 Revenue variance.

	Town Center	Beach Front
Revenue this year	$8,500,000.00	$4,500,000.00
Revenue last year	$8,000,000.00	$4,000,000.00
Variance in revenue	$500,000.00	$500,000.00

Compiled by Miguel Gassman from multiple sources.

24.2 Variance

The variance in revenue is the difference in revenue earned between two different time periods (the following equation):

Equation 24.3: Revenue variance

Revenue this year − Revenue the previous year = Revenue variance

Compiled by Miguel Gassman from multiple data sources.

In the example of Gibby's French Burgers above, for the Town Center restaurant, we can calculate the variance in revenue from 2017 to 2018 as

Equation 24.4: Variance in revenue 2017/2018.

Revenue this year	−	Revenue the previous year	Revenue variance
$8,500,000.00	−	$8,000,000.00	$500,000.00

Compiled by Miguel Gassman from multiple data sources.

24.2.1 Gibby's French Burger Town Center revenue variance

To calculate the revenue variance %, we utilize one of the following three formulas (Equation 24.5):

Equation 24.5: Revenue variance %.

$$\frac{\text{Revenue variance}}{\text{Revenue last year}} \times 100 = \frac{\$500,000}{\$8,000,000} \times 100 = 0.0625 \times 100 = 6.25\%$$

$$\frac{\text{Revenue this year} - \text{Revenue last year}}{\text{Revenue last year}} \times 100 = \frac{\$8,500,000 - \$8,000,000}{\$8,000,000} \times 100 = 0.0625 \times 100 = 6.25\%$$

$$\left(\left(\frac{\text{Revenue this year}}{\text{Revenue last year}}\right) - 1\right) \times 100 = \left(\left(\frac{\$8,500,000}{\$8,000,000}\right) - 1\right) \times 100 = 0.0625 \times 100 = 6.25\%$$

Compiled by Miguel Gassman from multiple data sources.

24.2.1.1 Gibby's French Burger Beach Front revenue variance

Using one of these formulas, we can see the increase for the Beach Front outlet this year increased by 12.5% over last year (the following equation):

Equation 24.6: Beach Front revenue variance %.

$$\frac{\text{Revenue variance}}{\text{Revenue last year}} \times 100 = \frac{\$500,000}{\$4,000,000} \times 100 = 0.125 \times 100 = 12.5\%$$

Compiled by Miguel Gassman from multiple data sources.

Using the revenue variance for both outlets for this year in comparison to last year, we can see that the Beach Front outlet of Gibby's French Burgers has done much better that the Town Center outlet.

24.3 Average check

Another term that must be understood before moving on is the average check, which refers to the average amount of money spent by any given guest during a particular meal or period: The average check is calculated as follows (Equation 24.7):

Equation 24.7: The average food check

$$\text{Average check} = \frac{\text{Total revenue earned}}{\text{Total number of guests served}}$$

Therefore, the average food check is calculated as follows:

$$\text{Average food check} = \frac{\text{Total food revenue}}{\text{Total number of guests served}}$$

Compiled by Miguel Gassman from multiple data sources.

The average beverage check is calculated as follows (the following equation):

Equation 24.8: The average beverage check

$$\text{Average beverage check} = \frac{\text{Total beverage revenue}}{\text{Total number of guests served}}$$

Compiled by Miguel Gassman from multiple data sources.

For example, Gibby's served 54 people during lunch today. The revenue for lunch is shown in Table 24.2:

Broken down, this equates to that found in the following equation (Equation 24.9):

Equation 24.9 Total food and beverage variances broken down.

$$\text{Average food check} = \frac{\text{Food revenue}}{\text{No. of guest served}} = \frac{\$5,400.00}{54} = \$100.00 \text{ per guest}$$

$$\text{Average beverage check} = \frac{\text{Beverage revenue}}{\text{No. of guest served}} = \frac{\$2,700.00}{54} = \$50.00 \text{ per guest}$$

$$\text{Average food and beverage check} = \frac{\text{Food and beverage revenue}}{\text{No. of guest served}} = \frac{\$8,100.00}{54} = \$150.00 \text{ per guest}$$

Compiled by Miguel Gassman from multiple data sources.

TABLE 24.2 Total food and beverage variances total.

Food revenue	$5400.00
Beverage revenue	$2700.00
Total revenue	$8100.00

Compiled by Miguel Gassman from multiple data sources.

24.4 Forecasting

When we are planning for future results, it is difficult to predict what will happen. Very often managers predict the feeling with what they feel or expect will happen. This is called a gut feeling. However, one of the characteristics of being human is that we very often repeat ourselves. As such and taking into consideration that we tend to do the same things the same way, in essence what has happened in the past is a very good guideline for helping predict what will happen in the future.

For this we use the variance in revenue growth from past years to help us predict the revenue growth for the following year.

Take for example Gibby's French Burger Town Center outlet, which can be found in the following Equation 24.10:

Equation 24.10: Gibby's French Burger Town Centre revenue variance.

$$\frac{\text{Revenue variance}}{\text{Revenue last year}} \times 100 = \frac{\$500,000}{\$8,000,000} \times 100 = 0.0625 \times 100 = 6.25\%$$

Compiled by Miguel Gassman from multiple data sources.

Here we can see that the revenue increase for this year over last year was 6.5%. This does not mean that next year we will have an increase in revenue of 6.5%! What this does indicate is that as we had an increase of 6.5% or revenue this year over last year, the chances of once again having an increase of 6.5% next year over this year are good.

As a manager, you might find that 6.5% is too high or even too low. The variance in revenue from one year to another is just a guide; it is not set in stone. If you feel that business is improving, you might feel that business will be even better next year, so you feel that business will increase by 7% or even 7.5%. Alternatively, you could feel that this year was an exceptional year and that business should stabilize next year so there will only be an increase in revenue of 3% over the revenue of this year.

The more years of financial history that we have, the better it is as we get a better feel for what happened. If we have the financial records for the past 3–4 years, we can calculate the variance in revenue for each of the years recorded. With the additional historical variances, we can get a better feel of how business is going and we can therefore make a better calculated guess of what will happen in the future.

To calculate the forecasted revenue for the following year, we use the following formula:

Equation 24.11: The forecasted revenue.

$$\text{Forecasted revenue for next financial period} = \text{Revenue from actual financial period} \times (1 \pm \text{expected increase or decrease in revenue})$$

Compiled by Miguel Gassman from multiple data sources.

For Gibby's Town Center restaurant, the manager feels that next year will be a bumper year, so he is predicting an increase of revenue of 9% for next year as shown in the following equation:

Equation 24.12: Forecasted revenue for next year

$$= \text{Revenue for this year} \times (1 + \text{forecasted increase in revenue})$$
$$= \$8,500,000.00 \times (1 + 9\%)$$
$$= \$8,500,000.00 \times 1.09$$
$$= \$9,265,000.00$$

Compiled by Miguel Gassman from multiple data sources.

24.5 The Profit and Loss (P&L) Statement

One of the most important reports that any F&B manager will have to understand, if not the most important, is the Profit and Loss (P&L) Statement.

Also known as the Income Statement, the P&L Statement is an accounting tool that details revenue, expenses, and profit for a given period of time. It lists all the revenues as well as the respective costs, labor costs, and other expenses that a restaurant may incur in its normal operation.

The revenues earned in a food and beverage operation normally come from the sale of food and beverage items. Three are some restaurants that get a substantial amount of other revenue from sources such as the sale of merchandised items for example.

For simplification purposes, we will assume that Christy's Burrito's gets all their revenue from the sale of food and beverage. See Christies P&L Statement for the previous 2 years below.

When analyzing the P&L statement, we must not only be able to read the report but also more importantly be able to understand and share the results of the report. To be able to give people a clearer understanding of our results, we will very often need to explain in percentage (%) terms. Most P&L Statements not only give the financial figures, they also indicate the percentage (%) of each figure in relation to the total revenue.

Every percentage (%) figure on the P&L Statement is based on total revenue for the financial period, except for two: (1) the cost of food % and (2) the cost of beverage %. The food cost % and the beverage cost % are explained in the next subsection. All other expense percentages are calculated as follows (the following equation):

Equation 24.13: Calculating expenses

Expense %	=	$\frac{\text{Expense}}{\text{Total revenue}}$

Compiled by Miguel Gassman from multiple data sources.

Take a look at Christy's P&L Statement for the past 2 years shown in the following table (Table 24.3):

It is also important to know that each of the categories on the P&L Statement is a cumulative total of other expenses that fall within that category. Take for example utilities.

TABLE 24.3 P&L Statement for Christy's Burrito's for the past 2 years.

	Last year	Percentage (%)	This year	Percentage (%)
Food revenue	$9,852,300.00	75.9	$11,034,576.00	76.4
Beverage revenue	$3,122,780.00	24.1	$3,403,830.00	23.6
Total revenue	**$12,975,080.00**	**100.0**	**$14,438,406.00**	**100.0**
Cost of food	$3,255,620.00	33.0	$3,841,411.00	34.8
Cost of beverage	$625,870.00	20.0	$637,727.00	18.7
Total cost of food and beverage	**$3,881,490.00**	**29.9**	**$4,479,138.00**	**31.0**
Cost of labor	$4,756,541.00	36.7	$5,094,173.00	35.3
Employee benefits	$342,555.00	2.6	$355,824.00	2.5
Total cost of labor	**$5,099,096.00**	**39.3**	**$5,449,997.00**	**37.7**
Other expenses				
Marketing	$758,984.00	5.8	$795,841.00	5.5
Entertainment	$310,516.00	2.4	$348,122.00	2.4
Utilities	$77,899.00	0.6	$82,754.00	0.6
Rent	$1,200,000.00	9.2	$1,400,000.00	9.7
Other operating expenses	$350,894.00	2.7	$395,358.00	2.7
Total other expenses	**$2,698,293.00**	**20.8**	**$3,022,075.00**	**20.9**
Operating income	**$1,296,201.00**	**10.0**	**$1,487,196.00**	**10.3**

Compiled by Miguel Gassman from multiple data sources.

Utilities could be a combination or water, electricity, and gas to name but a few. Below is the breakdown of utilities for Christy's Burrito's (Table 24.4):

24.6 The food and beverage cost percentage on the P&L statement

The cost of food sold percentage is based on the total food revenue and not the total revenue. This is because we want to know how much of the revenue we earned in the sale of food did we spend to be able to make the food we sold.

TABLE 24.4 Breakdown of utilities expenses for Christy's Burrito's.

Water	$8,499.00
Electricity	$54,222.00
Gas	$15,178.00
Total utilities	**$77,899.00**

Compiled by Miguel Gassman from multiple data sources.

24.6.1 Food cost percentage (%)

The food cost consists of the costs of all ingredients required to prepare and serve the dishes that were sold. If an ingredient has been prepared but has not yet been sold, it is not considered part of the food cost.

24.6.2 Beverage cost percentage (%)

The same principle applies to the beverage cost %.

The beverage cost consists of all the ingredients, both alcoholic and nonalcoholic, that are used to prepare the drinks that are sold. If a bottle of whiskey has been opened and only two portions of that bottle have been sold, only the cost of the two portions is added to the cost of beverage sold. The remaining liquor in the bottle is not added as it has not been used yet.

The formulas for calculating the food cost % and the beverage cost % are as follows:

Equation 24.14: Calculating food and beverage costs

$$\text{Food cost \%} = \frac{\text{Food cost}}{\text{Total food revenue}}$$

$$\text{Beverage cost \%} = \frac{\text{Beverage cost}}{\text{Total beverage revenue}}$$

Compiled by Miguel Gassman from multiple data sources.

Therefore, to calculate the food and beverage cost percentages for Christy's Burrito's, we calculate the figures as follows (Equation 24.15):

Equation 24.15: Christy's Burrito's food cost % and beverage cost %

$$\text{Food cost \% this year} = \frac{\text{Food cost}}{\text{Total food cost}} = \frac{\$3,841,411}{\$11,034,576} = 34.8\%$$

$$\text{Beverage cost \% this year} = \frac{\text{Beverage cost}}{\text{Total beverage revenue}} = \frac{\$637,727}{\$3,403,830.20} = 18.7\%$$

Compiled by Miguel Gassman from multiple data sources.

24.7 Analyzing the P&L statement

You need to be able to analyze the P&L Statement. Very often you will be measured on your ability of a manager to see if you are ready for promotion on how you can read and analyze the P&L Statement.

When analyzing the P&L statement, we start at the top and work our way downward.

24.7.1 Analyzing the revenue

When analyzing the revenue, as indicated before, we have to analyze the sales variance from one year to the next. We analyze the total sales variance as well as the individual food sales and beverage sales variances as is shown in the following equations:

Equation 24.16: Christy's Burrito sales variance for this year in comparison to last year:

$$\text{Food sales variance \%} = \frac{\text{Food sales this year} - \text{Food sales last year}}{\text{Food sales last year}}$$

Therefore,

$$\text{Food sales variance \%} = \frac{\$11,034,576 \quad - \quad \$9,852,300}{\$9,852,300} = 12.0\%$$

Compiled by Miguel Gassman from multiple data sources.

There could be four reasons for this increase:

1. More guests were served with the same average check
2. The same amount of guests was served but with a higher average check
3. More guest were served with a higher average check
4. Fewer numbers of guests were served but they had a much higher average check.

24.7.2 Analysis of food and beverage expense

When analyzing any expense that is not a fixed expense, i.e., it is a variable expense, it is important to remember that the expense incurred is directly related to the amount of revenue that was earned. For example, serving $10,000.00 worth of beverages would incur a much higher cost than that of sales of only $5000.00. In principle, the cost of the beverage for $10,000.00 of sales should be double of the cost for $5000.00 worth or sales.

Also, if you were told that your cost of beverage is $2,500,000, you might have a panic attack and immediately wonder what went wrong. But if you were told that you made $18,000.000 worth of sales from the $2,500,000, then you would be a lot more relaxed. As a result, whenever asked what your expenses were, you should always use percentage (%) to answer.

24.7.2.1 Analyzing the food cost

So, when analyzing the food cost for Christy's Burritos, we should compare the percentage of the last two respective years. In this case, we can see that this year, the food cost was 1.8% higher than the previous year's food cost % (Table 24.5).

To see the problem, we will have to delve deeper into the different categories that make up this 1.8% difference (Table 24.6).

By looking at these two reports, we can immediately pick up a few points of concern that should be looked into. Why is this year's meat figures so much higher than last year? In the opposing direction is the value for seafood, this year is much higher than last year. Was there

TABLE 24.5 Food cost for two years.

	Last year	Percentage (%)	This year	Percentage (%)
Food revenue	$9,852,300.00		$11,034,576.00	
Cost of food	$3,255,620.00	33.0	$3,841,411.00	34.8

Compiled by Miguel Gassman from multiple data sources.

TABLE 24.6 Breakdown of two years food costs.

Categories	Last year	Percentage (%)	This year	Percentage (%)
Meat	$1,256,897.00	12.8	$1,229,252.00	11.1
Seafood	$675,887.00	6.9	$998,767.00	9.1
Vegetables	$675,552.00	6.9	$845,111.00	7.7
Fruit	$190,504.00	1.9	$230,485.00	2.1
Dairy	$87,613.00	0.9	$103,716.00	0.9
Dry ingredients	$114,796.00	1.2	$130,607.00	1.2
Others	$254,371.00	2.6	$303,473.00	2.8
Total cost of food	$3,255,620.00	33.0	$3,841,411.00	34.8
Total food revenue	**$9,852,300.00**		**$11,034,576.00**	

Compiled by Miguel Gassman from multiple data sources.

an accounting error? What is the reason for the fish to be so high this year in comparison to last year?

You should then breakdown the fish categories even further and find out what the reason is for such high cost. It is possible your suppliers overcharged you or the cost of fish has substantially increased and you have not changed your pricing structure. You could have had a fridge that broke down and you had to throw a lot of food away.

You will have to go through each category one by one until you find the justification for the increase. Once you have found the reasons, you will have to take corrective measures and make sure they do not occur again.

24.7.2.2 *Analyzing the beverage cost*

The beverage cost % is analyzed in much the same way as the analysis for food was done (Table 24.7).

TABLE 24.7 Analyzing the beverage cost.

	Last year	Percentage (%)	This year	Percentage (%)
Beverage revenue	$3,122,780.00		$3,403,830.00	
Cost of beverage	$625,870.00	20.0	$637,727.00	18.7

Compiled by Miguel Gassman from multiple data sources.

Here we can see that this year we have controlled our beverage cost much better than the previous years. It might be required to investigate if the difference in beverage cost was very large. Also, we would have to compare our beverage cost with our budgeted beverage cost.

24.7.3 Analyzing the labor expense

To calculate the labor expense %, we use the following formula (the following equation):

Equation 24.17: Calculating the labor cost

$$\text{Labor cost \%} = \frac{\text{Cost of labor \$}}{\text{Total revenue}}$$

Compiled by Miguel Gassman from multiple data sources.

When analyzing the labor cost %, we can see that this year we are doing better than last year. We are 1.4% lower this year compared to last year, although we spent more money on labor. The reason for this is that revenue increased substantially as well (Table 24.8).

As many of your wagers are a fixed expense, as long as labor expenses do not make up 100% of your revenues, an increase in your sales revenue normally helps decrease your labor cost %.

Variable labor costs increase monetary-wise as long as revenue increases. However, the % value tends to remain the same.

Fixed labor cost, the monetary value remains the same whether you have an increase or a decrease in revenue as it is a fixed cost. As such the labor cost % will decrease as revenue increases. Vice versa, the labor cost % will increase as revenue decreases (the following equation) (Table 24.9).

Note a drastic decrease in labor cost is not necessarily a good sign. Remember we work in a labor-intensive industry. If your labor costs decrease too much, you might be providing very bad service to your customers.

TABLE 24.8 The cost of labor.

	Last year	Percentage (%)	This year	Percentage (%)
Total revenue	**$12,975,080**		**$14,438,406**	
Cost of labor	$4,756,541	36.7	$5,094,173	35.3
Employee benefits	$342,555	2.6	$355,824	2.5
Total cost of labor	**$5,099,096**	**39.3**	**$5,449,997**	**37.7**

Compiled by Miguel Gassman from multiple data sources.

TABLE 24.9 Labor cost movement.

	Decrease in sales	Percentage (%)	Normal	Percentage (%)	Increase in sales	Percentage (%)
Total sales	**$400,000**		**$500,000**		**$600,000**	
Fixed labor cost of $120,000	$120,000	30.0	$120,000	24.0	$120,000	20.0
Variable labor cost of 8.5%	$34,000	8.5	$42,500	8.5	$51,000	8.5
Total labor cost	$154,000	38.5	$162,500	32.5	$171,000	28.5

Compiled by Miguel Gassman from multiple data sources.

24.7.4 Analyzing other expenses

Other expenses are normally the minor costs; however, it is important to keep a close eye on your other expenses as they can build up quickly. In addition, many of the other expenses are uncontrollable expenses such as rent, maintenance, laundry. It is better to focus on items that can be easier controlled such as food cost.

Analyzing other expenses is done in the same way as the labor and food and beverage costs—we analyze the % figure and compare to previous years. When there is a discrepancy, we will have to find out what happened and take corrective action when necessary.

Calculating your other expense % is done using the following formula:

Equation 24.18: Calculating other expenses

$$\textit{Other expenses } \% = \frac{\text{Other expense}}{\text{Total sales}}$$

Compiled by Miguel Gassman from multiple data sources.

24.7.5 Analyzing the profit

When analyzing profit, it is important to remember we always want to take more profit to the bank. The more money we earn, the more our owners are satisfied with our performance. As such we compare profit from one financial period to another using the following formula:

Equation 24.19: Analysing the profit

$$\text{Profit variance } \% = \frac{\text{Net income from this period} - \text{Net income from last period}}{\text{Net income from last period}}$$

Compiled by Miguel Gassman from multiple data sources.

If we look at Christy's Burrito, we can see that net profit has increased this year in comparison to last year by $190,995.00. There was also an increase of 0.35 in the profit percentage. But how much better did they really do (the following equation):

Equation 24.20: Calculating profit variance

$$\text{Profit variance \%} = \frac{\text{Net income this year} - \text{Net income last year}}{\text{Net income last year}}$$

Therefore,

$$\frac{\$1,487,196 \;-\; \$1,296,201}{\$1,296,201} = \$190,995 \quad \text{PV} = \; 14.70\%$$

Compiled by Miguel Gassman from multiple data sources.

Now we can see how much Better Christy's Burrito really did; 14.7% increase in profit is a really good result.

24.8 Cost/volume/profit analysis

Every food and beverage operation wants to know how much business they need to do to start making money. This could be on an annual basis, monthly basis, or even on a meal basis. They all want to know when they will start making money.

The point where the restaurant will start to earn money is called the break-even point. At break-even point, the revenue earned is exactly equal to the expenses. There is zero profit at zero loss at this point.

Only once the restaurant earns more money that the break-even point will it start to earn a profit. Once you have passed the break-even point, you will want to know how much business you have to do to make a certain amount of money. To find out your break-even point and how much money you need to earn to make a certain amount of profit, you will need to the cost/volume/profit analysis.

24.8.1 Fixed and variable expenses

Before we go on, it is important to understand the difference between a fixed and a variable expense.

A fixed expense is an expense that does not change irrespective of how much money you have earned. An example would be rent: if you have signed a rental agreement with your landlord that you will pay him a fixed fee of $10,000.00 per month, then your rent is a fixed expense.

A variable expense is an expense that varies with the amount of revenue you have earned. Once again let us use the example of rent. If you have signed a rental agreement with your landlord that you will pay 10% of your revenue as a rental fee, then your rental fee will increase as your revenue increases.

You then have a mixed expense, which is a combination of fixed and variable expenses. Once again let us refer to your rent. You could have signed an agreement with your landlord that you will pay him a fixed rental fee of \$3500 per month plus an additional 4% of your revenue.

See the table below (Table 24.10) for an example of how much you will pay for all three different types of expense.

24.8.2 Contribution margin

The contribution margin is defined as the dollar amount that contributes to covering fixed costs and providing for a profit. The contribution margin is calculated as follows:

$$\text{Total sales} - \text{Variable costs} = \text{Contribution margin}$$

Gulantoo who runs an Indian Bistro in Taipa has rearranged P&L Statement to become a Contribution Margin P&L Statement as shown below in Equation 24.21

$$\text{Equation 24.21: Contribution margin P\&L Statement}$$
$$(\text{Total sales} - \text{Variable costs gives contribution margin}) - \text{Fixed costs} = \text{Profit}$$
$$\text{Therefore}$$
$$(\$650,000 - \$227,500 \text{ gives } \$422,500) - \$200,000 = \$222,500 \text{ profit}$$

Compiled by Miguel Gassman from multiple data sources.

TABLE 24.10 An example of fixed, variable, and mixed rents.

Example of fixed rent		Example of variable rent	
Fixed rent:	\$10,000.00	Variable rent:	10% of rent
Revenue	Rent due	Revenue	Rent due
\$	\$10,000.00	\$	\$
\$50,000.00	\$10,000.00	\$50,000.00	\$5000.00
\$100,000.00	\$10,000.00	\$100,000.00	\$10,000.00
\$150,000.00	\$10,000.00	\$150,000.00	\$15,000.00
\$200,000.00	\$10,000.00	\$200,000.00	\$20,000.00
Example of mixed rent			
Mixed rent = fixed rent + variable rent, i.e., \$3500 + 3.50%			
Revenue	Fixed rent	Variable rent	Rent due
\$ = \$0	\$3500.00	\$ = 3.5%	\$3500.00
\$50,000.00	\$3500.00	\$1750.00	\$5250.00
\$100,000.00	\$3500.00	\$3500.00	\$7000.00
\$150,000.00	\$3500.00	\$5250.00	\$8750.00
\$200,000.00	\$3500.00	\$7000.00	\$10,500.00

It will also be helpful for Gulantoo to view his contribution margin income statement in terms of per guest (or per unit). We can calculate the average check (AC) per guest by the following equation:

Equation 24.22: Average check (AC) per guest

$$\text{Average check (AC)} = \frac{\text{Total revenue}}{\text{No. of guests served}}$$

$$\text{Equates to}\,\frac{\$650,000}{5200\text{ guests}} = \$125\text{ per guest}$$

Compiled by Miguel Gassman from multiple data sources.

Furthermore, knowing the average check ($125) and the number of guests served ($5200), we can calculate the variable costs (VC) per guest (the following equation)

Equation 24.23: Calculating the variable costs per guest

$$\text{Variable cost per guest (VC/gst)} = \frac{\text{Variable cost}}{\text{No. of guests served}}$$

$$= \frac{\$227,500}{5200\text{ guests}} = \text{(VC) }\$43.75/\text{gst}$$

Compiled by Miguel Gassman from multiple data sources.

And lastly, we can calculate the contribution margin (CM) on a per guest basis by the following equation:

Equation 24.24: Calculating contribution margin per guest

$$\text{Contribution margin (CM) per guest} = \frac{\text{Contribution margin}}{\text{No. of guests served}}$$

Therefore

$$\text{Contribution margin (CM) per guest} = \frac{\$422,500}{5200\text{ guests}} = \text{(CM) }\$81.25/\text{gst}$$

Compiled by Miguel Gassman from multiple data sources.

From this point we can further calculate the percentages for the average check, contribution margin, and variable cost (see the following Equation 24.25):

Equation 24.25: Calculating contribution margin and variable cost

$$AC\% = 100\% = \$125.00$$

$$VC\% = \frac{\text{Variable cost/gst}}{\text{Average check}} = \$43.75 = 35\%$$

$$CM\% = \frac{\text{CM/gst}}{\text{Average check}} = \frac{\$81.25}{\$125.00} = 65\%$$

Furthermore from a P&L Statement to a contribution margin statement

$$AC\% - VC\% = CM\%$$

Ergo

$$100\% - 35\% = 65\%$$

Compiled by Miguel Gassman from multiple data sources.

Now that Gulantoo's P&L Statement has been converted to a Contribution Margin P&L Statement and he has calculated his per unit values as well as his percentages, he is now capable of calculating his break-even point.

To calculate the break-even point, the following formula is used:

Equation 24.26: The break – even point %

$$\frac{\text{Fixed costs}}{\text{Contribution margin \%}} = \text{Break – even point in sales}$$

Compiled by Miguel Gassman from multiple data sources.

So Gulantoo's Indian Bistro would reach its break-even point at

Equation 24.27: The break – even point $

$$\frac{\$200,000}{65\%} = \$307,692.31$$

Compiled by Miguel Gassman from multiple data sources.

From this we can see that Gulantoo will have to make at least $307,692.31 before he can earn any profit. Any sales less that this amount will result in a loss for the operation.

However, how many guests will Gulantoo have to serve before he breaks even, to calculate the break-even point in number of guests served, we use the following formula:

Equation 24.28: The break – even point in number of guests

$$\text{Fixed costs} = \frac{\text{No. of guests to serve to break even}}{\text{Contribution margin per guest}}$$

Therefore

$$\frac{\$200,000}{\$81.25} = 2,461.54 \text{ guests}$$

Compiled by Miguel Gassman from multiple data sources.

To break even, Gulantoo will have to serve a minimum of 2462 guests. If he serves less than this, he will be running at a loss.

Now let us assume Gulantoo wants to make $222,500 profit from this year's operation. How much revenue will he have to make to achieve this profit and how many guests will he have to serve? We use the following formula to calculate how much revenue Gulantoo will have to make to reach the desired profit:

Equation 24.29: Creating revenue to reach a desired profit

$$\frac{\text{Fixed costs} + \text{Desired profit}}{\text{Contribution margin \%}} = \text{Sales to achieve desired profit}$$

Therefore

$$\frac{\$200,000 + \$222,500}{65\%} = \$650,000$$

Following on from this:

$$\frac{\text{Fixed costs} + \text{Desired profit}}{\text{Contribution margin per guest}} = \text{Guests to serve to achieve desired profit}$$

$$\frac{\$200,000 + \$222,500}{\$81.25/\text{guest}} = 5200 \text{ guests}$$

Compiled by Miguel Gassman from multiple data sources.

From these calculations, we can see that to achieve a desired profit of $222,500, Gulantoo will have to achieve a minimum total sales of $650,000.00 and serve a minimum of 5200 guests.

Further reading

Dittmer, P.R., Heefe Iii, J.D., 2009. Principles of Food, Beverage, and Labor Cost Controls. John wiley and Sons, INC, USA.

Dopson, L.R., Hayes, D.K., 2011. Food and Beverage Cost Control, fifth ed. Willey and Sons, INC.

Keiser, J., Demicco, F.J., Cobanoglu, C., Grimes, R.N., 2008. Analyzing and Controlling Foodservice Costs, fifth ed. Pearson Education Ltd.

CHAPTER

25

Controlling costs and purchasing food

Miguel Gassmann de Oliveira

Institute for Tourism Studies, Macao (IFTM): Colina de Mong-Há Macao, China

OUTLINE

Food and Society
https://doi.org/10.1016/B978-0-12-811808-5.00025-8

It is all very nice and good knowing how to read a P&L statement as well as knowing how much money we need to make to break even. However, what happened when we do not break even. Very often, the first thing we say is that your pricing is too low.

This is, however, the cause for downfall of many restaurant operations. What needs to be done is to ensure that they are getting the best products at the best price and then producing the food the customer wants at a price the customer is willing to pay! To ensure that this is done, there is a seven-step process that needs to be followed. The steps are listed below. Each of these steps will then be further developed and explained.

- Maintain sales histories
- Forecast future sales levels
- Purchase and store needed food and beverage supplies
- Plan daily production levels
- Issue needed products to production areas
- Manage the food and beverage production process
- Price the products at a price so that customers feel it is a value for money operation

25.1 Maintaining sales histories

One of the first things you learn as an F&B manager is to ask yourself how much business am I going to do today, tomorrow, next week, and so on. This is so that you can produce just the right amount of food.

You don't want to produce too much because if you do you will have too many leftovers, which you might have to throw away. Any time you throw food away you throw money away, so producing too much is never a good sign.

Also, you don't want to produce too little as you might lose the opportunity to make a sale. You might also get a customer upset if he comes all the way to your restaurant only to find the food he wanted has been sold out.

You want to be able to produce enough so that you can reduce the amount of leftovers, while at the same time ensuring that you do not run out too early. To avoid both running out and making too much, we need to forecast (predict) how much sales we will be doing on a daily basis and we will then prepare quantity of food according to this prediction.

As indicated before, to predict what will happen in future, we will normally refer to what has happened in the past. For this reason, it is important that you keep your sales histories for a few years.

The sales histories should be kept not only in revenue and expenses but also in items sold, i.e., volume of sales.

A sales history is a systematic recording of sales achieved during a predetermined time period (Dopson and Hayes, 2011). Below (Table 25.1) is an example of sales volume history for 1 weeks' worth of sales.

TABLE 25.1 One week's sales history.

Week 1	Monday	Tuesday	Wednesday	Thursday	Friday
Hamburger	15	18	15	19	17
Cheeseburger	22	19	19	21	22
Hot dog	17	21	18	17	18
Fried rice	12	15	16	17	16
Beef noodles	25	22	18	23	25
Total sales	91	95	86	97	98

Compiled by Miguel Gassman from Multiple Data Sources.

25.2 Forecasting future sales

When we have the sales histories, we can convert the sales to averages. When we know what out average sales are, we can then use this figure to more accurately predict what our future sales will be. Using the above example, we can determine this as shown in Table 25.2:

From the above table we can see that an average of 16.8 hamburgers was sold during week 1. We would traditionally round off the numbers to make it easier to use (Table 25.3).

We can now see that on average, around 17 hamburgers were sold. We can therefore predict that the following week we will sell around 17 hamburgers per day. However, to ensure that we do not run out, we will normally produce a little extra of each just in case we get a few more customers asking for that particular item. This little bit extra that we produce is normally called a safety margin. Safety margins are normally not large amounts, normally 5%–10% maximum.

TABLE 25.2 Sales history averages.

Week 1	Monday	Tuesday	Wednesday	Thursday	Friday	Total sold	Average
Hamburger	15	18	15	19	17	84	16.8
Cheeseburger	22	19	19	21	22	103	20.6
Hot dog	17	21	18	17	18	91	18.2
Fried rice	12	15	16	17	16	76	15.2
Beef noodles	25	22	18	23	25	113	22.6

Compiled by Miguel Gassman from Multiple Data Sources.

TABLE 25.3 Rounded-off figures.

Week 1	Average	Rounded off
Hamburger	16.8	17
Cheeseburger	20.6	21
Hot dog	18.2	18
Fried rice	15.2	15
Beef noodles	22.6	23

Compiled by Miguel Gassman from Multiple Data Sources.

For small quantities such as the figures in the table above, a safety margin of 10% is not a lot; however, when you are working in large scales, 10% can be way too much. So, in every situation, the manager will have to estimate what his safety margin will be.

For this situation we will use a safety margin of 10%. We add the safety margin amount to the historical figure to get a new figure. We then round off this figure to get the actual amount that should be produced on a daily basis for week 2.

It is important to remember that these figures are estimates and consequently may not be 100% correct. This is why one must also, at times, rely on a gut feeling, i.e., what does one want as a manager expect will happen in the next financial period? One could increase or decrease the production of each item by as many as one perceives is needed. All this is not just theoretical—see Table 25.4 for a real-life example of forecasting sales volumes.

Forecasting the number of guests to be served is executed in the same way. Historical figures are used to give an idea of what has been happening. Then one will increase the amount of guests to be served by a figure of expect growth. It can either be that average growth from the sales history one has maintained or it could be an estimated figure.

TABLE 25.4 Forecasting sales volume.

Week 1		Week 2 daily forecast			
	Sales history	Safety margin (%)	Safety margin amount	Total amount	Amount to produce on daily basis
Hamburger	17	10	1.7	18.70	19
Cheeseburger	21	10	2.1	23.10	23
Hot dog	18	10	1.8	19.80	20
Fried rice	15	10	1.5	16.50	17
Beef noodles	23	10	2.3	25.30	25

Compiled by Miguel Gassman from Multiple Data Sources.

Some factors are to be taken into consideration when you are planning a sales forecast. Sales can be affected by

- Competition
- Weather
- Special events in your area
- Holidays
- Facility occupancy (for example, a full hotel will have more customers for breakfast than an empty hotel)
- Your own advertising and promotional offers
- Your competitors advertising and promotional offers
- Quality of service
- Changes in operation hours
- Operational consistency

Dopson and Hayes (2011).

25.3 Purchase and store-needed food and beverage supplies

Once you know how many of each item you will need to produce, it is time to order the ingredients to be able to produce those items. To understand which items and how many of each item to order, it is necessary to use standardized recipes.

25.3.1 Standardized recipes

A standardized recipe is a recipe that should be used by everyone producing the same dish in your restaurant. It is the recipe that has been designated as the correct one to use in the restaurant (Dittmer and Heefe, 2009).

Standard recipes are useful to a foodservice operation for the following reasons:

- It does not matter who makes it, it should always look and taste the same
- The portion size should always be the same
- There is a guide for chefs, so therefore less supervision is required during the production process
- The cost of the dish should be more or less the same every time it is produced.

Keiser et al. (2008).

A standardized recipe should contain the following information:

- Menu item name
- Total yield (number of servings)
- Portion size
- Ingredient list
- Preparation/method section

- Cooking time and temperature
- Special instruction (if any)
- Recipe cost (this is optional)

Dopson and Hayes (2011).

It is important that the same recipe is used every time no matter how many guests are being served. For instance, if you have a recipe for four persons, the same recipe would be used as if the chef was making enough food for 50 persons. However, to ensure consistency, you should convert the recipe to 50 persons. The method used for converting the recipe up or down is called the factor method.

To use the factor method, you need to calculate the conversion factor, which is calculated as follows (Dopson and Hayes, 2011):

Equation 25.1: The factor method:

$$\frac{\text{Desired yield}}{\text{Current yield}} = \text{Conversion factor.}$$

Compiled by Miguel Gassman from Multiple Data Sources.

Take for example the recipe mentioned above, the current yield is 4, we, however, need to make the same recipe for 50 persons, the conversion factor would be

Equation 25.2: The factor method example:

$$\frac{50}{4} = 12.5$$

Compiled by Miguel Gassman from Multiple Data Sources.

Once we have obtained the conversion factor, we will multiply all the ingredients in the recipe by the conversion factor.

25.3.1.1 *Product yield*

When ordering food for a recipe, it is important to remember that it is not possible to use 100% of the food that is purchased. Normally it is necessary to trim or peel the food before the food is ready to be cooked or served. Whenever we trim or peel the food, the part we discard is waste and is usually disposed. It is therefore necessary to remember that if we need 25 kg of potatoes, we will need to order more than 25 kg of potatoes to ensure that after we have peeled and cleaned them, we have at least 25 kg of ready-to-prepare potatoes.

To find out how much of an item we need to order to ensure we get enough of the ingredient to prepare the food, we will do a *yield test or butchers test*. Most products when they are delivered, they are in the *AP* or *as purchased state*. This refers to the weight or count that we receive or pay for. It is then necessary to trim, peel, cook, or portion the product so that we only have the part we will serve to the guest left; this is known as the *EP* or *edible portion*. The parts of the product that we peel, trim, and loose due to cooking loss or portioning are known as *product loss* or *waste*. Therefore, the AP is what the foodservice operation receives from the supplier, and the AP is what the guest receives from the foodservice operator (Dopson and Hayes, 2011):

Equation 25.3: The concept of the Edible Portion:

Yield Test Results		
Item name	**Leg of Lamb (bone in)**	
AP amount tested	3.2kg	
AP price per Kg	$75.80	
As Purchased	3.2kg	100%
Fat Loss	0.4kg	12.5%
Bone Loss	0.8kg	25.0%
Total product loss	1.2kg	37.5%
Ep portion	2kg	
Product yield	62.5%	
EP Cost per KG	$121.28	

Compiled by Miguel Gassman from Multiple Data Sources.

Yield % is an important area of recipe costing as it is necessary to take into consideration the EP and AP costs for ingredients: Recipes normally have the EP price on them, so it is necessary to calculate the EP price for the ingredients. Most good suppliers will be able to give you the standard yield % for their products. However, it is still advisable to do a yield test every once in a while to ensure the costing remains the same. To calculate the *waste* % of a product we use the following formula:

Equation 25.4: Calculating the **Waste**(%) of Product

$$\text{Waste }\% = \frac{\text{Product Loss}}{\text{AP Weight}}$$

Therefore

$$\text{Waste }\% = \frac{1.2\text{ kg}}{3.2\text{ kg}}$$

finally

$$\text{Waste }\% = 37.5\%$$

Compiled by Miguel Gassman from Multiple Data Sources.

Once we have determined waste %, it is possible to determine the yield %. The yield % is the percentage of a product that remains after it has been trimmed, peeled, cooked, or portioned.

Equation 25.5: Calculating the **Yield** (%) of Product (Number 1)

Yield% = (1−Waste%)

Therefore

Yield% = 1−37.5%

Which

= **62.5%**

Compiled by Miguel Gassman from Multiple Data Sources.

Another way to calculate the yield % is by dividing the *EP* weight by the *AP* weight as such:

Equation 25.6: Calculating the **Yield** (%) of Product (Number 2)

$$\text{Yield \%} = \frac{\text{EP Weight}}{\text{AP Weight}}$$

$$\text{Thus Yield \%} = \frac{2.0\ \text{kg}}{3.2\ \text{kg}}$$

$$\text{Ergo} = 62.5\%$$

Compiled by Miguel Gassman from Multiple Data Sources.

Once you know how much *EP* weight you need to a recipe, you can calculate how much *AP* weight you need to order to be able to produce the correct amount of portions. We calculate the *AP* weight as follows:

Equation 25.7: Calculating the AP weight Required

$$\text{AP Required} = \frac{\text{EP Weight}}{\text{Yield \%}}$$

Therefore

$$\text{AP Required} = \frac{2\ \text{kg}}{62.2\ \%} = 3.2\ \text{kg}$$

Compiled by Miguel Gassman from Multiple Data Sources.

So, to recap, to calculate the EP cost of lamb required to get 2 kg per kg after all the waste has been removed, we take, for example, the leg of lamb above. It costs $75.8 per kilogram to purchase, so the total paid for the leg was

Equation 25.8: Total Cost Required to Meet Sufficient Edible Portions

$$3.2\ \text{kg} \times \$75.8 = \$242.56$$

Compiled by Miguel Gassman from Multiple Data Sources.

Once we have removed all the waste, the total cost of the leg of lamb is still $242.56, but there is only 2 kg of edible portion. So, to calculate the *EP cost* (EP price per kg), we do it as follows (Equation 25.9):

Equation 25.9: Calculating the EP cost

$$\text{EP cost per kg} = \frac{\text{AP Price per KG}}{\text{Yield \%}}$$

Therefore

$$= \frac{\$75.80}{62.5\%}$$

Which

$$= \$121.28$$

Compiled by Miguel Gassman from Multiple Data Sources.

To calculate the recipe costs, you will always use the EP cost.

Once you know how much of a particular item you will need to purchase, you can go ahead and place your order.

25.3.2 Purchasing the ingredients

Once you know how much you need to purchase, you need to remember that when purchasing you are trying to determine the following:

- What should be purchased?
- What is the best price to pay?
- How can a steady supply be ensured?

25.3.2.1 What to purchase?

When the chef order steak, what steak does he want? If you order the wrong steak, he will not accept it and maybe the supplier will not accept that you return it. Therefore, when the chef orders a steak, you need to identify exactly what steak he wants in terms of quality, type, size, and quantity. To ensure we get the correct product for the correct recipe, a product specification needs to be established, also known as a standard purchase specification.

A *product specification* contains carefully written description listing in appropriate detail the specific and distinctive characteristics that best describe the items to be purchased. Product specifications commonly identify product name, pricing unit, grade, size, weight, degree of freshness, color, intended use, and other similar characteristics of a product (Dittmer and Heefe, 2009).

These product specifications are to be given to the suppliers so that when you order the particular steak the chef wants, you can order according to your product specification number. This way the supplier knows exactly what you want and can deliver accordingly.

25.3.2.2 *What is the best price to pay?*

Once you know the required products needed, it is time to figure out what is the best price to pay. *Take note, the best price to pay is usually not the cheapest price!*

The *best price* is, in fact, the lowest price that meets both the long-term goals of both the foodservice operation and the vendor (Dopson and Hayes, 2011). This means that it is a win-win situation for both the foodservice operator and the vendor. The foodservice operator is happy to pay the price for a product/service that the vendor is happy to offer.

Remember, vendors have more than one price. They will offer the same product with different prices to different customers, but more than the product itself is the service the vendor offers. If you manage to create a good working relationship with a vendor, the vendor will not only offer you good prices but also offer many guarantees that other suppliers might not offer. It is important to build a good working relationship with your vendors.

At the same time, it is important not to limit yourself to only one vendor per product whenever possible. The more vendors you have, the better chance you will have of getting a better deal, as you will need to get the vendors to fight for your business. Having too many vendors, on the other hand, will make it difficult to build a good working relationship and will also create too much paperwork. Therefore, it is a good idea to get around two to three vendors per product type. This way you can spread your business around them as well as ensure that you can get the best prices and best service constantly.

To ensure you get the best prices from your suppliers, you will regularly have to send the vendors a "*Bid Sheet*" for them to list the prices they can offer for certain products. The bid sheet will be sent to various suppliers.

Once all the suppliers have returned the completed bid sheets, it is then necessary to create a *Price Comparison Sheet*. A price comparison sheet is a sheet where you list down the different prices quoted by different vendors from a particular product; see both Tables 25.5 and 25.6.

25.3.2.3 *How much is required to be purchased?*

We have now managed to get the right product at the best price to ensure the right quality. We now have to find out how much we need to purchase. Before we purchase anything, we need to know how much of the products we have in our inventory. We do this by doing a daily inventory. This is done by using a daily inventory sheet as below (Table 25.7).

The par value is a preestablished level that a company needs to carry or have in stock of a particular item. The on-hand amount is the actual amount that is actually available. The special-order column is for when it is necessary to order an extra amount of a particular ingredient as more is required for the operation on the day. To calculate the order amount, the following formula is used:

Equation 25.10: Calculating the Order Amount

$$\text{Par Value} - \text{On Hand} + \text{Special Order} = \text{Order Amount}$$

So, the quantity of macaroni to order is $(10 - 4) + 6 = 12$

Compiled by Miguel Gassman from Multiple Data Sources.

Once we know how many of each ingredient is required, we can go ahead and prepare the purchase order.

TABLE 25.5 A comparison sheet.

Bid Sheet

Supplier: ____________________
Supplier address: ____________________
Supplier contact no: ____________________
Supplier email: ____________________

Item Number	Item Description	Unit of measure	Price per unit
201	Purple Cabbage	Kg	
202	Chinese Cabbage	Kg	
203	Cabbage	Kg	
204	Iceberg Lettuce	Kg	
205	Romaine Lettuce	Kg	
206	White Onion	Kg	
207	Purple Onion	Kg	

Bid prices guaranteed from __________ to __________

Salesperson: ____________________

Vendor Chop: ____________________

Compiled by Miguel Gassman from Multiple Data Sources.

TABLE 25.6 A price comparison sheet.

Price Comparison Sheet

Vendors **Category:** Produce

A) Gerome's Primary Veggies
B) Junior's organic plants
C) Dell's Fist Choice Greens

Price Comparison Sheet

Item Number	Item Description	Unit of measure	A	B	C	Best Bid ($)	Best Company	Last Price Bid
201	Purple Cabbage	Kg	$ 14.50	$ 14.00	$ 13.50	$13.50	C	$ 13.00
202	Chinese Cabbage	Kg	$ 12.50	$ 13.00	$ 13.00	$12.50	A	$ 13.00
203	Cabbage	Kg	$ 13.50	$ 13.75	$ 13.80	$13.40	A	$ 13.50
204	Iceberg Lettuce	Kg	$ 15.00	$ 14.00	$ 14.50	$14.00	B	$ 14.00
205	Romaine Lettuce	Kg	$ 14.00	$ 14.00	$ 13.50	$13.50	C	$ 13.00
206	White Onion	Kg	$ 9.00	$ 8.70	$ 9.50	$ 8.70	B	$ 9.00
207	Purple Onion	Kg	$ 7.00	$ 8.00	$ 6.50	$ 6.50	C	$ 8.00

Compiled by Miguel Gassman from Multiple Data Sources.

TABLE 25.7 The daily inventory sheet.

Daily Inventory Sheet

Item	Description	Unit	Par Value	On hand	Special Order	Order Amount
Ketchup Bottle	330g	Bottle	10	6		4
Mustard	250g	Jar	10	4		6
Chilli Sauce	250g	jar	10	9		0
Sardinnes (canned)	180g	can	24	10	12	26
Chick peas (canned)	800g	can	12	2		10
White Beans (canned)	800g	can	12	5	12	19
Spaghetti	500g	pkt	10	5		5
Macaroni	500g	pkt	10	4	6	12

Compiled by Miguel Gassman from Multiple Data Sources.

25.3.2.4 *Preparing the purchase order*

The purchase order is a formal agreement that a product is going to be bought for a specific price from a specific supplier (Dittmer and Heefe, 2009). This document is then sent to the selected supplier that contains the following information; see also Table 25.8.

TABLE 25.8 An example of a purchase order.

Supplier: ____________ **Purchase order #:** ________
Supplier Address: ____________ **Delivery Date:** ________
Suppliers Contact Person: ________
Supplier Contact: ____________

Item Number	Item Description	Unit of measure	Quantity ordered	Quoted price	Extended Tprice

Total Value of order ________

Order date: ____________ **Comments**
Order by: ____________

Received by: ____________
Delivery Instuctions: ____________

Compiled by Miguel Gassman from Multiple Data Sources Dopson, L.R., Hayes, D.K., 2011. Food and Beverage Cost Control, fifth ed. Wiley and Sons, INC.

Purchase order information should contain the following:

- Item name
- Specification number
- Quantity ordered
- Quoted price
- Extended price
- Total value of the order
- Supplier information
- Date of order
- Date of expected delivery
- Special delivery instructions
- Comments

25.3.3 Receiving the goods

Now that the items have been purchased, it is time to receive all the items. To properly receive the items ordered, it is necessary to have the following:

1. Proper location: It is necessary to have enough space to be able to check all the items ordered.
2. Scales: Never trust the supplier word for the amount delivered. Everything should be counted or weighed to verify the invoice is correct.
3. Thermometer: Some items should be delivered at a certain temperature to ensure their quality, so have a thermometer available to ensure the product is at the right temperature.
4. Calculators: Never trust the extended price on the invoice, although it is a computer-printed invoice as sometimes the totals are manually input.

If the products received are not to the required standard or are in fact the wrong product, they should be returned. If the quantity received is not the same as the quantity delivered, a credit memo is issued, which is signed by both the delivery person and the receiving person. This credit note is then sent to the supplier to revise the invoice.

Always reconfirm the invoice price with the quoted price on the purchase order to ensure they are the same. Very often many establishments request the delivery invoice be accompanied by a copy of the purchase order received by the supplier.

25.3.3.1 Placing the items in storage

Once the items are confirmed as the correct items and they are in a good quality, it is necessary to store the items properly. The best way to ensure that products do not pass their validity date is to use the FIFO system. FIFO stands for first in first out, which means that the older items are used first and the new items are placed at the bottom of the pile. This method helps reduce spoilage as well as products passing their use-by dates. In small operations with small storage areas, most products are bought daily whenever possible to ensure a little spoilage as possible except items with a long shelf life such as canned and bottled goods.

For larger operations such as restaurants in hotels, normally all items are stored in the storeroom. These storerooms have different areas for storing different products. They will also very often buy in bulk to try to get a better purchase price. Different types of storage areas that will exist in these storerooms are

5. Dry storage: It is the space where dry goods, canned, and bottled goods are normally stored. Normally, the temperature is between 18 and 24°C.
6. Refrigerated storage: Here, temperature normally ranges between 2 and 4°C. Different fridges are used for storing different types of food, such as
   - Fish and seafood
   - Meats
   - Vegetables and fruit
   - Dairy products.
   - Freezer storage: Temperature normally varies between −18 and −23°C.

Whenever storing any products, there are a few basics that need to be observed (Dittmer and Heefe, 2009).

- Storage containers: Products must be stored in the appropriate containers. All containers should be tight, sealed, and insect-proof containers (for dry goods).
- Shelving: Shelves should be slated to help the circulation of air. At no time should items be stored directly on the floor. For large heavy items such as rice and flour, they should be stored on crates on the floor. All shelves should be a minimum of 15 cm off the bottle from the floor.
- Cleanliness: Absolute cleanliness should be maintained at all times in the storeroom. Storerooms are great places for vermin and insects to breed. So, a regular cleaning schedule should be implemented.

25.3.3.2 Determining inventory value

In most establishments, the value of the food inventory on hand is done at the close of the accounting period, week, month, or year. Most places do a monthly inventory. Doing the inventory check is the only way to know exactly what is available in the storeroom. This is done by doing a physical inventory.

Taking physical inventory required counting the actual number of units (pieces) on hand for each individual item (Dittmer and Heefe, 2009) and recording that amount on a specially prepared inventory form. An inventory form lists all the items available in the storeroom by storage area so that when doing the inventory, you do not have to move around to look for the items on the list. The inventory list identifies the following:

- Product name
- Product code
- Measuring unit
- Price per unit of measure
- Quantity that should be in stock
- Space for the actual amount of items on stock.

TABLE 25.9 A costed inventory stock list.

Date of purchase	Quantity	Unit cost	Total value
Opening inventory	10	$9.50	$95.00
June 09	36	$9.80	$352.80
June 18	36	$9.70	$349.20
June 25	24	$10.00	$240.00
Total available for sale	**106**		**$1037.00**

Compiled by Miguel Gassman from Multiple Data Sources.

Calculating the value of the inventory is then a question of using the formula

Number of items on hand × Unit cost of item = Inventory value for item

Before determining the inventory value of goods, it is necessary to determine the unit cost value of the inventory items on hand because sometimes the same item will have a different cost. We will be using cans of tomato paste to demonstrate how to calculate the unit cost of each stock item.

For the month of June, the inventory record for cans of tomato paste shows the following (Table 25.9):

On the 30th of June, an inventory was conducted and it was found that there were 15 cans left in the store room. It is now necessary to assign a value to these cans.

The three most common methods of calculating the cost price of inventory items are

- Actual purchase price method
- Weighted-average purchase price method
- Latest purchase price method

25.3.3.2.1 Actual purchase price method

This method involves using the actual price for each item that is in the storeroom. This is one of the better methods but requires that each item being received is marked down with the actual price of purchase (Table 25.10).

Note that there are still cans listed from the opening inventory and all the other times items were purchased, so the stock rotation was not properly conducted.

TABLE 25.10 Tabulated inventory value.

No. of cans	Purchased price	Total value
4	$9.50	$38.00
4	$9.80	$39.20
5	$9.70	$48.50
5	$10.00	$20.00
Total value		$145.70

Compiled by Miguel Gassman from Multiple Data Sources.

25.3.3.2.2 Weighted-average purchase price method

When we cannot guarantee that the stock has been properly rotated, the weighted-average purchase price offers a good solution. You can calculate the weighted-average price of the cans by dividing the total value of the good available for sale by the number of good available for sale (Equation 25.11).

Equation 25.11: The weighted average price of Inventory

$$\frac{\text{Total value of good available for sale}}{\text{Number of goods available for sale}}$$

equals

$$= \frac{\$1,037.00}{106}$$

which then

$$= \$9.78/\text{can}$$

Compiled by Miguel Gassman from Multiple Data Sources.

We can then calculate the inventory value of the item by multiplying the number of items on hand. In this case, the cans of tomato paste, as follows (Equation 25.12):

Equation 25.12: Calculating the Inventory Value of the Item

$$\text{Total Stock on Hand} = \text{Number of Units on Hand} \times \text{Weighted Average Price/Unit}$$

which

$$= 15 \times \$9.78$$

Total stock on hand

$146.70

Compiled by Miguel Gassman from Multiple Data Sources.

25.3.3.2.3 Latest purchase price method (most recent price)

A simpler and faster method to calculating the value of the stock item is to use the latest price which the goods were purchased for. In the above example, the last price paid for the cans of tomato was $10.00 each. Therefore, the inventory value would be calculated as follows:

Equation 25.13: Alternative Method of Calculating Inventory Value

$$\text{Total Value of Stock on Hand} = \text{Number of Units on Hand} \times \text{The most Recent Price}$$

which

$$= 15 \times \$10.00$$

$$= \$150$$

Compiled by Miguel Gassman from Multiple Data Sources.

25.3.4 Determining the food cost

Once we have calculated the total value of all the stocks on hand, we can now turn out attention to calculating the total food cost for the month. This method is applicable to all foodservice operations:

Equation 25.14: The basic formula to calculate the cost of food issued is:

Opening inventory + Purchase = Total Goods Available for Sale – Closing inventory
= Cost of food issued

Compiled by Miguel Gassman from Multiple Data Sources.

where the *Opening Inventory* is the value of goods that are available for sale at the beginning of the accounting period, *Purchase* is the value of all the goods purchased during the accounting period that are for sale, *closing inventory* is the total value of goods in the storeroom at the end of the accounting period that have not yet been used, i.e., are still in the storeroom, and *cost of food issued* refers to all the food that was used to generate the sales achieved.

This formula is fine when you are talking about a single standalone restaurant. In this case the cost of food issued would be considered as the cost of food sold. However, when you are working in an establishment with multiple restaurants such as a hotel, very often food is transferred from one restaurant to the other. Also, very often the hotel will offer its employees complimentary meals as a benefit to the staff.

When Restaurant A transfers some food to Restaurant B for sale in Restaurant B, it is only fair that the cost of the food also be transferred to Restaurant B. If the cost is not transferred to Restaurant B, Restaurant A's food cost will be unfairly high.

Similarly, a meal offered to an employee should not be charged to the restaurant as it is not a cost that was created to the sale of food, it is a staff benefit so it is a Human Resources cost and should be charged to Human Resources accordingly.

When these two factors mentioned above come into play, we revise the formula for calculating the cost of food sold to the following:

Equation 25.15: Calculating the Cost of Multiple Outlets/Conditions

Opening inventory + Purchases – Closing Inventory ± Transfers – Employee meals
= Cost of Food Sold

Compiled by Miguel Gassman from Multiple Data Sources.

We can display this in the following formula:

Equation 25.16: Transfers Effects on Food Costs

Opening Inventory + Purchases = Goods available for sale – Closing Inventory = Cost of food issued – Transfers out + Transfers in – Employee meals = Cost of food sold

Compiled by Miguel Gassman from Multiple Data Sources.

TABLE 25.11 Hugo's Big Restaurant costs.

Total sales	**$1,116,800.00**
Opening inventory	$78,550.00
Purchases	$455,100.00
Closing inventory	$101,250.00
Transfer to Restaurant A	$5200.00
Transfer to marketing	$4300.00
Transfer from Restaurant B	$1200.00
Transfer from bar	$2100.00
Employee meals	$2550.00

Compiled by Miguel Gassman from Multiple Data Sources.

When we look at the figures for Hugo's Big Restaurant below, for instance (Table 25.11): we can then calculate Hugo's Cost of food sold by the following equation:

Equation 25.17: Cost of Food Sold at Hugo's Place

$$\text{Opening Inventory} + \text{Purchases} = \text{Goods available for sale} - \text{Closing Inventory} = \text{Cost of food issued} - \text{Transfers out} + \text{Transfers in} - \text{Employee meals} = \text{Cost of food sold}$$

$$\$78,550.00 + \$455,100.00 + \$533,650.00 - \$101,250.00 = \$\ 432,400.00 - \$9,500.00 + \$3,300.00 - \$2,550.00 = \$423,650.00$$

Compiled by Miguel Gassman from Multiple Data Sources.

With the cost of food sold we can now calculate Hugo's cost of food sold in percentage (%) using the following formula (Equation 25.18):

Equation 25.18: Cost of food sold in % at Hugo's Place

$$\text{Food Cost \$} = \frac{\text{Cost of Food Sold}}{\text{Total Sales}}$$

Which equals

$$= \frac{\$423,650}{\$1,116,800}$$

Ergo Hugo's Food Cost %

$$= 37.93\%$$

Compiled by Miguel Gassman from Multiple Data Sources.

25.4 Plan daily production levels

One of the first questions a chef and foodservice manager need to ask themselves is "How much of each dish to prepare today?" This is a question that needs to be answered every day. Not only do they need to ensure they have enough ingredients to prepare the food, but they also need to schedule enough staff to prepare and serve the food.

The process for determining how much of each menu item needs to be prepared is based on the following formula (Dopson and Hayes, 2011):

Equation 25.19: Cost of food sold in % at Hugo's Place

$$\text{Prior Day's Carryover} \pm \text{Today's Production} = \text{Today's Sales forecast} \pm \text{Margin of Error}$$

Compiled by Miguel Gassman from Multiple Data Sources.

where *Prior Day's Carryover* is the amount of food that was produced for the day before that was not sold and can be carried over to be sold today without affecting the quality of the dish and *Margin of Error* should be a small sum because forecasting food sales is an imprecise science, so usually foodservice managers will find they need to produce a few more of each item.

From the above example (Table 25.12), it can be seen for the African chicken dish that five extra portions are produced as the safety margin, that is why the new production is 45 portions for the day.

For the sautéed vegetables, there were zero items carried over. This is because the quality of vegetables is greatly reduced from 1 day to the next, so it cannot be kept from 1 day to the next. This does not mean that if not all portions are used, the vegetables are thrown away. The vegetables will be used to prepare another dish such as soup or stews.

For the Serradura, there are 120 portions carried over, which is a lot more than the forecasted daily average. This is because some items when produced are in bulk, as producing one item only takes a lot of manpower, and the items can be stored for a while without the quality deteriorating. The same happens for the lamb curry, which is a dish that takes a long time to prepare from scratch; as such, Chefs normally produce large batches and

TABLE 25.12 Example of production schedule.

Menu item	Sales forecast	Prior Day's carryover	New production	Total available	Number sold	Carryover
African chicken	55	15	45	60		
Lamb curry	52	12	100	112		
Sautéed vegetables	78	0	100	100		
Serradura	45	120	0	120		

Compiled by Miguel Gassman from Multiple Data Sources.

TABLE 25.13 Example of a food requisition order.

Item requisitioned	Unit	Quantity requested	Quantity issued	Unit cost	Total cost
Rice	Kg	10	10	$12.00	$120.00
Beef chuck	Kg	12	15	$85.00	$1275.00
Cabbage	Kg	3	2.5	$8.00	$20.00
Canned tomato	Can	10	10	$12.00	$120.00
Total					**$1535.00**

Compiled by Miguel Gassman from Multiple Data Sources.

then extra batches and then store them in individual portions; they will then just finish the cooking once the food is ordered. Also, some dishes get better with time if properly stored.

25.5 Issue needed products to production areas

Requisitioning ingredients from the storeroom should not be an overly complicated procedure. It, however, cannot be so easy that anyone can go into the storeroom at any time to collect the ingredients. It is important that all food collected from the storeroom should always be requisitioned before pickup. This requisition should then be preapproved by management before any food is collected. Although it is important to ensure that all products are requisitioned, it is a relatively easy procedure to ensure product security, if the following steps are followed (Dopson and Hayes, 2011):

- Food and supplies should only be requisitioned on an as-needed basis
- Required items can only be issued with prior management approval
- Each person removing any item from the storeroom should sign for it
- Products that do not get used should be returned to the storeroom whenever possible.

It is important to remember that what is issued is very often not exactly what was requested; take for example the requisition order below (Table 25.13):

For the ingredients above, for the rice and canned tomatoes, the exact same amount of food ordered is issued as it is easy to measure or count and issue.

For the beef chuck, it is difficult to get a piece of beef that weighs exactly 12 kg, as such whole portions of beef chuck are issued until the minimum order is completed.

For the cabbage, only 2.5 kg is issued as may be only 2.5 kg is available in the storeroom.

As for the amount charged, it is based on the amount issued, not the amount ordered.

25.6 Manage the food and beverage production process

Managing the food production area is a process of controlling five areas (Dopson and Hayes, 2011):

- Waste
- Overcooking

- Overportioning
- Improper carryover utilization
- Inappropriate make or buy decisions

25.6.1 Waste

Waste is very often thought of a minor cost; however, it is important to remember that every time something is thrown away, money is being thrown away. Not cleaning a pot properly when a dish is prepared, the food left over in the pot is then thrown away. Little costs like these add up and, in the end, form a large and important amount.

25.6.2 Overcooking

Overcooking affects the quality of the food as most foods are very heavy in moisture, so when we overcook the food, more water than is planned for evaporates, so the yield is a lot less. Take for example a piece of roast beef, which should be cooked for 1 h. If we let the beef cook for 1.5 h, the size of the beef will be a lot smaller than a similar piece cooked for only 1 h. Also, the meat will be a lot dryer, tougher, and have a lot less flavor. Also, often when we overcook, we burn the food. The only solution for burned food is normally to throw it away, which means we are creating more waste. Overcooking is especially critical in the pastry, as when the food is overcooked, most of the time the only solution is to make a new batch.

Therefore, in an attempt to control overcooking, it is necessary to strictly enforce the use of standardized recipes. If everyone follows the standardized recipe, the chance of overcooking is greatly reduced.

25.6.3 Overportioning

Over portioning is a very widely talked about subject, not only because of the cost implications but also because of operational standardization of the dishes. If 6 kg of a dish has been prepared to serve 100 gr portions to 60 guests, it is important to make sure everyone get correct portion. If we start serving everyone 120 g, then after 50 guests have been served, we will run out of food. Normally the cost of the menu has been planned on 100 g per person; if they are served portions of 120g, then the food cost is 20% higher than expected.

Also, a person is always served a larger portion than what the actual portion is; when the guests come to the restaurant and are served the actual proper size, they will feel as if they are being robbed as the portion is so much smaller.

It is therefore important to ensure that all the necessary tools are made available to the staff to be able to quickly and efficiently prepare and serve the correct size portion every time.

It is important to establish the *standard portion size*, which is defined as the quantity of any item that is to be served each time that item is ordered. It is basically the exact amount of food to be served for each particular dish to each guest at any given time. Every item on the menu can be quantified by one of the three ways: weight, volume, and count (Dittmer and Heefe, 2009).

1. *Weight*: Normally expressed in grams, it is frequently used to measure sizes of portions, for example, steak.
2. *Volume*: Usually measured in milliliters, it is normally used to measure liquids such as soups, coffee, juices, etc.
3. *Count*: Used by operators to ensure all guests get the same quantity of an item, for example, eggs or sausages: Some items are also purchased by count, which defines their size such as prawns: prawn size at 60–70 pieces per kilogram are a lot smaller that prawns that are 20–30 pieced per kilogram.

25.6.4 Improper carryover utilization

When food that is kept from one day to the next as it was not sold is not properly packed and stored, the quality of the food will not be good enough to reuse. In this case, there is no other option than to throw it away. It is therefore necessary to ensure that if food is to be carried over to the next day, it is properly stored. If food cannot be carried over to the next day, then it is important that not too much is produced—only produce the amount that is required, based on your predictions.

25.6.5 Inappropriate make or buy decisions

In an effort to reduce labor costs, many operators decide to purchase food that is preprepared, also known as convenience foods or ready foods. This is not only done to save food costs but also because sometime a particular item requires very specialized equipment so it cannot be made in house. When deciding if a product should be made from scratch or should be bought preprepared, the following four points should be taken into consideration:

1. Is the quality acceptable? If the quality is not good enough, it should not be used.
2. Will the product save labor? If it does not save labor, then what is the use? Normally convenience foods are more expensive.
3. Would it matter if the guest knew that you are using preprepared products? Do you mind if they know that the food you are selling was not 100% made in your restaurant?
4. Does the product come in an acceptable package size? If it come in large packages of five portions each, and you only need one portion, what do you do with the other four portions?
5. Is the appropriate storage space available? Very often these foods come frozen, so is there enough freezer space?

25.7 Price the products at a price so that customers feel it is a value for money operation

We have so far discussed various important points on how to control your costs. The process started with maintaining your sales histories so that it could better forecast your sales, once we know more or less how much business we will be doing we can order the correct amount of ingredients and store them accordingly. We then plan our daily production levels and ensure that we have not only enough ingredients but also enough staff to prepare and serve the food. The food is then requisitioned, picked up from the storeroom and prepared, and ready to serve to the guest.

Now it is the next step and very possible the most important step, calculating the selling price. After all the hard work has been put into the preparation of the food, if the guests feel that the value is not there, they will not buy the product. So, it is important that the items are priced correctly.

Very often when the company is making a loss, although all the control measures have been put in place properly, the first thing the managers say is that the prices are too low. The goal of any company is to increase revenue, which is calculated by

Equation 25.20: Calculating Revenue

$$\text{Price} \times \text{Number of items sold} = \text{Revenue}$$

Compiled by Miguel Gassman from Multiple Data Sources.

From the above formula we can see that two factors that affect total revenue are *price* and *number of items sold*. Following the theory of law and demand, the higher the price, the lower the number of items sold and vice versa, the lower the price, the higher the volume of sales. So, whenever you increase your prices, the volume of sales will also go down.

Customers are very conscious of the *price/value relationship*. Should they feel that the value is there, they will not mind paying whatever price is stipulated. It is therefore necessary to ensure that you get the correct pricing strategy. Before you decide on the prices you will be charging, it is necessary to take into consideration the following factors that significantly affect your level of business.

1. *Competition*: What your competitors charge is definitely a factor that affects your price. If two competitors offer a similar product, customers usually go for the cheapest. However, if the quality of one of the products is better and more expensive than the other and the customers feel the price is justified, they will choose the more expensive product. It is also important to remember that it is very easy for a competitor to lower their prices by offering a lower quality product. Never assume because your competitor offers a cheaper product, you also have to offer the product at the same price.
2. *Service levels*: When the service levels are high, guests naturally expect to pay more. However, it is important to make sure the service level is not only high but also good. Having all the necessary hardware to provide great service but then having staff who are unqualified will degrade the experience for the customer drastically. They might not want to return if this happens. So, it is important to have a very high level of consistency in your service.

3. *Guest type*: It is important to identify who your customers are and what they value, as well as how much can they afford to pay. Some guests are more sensitive to prices than others. Once you know what type of customer you are going to serve, then you can plan your pricing accordingly.
4. *Product quality*: Generally speaking, most guests think the higher the quality, the higher the price. It is necessary to remember that not all guests think that all products are of a high quality. It is important then that you be able to identify the best quality product for the right price that the customer will be willing to pay for. The quality will have to be of a consistent quality as well.
5. *Portion size*: It is important to remember that guests eat with their eyes first. So, if you give them dish that looks very small on a very large oversized plate, they might think the portion is tiny. If you give the same portion on a regular size dish, taking this into consideration, it is also important to note that the larger the portion, the higher the cost. So, it is important to ensure that you prepare the correct portion using the standardized recipes which will have more or less standardized cost.
6. *Ambience*: In most cases, creating the right ambience in a restaurant costs money. It could be the decoration or the entertainment. In some cases, guests do not like paying more for the ambience.
7. *Meal period*: As a rule, people are willing to spend more at diner than at lunch. This is because at dinner they normally have more time to relax and enjoy the meal. At lunch time most people only have a limited time to eat so they prefer not to have to spend too much money on the meal.
8. *Location*: The location of your establishment also very much affects the prices you charge. If you are in a location that has a lot of competition in the area, you will need to charge more or less what the competition charges. If you are in a location where the customers have no choice but to use your establishment, then you might be able to charge premium prices. Take for example the amusement parks. The operators in these venues can charge premium prices because they know that the customers do not really have much of a choice.
9. *Sales mix*: It is, most probably, the most important factor when it comes to pricing. It is important to remember that guests do not only buy one item in your restaurant, but also they buy a variety of items. It is therefore important to have some items with a low price and some items with a higher price. The goal is at the end, the total revenue is what is expected.

25.7.1 Assigning menu prices

There are two main methods of assigning menu prices:

1. Product cost %
2. Product contribution margin

25.7.1.1 *Product cost %*

This is one of the most common methods of assigning menu prices. The formula for calculating the selling price of the item is as follows (the following equation):

Equation 25.21: Product cost

$$\text{Selling price in \$} = \frac{\text{Food cost of the dish(\$)}}{\text{Desired food cost\%}}$$

So for example, if we are selling hotdogs. The food cost of the hot dog is \$15.00.

We have established that we want a food cot % of 35%, the selling price of the hotdog will be:

$$\text{Selling price in \$} = \frac{\$15.00}{35\%} = \$42.86$$

Compiled by Miguel Gassman from Multiple Data Sources.

We can then round off the price to any of the following:

- Round up to either \$43.00 or \$45.00
- Round down to \$42.00 or \$40.00

25.7.1.2 *Product contribution margin*

A contribution margin is defined as the amount that is left after the product cost of the menu item is removed from the sales price (Dopson and Hayes, 2011). Therefore, the formula for calculating the selling price using the contribution margin is seen in the following equation:

Equation 25.22: The contribution margin:

$$\text{Selling Price(\$)} = \text{Cost Price of item(\$)} + \text{Desired Contribution Margin}$$

So, for the case of the hotdog above, if the restaurant wants to make \$20.00 contribution margin on every hotdog, they sell the selling price for the hot dog will be:

$$\text{Selling Price(\$)} = \$15.00 + \$20.00 = \$35.00$$

Compiled by Miguel Gassman from Multiple Data Sources.

References

Dittmer, P.R., Heefe III, J.D., 2009. Principles of Food, Beverage, and Labor Cost Controls. John wiley and Sons, INC, USA. John Dittmer, P. R. & Heefe Iii, J. D. 2009. Principles of Food, Beverage, and Labor Cost Controls, USA.

Dopson, L.R., Hayes, D.K., 2011. Food and Beverage Cost Control, fifth ed. Willey and Sons, INC.

Keiser, J., Demicco, F.J., Cobanoglu, C., Grimes, R.N., 2008. Anlyzing and Controlling Foodservice Costs, fifth ed. Pearson Education Ltd.

CHAPTER

26

F&B spatial design and functionality

Marta Hoi Yan Lam

Institute for Tourism Studies, Macao (IFTM): Colina de Mong-Há Macao, China

OUTLINE

26.1 Introduction

Food and beverage industry is not just about eating and drinking. It is a matter of how we can deliver "an experience" to the customers. This experience is perceived through customers' five senses: seeing, smelling, feeling, hearing, and tasting (Spence and Piqueras-Fiszman, 2014). Tasting, by all means, depends on the food which is to be served on the table. The other senses are somehow also related to customers' encounter through the servicescape[1] in and around the restaurant.

[1]Servicescape is the environment where service takes place, which if designed successfully will change people's behavior to enhance the overall experience.

Food and Society
https://doi.org/10.1016/B978-0-12-811808-5.00026-X

One of the key elements of servicescape within foodservice outlets is spatial arrangements. A logical progression from one space to another; from exterior to entrance to dining room; and from kitchen to dining room are all part of the deal to create an aesthetic yet functional environment to staff and customers. Developing a restaurant design and floor plan involves knowing every anthropometric detail of human, equipment, and movement path, that is, the space required for standing, sitting, and working to make the optimum space utilization together with the requirements of the restaurant operations. Ideally, 60% area in a floor plan must go to the dining area and the other 40% to the kitchen, storage, freezer, etc. (The Restaurant Times, 2017). This proportion of front of house serving area versus the back of house kitchen area varies differently among different concepts of foodservice outlets. But the ultimate emphasis on creating an amenity design lies on correct positioning within the market with functional use of space, yet with energy efficiency achieved in all equipment installations. With every nation in quest on new technology to create a carbon neutral operation, foodservice designs are also heading toward this sustainability mission.

26.2 Foodservice design factors: front and back of house—location and site characteristics

Choosing the best location to start up a foodservice business is the first key to success. But not all locations suit the owner's expectations, or sometimes the site itself is technically unfit for the establishment. Some newly developed sites lack the extensive coverage on building services (electricity, water, or gas supply), which results in inadequate support for equipment installations and might cause restrictions in future operations. Thorough checking on local regulations and utility provisions at the desired location is important before putting a finalized decision to run the business at the location.

The size and shape of the foodservice space, of course, depends on the original characteristics of the site. However, designing the space with higher ceilings or lower ceilings greatly affect the perception of people (Love and Grimley, 2007). Higher ceiling creates an airy space for diners and will also diminish the apparent width of the space (Ching and Binggeli, 2004), which engages people into the dramatic atmosphere. While a low ceiling space visualizes a huge area into a small enclosure, it helps to create a cozy and warm feeling in a restaurant (Baraban and Durocher, 2010). Shape of the dining space always has psychological impact. Most people are attracted to curvature forms, but curve walls and lines are difficult to layout table settings, losing the flexibility to combine tables for larger groups.

The geographical location also affects the back of house kitchen design. For instance, food can be purchased daily in urban restaurants; however, in more remote places, larger storage facilities within the restaurant are needed to accommodate for longer period before procurement and delivery can arrive (Walker, 2014). Thus, extra storage space must be planned in the design according to the logistics of food delivery.

If renovations are required to fit to an existing structure in a desired location, it is always necessary to seek design solutions to embrace the current elements of the existing structure in the design rather than cover them up (Baraban and Durocher, 2010). Taking an example of traditional Chinese architecture, legislation of keeping the shop front icon may not be a total fit to the concept and theme of a western cuisine foodservice. But with cross-over themes now coming into trend, a cultural and historical view at the outlook entrance can sometimes be an attractive gimmick to arise people's attention.

Visibility is the extent to which the restaurant can be seen for a reasonable amount of time, whether the potential guest is walking or driving by. The signs, parking area, landscaping, design and color of the building, windows, view of the inside, and lighting will all be closely scrutinized to maintain a visible sight to people. It is best to have the restaurant to be visible from both sides of the street, as far away as 400 feet (Katsigris and Thomas, 2009). This also means that checking local signage regulations is important so as to make sure you can or cannot put up your restaurant logo and signage up without affecting the neighborhood. Some worst-case scenarios, the size and height of the restaurant name badge might have restrictions in urban areas, making sure it complies with the local law and appropriate brightness level would definitely help in selecting the best foodservice location.

26.3 Spatial allocation with types of service

The restaurant design and layout will very much depend on whether the staff would serve the food or if it is a self-service restaurant. This refers to the "type of service" or concept in foodservice sector and is the start of the journey in developing a new restaurant. Examples of types of service include fine dining, casual dining, fast food, cafes, pubs and bars, buffets, etc. (Birchfield, 2008). Fine dining restaurants can be categorized with generous space in seating and in-service aisles. The ample space in fine dining will not only provide an upscale atmosphere to the customers but also allow tableside tools to move around the space, where casual dining and fast food services would not provide tableside service. Cafes, on the other hand, will need extra space on display refrigeration units for high visual impact to the customers to trigger their desire on cakes, sandwiches, and confectionary. Each and every type of service would have their own requirements on space allocation to equipment, staff, and customers.

According to the North American Association of Food Equipment Manufacturers (NAFEM), the following chart lists the average allotted square feet per customer by service type (Table 26.1).

TABLE 26.1 Space allocation for different type of foodservice operations.

Type of operation	Space allocation—per seat (Sq. Feet)
School lunchroom/cafeteria	9–12
Banquet room	10–11
Table service (minimum)	11–14
College or business and industry cafeteria	12–15
Table service at a hotel, club, or restaurant	15–18
Commercial cafeteria	16–18
Counter service restaurant	18–20

Source: Marta Lam.

26.4 The space layout

Design focuses on the overall space planning and includes defining size, shape, style, and decoration of a space. Layout, on the other hand, is the detailed arrangement of the equipment, floor space, and counter space (Gregoire, 2016).

26.4.1 Front of house

As mentioned in the beginning of the chapter, dining out is hardly just about the food. The scenes, artwork, decoration, and socializing with the servers are equally important. When these are done right, these design elements will alleviate the whole dining experience. A simple meal would seem to be much more scrumptious when paired with fascinating furniture, alluring lighting, and stunning artifacts.

From the moment when one enters the restaurant, all designers are hoping to acclimatize the customers through human's senses. First impression is the key to cognitive thought which will induce to the next behavioral response (Fitzsimmons, 2014). Gentle music, soft lighting, and a welcoming host are all part of the deal to facilitate the foodservice delivery starting point. When customers are adapted to this first acclimatization, the journey to lure into the design and theme commences.

Aside from the decoration, the seating layout is probably the next most concerning issue to customers' affection. Booth seating against the wall is, by all means, mostly favored by customers. This is because of the cozy and private sensation that one can get when sitting in booths. However, anchored booth seating has a restriction in movement, and the construction space needed is higher than typical rectangular and square tables. Where upscale restaurants contain more area to accommodate different types of table settings, the furniture used in upscale restaurants is normally larger and bulkier, thereby also forbidding the flexibility to accommodate changes. Round tables on the other hand are preferable for group dining because of easier socialization. Chinese cuisine restaurants, for instance, are designed with lots of round tables for the ability to share dishes in the Chinese food culture. But again, round tables also lack the flexibility to combine tables and pushing toward corner areas.

All types of foodservice space would require the provision of restrooms—no matter if it is a counter service, sit in service, or even a takeaway service. Careful consideration on where the restrooms are to be located depends on, majority, where the building services like plumbing, drainage, and water supply pipes are laid. If a total new makeover of the restaurant is to be done, new pipelines can be installed and can give alternative positioning of where the restrooms can locate within the dining space. The best to place the restrooms is far away from immediate dining seating area, although of course in some instances this cannot be avoided. In case of this, partitions should be installed in between the dining space and the restroom entrance to make sure a visual separation is maintained.

26.4.2 Back of house

The back of house area of a foodservice space is the kitchen, which should be well-designed to a happy, sanitized, and efficient working area for food preparation. It links to the importance of workflow, equipment positioning, and accommodating the required staff number to facilitate a smooth operation. It is always better that fewer steps are required to complete a task. Careful and well-considered kitchen design will save money and time during the construction phase and increase the profitability over the lifetime of the kitchen.

Kitchen design begins with an analysis of the individual task that must be performed in each area. The menu will govern the quantity of food and supply items required in the preparation and cooking processes. Correlating to the number of seating and turnover rate in the front dining space, the amount of storage space, size and type of production equipment needed at concurrent time can be evaluated. Below (Fig. 26.1) are a few steps to consider when performing kitchen space planning.

The last item listed in Fig. 26.1 refers to the "workflow," which is the movement of people in an operation. There are normally two types of flow in foodservice designs, namely product flow and traffic flow (Gregoire, 2016).

According to Gregoire (2016), product flow is the movement of food and ingredients from receiving through production to the customer and then through trash removal. On the other hand, traffic flow is the movement of staff through the operation from start to completion of their work. The best scenario is to a have a straight line flow from kitchen to serving space, or even best from receiving food delivery through warewashing (i.e., receiving → storage → preparation → serving → cleaning), which minimizes backtracking and cross-over movement of food and people. However, this rarely happens because the kitchen space shape is never a long rectangular dimensional space. According to Birchfield (2008), the flow of a foodservice design could ideally be demonstrated in Fig. 26.2.

In Payne-Palacio and Theis (2016), the typical diagram showing desirable work area relationships and progression of work from receiving goods to serving without backtracking and with little cross traffic is shown in Fig. 26.3.

Almanza et al. (2000) suggested the understanding of work centers in kitchen area to implement a smooth design fitting into the kitchen plan shape. Work centers are areas where a group of closely related tasks are done by an individual or individuals. According to Almanza et al. (2000), a work center is all about maximum reach and normal reach distance to obtain tools and materials for quick, convenient operation. Each work center should be

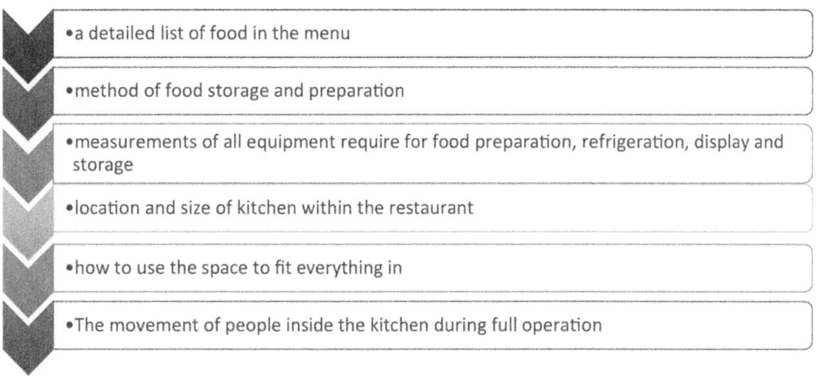

FIGURE 26.1 Simple steps in defining the space planning in kitchens. *Source: Marta Lam.*

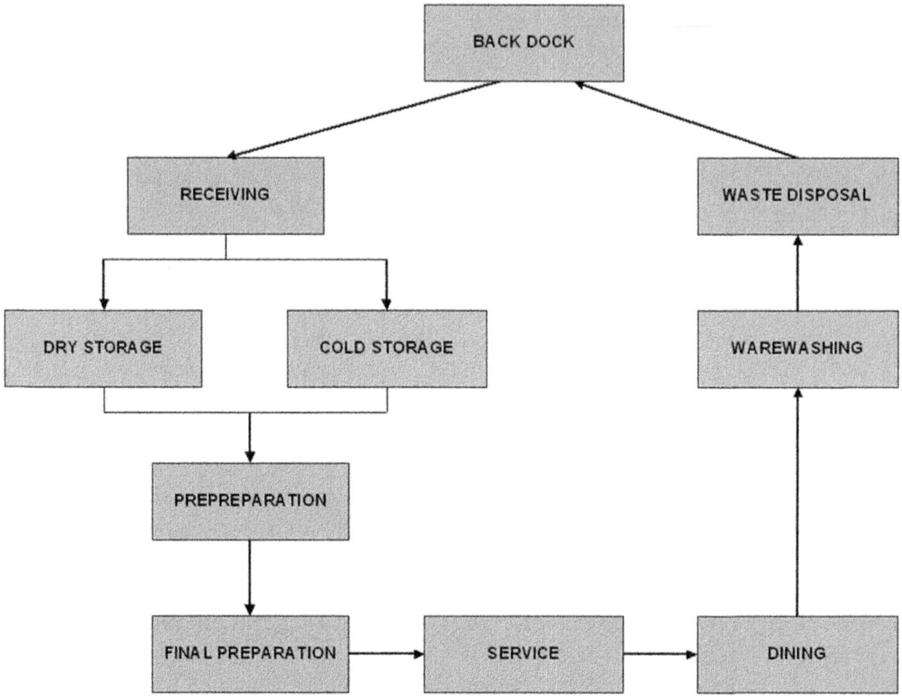

FIGURE 26.2 An example of foodservice design flow by Birchfield (2008). *Adapted from Marta Lam.*

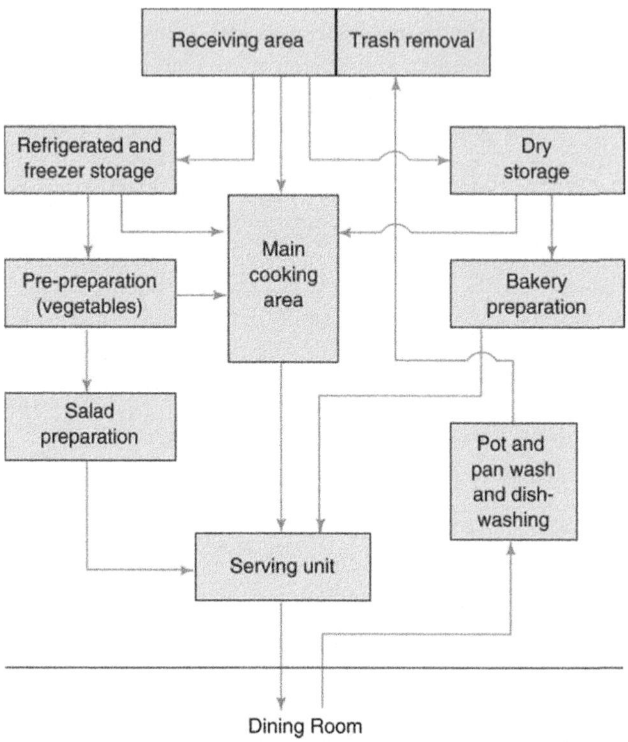

FIGURE 26.3 An example of foodservice design flow by Payne-Palacio and Theis (2016). *Adapted from Marta Lam.*

located at different points in the kitchen space so that staff and chef will not need to outreach places while performing their tasks. Thus, efficiency will be achieved. Each individual's work center will be combined together to form the total work flow inside the kitchen. These work centers do not need to be ideally in a straight line dimension, but they can be fitted according to the shape and size of the kitchen space. The aim is to eliminate unnecessary travel and energy consuming crisscrossing and backtracking and to obtain the smoothest flow line within the restricted kitchen plan shape.

Birchfield (2008) stated that flow is an important consideration in efficient design throughout a foodservice facility and listed out some flow considerations, quoted as below:

- The movement of employees from one section of the kitchen to another
- The flow of dishes through dishwashing system and back to the service area
- The flow of raw food ingredients through the main traffic aisles of the kitchen to the preparation area

In the design process, it will be very useful for the designer to diagram the flow patterns on the preliminary floor plan, showing the movement of chefs, servers, customers, food, dishes, and trashing rubbish. Color coding the flow lines of each party and equipment makes the patterns easier to distinguish and assists the designer in arriving at a design solution that accommodates the proper flow of materials and personnel. Rough sketches called "bubble diagrams" are used by designers to help in the preliminary design process. A bubble diagram quickly shows the spatial relationships among major areas of a restaurant. An example of a bubble diagram is shown in Fig. 26.4. Sometimes, these bubble diagrams can be expanded in details, thereby showing every equipment placement so it can be shared with different employees.

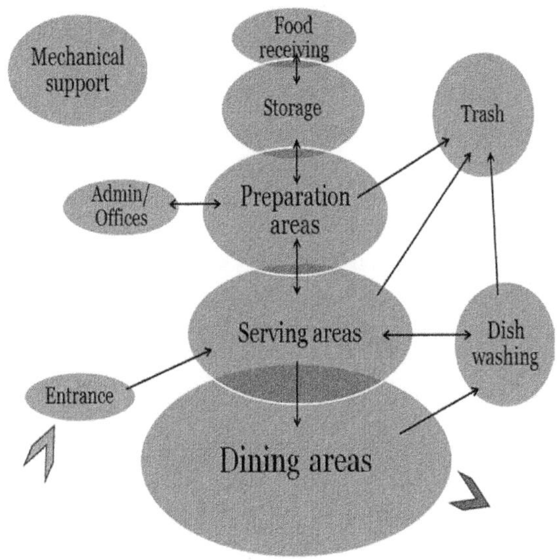

FIGURE 26.4 An example of a bubble diagram to show the schematic relationships between the various working areas (Mion, 2017). *Adapted from Marta Lam.*

26.5 Features of a comfortable restaurant design

Foodservice interior design is not only about how exquisite the decorations are or on the success purely laid on how exciting the menu is. People asked for comfort. According to the United States Environmental Protection Agency (EPA), an average American spent 90% of his time indoors. This posed a great challenge to architects, designers, and engineers to come up with eye-catching and comfortable and financially viable designs to implement in the current days. Hence, if your restaurant interior design is comfortable and if it gives the customers a sense of comfort, then there are high chances that your restaurant will top your customers' favorite list.

According to the Chartered Institution of Building Services Engineers (CIBSE), "the indoor environment should be designed and controlled so that occupants' comfort and health are assured." The overall layout of space and the accompanying environmental conditions of climate influence how people feel in a space.

26.6 Thermal

When the word thermal comes up, temperature is always considered as the relationship with it. Actually, thermally comfort is not just about the right temperature. It is all linked together with the relative humidity, the ambient temperature, and also the activity of oneself. Sometimes with the same ambient temperature at different humidity levels, humans experience comfort discrepancy. Removing excess moisture to lower the relative humidity (but at the same temperature) can, in some circumstances, achieve a thermally comfortable environment.

Indoor comfort can be achieved through air conditioning, which is either through heating or cooling. Outdoor comfort is about shelter and about external air modification through localized equipment. With alfresco dining becoming more and more common and with unbearable hotter summers, a lot of outdoor dining space has installed mechanical fans, localized cooling units, or even evaporative cooling fans by spraying jets of water to cool down customers outdoor.

The internal heat gains in kitchens are extremely high because of cooking, lighting, electrical equipment, and the number personnel, whereas in the front dining space it is much cooler to the customers because of steady activity, such as sitting down and enjoying the meal. Thus, the thermal requirements in each part of the restaurant could be totally different. Adjusting each zone into the required standard of thermally comfort level to occupants is essential for engineers to implement at the design stage.

26.7 Ventilation

Ventilation in foodservice designs is all about capturing and exhausting odors, grease, and heat from the kitchen and keeping them from entering the dining area. Controlling all these are never an easy task. Cooking equipment generates moisture and heat, and food preparation generates smoke and grease to the air. Mechanical ventilation must be installed in the kitchen area to force supply air into the space; and if the air properties are changed (cooler or hotter) before entering the kitchen, it will be known as air-conditioned.

Extract fans and exhaust hoods in the kitchen will remove smoke and grease to the exterior, which the airflow will induce changes of pressure in the space. At the same time, makeup air is added to the kitchen, the overall design promotes a negative pressure that prevents kitchen air pollution from reaching dining area, whereas in the dining area a positive pressure is to be maintained through mechanical ventilation systems such that outdoor dust and kitchen smell will not flow over to the dining space (Baraban and Durocher, 2010). In case there is a smoking bar next to the dining area, further considerations are needed so that the cigarette smell can be maintained and expelled out, while at the same time not affecting diners next to them.

Positioning the ventilation grilles (air outlets) can sometimes be a very hard job for the designers and engineers. Locating grilles far away from the dining tables may not be able to provide the required thermal comfort level. On the other hand, placing these air outlets near to dining seats can result in discomfort with high wind turbulence onto the diners. Thus, an efficient ventilation system in a restaurant should make all the above merits happen, within the space allocation of the ventilation equipment installation.

26.8 Lighting and color

Lighting is, of course, one of the major factors in designing restaurant as it enables visibility and also spontaneously affects the mood of the customers. Hence, choosing the proper lighting level and the lighting color is based on the restaurant theme, concept, and target market. In case of targeting young people like teenagers, then choosing funky lights could be appropriate. A quick service restaurant, for example, should be brightly lit so as to let patrons move through ordering and the pick-up process. The most important issue in this case is to prevent lingering to promote the turnover. On the other hand, if planning to target families or couples, using lights that create an intimate environment would be more suitable. Lighting must not be too bright or too dim in the dining area. A proper lighting level can enhance the decoration and increase the appeal to customers.

Large-sized windows capturing natural daylight could help in lighting up the space, giving a sense of well-being to customers. Another advantage of harnessing natural daylighting is to lower the energy bills. Brightly lit up restaurants typically spend from 12% to 15% of the annual budget on lighting (Katsigris and Thomas, 2009). However, dining tables located adjacent to the solar bearing windows receive a lot of heat gains, which brings to the dilemma of enjoying sunlight or becoming too hot for the discomfort diners. A well-balanced window area, with effective shading devices and heat reflective glasses, is somehow a key way to get a view and maintain comfort to the customers. The challenge while designing lighting is to understand under what circumstances certain visual conditions might be better than others and to use that knowledge to design lighting systems that improve the overall performance and human comfort (The Restaurant Times, 2017).

Color has a great psychological impact on guests. Different colors stimulate different emotions and can heavily impact feelings of hunger, thirst, and even space in customers

(The Restaurant Times, 2016). Tacky colors or the ones that hurt the eyes will not be much appreciated by humans. On the contrary, using proper color can help making the restaurant look spacious; similarly, a wrong choice of color can shrink the space. Colors are known to affect the eating habits of the customers; for instance, quick service restaurants more commonly choose bright colors such as red and yellow (such as McDonald's), as it inspires a feeling of excitement and makes customers order and eat quickly. Using warmer colors at the entrance of a restaurant makes the customers feel that the temperature is a few degrees higher than normal, providing a comfy atmosphere. In contrast, cooler shades send a psychological stimulus that makes people feel that the temperature is actually cooler (The Restaurant Times, 2016).

26.9 Acoustics

Acoustics is the design of sound within a space. In foodservice designs, control of the auditory environment is both important in dining space and in the kitchen. One of the greatest challenges is to safeguard perceptible conversations at the dining seats without being overpowered by the background noise and music. Thus, installing ceiling speakers or wall-mounted televisions or any music producing equipment must be carefully considered to each of the dining seats.

According to Baraban and Durocher (2010), the average noise level of a typical restaurant during a dining rush is around 80 dB (some could reach up to 110 dB). Within the dining space alone, an unsuccessful acoustics environment can also induce shouting and hearing problems. Most likely it is because the sound generated within the space is not absorbed within the decoration materials, causing echo and noisy atmosphere. Without the need to change the space layout, some acoustics solutions are, for example, to treat the dining ceilings with sound baffles; to treat the floors with carpet; to replace with soft furnishing and upholstered chairs; to use table cloths (reducing utensils noises on dining tables); and to hanging up curtains, which all can provide sound absorbing qualities to reduce the echo and noise reverberation problems.

Kitchens are obviously noisy spaces made louder by communication between chefs and service staff. Sound level above 65 dB will force chefs and servers to shout so that they can be heard over background noises (Baraban and Durocher, 2010). With remote printers in kitchens now linked with point of sale system, the need for conversation between production and service staff is significantly reduced. In case of designs with display kitchens, the carrying over of cooking and utensil noises must be minimized because of the proximity to the dining areas. These kinds of concerns are also vital in ensuring an appropriate acoustics environment in any foodservice space.

26.9.1 The trend in foodservice design

Foodservice sector is facing the same challenge to all industries in the nation, which is sustainability (Smith, 2001). It is not only about how delicious the food tastes nowadays, but also customers and owners are becoming more concerned about environment they are situating.

They would question about where the ingredients are from and if they locally sourced or grown from sustainable agriculture. Some people are more concerned with how food waste is being treated or whether there is a recycling facility available within the foodservice sector. Some customers will even query about the type of wood used in decoration and what preservatives have been used on the wood. Owners may request designers to use a mix use of natural lighting and artificial lighting in the dining space and urge the replacement of energy efficient luminaries irrespective of higher installation costs. All these issues are happening all around us as people are now leading a more environmental conscious lifestyle. The designs in all types of infrastructure, including those of foodservice spaces, are not only needed to fulfill the functional and esthetic qualities but also to aim for higher standards of sustainable designs to endure the future.

Accreditation points have been developed for energy conservation in foodservice designs. The most known system is the LEED certification process (Leadership in Energy and Environmental Design). It is a standard that architects and construction firms competing to market their buildings as green (Payne-Palacio and Theis, 2016). The LEED rating systems include groups of requirements for projects that want to achieve LEED certification. Projects satisfy different areas to earn points within the LEED credit categories so as to satisfy the green building requirements. Designs which gained 40 points will be LEED certified—50 points will be certified as silver level; 60 points as gold level; and 80 points as platinum level. The areas of scaling will include design and construction; site location; water efficiency; energy and atmosphere; material and resources; indoor quality; and innovation in design.

Another design trend in the foodservice sector is to become more universal design capable. The American Disability Act (ADA) was enforced to ensure that the environments to be designed nowadays are catered for everyone without much extra effort for people to use, termed as "Universal Design." Restaurant designs must comply with the ADA requirements, in particular to ingress and egress, size allocations, and perceptible sensory facilities. These are applied not only to patrons but also to each and every employee working inside the foodservice outlet, for example, providing slope access, handrail support, effective communication tools for people with hearing, vision, or speech disabilities and removing barriers, etc. This movement toward Universal Design will continue to strengthen as worldwide population continues to age. Architects and designers are in need to design and refurbish more environments to accommodate our aging community, providing easier access and alleviating use to all types of people.

References

ADA. American with Disabilities Act. https://www.ada.gov/ [Online].

Almanza, B.A., Kotschevar, L.H., Terrell, M.E., 2000. Foodservice Planning: Layout, Design, and Equipment, fourth ed. Prentice-Hall.

Baraban, R.S., Durocher, J.F., 2010. Successful Restaurant Design, third ed. John Wiley & Sons.

Birchfield, J.C., 2008. Design and Layout of FoodService Facilities, third ed. John Wiley & Sons.

CIBSE. The Chartered Institution of Building Services Engineers, 2015. Guide A: Environmental Design. CIBSE.

Ching, F.D., Binggeli, C., 2004. Interior Design Illustrated, second ed. John Wiley & Sons.

EPA. Environmental Protection Agency. Indoor Air Qual. [Online] https://cfpub.epa.gov/roe/chapter/air/indoorair.cfm.

Fitzsimmons, J.A., 2014. Service Management: Operations, Strategy, Information Technology, eighth ed. McGraw-Hill.
Gregoire, M.B., 2016. Foodservice Organizations: A Managerial and Systems Approach, ninth ed. Pearson.
Katsigris, C., Thomas, C., 2009. Design and Equipment for Restaurants and Foodservice: A Management View, third ed. John Wiley & Sons.
LEED. Leadership in Energy & Environmental Design. https://new.usgbc.org/leed [Online].
Love, M., Grimley, C., 2007. Color, Space, and Style: All the Details Interior Designers Need to Know but Can Never Find. Rockport Publishers.
Mion, E.G., 2017. Whole Building Design Guide: Dining Facilities [Online]. Available: https://www.wbdg.org/building-types/community-services/dining-facilities.
NAFEM. North American Association of Food Equipment Manufacturers. https://www.nafem.org/ [Online].
Payne-Palacio, J., Theis, M., 2016. Foodservice Management: Principles and Practices, thirteenth ed. Prentice-Hall.
Smith, P.F., 2001. Architecture in a Climate of Change – a Guide to Sustainable Design. Oxford Architectural Press.
Spence, C., Piqueras-Fiszman, B., 2014. The Perfect Meal: The Multisensory Science of Food and Dining. Wiley-Blackwell.
The Restaurant Times, 2016. How to Use Restaurant Interior Colours to Increase Your Restaurant Sales [Online]. Available: https://www.posist.com/restaurant-times/restro-gyaan/know-restaurant-interior-colors-can-affect-sales.html?utm_source=rt&utm_medium=blog&utm_campaign=restaurant_design_tips&utm_term=restaurant_interior_colors.
The Restaurant Times, 2017. How to Design a Restaurant Layout & Floor Plan [Online]. Available: https://www.posist.com/restaurant-times/resources/restaurant-design-guidelines-to-achieve-the-optimum-utilisation-of-floor-area-in-restaurants.html.
Walker, J.R., 2014. The Restaurant: From Concept to Operation, seventh ed. John Wiley & Sons.

PART FIVE

Summary

How little we know of what there is to know
Ernest Hemingway
For Whom the Bell Tolls

Summary: What role food?

In the first part of this book, we examined, in a fairly detailed manner, the varying contemporary issues surrounding food and society. Indeed, in this undertaking, we covered many topics from historical and contemporary perspectives, especially in the areas of nutritional guidelines and how they came to be. This section also introduced notions of food and health and the socioeconomics of food and political governance. In fact, there were few areas untouched by the phenomenon of food. The 20th century was indeed a very thought-provoking period giving the reader important background or contextual reading in preparation for the rest of the book. From these early sections it becomes clear that food and society were in fact inextricably linked to matters of political issues including, and not limited to, governance and development theory, especially in agriculture as a means of economic growth and international trade. Further digging into food and society, one is drawn to social issues of hunger and food security, briefly describing the many issues of the "have" and the "have not" and the vast gap that exists between those that remain hungry throughout the world and obesity, which is a paradox of modern economic development whereby both often coexist in the same country or region at the same time. Food—its production, consumption, and waste—not surprisingly, is also a major consideration of environmental and sustainability issues. No wonder then that much time and effort is given to policies of waste management, reusability, the global food chain, in the form of legislation, corporate social responsibility, or public awareness campaigns. Food in this and many other aspects has been politicized and the editor believes that it will only become more entrenched in sociopolitical governance.

In Part 2 and 3, ideas that describe and evaluate issues surrounding the anthropology of food are further developed—additionally developing ideas of food trends; exploitation; fairtrade; the rising tide of food rights; a growing international body of non-governmental organizations (NGOs); the organic movement; slow food, enhanced governance; and more importantly society's culture. Indeed, when it comes to food and society, this is not a modern phenomenon—for millennia, food has been the center of history's most auspicious and some of the most disastrous moments, through practiced rituals, festivals, social gathering tourism (and gastro-tourism), and other similar societal events.

This section leads nicely into some practical issues of the food service industry, particularly surrounding hospitality and hospitality-related fields. Not surprisingly, food and beverage have become large components of a huge global market.

As food is all pervasive and touches every corner of peoples' lives and every ounce of the earth, it is no wonder so much has been written on the topic. Indeed, the breadth and scope of the above topics might seem somewhat disparate at first glance; however, there has been a great deal of thought given to what was to be included in this book and what would not.

There have been agrarian revolutions, industrial revolutions, political revolutions, and innovation and technology; there have been different schools of thought over food and agriculture's development potential.

Muslim agricultural revolution, followed by the Columbian exchange, colonial agriculture, and a period of enlightenment and an ever-increasing social awareness and the latter half of this great evolutionary period ushered in the game-changing industrial revolution at large and to a great extent.

General economic theory

A2.1 Economics of food: general background

How society uses its resources to satisfy its wants and needs is the basis of all economic activities. Essentially there are three basic varieties of economic systems—traditional, command, and market—although some suggest a fourth variety, a mixed economy. A *traditional* economy relies very much on the status quo—what has been will be—so to speak. It is an economic system largely based on custom and tradition and is thought of as a primitive economic system. As a system too, it is often relying on such practices as subsistence farming, etc. A *command* economy on the other hand is presided over and controlled at the government level. Such systems are particularly prevalent in communist political systems such as Korea and Cuba. The *market* economic system is based on capitalist values and seeks to leave the market mechanism alone to determine supply and demand unimpeded by interventionist policies. This type of economic system naturally encourages change and growth. It is thought of as the most efficient economy model, and its most famous proponent Adam Smith promoted it as a natural and beneficial system in which people could trade fairly based on their comparative advantages. Lastly a mixed economy is one in which both command and market forces are at work. This socialist approach favors the government owning and controlling basic productive resources while the rest of businesses are privately owned. This model is predicated on big government involvement where they have more say in the economy than in the capitalist model but less so than in communism. In theory, the mixed economic model should allow for everyone to be provided for, particularly those who are not able to equitably share in society's benefits. In this way, certain benefits such as health care, welfare, and other social safety nets are catered for at the state level. While there are few downfalls to this type of socialism, it is not usually seen as productive or as efficient as capitalism. As a result, another form of mixed economy favors a mixture of capitalism and socialism. This mixed model is thought to allow some freedom of market forces that facilitate trade while also ensuring that government intervention takes care of the welfare side of society.

A2.2 Macro- and microeconomic principles

Whatsoever the economic model is, any economy can be subdivided into two components: macroeconomics and microeconomics. In macroeconomics, analysts and policy-makers are concerned with the overall growth of the economy; inflation issues; development economics; the business cycle; monetary policy and policy regulation; and the overall balance of payments (BoP) among others. Microeconomics on the other hand is concerned with ensuring the well-oiled operational components of a national economy—these include issues dealing with the markets, individuals, and businesses, etc. There are several economic areas relevant to the global food sector including the market mechanism to development economics.

A2.3 Balance of payments

Perhaps the overriding economic concern at the macrolevel is that of the BoP. The BoP is a record of all the monetary transactions of countries' consumers, their businesses, and government activity with the rest of the world—it is the country's balance sheet. The BoP is made up of three accounts: the current, capital, and reserve accounts, and by definition they must balance; i.e., Current Account Balance + Financial Account Balance + Official Reserve Account Balance = 0. In short, a country's BoP is concerned with balancing payments for such things as a country's exports and imports (goods, services, and financial capital), as well as its financial transfers.

A2.4 Current account

A country's current account tracks all the money going in or out of that country. In turn the current account is composed of three named accounts: the balance of trade; net investment income; and other transfers. It is in the current account that trade surpluses and deficits are identified. The trade deficit essentially means that more goods and services are being imported into a country than are being exported resulting in a net outflow of finance. A trade surplus on the other hand comes about when there are more exports than imports. The same is true of investment or net investment flows to be precise; this can be negative with net investment outflow or positive with net investment inflow.

A2.4.1 Capital account

While the current account tracks changes in the income of a country on the BoP, the capital account can be thought of as tracking the assets of a particular country. It is composed of four accounts: foreign direct investment (FDI) account; portfolio investment account; other Investment account; and the Reserve account. While the FDI refers to long-term capital investment in such things as the purchase of machinery, buildings, and businesses, the portfolio investment account looks after the purchase of shares and bonds. Besides these two accounts, the other investment account records net capital flows into banks or money provided as loans as well as taking advantage of fluctuations in interest or exchange rates between currencies as a result of the common practice of short-term financial flows between nations.

A2.4.2 Reserve account

Operated by a nation's central bank, the reserve account, also known as the official settlements account, is the balancing mechanism of the BoP. Everything else being equal, trade between two countries is financed by the country's reserve account. In this way, imports into country create a net outflow of the reserve account while exports result in inflows of foreign exchange. As a result, and in principle, the BoP account should balance, although there are almost always discrepancies.

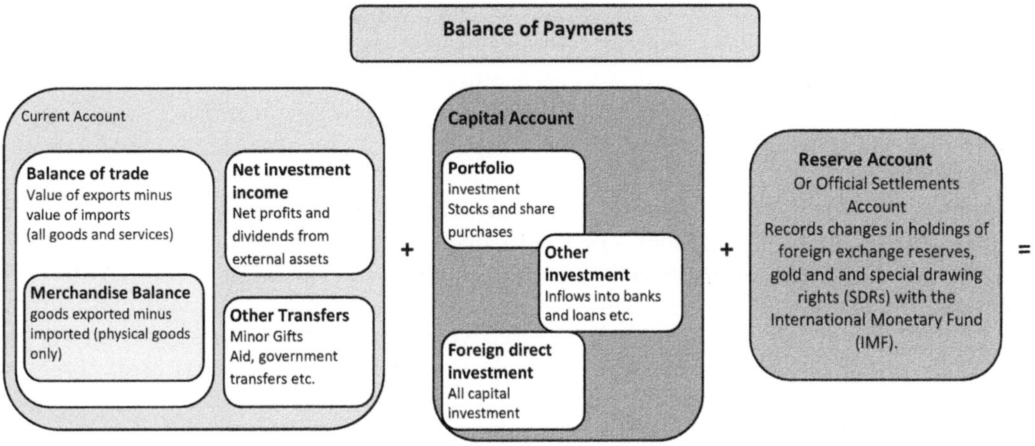

FIGURE A2.1 Balance of payments accounting system. *Source: Mark Gibson.*

Overall the wealth of a country or its economic development then is calculated using the BoP. For instance, while initially the net inflow of funds from favorable export volumes results in financial inflows into a country, so then foreign exchange, gold, and other reserves quickly build up—effectively increasing a country's assets. While this might be good news for some, such as making imports cheaper, overall this becomes unattractive as it can place pressure on domestic currencies. The central bank in these situations then comes under pressure to intervene to stabilize or devalue the local currency. If it fails to do so, the financial markets automatically and effectively adjust terms of trade accordingly through foreign exchange rates instead. Also, if unaddressed, strong local currencies can sometimes result in currency appreciation as expressed through inflation, thus restoring equilibrium in terms of trade (Fig. A2.1).

A2.4.3 Exchange rates

Trade is the cornerstone of economic activity. In early civilization, traditional bartering in exchange for goods sufficed for the longest time; however, as progress was made, this form of exchange became inefficient and inflexible. Eventually coinage was introduced based on precious metals such as gold and silver, etc. However, this had fundamental limitations in that such metals were scarce, and as economies expanded, an alternative was therefore needed to be found. The solution was fiat money yet, unlike the intrinsic value of gold the value of fiat money was illusional; that is, it thrived solely on the belief of its worth. Ultimately, acceptance and growth in fiat money allowed governments' flexibility and leverage to manage the money in circulation.

However, this growth created problems of its own; through excessive leverage over the years too much credit was being introduced into the economy with the effect of destabilizing it. Controls had to be put in place and one way of achieving this was the introduction of gold convertibility. By 1821, England adopted a gold standard, although by 1871 with the advent of a new international standard much of the developed world pegged their currencies to gold. Thus, the gold standard allowed countries to values their currencies against their stockholding of gold and by extension to each other. This became known as the fixed exchange rate system. By the 20th century, all major industrial powers were using the new gold standard and ensured a good deal of stability shepherding a new era of international trade among nations.

However, with the advent of the First World War, it surfaced that there was insufficient gold to back the amount of money needed to finance the war effort resulting in many countries abandoning or suspending the gold standard. This led to much speculation on currency markets as countries no longer defined the values of their currencies in terms of a gold standard. Instability and speculation caused volatility in the markets—the answer to stability in the form of a new version of the gold standard was introduced in the Bretton Woods Agreement. In this incarnation, countries pegged their domestic currency to that of the dollar, which in turn was pegged to gold. This was fine for close to three decades by which time inflationary pressures in 1971 saw the dollar devalue and breakaway from the agreed convertibility standard. Others quickly followed allowing for a new floating rate exchange system where exchange rate values were determined in the market place through supply and demand. A floating exchange rate by its very nature is constantly in change and helps promote efficiency and self-correction. Nowadays, while today the use of gold standard is long gone, in reality no currency is either fully floating or indeed fully fixed. In this sense it is left up to national governments to choose to have a "floating" or "fixed." Of course, there are inherent benefits and weaknesses in both fixed and floating exchange rate systems. The main reason to "fix" or peg a currency is stability; however, a peg is difficult to maintain in the long run and can sometimes lead to financial crises. As such we often see this kind of monetary policy in less sophisticated financial markets characterized by weak regulatory institutions. By contrast, a floating currency is seen as a more efficient exchange rate mechanism associated with strong, mature economies.

Millennium Development Goals (MDG)

The official list of Millennium Development Goals (MDGs) as listed below comprise indicators that emerged out of the Summit are further expanded upon in Table A3.1.

Goal 1: Eliminate the worst excesses of poverty and hunger
Goal 2: The achievement of universal primary education
Goal 3: Promote gender equality and empower women
Goal 4: Reduce child mortality
Goal 5: Improve maternal health
Goal 6: Combat HIV/AIDS, malaria, and other diseases
Goal 7: Ensure environmental sustainability
Goal 8: Develop a global partnership for development

TABLE: A3.1 The official list of Millennium Development Goals (MDG) indicators that emerged out of the Millennium Summit.

Targets and goals that emerged from the Millennium Declaration	Suggested indicators for monitoring progress of MDGs
Goal 1: Eliminate the worst excesses of poverty and hunger	
Target 1.A: Between 1990 and 2015. The target of the first goal was to proportionally reduce the number of people by half who earned less than one dollar a day	To achieve this target the suggested indicators to watch were several including the proportion of population who earned less than the $1 (PPP) per day; the proportion of national consumption; and of course the poverty gap ratio.
Target 1.B: Target 1B looks toward achieving not only full and productive employment for all, but also decent work for women as well as the younger generation too.	There are several indicators used in the assessment of Target 1.B: The first and perhaps the most commonly used is that of per person GDP growth rate followed by the employment-to-population ratio; the proportion of increasing owner account holders; together with family workers prevalence in total employment proportion of employed people living below $1 (PPP) per day.
Target 1.C: This target concerns itself with the halving (between 1990 and 2015), the proportion of the population who suffer from hunger	This target relies on such indicators as the prevalence of underweight children (under-5's), as well as the proportion of the population whose daily diet falls below the considered "minimum" level of dietary energy consumption.

(*Continued*)

TABLE: A3.1 The official list of Millennium Development Goals (MDG) indicators that emerged out of the Millennium Summit.—cont'd

Targets and goals that emerged from the Millennium Declaration	Suggested indicators for monitoring progress of MDGs
Goal 2: The achievement of universal primary education	
Target 2.A: Recognizes the fact that in terms of future governance among other things – will, at some point, be governed by those we seek to help. As such Target 2.A is responsible for ensuring that, by 2015, children everywhere (without discrimination), should be able to complete a full course of primary education	Ensuring Target 2.A is on track, the MDGs look to include components of net enrollment ratio in primary education; Proportion of pupils starting grade 1 who reach last grade of primary as well as the literacy rate; literacy rate of 15–24-year-old women and men.
Goal 3: Promote gender equality and empower women	
Target 3.A: Aims to eradicate gender inequality or disparity in both primary and secondary educational facilities. This is to happen primarily by 2005, with a 2nd remit that covers all other levels of education no later than 2015	This target is tracked through the assessment rubrics of ratios of girls to boys in primary, secondary, and tertiary education; the share of women employment in the nonagricultural sector; and the ambitious proportion of seats occupied by women in local and national councils/ parliaments.
Goal 4: Reduce child mortality	
Target 4.A: Concentrates on the reduction of under-five mortality rate by two-thirds, between 1990 and 2015	This target is easily measured through the under-5 mortality rates; the infant mortality rate; as well as the proportion of 1-year-old children who have been immunized against measles.
Goal 5: Improve maternal health	
Target 5.A: Between 1990 and 2015 Target 5A. Aims to reduce by three quarters, overall maternal mortality ratio	Target 5A seeks to reduce a region or country's maternal mortality ratio, as well as increasing the proportion of births attended by skilled health professionals.
Target 5.B: The target of 5.B was to accomplish, by 2015, not just access but universal access to reproductive health healthcare	This was measured through such indicators as adolescent birth and contraceptive prevalence rates, antenatal care, etc.
Goal 6: To combat HIV/AIDS, malaria, and other diseases	
Target 6.A: To have halted and begun to reverse the spread of HIV/AIDS by 2015	By monitoring such statistics as the prevalence of HIV among the population of 15–24-year old's. Condom use at last-known high-risk sex; the percentage of 15–24-year old's with comprehensive knowledge of HIV/AIDS. A healthier ratio of school attendance of orphans to school attendance of nonorphans aged 10–14 years of age
Target 6.B: Goal is to achieve, widespread access (by 2010) to treatment for HIV/AIDS for all those who require it	The main designated indicator of this target is the proportion of population that has advanced HIV infection rates and with access to antiretroviral drugs.
Target 6.C: To have (by 2015) stopped and started to reverse the prevalence of malaria and other major diseases	The best benchmarks of Target 6.C are the number of under 5's sleeping beneath insecticide-treated bed nets; the Frequency and death rates resulting from malaria, as well as the numbers of tuberculosis death rates and detected cases cured under supervision.

(*Continued*)

TABLE: A3.1 The official list of Millennium Development Goals (MDG) indicators that emerged out of the Millennium Summit.—cont'd

Targets and goals that emerged from the Millennium Declaration	Suggested indicators for monitoring progress of MDGs
GOAL 7: ENSURE ENVIRONMENTAL SUSTAINABILITY	
Target 7.A: Aims to integrate sustainable development principles into a country's policies and programs in order to arrest and reverse the loss of environmental resources. Target 7.B: In unison with the above aim, the MDG's look to slow-down and stop the loss of biodiversity loss, aiming for a substantial reduction in the rate of loss by 2010	These two targets utilize many of the indicators below: the proportion of species threatened with extinction; the total emissions of CO_2, per capita and per \$1 GDP (PPP); ensuring the proportion of fish stocks remain in safe biological limits; the total water resources used; the amount of land covered by forest; and finally, the amount of terrestrial and marine areas that are currently protected.
Target 7.C: To halve (by 2015), the quantity of people who do not have sustainable access to safe drinking water and basic sanitation	Indictors include the proportion of population using improved/treated water sources as well as the numbers/proportion of people who have access to an improved sanitation facility.
Target 7.D: By 2020, the MDG's target is to have improved the lives of at least 100 million slum dwellers	Without specific indicators the only relevant measure is the prevalence of urban populations living in slums.
Goal 8: Develop a global partnership for development	
Target 8.A: In developing further global development the, MDG board advocates an accessible, predictable, non-discriminatory-financial as well as a global equitable trading platform. On top of this a commitment to good governance, is required. Target 8.B: Address the special requirements of the least developed countries. Target 8.C: Address the special needs of landlocked developing countries as well as small island developing States. Target 8.D: Deal comprehensively with issues of debt problems of developing countries	Some of the indicators listed below are monitored separately for the least developed countries, Africa, landlocked developing countries, and small island developing states; official development assistance. Others are measured in aggregate; thus, it is difficult to separate goals or indicators for this particular target. A few multiple target indicators include items such as the amount of overseas development assistance received by landlocked developing countries as a proportion of their gross national income; quota and/or tariff free access for the least developed countries' exports; market access; proportion of a country's total developed imports from developing countries and least developed countries, freely entering such countries free of duty. Similar to the above, this particular target (8.D) invokes a wide-ranging suit of indicators including the forward-looking and far-reaching cancellation of the country in question's official bilateral debt. Growth through agricultural development channeled through the agricultural support estimate for OECD countries proportional to their gross domestic product; debt sustainability.
Target 8.E: Working with large multinational pharmaceutical companies, access to affordable essential drugs is provided in developing countries	Proportion of the population with access to essential affordable drugs on a sustainable basis.
Target 8.F: Once again in cooperation with the private sector, this target makes available the benefits of new technologies, particularly in the information and communications sector	Telephone lines per 100 population; cellular subscribers per 100 population; Internet users per 100 population.

Source: Official list of The Millennium Development Goals (MDG) and targets which became the official outcome forming the Millennium Declaration. Signed by 189 countries, orchestrated in 2000 by the UN (2000b) and from further agreement by member states at the follow-up summit in 2005 in the form of the instrument of the General Assembly – A/RES/60/1.

The MDGs working on popular politically acceptable targets did make some remarkable inroads regarding the targets however, their mandate was only effective until 2015, after which another instrument (also via the United Nations) was put into play: The 2030 Agenda for Sustainable Development with the Sustainable Development Goals (SDGs).

The 2030 Agenda for Sustainable Development

The sustainable development agenda and its 17 goals are very similar to the MDGs and essentially carried on where the MDGs left-off. They were unanimously adopted by all United Nations Member States in 2015 and effectively provided a framework for peace, prosperity and a sustainable future for the planet and its people.

The target achievements to 2030 became known as the Sustainable Development Goals (SDGs). Seventeen in total the SDGs collectively outlined the entire member state's objectives for sustainable development. This was essentially an urgent call for action by both the developed and developing countries as a coherent and collective worldwide partnership. One noticeable difference - while the interwoven nature of the goals of the MDGs were generally implicit in the MDGs, it was made overtly explicit in the SDGs of 2015. Through the recognition of the notion that ending poverty and other similar targets were often aligned - or went hand in hand with other targets. For instance, it was noted that strategies which improved say, health and education, might also reduce inequality, and spur economic growth and which in-turn could potentially affect climate change; this emphasized the realization of the interwoven character of certain goals which was ultimately expressed in the often mention mantra: that the goal's should not be tackled in isolation.

In and of itself the SDGs are the sum of countless decades of work by individuals, countries as well as governmental and civil societal bodies - especially when it came to the work of the UN Department of Economic and Social Affairs. Kick-starting the targets and goals recent incarnation was the United Nations who, in 1992 held the "Earth Summit" in Rio de Janeiro, Brazil. Building on a solid base of agreement of more than 178 countries, and after countless debates and discussions, the outcome - the Agenda 21 plan was adopted. Agenda 21 was essentially a plan of action that pushed for sustainable development to improve the lot of human lives whilst protecting the environment. It was a good beginning and years later paved the way for the unanimous adoption in 2000 of the Millennium Summits' Millennium Declaration (*see above*). This led to the above mentioned eight Millennium Development Goals (MDGs) with the main goal to reduce and if possible eliminate, extreme poverty by 2015. This was further endorsed and improved upon in the 2002 World Summit on Sustainable Development in South Africa - officially the Johannesburg Declaration on Sustainable Development. This effectively reaffirmed the global community's commitment, through emphasis on the joint cooperation of achievement of the goals both the Agenda 21 and the MDG's through greater collaboration and stronger multilateral partnerships.

At the UN Conference on Sustainable Development in June 2012 (also held in Rio de Janeiro, Brazil (Rio+20)), and after various negotiations and discussions, Member States finally adopted the 'plan of action' or the outcome document entitled "The Future We Want" in which Members decided, among other things to develop a set of Sustainable Development Goals that built on the MDGs. By 2015, the UN General Assembly started the ball rolling on the post-2015 development agenda, this resulted in the UN Sustainable Development Summit in September 2015s final adoption of the 2030 Agenda for Sustainable Development and its 17 goals.

Nutritional guidelines

Regarding nutrients, whether macro-, micro-, or trace elements, it just remains to determine the actual guidelines. As with previous discussions on this, while there are many nutritional guidelines available to the general public, some guidelines are more acceptable than others, especially at the international level. These include those of the World Health Organization (WHO), the United States, and the European Union. The following table summarizes the various nutrient guidelines in question (Table A4.1).

TABLE A4.1 Macro- and micronutrient guidelines of the European Union (EU), United States (USA), and World Health Organization (WHO) for adult men and women.

			Energy																
			Carbohydrate			Fat				Protein[b]									
	Water	Daily Calories	Total	Sugar	Fiber	Total	Saturated Fat	n-6 PUFA Linoleic acid	n-3 PUFA α-Linolenic acid	Total	Vitamin A[c]	B_1 Thiamine	B_2 Riboflavin	B_3 Niacin[d]	B_5 Pantothenic Acid	B_7 Biotin	Vit B_6 Pyridoxine	B_9 Folate	Vit B_{12} Cobalamin
Body	(L/d)	kcal	(g/d)		(g/d)	(g/d)	g/d	(g/d)	(g/d)	(g/d)	(μg/d)	(mg/d)	(mg/d)	(mg/d) e	(mg/d)	(μg/d)	(mg/d)	(μg/d)	(μg/d)
USDA[1]	2.7-3.7[a,*]	**W: 2000[9] M: 2400[9]**	>130[3] **W: 225-325 M: 270-390**		25-38*	ND **W: 44-67 M: 53-93**		12-17*	1.1-1.6*	46-56 **W: 50-175 M: 60-210**	7-900	1.1-1.2	1.1-1.3	14-16	5*	30*	1.3	400	2.4
FAO/WHO[2]	1.4-2 (3[a])	**W: 2000[10] M: 2550[10]**	**W: 275-375 M: 350-480**			**W:33-66 M:42-84**				54-58[7] **W: 50-75 M: 64-96**	5-600	1.1-1.2	1.1-1.3	14-16	5	30	1.3	400	2.4
EU	2-2.5[a,*,5]	**W: 2000[11] M: 2500[11]**	230[4] W: 230-270[6] M: 300-340[6]	90[6] W: -90[6] M: 110-120[6]		70[4] W:65-70[6]; M:80-95[6]	20[4] W: 20[6] M: 30[6]			W: - M:	800	1.1	1.4	16	6	50	1.4	200	2.5

Body	Vitamin C (mg/d)	Vitamin D[e] (μg/d)	Vitamin E (mg/d)	Vitamin K (μg/d)	Calcium (mg/d)	Chloride (g/d)	Choline (mg/d)	Chromium (μg/d)	Copper (μg/d)	Fluoride (mg/d)	Iodine (μg/d)	Iron (mg/d)	Magnesium (mg/d)	Manganese (mg/d)	Phosphorous (mg/d)	Potassium (g/d)	Selenium (μg/d)	Sodium (g/d)	Zinc (mg/d)
USDA[1]	75-90	5*	15	90-120*	1000*	2.3*	425-550*	25-35*	900	3-4*	150	8-18	310-400	43-45	700	4.7*	55	1.5*	8-11
FAO/WHO[2]	45	5	7.5-10	55-65	1000		750*	-	1100	3-4*	150	9-59[8]	220-260	-	-	2*	26-34	0.5*	3-14[8]
EU	80	5	12	75	800	0.8	425-550*	40	1000	3.5	150	14	375	2	700	2	55	0.575-3.5	10

Notes:

[1]*USDA adult values 19-30 years based on Institute of Medicine tables;*

[2]*Adults 19-50 years (approx);*

[3]*The RDA for carbohydrate is set at 130 g/d for both adults and children based on the average minimum needed to keep brain activities functional. This level is typically exceeded with median intakes of 220 to 330 g/d for men and 180 to 230 g/d for women;*

[4]*Taken from the labeling Reference Intake at 2000 kcal values and based on scientific opinion undergoing consultation and subject to change;*

[5]*European Food Safety Authority (EFSA) water intakes are preliminary based on scientific opinion; undergoing consultation and are regularly subect to change;*

[6]*Based on EU's current labeling Intake Values;*

[7]*Based on a 65-70kg adult;*

[8]*Depending on Bioavailability;*

[9]*USDA's domestic reference energy intakes of sedentary adults;*

[10]*FAO's sedentary adult requirements (for average bodyweights of 70 kg);*

[11]*Based on EU labeling Reference Intake guides;* ***Bold 'Type'*** *is calculated using stated domestic or industrialized calorie requirements for each institution and recommended energy balance portfolios (table 2.5);*

[a]*Total water includes all drinking water and water contained in food and beverages;*

[b]*Based adults 0.8 g/kg body weight for the reference body weight;*

[c]*USDA figures=Retinol activity equivalents (RAEs);*

[d]*Niacin equivalents;*

[e]*In the absence of adequate sunlight;*

* *In the absence of firm recommendations in the form of RDAs, RNIs and PRIs the relevant institutions own Adequate Intakes (AIs) or equivalent are used based on estimations from balance studies, factorial analysis, daily intakes and/or biochemical indicators.*

Source: Compiled from multiple data sources including Gibson (2016) and SCF (1993), SCF (2000), FAO/WHO (2001), WHO/FAO (2002a,b), WHO/FAO (2002a,b), Europa (2003), WHO/FAO (2003), FAO/WHO/UNU (2004), WHO/FAO (2004), IoM (2005), EUFIC (2006), WHO/FAO/UNU (2007), EFSA (2009), Lagiou, Lavik et al. (2009), Agostoni, Bresson et al. (2010), CIAA (2010), EFSA (2010a,b,c), EFSA (2010a,b,c), EFSA (2010a,b,c), USDA (2010), EFSA (2012), Gibson (2012), EFSA (2013), FDF (2013), WHO (2015), Gibson (2016), IoM of The National Academies (2017), NAS (2017), WHO (2017).

References

Agostoni, C.V., et al., 2010. Scientific opinion on dietary reference values for water. EFSA Journal 8 (3), 1–48.

CIAA, 2010. GDA's: Guideline Daily Amounts. From: https://www.fooddrinkeurope.eu/uploads/publications_documents/GDAs_-_Guideline_Daily_Amounts.pdf.

EFSA, 2009. Scientific Opinion of the Panel on Dietetic Products, Nutrition and Allergies on a request from the Commission related to the review of labelling reference intake values for selected nutritional elements. Parma, European Food Safety Authority, p. 1008.

EFSA, 2010a. Dietary reference values and dietary guidelines. Retrieved from: https://www.efsa.europa.eu/en/topics/topic/dietary-reference-values-and-dietary-guidelines.

EFSA, 2010b. Scientific Opinion on Dietary Reference Values for carbohydrates and dietary fibre. Parma. European Food Safety Authority: 5.

EFSA, 2010c. Scientific Opinion on principles for deriving and applying Dietary Reference Values. EFSA Journal EFSA Journal 8 (3), 1458.

EFSA, 2012. Scientific Opinion on Dietary Reference Values for protein. EFSA Journal 10 (2). 2557-n/a.

EFSA, 2013. Scientific Opinion on Dietary Reference Values for Energy. EFSA Journal 11 (1). 3005-n/a.

EUFIC, 2006. Minerals: what they do and where to find them. Retrieved from: http://www.eufic.org/article/en/page/MARCHIVE/expid/miniguide-minerals/#6 http://www.eufic.org/en/.

Europa, 2003. Opinion of the Scientific Committee on Cosmetic Products and Non-Food Products Intended for Consumers Concerning – CHLORIDE. Parma, Europa, p. 9.

FAO/WHO, 2001. Report of the Joint FAO/WHO Expert Consultation on Human Vitamin and Mineral Requirements. FAO/WHO, Bankok, Thailand, p. 303.

FAO/WHO/UNU, 2004. Human energy requirements Report of a Joint FAO/WHO/UNU Expert Consultation: Rome, 17–24 October 2001. UNU/WHO/FAO, Rome.

FDF, 2013. Food and Drink Labelling: A Tool to Encourage Healthier Eating. The Food and Drink Federation (FDF), UK, p. 16.

Gibson, M., 2012. The Feeding of Nations: Re-defining Food Security for the 21st Century. CRC Press, Boca Raton, Florida.

Gibson, M., 2016. The Feeding of Nations: Re-defining Food Security for the 21st Century. CRC Press, Boca Raton, Florida.

IoM, 2005. Dietary Reference Intakes for Energy, Carbohydrate, Fiber, Fat, Fatty Acids, Cholesterol, Protein, and Amino Acids. National Academy Press, Washington, D.C.

IoM of The National Academies, 2017. Dietary Reference Intakes Tables and Application. Dietary Reference Intakes. Retrieved from: http://www.nationalacademies.org/hmd/Activities/Nutrition/SummaryDRIs/DRI-Tables.aspx.

Lagiou, P., et al., 2009. Labelling reference intake values for n-3 and n-6 polyunsaturated fatty acids Scientific Opinion of the Panel on Dietetic Products, Nutrition and Allergies on a request from the Commission related to labelling reference intake values for n-3 and n-6 poly. EFSA 1176, 1–11.

NAS, 2017. Website of the Institute of Medicine National Academies of Sciences. Retrieved from: http://www.nationalacademies.org/hmd/Home/Global/News%20Announcements/DRI.

SCF, 1993. Nutrient and energy intakes for the European Community. Reports of the Scientific Committee for Food, Thirty First Series. Luxembourg, European Commission: Scientific Committee for Food.

SCF, 2000. Guidelines of the Scientific Committee on Food for the Development of Tolerable Upper Intake Levels for Vitamins and Minerals. SCF/CS/NUT/UPPLEV/11 Final. Luxembourg, European Commission: Scientific Committee for Food.

USDA, 2010. Dietary Guidelines for Americans. U.S. Department of Agriculture and U.S. Department of Health and Human Services, Washington DC, p. 99.

WHO, 2015. Sugars Intake for Adults and Children Guideline. World Health Organisation, Geneva, Switzerland, p. 49.

WHO, 2017. Website of the World Health Organisation. Retrieved from: http://www.who.int/nutrition/topics/5_population_nutrient/en/.

WHO/FAO, 2002a. Human Vitamin and Mineral Requirements. Report of a joint FAO/WHO expert consultation Bangkok. WHO/FAO, Thailand.

WHO/FAO, 2002b. Vitamin and Mineral Requirements. WHO/FAO, Rome.

WHO/FAO, 2003. Diet, Nutrition and the Prevention of Chronic Diseases. WHO Technical Report Series 916. WHO/FAO, Switzerland.

WHO/FAO, 2004. Vitamin and Mineral Requirements in Human Nutrition (Draft Version). WHO/FAO, Sun Fung, China.

WHO/FAO/UNU, 2007. Joint FAO/WHO/UNU Expert Consultation on Protein and Amino Acid Requirements in Human Nutrition. WHO Technical Report Series 935. World Health Organisation, Geneva.

Glossary

Acetic acid: an effective agent of preservation in which a weak solution (0.1%) is able to inhibit the growth of many microbes
Agenda 21: the program of action arising from the Earth Summit in 1992
Agribusiness: any business that includes a component of agriculture
Agricultural revolution: a movement of increased innovation and industrialization in agriculture during the 18th and 19th centuries
Amorphous: entities that are without a clearly defined shape or form
Antibody: an antibody—also known as immunoglobulin; it is a protective protein produced by the immune system in response to the presence of a foreign substance in the body—called an antigen
Antioxidant: helps to fight off free radicals
Aquaculture: the farming of aquatic organisms
Aquifer: an underground source of water
Biennials: (not to be confused with biannuals—which occur twice a year), either live for two years or their lifecycle is a 2-year long process
Bioenergy: biomass used to produce energy
Biogenic: produced by natural processes such as those produced by plants and animals
Biotechnology: any technology that utilizes biological or living organisms
Black tea: the strong "black" tea that is most familiar in the West today is the result of the processing the Chinese developed in the 1840s specifically for export to the West
Bog: poorly drained wetland
Cacao nib's: raw cacao nibs are simply cacao beans that have been ground into smaller pieces
Caking: the tendency of some dry goods like flour and starch to clump together and cause lumps in sauces or in very humid places
Carbohydrates: provide body to the coffee
Carbon dioxide: a colorless, odorless occurring gas found in the atmosphere as well as resulting from the use of fossil fuel combustion
Carrying capacity: the theoretical equilibrium of a sustainable population vis-à-vis the replenishable resources natural resources
Catalyst: to catalyze something is to encourage a chemical reaction to happen faster
Caviar: although "caviar" should refer only to the salted eggs of sturgeon, it has come to mean any lightly salted eggs
Cereals: generally (but not always) referred to as annual plants of the gramineous family
Chemesthesis: arises when certain chemical compounds activate receptors normally associated with other senses that mediate pain, touch, and thermal perception, i.e., false heat and false coolness
Chlorophyll: a green pigment, present in green plants and in cyanobacteria, which is responsible for the absorption of light that provides energy for photosynthesis
Cow's milk: contains more than double the protein and minerals found in mother's milk
Cultivars: plant varieties that have been propagated not only from seeds but also from bulbs and stem cuttings too
Distillates: of wine are called brandies
Energy: in mammals comes from the glucose that has either been ingested (eaten) directly or synthesized from the food eaten. If it is not needed for energy use straight away, it is converted into glycogen and stored as a form of energy primarily in the cells of the liver and the muscles. It functions as the secondary long-term energy storage deposit after adipose tissue (tissue containing fat cells)
European Commission: a principal body of the European Union
European Union (EU): integration of various European countries
Exogenous: in this sense means developed from external substances

Fatty acids: the building blocks of fat within the human body as well as within the food we eat. So, while we consume whole fats, the human digestive system breaks down these fats into their component fatty acids which can then be absorbed into the blood. Fatty acid molecules are usually joined together in groups of three, forming a triglyceride. Triglycerides are also made in our bodies from the carbohydrates that we eat too. Fatty acids have many important known functions in the body, including energy storage. Furthermore, fatty acids act as fuel for the cells after the other macronutrients

Flash flood: high peak discharge of water in a short period of time

Flood plain: a low area along a river considered to be at constant risk of flooding

Flora: plant life

Freezer burn: the drying out of produce caused by the evaporation of frozen water molecules (sublimation), which in turn develop 'burned' patches leaving a tough texture and stale flavor

Gelatinization: of starch means that it absorbs moisture, swells, and then becomes semifirm

Genotype: in a very broad sense the term "genotype" refers to the genetic makeup of an organism. That is to say it describes an organism's complete set of genes

Globalization: a word used to describe the increasing internationalization of goods and services characterized by increased mobility of people and capital

Grains or cereals: used more often than not interchangeably. Cereals are plants in the grass family, the Gramineae

Green revolution: the culmination of technologies mostly in the 20th century that resulted in increased productivity

Hemoglobin: contains in excess of 10,000 atoms of hydrogen, carbon, nitrogen, oxygen, sulfur, and 4 atoms of iron. It is the iron in hemoglobin that is more important as it gives blood its ability to combine with oxygen

Herbaceous: plants with little or no woody tissue and which generally persist for only one growing season

Heterogeneous: populations are diverse, varied, disparate, and multifarious. This differing nature means recommended daily intakes of nutrients vary quite differently

Human Development Index: composite indices describing the well-being of the country's people as opposed to economic development

Hunger: the physiological symptoms feeling hungry

Hydrogenation: see trans-fats

Hydrolysis: is the breakdown of a compound by a chemical reaction with water

Hydropower: electrical energy produced by converting the power of water

Industrial enzymes: complex molecules or proteins that are manufactured and which aid in catalysis. They are sourced from plant or animal tissue and microbes

International Monetary Fund: a World Bank body that promotes international monetary cooperation as well as the advancement of trade through the promotion of stability

Lactic acids: many of the lactic acids are the same species found in fermented dairy products and vegetables

Lactose: peculiar to milk and only a handful of plants and therefore very few organisms possess lactose-breaking enzymes

Lipophilic: the ability of chemical compounds to dissolve in fats, oils, lipids, and nonpolar solvents such as hexane or toluene

Metabolism: breaking down (catabolism) or the building up (anabolism) of food in the body. Metabolism can be considered as functionally active in three main areas—firstly, the transformation of food into energy/fuel to operate cellular processes; secondly, the transformation of food into building blocks for proteins, lipids, some carbohydrates, and nucleic acids; and lastly, the transformation of food to be used as vehicles for the safe disposal of nitrogenous wastes, which would otherwise tend to form toxic ammonia

Metabolite: It is any substance produced during metabolism, although the term might also be used to refer to the product that remains after the process of metabolism by the body

Milk fat: accounts for about half the calories of whole milk as well as the fat-soluble vitamins (A, D, E, K)

Monounsaturated and polyunsaturated fats: known by other names—omegas. There are three types of omega fatty acids: omega-3, omega-6, and omega-9

Morbidity: the incidence of disease in a population

Musk: a strong-smelling substance used predominantly in perfumes and is either artificially made or obtained from a gland of an Asian (musk) deer

Noble rot: is a gray fungus *Botrytis cinerea* that affects wine grapes, which in the wrong conditions can whole destroy crops. However, under the right conditions, i.e., dry, less humid conditions, the gray rot becomes beneficial, and as the grapes become a little dried, like raisins, this type of infection is known as noble rot. This produces particularly fine and sweet concentrated wines
Nucleotides: form the basic structural unit of nucleic acids such as DNA
Oats: are inherently gluten-free, but are frequently contaminated with wheat during growing or processing
Oenology: the science and study of wine and winemaking; it is separate from viticulture, the science of vine-growing and of grape harvesting. Oenology is the overarching field of study
Pellagra: a condition brought about by deficiency in vitamin b_3 niacin
Pepsins: the main digestive enzyme produced in the human stomach responsible for breaking down proteins
Peptides: biologically occurring short chains of amino acids
Per capita: a substitute term to describe per person
Perennial: plants that live for more than two years
Phosphate: a salt or ester of phosphoric acid
Phospholipids: the human body is made up of cells, including phospholipids, which are important molecules providing structure and protection to cells. Phospholipids consist of a hydrophilic (water loving) and a hydrophobic (water fearing) tail. Phospholipids line up arranging themselves into two parallel layers, a "phospholipid bilayer." This layer makes cell membranes and is critical to a cell's ability to function
Phosphoproteins: common in living organisms in which they play a role in various metabolic processes such as in the regulation of cell nuclei; in ion transport; and in the oxidation processes in a cell's mitochondria
Phytochemicals: nonessential nutrients, meaning that they are not required by the human body for sustaining life
Polysaccharides: carbohydrates (e. g. starches, cellulose, or glycogen) whose molecules consist of a number of sugar molecules bonded together
Population density: a number describing the total population living in a per square kilometer
Protease: refers to a group of enzymes whose catalytic function is to hydrolyze (breakdown) proteins. They are also called proteolytic enzymes or systemic enzymes
Proteins: in adding protein to hot liquids one must be careful not to coagulate the proteins on contact. It is generally more prudent to add a little of the hot liquid into the protein to gently bring up the heat before adding the protein mix to the sauce
Proteolysis: the breakdown of proteins or peptides into amino acids by the action of enzymes
Protoplasm: in biology protoplasm is the living contents of a cell surrounded by a plasma membrane that effectively separates the interior of the cell from the outside environment
Purees: the solid particles in a puree are plant cell walls—an agglomeration of many thousands of carbohydrate and protein molecules
Rheology: applicable to nonfood items too
Rhizopus: the major mold in soybean tempeh—a soy product from Indonesia. It is made by natural culturing and controlled fermentation. This binds soybeans into a cake form
Robusta: and Arabica plants are sometimes used for cross-breeding of coffea species
Robusta: not all Robusta's are of inferior quality, some are valued especially in espressos for their deep flavor and good crema
Roes: consist of separate eggs held loosely together in a weak protein solution all encased in a thin, fragile membrane
Round fish: a general term used to describe the oval/rounded fish as compared to flatfish
Rural population: that is left after taking out urban population from total population figures
Saccharides: any number of sugar compounds from simple monosaccharides to polymerized disaccharides and polysaccharides
Saturated and unsaturated compounds: might comprise simple straight or branched or ring arrangements or they may also contain functional groups
Seaweeds: all seaweeds are algae but not all algae are seaweeds. Seaweeds tend to be easily visible, comprise many cells, and grow in the sea. There are red, brown, and green seaweeds. However, many algae are microscopic single-cell diatoms consisting of a single cell
Sequestration: a process of removing carbon dioxide from the atmosphere through utilization of natural and man-made things that naturally absorb airborne CO_2—these include land-use change, afforestation, reforestation, etc. Shellfish can be a little misleading as not all members have a shell or an exoskeleton

Silage: or ensilage is a method of green fodder preservation through which fermentation is used to slow spoiling
Skin: good for fish stock as the flesh can be as little as between 0.3%–3% collagen
Solera: the process of aging liquids such as wine, beer, vinegar, and brandy, by means of fractional blending. The purpose of this intensive process is the creation and maintenance of reliable qualities in the beverage over time
Soybeans: are legumes mainly cultivated in the northern and southern parts of America. Soybeans are a major source of vegetable oil as well as feed
Specialized agencies: in terms of the United Nations specialized agencies are autonomous organizations related to the United Nations through individual agreements
Sturgeon: a fish from which caviar is made—there are 27 varieties of Sturgeon
Sugar crops: cultivated for their sugar content. Usually from sugarcane and sugar beets
Swamp: a sort of fresh or saltwater, tidal, or nontidal wetland replete with woody vegetation while lacking peat deposits
Swim bladders: balloons of connective tissue which fill with air or gas to adjust the fishes' buoyancy
Syneresis: the drawing together of the protein particles of a gel, with some separation of the liquid medium and shrinkage of the gel, i.e., weeping
Tastant: a tasting is the thing being tasted
Tea: one of the highest flavonoid contents of all plants taking up as much as 15% of the dry leaf weight
Trans-fats and hydrogenation: in their natural state, unsaturated fats can be oxidized creating a rancid, stale, or unpleasant smell or flavor. By saturating the fat (fixing more hydrogen to natural unsaturated fats), the number of double bonds is decreased in a process called hydrogenation; this slows down the process of rancidity
Transgenics: a branch of research that splices genes from one organism into another resulting in an organism whose genetic makeup is a combination of both organisms
Treaty: an international agreement between states/countries which in turn is governed by international law
Triglycerides: derived from glycerol and three fatty acids. They are the main component of body fat in animals and humans as well as vegetable fat
Triglycerides: the collective name for all lipids, fats, and oils
Triple point of water: the point when temperature and pressure are in equilibrium to allow the three phases (solid, liquid, and gas) to coexist
Triticale: a cereal hybrid cross between wheat and rye
Tropical climate: typical of equatorial regions that display continuous high temperatures with appreciable precipitation
Tubers: short, thickened, fleshy parts of an underground stem
Urea: is what most animals turn protein waste into before they excrete it
Vinegar: made from wine as the ethanol oxidizes into acetic acid in the presence of air
Viticulture: the science, production, and study of grapes in general while viniculture refers same thing only for grapes specifically used for wine
Volatility: in chemistry volatility refers to a substance's tendency to vaporize
Water conservation: the conservation of both ground and surface water
Water vapor: often overlooked as a greenhouse gas, yet it is one of the most abundant of the greenhouse gases
Wheat: one of the most important cereals
Whisky: the anglicized term of the Gaelic "water of life"
Wild yeasts: many things like sweet liquids, sour bread, and wine, etc., will spontaneously ferment and eventually sour naturally, some wild yeasts will give certain "off" flavors to the final product. So, whenever yeasts are required in a product, most people will use specific "cultured" yeasts for specific purposes
Winnowing: the practice of blowing a current of air through (usually grain) to remove the chaff or outer papery coverings
World Bank: an international lending body comprising several institutions with the remit to reduce poverty and strengthening economies through the promotion of sustainable development
World Trade Organization (WTO): an international organization that came after the general agreement on tariffs and trade (GATT)
Yeasts: for brewing cannot tolerate more than approximately 20% alcohol concentration
Zooplankton: microscopic animals found in open water

Index

Note: 'Page numbers followed by "*f*" indicate figures and "*t*" indicate tables'.

B

C

D

F

J

K

L

M

T

Y

Z

A7/1019613/9/10/21

CPI Antony Rowe
Eastbourne, UK
March 05, 2020